Beyond Natural Resources to Post-Human Resources

Beyond Natural Resources to Post-Human Resources

Towards a New Theory of Diversity and Discontinuity

By

Peter Baofu

CAMBRIDGE SCHOLARS PUBLISHING

Beyond Natural Resources to Post-Human Resources: Towards a New Theory of Diversity and Discontinuity, by Peter Baofu

This book first published 2013

Cambridge Scholars Publishing

12 Back Chapman Street, Newcastle upon Tyne, NE6 2XX, UK

British Library Cataloguing in Publication Data
A catalogue record for this book is available from the British Library

ISBN (10): 1-4438-4453-5, ISBN (13): 978-1-4438-4453-6

To Those in the Future World beyond Diversity and Discontinuity

Books also by Peter Baofu

Category I: Social Sciences and Related Fields

- 54. *The Future of Post-Human Waste* (2012)
- 53. *The Future of Post-Human Migration* (2012)
- 52. *The Future of Post-Human Criminality* (2012)
- 51. *The Future of Post-Human Geography* (2011)
- 50. *The Future of Post-Human Gambling* (2011)
- 49. *The Future of Post-Human Education* (2011)
- 48. *The Future of Post-Human War and Peace* (2010)
- 47. *The Future of Post-Human Law* (2010)
- 46. *The Future of Post-Human Organization* (2009)
- 45. *The Future of Post-Human Mass Media* (2009)
- 44. *The Rise of Authoritarian Liberal Democracy* (2007)
- 43. *Beyond the World of Titans, and the Remaking of World Order* (2007)
- 42. *Beyond Capitalism to Post-Capitalism* (2005)
- 41. Volume 1: *Beyond Democracy to Post-Democracy* (2004)
- 40. Volume 2: *Beyond Democracy to Post-Democracy* (2004)
- 39. *The Future of Capitalism and Democracy* (2002)

Category II: Natural Sciences and Related Fields

- 38. *The Future of Post-Human Transportation* (2012)
- 37. *The Future of Post-Human Architecture* (2012)
- 36. *The Future of Post-Human Chemistry* (2011)
- 35. *The Future of Post-Human Acoustics* (2011)
- 34. *The Future of Post-Human Geology* (2010)
- 33. *Beyond Cosmology to Post-Cosmology* (2010)
- 32. *The Future of Post-Human Personality* (2010)
- 31. *The Future of Post-Human Sexuality* (2010)
- 30. *The Future of Post-Human Urban Planning* (2009)
- 29. *The Future of Post-Human Engineering* (2009)
- 28. *The Future of Post-Human Unconsciousness* (2008)
- 27. *The Future of Complexity* (2007)
- 26. *Beyond Nature and Nurture* (2006)
- 25. *The Future of Post-Human Space-Time* (2006)
- 24. *The Future of Post-Human Consciousness* (2004)

Category III: The Humanities and Related Fields

- 23. *The Future of Post-Human Culinary Art* (2012)
- 22. *The Future of Post-Human History* (2012)
- 21. *The Future of Post-Human Performing Arts* (2012)
- 20. *The Future of Post-Human Literature* (2011)
- 19. *The Future of Post-Human Humor* (2011)
- 18. *Beyond Ethics to Post-Ethics* (2011)
- 17. *The Future of Post-Human Religion* (2010)
- 16. *The Future of Post-Human Chess* (2010)
- 15. *The Future of Post-Human Martial Arts* (2009)
- 14. *The Future of Post-Human Creative Thinking* (2009)
- 13. *The Future of Post-Human Knowledge* (2008)
- 12. *The Future of Aesthetic Experience* (2007)
- 11. *Beyond Civilization to Post-Civilization* (2006)
- 10. Volume 1: *The Future of Human Civilization* (2000)
- 9. Volume 2: *The Future of Human Civilization* (2000)

Category IV: Formal Sciences and Related Fields

- 8. *The Future of Post-Human Semantics* (2012)
- 7. *The Future of Post-Human Data Analysis* (2011)
- 6. *The Future of Post-Human Computing* (2011)
- 5. *The Future of Post-Human Formal Science* (2010)
- 4. *The Future of Post-Human Language* (2009)
- 3. *The Future of Post-Human Geometry* (2009)
- 2. *The Future of Post-Human Mathematical Logic* (2008)
- 1. *The Future of Information Architecture* (2008)

Contents

Part Three: Discontinuity

Part Four: Conclusion

Tables

Category III: Visions on the Mind

Category IV: Visions on History

Category VII. Visions on Society (Socio-Political)

Category VIII. Visions on Society (Socio-Economic)

Category IX: Visions on Culture

ACKNOWLEDGMENTS

This book, like all other previous books of mine, is written to challenge conventional wisdom or, more precisely, in the current context, the one on natural resources—and to replace it with an original way of thinking about its future fate (in relation to diversity and discontinuity).

Consequently, this book receives no external funding nor help from any formal organization or institution, because of its political incorrectness—as this is something that I often stressed (and repeated) in all my previous books.

My only reward is those wonderful moments of conceiving something original in the intellectual history of natural resources.

In any event, I bear the sole responsibility for the ideas presented in this book.

ABBREVIATIONS

ALD = Peter Baofu. 2007. *The Rise of Authoritarian Liberal Democracy: A Preface to a New Theory of Comparative Political Systems*. Cambridge, England: Cambridge Scholars Publishing, Ltd.

BCIV = Peter Baofu. 2006. *Beyond Civilization to Post-Civilization: Conceiving a Better Model of Life Settlement to Supersede Civilization*. NY: Peter Lang Publishing, Inc.

BCPC = Peter Baofu. 2005. *Beyond Capitalism to Post-Capitalism: Conceiving a Better Model of Wealth Acquisition to Supersede Capitalism*. NY: The Edwin Mellen Press.

BCOS = Peter Baofu. 2010. *Beyond Cosmology to Post-Cosmology: A Preface to a New Theory of Different Worlds*. Cambridge, England: Cambridge International Science Publishing, Ltd.

BDPD1 = Peter Baofu. 2004. Volume 1. *Beyond Democracy to Post-Democracy: Conceiving a Better Model of Governance to Supersede Democracy*. NY: The Edwin Mellen Press.

BDPD2 = Peter Baofu. 2004. Volume 2. *Beyond Democracy to Post-Democracy: Conceiving a Better Model of Governance to Supersede Democracy*. NY: The Edwin Mellen Press.

BEPE = Peter Baofu. 2011. *Beyond Ethics to Post-Ethics: A Preface to a New Theory of Morality and Immorality.* Charlotte, NC: Infomration Age Publishing.

BNN = Peter Baofu. 2006. *Beyond Nature and Nurture: Conceivng a Better Way to Understand Genes and Memes.* Cambridge, England: Cambridge Scholars Publishing, Ltd.

BNR = Peter Baofu. 2012. *Beyond Natural Resources to Post-Human Resources: Towards a New Theory of Diversity and Discontinuity.* Cambridge, England: Cambridge Scholars Publishing, Ltd.

BWT = Peter Baofu. 2007. *Beyond the World of Titans, and the Renaking of World Order: A Preface to a New Logic of Empire-Building.* Cambridge, England: Cambridge Scholars Publishing, Ltd.

FAE = Peter Baofu. 2007. *The Future of Aesthetic Experience: Conceiving a Better Way to Understand Beauty, Ugliness and the Rest.* Cambridge, England: Cambridge Scholars Publishing, Ltd.

FC = Peter Baofu. 2007. *The Future of Complexity: Conceiving a Better Way to Understand Order and Chaos.* London, United Kingdom: World Scientific Publishing Co.

FCD = Peter Baofu. 2002. *The Future of Capitalism and Democracy.* MD: The University Press of America.

FHC1 = Peter Baofu. 2000. Volume 1. *The Future of Human Civilization.* NY: The Edwin Mellen Press.

FHC2 = Peter Baofu. 2000. Volume 2. *The Future of Human Civilization.* NY: The Edwin Mellen Press.

FIA = Peter Baofu. 2008. *The Future of Information Architecture: Conceiving a Better Way to Understand Taxonomy, Network, and Intelligence.* Oxford, England: Chandos Publishing (Oxford) Limited.

FPHA = Peter Baofu. 2011. *The Future of Post-Human Acoustics: A Preface to a New Theory of Sound and Silence.* Cambridge, England: Cambridge International Science Publishing, Ltd.

FPHARCH = Peter Baofu. 2012. *The Future of Post-Human Architecture: A Preface to a New Theory of Form and Function.* Cambridge, England: Cambridge International Science Publishing, Ltd.

FPHC = Peter Baofu. 2004. *The Future of Post-Human Consciousness.* NY: The Edwin Mellen Press.

FPHCA = Peter Baofu. 2012. *The Future of Post-Human Culinary Art: Towards a New Theory of Ingredients and Techniques.* England: Cambridge Scholars Publishing, Ltd.

FPHCHEM = Peter Baofu. 2011. *The Future of Post-Human Chemistry: A Preface to a New Theory of Substances and their Changes.* England: Cambridge Scholars Publishing, Ltd.

FPHCHESS = Peter Baofu. 2010. *The Future of Post-Human Chess: A Preface to a New Theory of Tactics and Strategy.* Cambridge, England: Cambridge International Science Publishing, Ltd.

FPHCOM = Peter Baofu. 2011. *The Future of Post-Human Computing: A Preface to a New Theory of Hardware and Software.* Cambridge, England: Cambridge International Science Publishing, Ltd.

FPHCRIM = Peter Baofu. 2012. *The Future of Post-Human Criminality: A Preface to a New Theory of Heroes and Villains.* Cambridge, England: Cambridge International Science Publishing, Ltd.

FPHCT = Peter Baofu. 2009. *The Future of Post-Human Creative Thinking: A Preface to a New Theory of Invention and Innovation.* Cambridge, England: Cambridge Scholars Publishing, Ltd.

FPHDA = Peter Baofu. 2011. *The Future of Post-Human Data Analysis: A Preface to a New Theory of Quantitative and Qualitative Research.* Cambridge, England: Cambridge International Science Publishing, Ltd.

FPHE = Peter Baofu. 2009. *The Future of Post-Human Engineering: A Preface to a New Theory of Technology.* Cambridge, England: Cambridge Scholars Publishing, Ltd.

FPHEDU = Peter Baofu. 2011. *The Future of Post-Human Education: A Preface to a New Theory of Teaching and Learning.* Cambridge, England: Cambridge International Science Publishing, Ltd.

FPHFS = Peter Baofu. 2010. *The Future of Post-Human Formal Science: A Preface to a New Theory of Abstraction and Application.* Cambridge, England: Cambridge Scholars Publishing, Ltd.

FPHG = Peter Baofu. 2009. *The Future of Post-Human Geometry: A Preface to a New Theory of Infinity, Symmetry, and Dimensionality.* Cambridge, England: Cambridge Scholars Publishing, Ltd.

FPHGAM = Peter Baofu. 2011. *The Future of Post-Human Gambling: A Preface to a New Theory of Risk and Caution.* Cambridge, England: Cambridge International Science Publishing, Ltd.

FPHGEOG = Peter Baofu. 2011. *The Future of Post-Human Geography: A Preface to a New Theory of Environments and their Interactions.* Cambridge, England: Cambridge International Science Publishing, Ltd.

FPHGEOL = Peter Baofu. 2010. *The Future of Post-Human Geology: A Preface to a New Theory of Statics and Dynamics.* Cambridge, England: Cambridge International Science Publishing, Ltd.

FPHH = Peter Baofu. 2011. *The Future of Post-Human Humor: A Preface to a New Theory of Joking and Laughing.* Cambridge, England: Cambridge International Science Publishing, Ltd.

FPHHIST = Peter Baofu. 2012. *The Future of Post-Human History: A Preface to a New Theory of Universality and Relativity.* Cambridge, England: Cambridge Scholars Publishing, Ltd.

FPHK = Peter Baofu. 2008. *The Future of Post-Human Knowledge: A Preface to a New Theory of Methodology and Ontology.* Oxford, England: Chandos Publishing (Oxford) Limited.

FPHL = Peter Baofu. 2009. *The Future of Post-Human Language: A Preface to a New Theory of Structure, Context, and Learning.* Cambridge, England: Cambridge Scholars Publishing, Ltd.

FPHLAW = Peter Baofu. 2010. *The Future of Post-Human Law: A Preface to a New Theory of Necessity, Contingency, and Justice.* Cambridge, England: Cambridge Scholars Publishing, Ltd.

FPHLIT = Peter Baofu. 2011. *The Future of Post-Human Literature: A Preface to a New Theory of Fiction and Non-Fiction.* Cambridge, England: Cambridge International Science Publishing, Ltd.

FPHMA = Peter Baofu. 2009. *The Future of Post-Human Martial Arts: A Preface to a New Theory of the Body and Spirit of Warriors.* Cambridge, England: Cambridge Scholars Publishing, Ltd.

FPHMIG = Peter Baofu. 2012. *The Future of Post-Human Migration: A Preface to a New Theory of Sameness, Otherness, and Identity.* Cambridge, England: Cambridge Scholars Publishing, Ltd.

FPHML = Peter Baofu. 2008. *The Future of Post-Human Mathematical Logic: A Preface to a New Theory of Rationality.* Cambridge, England: Cambridge Scholars Publishing, Ltd.

FPHMM = Peter Baofu. 2009. *The Future of Post-Human Mass Media: A Preface to a New Theory of Technology.* Cambridge, England: Cambridge Scholars Publishing, Ltd.

FPHO = Peter Baofu. 2009. *The Future of Post-Human Organization: A Preface to a New Theory of Communication, Decision-Making, and Leadership.*.Cambridge, England: Cambridge Scholars Publishing, Ltd.

FPHP = Peter Baofu. 2010. *The Future of Post-Human Personality: A Preface to a New Theory of Normality and Abnormality.* Cambridge, England: Cambridge International Science Publishing, Ltd.

FPHPA = Peter Baofu. 2012. *The Future of Post-Human Performing Arts: A Preface to a New Theory of the Body and its Presence.* Cambridge, England: Cambridge Scholars Publishing, Ltd.

FPHR = Peter Baofu. 2010. *The Future of Post-Human Religion: A Preface to a New Theory of Spirituality.* Cambridge, England: Cambridge International Science Publishing, Ltd.

FPHS = Peter Baofu. 2010. *The Future of Post-Human Sexuality: A Preface to a New Theory of the Body and Spirit of Love-Makers.* Cambridge, England: Cambridge Scholars Publishing, Ltd.

FPHSEM = Peter Baofu. 2012. *The Future of Post-Human Semantics: A Preface to a New Theory of Internality and Externality.* Cambridge, England: Cambridge Scholars Publishing, Ltd.

FPHST = Peter Baofu. 2006. *The Future of Post-Human Space-Time: Conceivng a Better Way to Understand Space and Time.* New York: Peter Lang Publishing, Inc.

FPHTRAN = Peter Baofu. 2012. *The Future of Post-Human Transportation: A Preface to a New Theory of Networks and Operations.* Cambridge, England: Cambridge Scholars Publishing, Ltd.

FPHU = Peter Baofu. 2008. *The Future of Post-Human Unconsciousness: A Preface to a New Theory of Anomalous Experience.* Cambridge, England: Cambridge Scholars Publishing, Ltd.

FPHUP = Peter Baofu. 2009. *The Future of Post-Human Urban Planning: A Preface to a New Theory of Density, Void, and Sustainability.* Cambridge, England: Cambridge Scholars Publishing, Ltd.

FPHW = Peter Baofu. 2012. *The Future of Post-Human Waste: Towards a New Theory of Uselessness and Usefulness.* Cambridge, England: Cambridge Scholars Publishing, Ltd.

FPHWP = Peter Baofu. 2010. *The Future of Post-Human War and Peace: A Preface to a New Theory of Aggression and Pacificity.* Cambridge, England: Cambridge Scholars Publishing, Ltd.

• PART ONE •

Introduction

CHAPTER 1
INTRODUCTION—THE AMBIENCE OF NATURAL RESOURCES

Earth provides enough to satisfy every man's need, but not every man's greed.
—Mahatma Gandhi (TE 2012)

A Question to a Fashionable Ideology

Are natural resources really so limited that, as Mahatma Gandhi once famously said, "Earth provides enough to satisfy every man's need, but not every man's greed"? (TE 2012)

This limiting view of natural resources can be contrasted with an opposing view by John Maynard Keynes, who "summarized Say's Law as 'supply creates its own demand'" but then "turned Say's Law on its head in the 1930s by declaring that demand creates its own supply," so whenever a demand exists, there will be resources to create the supply. (EN 2012)

Contrary to these opposing views (and other ones as will be discussed in the book), natural resources (in relation to both diversity and discontinuity) are neither possible (or impossible) nor desirable (or undesirable) to the extent that the respective ideologues (on different sides) would like us to believe.

Needless to say, this challenge to the opposing views of natural resources does not mean that natural resources are unimportant, or that those interdisciplinary fields (related to natural resources) like conservation biology, environmental management, ecological economics, political ecology, environmental ethics, deep ecology, adaptive management, genetic engineering, Malthusianism, and so on are not worth studying. Of course, neither of these extreme views is reasonable.

Rather, this book offers an alternative (better) way to understand the future of natural resources, especially in the dialectic context of diversity and discontinuity—while learning from different approaches in the literature but without favoring any one of them (nor integrating them, since they are not necessarily compatible with each other).

More specifically, this book offers a new theory (that is, *the resilient theory of natural resources*) to go beyond the existing approaches in a novel way.

If successful, this seminal project is to fundamentally change the way that we think about natural resources in relation to diversity and discontinuity from the combined perspectives of the mind, nature, society, and culture, with enormous implications for the human future and what I originally called its "post-human" fate.

Definition and History of Natural Resources

A good point of departure to start this inquiry is to provide a working definition of the term "natural resource" and its brief history.

Definition

The term "natural resource" has two words, namely, "natural" (from the noun "nature") and "resource."

In biology and ecology, "a resource is defined as a substance or object required by a living organism for normal growth, maintenance, and reproduction….Resources, such as food, water, or nesting sites, can be consumed by an organism and, as a result, become unavailable to other organisms. For animals, key resources include food, water, and territory. For plants key resources include sunshine, nutrients, water, and a place to grow." (WK 2012; R. Ricklefs 2005)

And the term "natural resource" refers to "anything obtained from the environment to satisfy human needs and wants." (WK 2012; G. Miller 2011) In other words, natural resources are resources which "occur naturally within environments that exist relatively undisturbed by mankind, in a natural form. A natural resource is often characterized by amounts of biodiversity and geodiversity existent in various ecosystems." (WK 2012a)

History

With this definition in mind—historically, "natural resources are derived from the environment. Some of them are essential for our survival while most are used for satisfying our wants." (WK 2012a)

Since pre-historical times, natural resources have been used by humans to promote their welfare. An excellent example is "resource extraction," which "involves any activity that withdraws resources from nature. This can range in scale from the traditional use of preindustrial societies, to global industry. Extractive industries are, along with agriculture, the basis of the primary sector of the economy. Extraction produces raw material which is then processed to add value. Examples of extractive industries are hunting and trapping, mining, oil and gas drilling, and forestry." (WK 2012a)

However, unlike preindustrial societies which used natural resources on a relatively smaller scale—contemporary societies in the 21st century consume them so much that, "in recent years, the depletion of natural resources has become a major focus of governments and organizations such as the United Nations (UN). This is evident in the UN's Agenda 21, Section Two, which outlines the necessary steps to be taken by countries to sustain their natural resources. The depletion of natural resources is considered to be a sustainable development issue." (WK 2012a; UN 2011; M. Schilling 2011)

And "the term sustainable development has many interpretations, most notably the Brundtland Commission's 'to ensure that it meets the needs of the present without compromising the ability of future generations to meet their own needs'; however, in broad terms it is balancing the needs of the planet's people and species now and in the future. In regards to natural resources, depletion is of concern for sustainable development, as it has the ability to degrade current environments and potential to impact the needs of future generations." (WK 2012a; UN 2011 & 2011a; M. Schilling 2011; L. Salvati 2008)

Diversity and Discontinuity in Natural Resources

With this definition and history of natural resources in mind—this book is to redefine the field in terms of the dialectic relationship between *diversity* and *discontinuity*, as explained below (and summarized in *Table 1.1*).

Diversity

If "natural resource is often characterized by amounts of biodiversity and geodiversity existent in various ecosystems" (as defined in the previous section), then it is important to find out what diversity is.

In accordance to the Merriam-Webster dictionary, "diversity" is defined as "the condition of having or being composed of differing elements" or "an instance of being composed of differing elements or qualities." (MWD 2012) For example, in biology, the term "diversity" refers to "the degree of variation of life forms within a given species, ecosystem, biome, or an entire planet." (WK 2012b)

In the context of natural resources, "diversity" can be understood in this book in relation to (a) conservation, (b) management, and (c) equilibrium, just to cite three good examples, as explained below.

Conservation

The first way to understand diversity in the context of natural resoruces has to do with "conservation," which, especially in biology, refers to "the scientific study of the nature and status of Earth's biodiversity with the aim of protecting species, their habitats, and ecosystems from excessive rates of extinction and the erosion of biotic interactions. It is an interdisciplinary subject drawing on sciences, economics, and the practice of natural resource management." (WK 2012c; S. Sahney 2008; B. Wilcox 1980; M. Soule 1986 & 198a; M. Hunter 1996; G. Meffe 2006; F. van Dyke 2008)

A relevant field of study is "conservation biology," which is also known as "discipline with a deadline," because of "the rapid decline of established biological systems around the world"—and "conservation biology is tied closely to ecology in researching the dispersal, migration, demographics, effective population size, inbreeding depression, and minimum population viability of rare or endangered species. To better understand the restoration ecology of native plant and animal communities, the conservation biologist closely studies both their polytypic and monotypic habitats that are affected by a wide range of benign and hostile factors." (WK 2012c; E. Wilson 2002; S. Sahney 2010; H. Theel 1948)

Management

The second way to understand diversity in the context of natural resoruces has to do with "management" in regard to "land, water, soil, plants and animals, with a particular focus on how management affects the quality of life for both present and future generations." (WK 2012a)

In general, "management of natural resources involves identifying who has the right to use the resources and who does not for defining the boundaries of the resource. The resources are managed by the users according to the rules governing of when and how the resource is used depending on local condition." (WK 2012a; N. Kommers 2005)

For instance, "a successful management of natural resources should engage the community because of the nature of the shared resources the individuals who are affected by the rules can participate in setting or changing them. The users have the rights to device their own management institutions and plans under the recognition by the government. The right to resources includes land, water, fisheries and pastoral rights. The users or parties accountable to the users have to actively monitor and ensure the utilisation of the resource compliance with the rules and to impose penalty on those peoples who violates the rules. These conflicts are resolved in a quick and low cost manner by the local institution according to the seriousness and context of the offence." (WK 2012a; N. Kommers 2005)

Equilibrium

And the third way to understand diversity in the context of natural resoruces has to do with "equilibrium," which refers to a return to an "initial state" after "periodic disturbances" in an ecosystem. (WK 2012d)

In ecosystem dynamics, "when an ecosystem is subject to some sort of perturbation, it responds by moving away from its initial state. The tendency of a system to remain close to its equilibrium state, despite that disturbance, is termed its resistance." (WK 2012d)

In ecological processes, "disturbance...plays an important role....F. Stuart Chapin and coauthors define disturbance as 'a relatively discrete event in time and space that alters the structure of populations, communities and ecosystems and causes changes in resources availability or the physical environment.' This can range from tree falls and insect outbreaks to hurricanes and wildfires to volcanic eruptions and can cause large changes in plant, animal and microbe populations, as well soil organic matter content. Disturbance is followed by succession, a 'directional change in ecosystem structure and functioning resulting from biotically driven changes in resources supply.'" (WK 2012d)

Indeed, "the frequency and severity of disturbance determines the way it impacts ecosystem function. Major disturbance like a volcanic eruption or glacial advance and retreat leave behind soils that lack plants, animals or organic matter. Ecosystems that experience disturbances that sever undergo primary succession. Less severe disturbance like forest fires, hurricanes or cultivation result in secondary succession. More severe disturbance and more frequent disturbance result in longer recovery times. Ecosystems recover more quickly from less severe disturbance events." (WK 2013d; F. Chapin 2002)

Discontinuity

If there is diversity, there is likewise a countervailing process known as "discontinuity."

In accordance to the Merriam-Webster dictionary, "discontinuity" is defined as the "lack of...cohesion" or "the property of being not mathematically continuous." (MWD 2012a)

In the context of natural resources, "discontinuity" can be understood in this book in relation to (a) extinction, (b) fragmentation, and (c) reduction (in range), just to cite three examples, as explained below.

Extinction

The first way to understand discontinuity in the context of natural resoruces has to do with "extinction," which, especially in biology and ecology, refers to "the end of an organism or of a group of organisms (taxon), normally a species." (WK 2012e)

In general, "the moment of extinction is...considered to be the death of the last individual of the species, although the capacity to breed and recover may have been lost before this point. Because a species' potential range may be very large, determining this moment is difficult, and is usually done retrospectively. This difficulty leads to phenomena such as Lazarus taxa, where a species presumed extinct abruptly 're-appears' (typically in the fossil record) after a period of apparent absence." (WK 2012e)

Extinction is better understood through the study of "evolution," which suggests that "new species arise through the process of speciation—where new varieties of organisms arise and thrive when they are able to find and exploit an ecological niche—and species become extinct when they are no longer able to survive in changing conditions or against superior competition. The relationship between animals and their ecological niches has been firmly established." (WK 2012e; S. Sahney 2010; M. Newman 1994)

For instance, "a typical species becomes extinct within 10 million years of its first appearance, although some species, called living fossils, survive virtually unchanged for hundreds of millions of years. Most extinctions have occurred naturally, prior to Homo sapiens walking on Earth: it is estimated that 99.9% of all species that have ever existed are now extinct." (WK 2012e; M. Newman 1994; D. Raup 1991)

Fragmentation

The second way to understand discontinuity in the context of natural resoruces has to do with "fragmentation," which, "as the name implies, describes the emergence of discontinuities (fragmentation) in an organism's preferred environment (habitat), causing population fragmentation." (WK 2012f; S. Sahney 2010a)

For instance, "habitat fragmentation can be caused by geological processes that slowly alter the layout of the physical environment (suspected of being one of the major causes of speciation), or by human activity such as land conversion, which can alter the environment much faster and causes extinctions of many species." (WK 2012f; S. Sahney 2010a)

Fragmentation is important, because "one of the major ways that habitat fragmentation affects biodiversity is by reduction in the amount of available habitat (such as rainforests, boreal forests, oceans, marshlands, etc.) for all organisms in an ecological niche. Habitat fragmentation invariably involves some amount of habitat destruction. Plants and other sessile organisms in these areas are usually directly destroyed. Mobile animals (especially birds and mammals) retreat into remnant patches of habitat. This can lead to crowding effects and increased competition." (WK 2012f)

This then means that "the remaining habitat fragments are smaller than the original habitat. Species that can move between fragments may use more than one fragment. Species which cannot move between fragments must make do with what is available in the single fragment in which they ended up. Since one of the major causes of habitat destruction is agricultural development, habitat fragments are rarely representative samples of the initial landscape." (WK 2012f)

Reduction (in Range)

And the third way to understand discontinuity in the context of natural resoruces has to do with "reduction" (in range), which, especially in biology, refers to the decrease of "the range or distribution of a species" in "the geographical area within which that species can be found." (WK 2012g)

There can be both natural and human causes of a change in range; for instance, a good instance of natural cause is for species to move to "different regions at different times of year," like "summer range and winter range." (WK 2012g)

But human agency can also play a role; for instance, a species' range can be "introduced by human agency (deliberately or accidentally), as well as where it has been re-introduced following extirpation" (or local extinction). (WK 2012g)

Range patterns can come in all shapes and sizes, and good examples include those which are "scattered/random (random placement)," "clustered/grouped (the majority are placed in one area)," "linear (their placements form a line)," "radial (placements form a ' x ' shape)," and "regular/ordered (they are not random at all, but follow a set placement, much like a grid)." (WK 2012g)

Multiple Aspects of Natural Resources

This introduction to "diversity" and "discontinuity" in natural resources thus shows their more complicated nature.

To illustrate this more complicated nature of natural resources further, consider the multiple aspects of natural resources below (as summarized in *Table 1.1*).

Source of Origin

The first aspect of natural resources to be illustrated here in regard to their more complicated nature concerns "source of origin."

On the basis of the "source of origin," natural resources can be divided into two main groups (as shown below): (WK 2012a)

- "Biotic" Resources—are those which "are obtained from the biosphere (living and organic material), such as forests, animals, birds, and fish and the materials that can be obtained from them. Fossil fuels such as coal and petroleum are also included in this category because they are formed from decayed organic matter."
- "Abiotic" Resources—are those "that come from non-living, non-organic material. Examples of abiotic resources include land, fresh water, air and heavy metals including ores such as gold, iron, copper, silver, etc."

Surely, there can be other examples in each group, so the ones indicated above are illustrative, not exhaustive.

Stage of Development

The second aspect of natural resources to be illustrated here in regard to their more complicated nature concerns "stage of development."

On the basis of the "stage of development," natural resources can be divided into the following groups: (WK 2012a)

- "Potential Resources"—are "those that exist in a region and may be used in the future. For example, petroleum may exist in many parts of India, having sedimentary rocks, but until the time it is actually drilled out and put into use, it remains a potential resource."
- "Actual Resources"—"are those that have been surveyed, their quantity and quality determined and are being used in present times. The development of an actual resource, such as wood processing depends upon the technology available and the cost involved."
- "Reserve Resources"—are those actual resources which "can be developed profitably in the future."
- "Stock Resources"—are "those that have been surveyed but cannot be used by organisms due to lack of technology. For example: hydrogen."

Of course, there can be other groups, but the ones here suffice for the purpose at hand.

Renewability

The third aspect of natural resources to be illustrated here in regard to their more complicated nature concerns "renewability."

On the basis of "renewability," natural resources can be divided into the following two categories: (WK 2012a; D. Cohen 2007; R. Bailey 2010)

- "Renewable Resources"—are those "that can be replenished naturally. Some of these resources, like sunlight, air, wind, etc., are continuously available and their quantity is not noticeably affected by human consumption. Though many renewable resources do not have such a rapid recovery rate, these resources are susceptible to depletion by over-use. Resources from a human use perspective are classified as renewable only so long as the rate of replenishment/recovery exceeds that of the rate of consumption."

- "Non-Renewable Resources"—are those "that form extremely slowly and those that do not naturally form in the environment. Minerals are the most common resource included in this category. By the human perspective, resources are non-renewable when their rate of consumption exceeds the rate of replenishment/recovery; a good example of this are fossil fuels, which are in this category because their rate of formation is extremely slow (potentially millions of years), meaning they are considered non-renewable. Some resources actually naturally deplete in amount without human interference, the most notable of these being radio-active elements such as uranium, which naturally decay into heavy metals"

In addition, some natural resources can be recycled, whereas others cannot; for instance, "the metallic minerals can be re-used by recycling them, but coal and petroleum cannot be recycled." (WK 2012a; D. Cohen 2007; R. Bailey 2010)

Earth's Spheres

And the foufth aspect of natural resources to be illustrated here in regard to their more complicated nature concerns "Earth's spheres."

On the basis of "Earth's spheres," natural resources can be studied in relation to the following spheres: (WK 2012h)

- The "atmosphere" (corresponding to air)
- The "cryosphere" (corresponding to ice)
- The "hydrosphere" (corresponding to water)
- The "lithosphere" (corresponding to rocks)
- The "pedosphere" (corresponding to soil)
- The "biosphere" (corresponding to life)

Surely, other related issues are also important, and good examples include those related to environmental science, geography (both human and physical), geology, and so on.

The Theoretical Debate

These multiple aspects of natural resources, together with the introduction to their "diversity" and "discontinuity," then lead us to a more important question concerning the theoretical debate on the very nature of natural resources. Indeed, this theoretical debate can also say much about the study of natural resources as a discipline over time.

Although there can be many different theoretical approaches in the debate—there are five main ones to be summarized hereafter, and they can be called, in the absence of better words, namely, (a) the *mental* argument, (b) the *natural* argument, (c) the *societal* argument, (d) the *cultural* argument, and (e) the *resilient* argument (which is my original contribution to the debate), to be elabroated hereafter, respectively.

Lest misunderstanding occurs, three clarifications should be made here, as this is something that I regularly emphasize in my previous books, whenever a theoretical debate is summarized for a topic, *almost verbatim*.

Firstly, there can be other arguments besides the first four here, that is, the "mental," the "natural," the "societal," and the "cultural" arguments (as indicated above). The advantage to select these four here has to do with their illustrative representation of the diverse theories in the literature (which are deemed sufficient for the purpose at hand in this book).

Secondly, the examples (theories) given to illustrate each argument are also supposed to be illustrative, not exhaustive. After all, the arguments will be addressed again in the rest of the book, with other examples (theories) to be introduced.

And thirdly, the four arguments are not necessarily mutually exclusive. For instance, those who advocate the "mental" argument can also consider the "cultural" argument, although they do not focus on the latter.

And the reverse is also true, in that those who make the "cultural" argument can also consider the "mental" argument, although they do not focus on the latter.

In other words, their disagreement is often one in degree, not in kind.

The Mental Argument

With these three clarifications in mind—the first major theoretical approach is called, in the absence of better words, the *mental* argument, which focuses, relatively speaking of course, more on the biological and psychological aspects of natural resources.

Thus, the word "mental" in the title of the argument is suggestive, because it refers to the focus on the physiological, biological and psychological aspects of natural resources so understood.

But the mental argument has different issues of concern, and different defenders of the argument offer their own distinctive stands on the issues, depending on their own interests.

For illustration, consider two major versions of the mental argument (with the first on diversity and the second on discontinuity), namely, (a) biodiversity and human health, and (b) extinction and overconsumption—to be addressed below, in that order.

Biodiversity and Human Health

The first version of the mental argument has to do with biodiversity and human health (in the context of diversity) in the study of natural resources.

An argument here is that there is a relationship between human health and biodiversity, because "many of the anticipated health risks of climate change are associated with changes in biodiversity (e.g. changes in populations and distribution of disease vectors, scarcity of fresh water, impacts on agricultural biodiversity and food resources etc.) This is because the species most likely to disappear are those that buffer against infectious disease transmission, while surviving species tend to be the ones that increase disease transmission, such as that of West Nile Virus, Lyme disease and Hantavirus, according to a study done and co-authored by Felicia Keesing, ecologist at Bard College, and Drew Harvell, associate director for Environment of the Atkinson Center for a Sustainable Future (ACSF) at Cornell University." (WK 2012i; K. Ramanujan 2010; CBD 2009)

However, for the critics, this study of biodiversity and human health has some major problems. Consider, for illustration, a few examples below.

Firstly, one main criticism of the study of biodiversity and human health is that there are many different causes of human health problems.

For example, for the critics, an individual may suffer from human health problems which are not necessarily related to biodiversity, and good examples include overconsumption, junk diet, personality disorders, etc.

And secondly, another main criticism of the study of biodiversity and human health is that there can be alternative resources for health improvement, even when biodiversity occurs.

For instance, for the critics, although it is true that "biodiversity provides critical support for drug discovery and the availability of medicinal resources" to improve human health—it is also important to point out that "the decline in output from the pharmaceutical sector since the mid-1980s can be attributed to a move away from natural product exploration ('bioprospecting') in favor of genomics and synthetic chemistry" for health improvement instead. (WK 2012i; E. Hawkins 1992)

Extinction and Overconsumption

And the second version of the mental argument has to do with extinction and overconsumption (in the context of discontinuity) in the study of natural resources.

An argument here is that there is a relationship between extinction and overconsumption, because human overconsumption can lead to the "overexploitation" of animals and other resources. (WK 2012i)

In other words, "overexploitation occurs when a resource is consumed at an unsustainable rate. This occurs on land in the form of overhunting, excessive logging, poor soil conservation in agriculture and the illegal wildlife trade. Joe Walston, director of the Wildlife Conservation Society's Asian programs, called the latter the 'single largest threat' to biodiversity in Asia. The international trade of endangered species is second in size only to drug trafficking." (WK 2012i; NYT 2010; B. Weirum 2007)

However, for the critics, this study of extinction and overconsumption has some major problems. Consider, for illustration, a few examples below.

Firstly, one main criticism of the study of extinction and overconsumption is that there are different threats to species, and overconsumption can be only one of them.

For instance, for the critics, there are other important threats like "habitat destruction," "genetic pollution/erosion," "climate change," "human overpopulation," etc. (WK 2012i)

And secondly, another main criticism of the study of extinction and overconsumption is that overexploitation does not necessarily lead to extinction of animals and other resources.

For instance, for the critics, "overexploitation need not necessarily lead to the destruction of the resource, nor is it necessarily unsustainable," although "depleting the numbers or amount of the resource can change its quality. For example, footstool palm is a wild palm tree found in Southeast Asia. Its leaves are used for thatching and food wrapping, and overharvesting has resulted in its leaf size becoming smaller," but not in extinction. (WK 2012j)

The Natural Argument

The second theoretical approach is called, in the absence of better words, the *natural* argument, which focuses, relatively speaking of course, more on the natural aspects of natural resources, especially in relation to the state of nature.

Therefore, the word "natural" in the title of the argument is suggestive, because it refers to the focus on the natural aspects of natural resources so understood.

Like the mental argument—the natural argument has different issues of concern, and different defenders of the argument offer their own distinctive stands on the issues, depending on their own interests.

For illustration, consider two major versions of the natural argument (with the first on diversity and the second on discontinuity), namely, (a) conservation biology and environmental protection, and (b) invasive species and reduction in range—to be summarized below, in that order.

Conservation Biology and Environmental Protection

The first version of the natural argument has to do with conservation biology and environmental protection (in the context of diversity) in the study of natural resources.

An argument here is that there is a relationship between conservation biology and environmental protection, in that conservation biology can contribute to the protection of "species, their habitats, and ecosystems from excessive rates of extinction and the erosion of biotic interactions," with the help of "sciences, economics, and the practice of natural resource management." (WK 2012c; S. Sahney 2008; B. Wilcox 1980; M. Soule 1986 & 198a; M. Hunter 1996; G. Meffe 2006; F. van Dyke 2008)

However, for the critics, this study of conservation biology and environmental protection has some major problems. Consider, for illustration, a few examples below.

Firstly, one main criticism of the study of conservation biology and environmental protection is that it does not give sufficient attention to the important problem of poor management.

For instance, for the critics, "the main issues for developing countries like Brazil and Mexico are that protected areas suffer from encroachment and poor management. In Brazil, protected areas are increasing, but there are significant challenges caused by human impacts. Logging and mining are potentially huge threats to protected areas." (WK 2012k)

And secondly, another main criticism of the study of conservation biology and environmental protection is that it often ignores the factors of power abuse and financial shortage.

For instance, for the critics, "African governments face several challenges in implementing environmental protection mechanisms. In Tanzania for example these include lack of financial resources to manage protected areas, poor governance and corruption...." (WK 2012k; D. Pallangyo 2007)

Invasive Species and Reduction in Range

And the second version of the natural argument has to do with invasive species and reduction in range (in the context of discontinuity) in the study of natural resources.

An argument here is that there is a relationship between invasive species and reduction in range, in that "biotic invasion is considered one of the five top drivers for global biodiversity loss," because "invasive species may drive local native species to extinction" (in most serious cases) or to reduction (in less serious cases) "via competitive exclusion, niche displacement, or hybridisation with related native species. Therefore, besides their economic ramifications, alien invasions may result in extensive changes in the structure, composition and global distribution of the biota of sites of introduction, leading ultimately to the homogenisation of the world's fauna and flora and the loss of biodiversity." (WK 2012l; L. Odendaal 2008)

However, for the critics, this study of invasive species and reduction in range has some major problems. Consider, for illustration, a few examples below.

Firstly, one main criticism of the study of invasive species and reduction in range is that more actual evidences are needed for the causal assertion.

For instance, for the critics, "concern over the impacts of invasive species on biodiversity must therefore consider the actual evidence (either ecological or economic), in relation to the potential risk," as sufficient solid scientific evidences are still lacking.

And secondly, another main criticism of the study of invasive species and reduction in range is that there are alternative causes to reduction in range.

For instance, for the critics, "it is difficult to unequivocally attribute extinctions" (in most serious cases) or "reduction" (in less serious cases) to "a species invasion, and the few scientific studies that have done so have been with animal taxa." (WK 2012l) After all, there can be other causes like "habitat destruction," "climate change," "human overpopulation," etc. (WK 2012i)

The Societal Argument

The third theoretical approach is called, in the absence of better words, the *societal* argument, which focuses, relatively speaking of course, more on the societal aspects of natural resources (e.g., organizational, institutional, structural, and technological issues).

In this sense, the word "societal" in the title of the argument is suggestive, because it refers to the societal aspects of natural resources so understood.

Like the mental and natural arguments—the societal argument has different issues of concern, and different defenders of the argument offer their own distinctive stands on the issues, depending on their own interests.

For illustration, consider two major versions of the societal argument (with the first on diversity and the second on discontinuity), namely, (a) the restoration of life, and the technology of cloning, and (b) the North-South divide, and environmental destruction—to be summarized in what follows, in that order.

The Restoration of Life, and the Technology of Cloning

The first version of the societal argument has to do with the restoration of life and the technology of cloning (in the context of diversity) in the study of natural resources.

An argument here is that there is a relationship between the restoration of life and the technology of cloning, in that, "by using DNA from the remains of an extinct species, through the process of cloning, the species may be 'brought back to life.' Proposed targets for cloning include the dinosaurs, the mammoth, thylacine, and the Pyrenean Ibex. For such a program to succeed, a sufficient number of individuals would have to be cloned, from the DNA of different individuals (in the case of sexually reproducing organisms) to create a viable population." (WK 2012e)

However, for the critics, this study of the restoration of life and the technology of cloning has some major problems. Consider, for illustration, a few examples below.

Firstly, one main criticism of this study of the restoration of life and the technology of cloning is that it is still technological infeasible in our time.

For instance, for the critics, "in 2003, scientists attempted to clone the extinct Pyrenean Ibex (*C. p. pyrenaica*). This initial attempt failed; of the 285 embryos reconstructed, 54 were transferred to 12 mountain goats and mountain goat-domestic goat hybrids, but only two survived the initial two months of gestation before they too died. In 2009, a second attempt was made to clone the Pyrenean Ibex; one clone was born alive, but died seven minutes later, due to physical defects in the lungs." (WK 2012e; S. Connor 2009; R. Gray 2009) And worse, "researchers have found several abnormalities in cloned organisms....The cloned organism may be born normal and resemble its non-cloned counterpart, but majority of the time will express changes in its genome later on in life." (WK 2012m; HEA 2001)

And secondly, another main criticism of this study of the restoration of life and the technology of cloning is that there remain some deep ethical issues to be resolved.

For instance, for the critics, there are deep "bioethical and philosophical objections" to "the cloning of extinct creatures," which have yet to be satisfactorily answered. (WK 2012e; A. Zitner 2000) Indeed, "many religious organizations oppose all forms of cloning, on the grounds that life begins at conception....From the standpoint of classical liberalism, concerns also exist regarding the protection of the identity of the individual and the right to protect one's genetic identity." (WK 2012m; M. Farley 2001)

The North-South Divide, and Environmental Destruction

And the second version of the societal argument has to do with the North-South divide, and environmental destruction (in the context of discontinuity) in the study of natural resources.

An argument here is that there is a relationship between the North-South divide and environmental destruction, in that "an increasing gap between rich and poor and the global North-South divide denies many access to basic human needs, rights, and education, leading to further environmental destruction," whereas those who are privileged enjoy better life instead. (WK 2012n; UN 2002; WB 2011; D. Dunphy 2007)

However, for the critics, this study on the North-South divide, and environmental destruction has some major problems. Consider, for illustration, a few examples below.

Firstly, one main criticism of the study on the North-South divide, and environmental destruction is that it is often driven by leftist ideology to close the divide.

For instance, for the critics, this leftist ideology does not answer the underlying cause for the divide, because "Global Capitalism...enhances social and economic gaps between countries, since it requires economies and societies to adapt in a very rapid manner, and because this almost never happens in an equal fashion, some nations grow faster than others.

Rich countries exploit poorer countries to a point where developing countries become dependent on developed countries for survival. The very structure and process of globalization perpetuates and reproduces unequal relationships and opportunities between the North and the South, it tends to 'favor the privileged and further marginalize the already disadvantaged.'" (WK 2012o; A. Kacowicz 2007)

And secondly, another main criticism of the study on the North-South divide, and environmental destruction is that there are alternative causes of environmental destruction.

For instance, for the critics, it is tempting to blame environmental destruction and thus does not give sufficient attention to alternative causes, and good examples include "climate change," "anthropogenic sources" (like war, mining, transport, etc.), and so on. (WK 2012n & 2012p)

The Cultural Argument

And the fourth theoretical approach is called, in the absence of better words, the *cultural* argument, which focuses, relatively speaking of course, more on the cultural aspects of natural resources (e.g., issues in relation to values and beliefs).

Therefore, the word "cultural" in the title of the argument is suggestive, because it refers to the focus on the cultural aspects of natural resources so understood.

Like the mental, natural, and societal arguments—the cultural argument has different issues of concern, and different defenders of the argument offer their own distinctive stands on the issues, depending on their own interests.

For illustration, consider two major versions of the cultural argument (with the first on diversity and the second on discontinuity), namely, (a) biodiversity and leisure activities, and (b) traditional knowledge and bioprospecting—to be summarized in what follows, in that order.

Biodiversity and Leisure Activities

The first version of the cultural argument has to do with biodiversity and leisure activities (in the context of diversity) in the study of natural resources.

An argument here is that there is a relationship between biodiversity and leisure activities, in that "biodiversity enriches leisure activities such as hiking, birdwatching or natural history study. Biodiversity inspires musicians, painters, sculptors, writers and other artists. Many cultures view themselves as an integral part of the natural world which requires them to respect other living organisms." (WK 2012i)

However, for the critics, the study of biodiversity and leisure activities has some major problems. Consider, for illustration, a few examples below.

Firstly, one main criticism of the study of biodiversity and leisure activities is that there are philosophical disagreements in regard to the use of biodiversity for leisure activities.

For instance, some people argue philosophically that "biodiversity has intrinsic aesthetic and spiritual value to mankind in and of itself," but others disagree, because they instead suggest that "tropical forests and other ecological realms are only worthy of conservation because of the services they provide." (WK 2012i)

And secondly, another main criticism of the study of biodiversity and leisure activities is that what is considered as a leisure activity in a culture may not be so at all in another culture, which may choose the use of different animals or natural resources instead.

For instance, for the critics, "the Chinese brought goldfish indoors during the Song Dynasty to enjoy them in large ceramic vessels," but "wealthy Romans kept lampreys and other fish in salt water pools" instead. (WK 2012q)

Traditional Knowledge and Bioprospecting

And the second version of the cultural argument has to do with traditional knowledge and bioprospecting (in the context of discontinuity) in the study of natural resources.

An argument here is that there is a relationship between traditional knowledge and bioprospecting, in that indigenous knowledge (or wisdom) in less advanced countries can be exploited by biopirates in developed countries for profits, so as to contribute to social fragmentation like the inequality of benefiting from biodiversity between the rich and the poor.

A good example is "biopiracy," which is "a situation where indigenous knowledge of nature, originating with indigenous people, is used by others for profit, without permission from and with little or no compensation or recognition to the indigenous people themselves. For example when bioprospectors draw on indigenous knowledge of medicinal plants which is later patented by medical companies without recognizing the fact that the knowledge is not new, or invented by the patenter, and depriving the indigenous community to the rights to commercial exploitation of the technology that they themselves had developed. Critics of this practice such as Greenpeace, claim these practices contribute to inequality between developing countries rich in biodiversity, and developed countries hosting companies that engage in 'biopiracy.'" (WK 2012r)

However, for the critics, the study of traditional knowledge and bioprospecting has some major problems. Consider, for illustration, a few examples below.

Firstly, one main criticism of the study of traditional knowledge and bioprospecting is that it can become quickly political in tone (without examining the actual evidences more carefully).

For instance, for the critics, "one common misunderstanding is that pharmaceutical companies patent the plants they collect. While obtaining a patent on a naturally occurring organism as previously known or used is not possible, patents may be taken out on specific chemicals isolated or developed from plants. Often these patents are obtained with a stated and researched use of those chemicals....As a result, even if the indigenous medical knowledge is taken..., that knowledge does not by itself make the active chemical compound 'obvious,' which is the standard applied under patent law." (WK 2012r)

And secondly, another main criticism of the study of traditional knowledge and bioprospecting is that there are legal protections of indigenous knowledge for fair use.

For instance, for the critics, a good case in point is the "Convention on Biological Diversity (CBD)," which "came into force in 1993. It secured rights to control access to genetic resources for the countries in which those resources are located. One objective of the CBD is to enable lesser-developed countries to better benefit from their resources and traditional knowledge. Under the rules of the CBD, bioprospectors are required to obtain informed consent to access such resources, and must share any benefits with the biodiversity-rich country." (WK 2012r) This is so, even though some opponents of CBD argued that "the CBD has failed to establish appropriate regulations to prevent biopiracy." (WK 2012r)

The Resilient Argument

And fifthly, unlike the previous four arguments (viz., the mental, the natural, the societal, and the cultural)—the fifth one to be introduced here can be called, in the absence of better words again, the *resilient* argument, which learns from all other theoretical approaches (which include the previous four arguments and also other examples as will be introduced in the rest of this book), without, however, favoring any of them, nor trying to integrate them (as they are not necessarily compatible with each other, as the introduction of the theoretical debate in the previous sub-sections has shown).

The Resilient Theory of Natural Resources

My *resilient* argument can be more precisely called *the resilient theory of natural resources*, since it is to propose a new theory of diversity and discontinuity in the study of natural resources.

My *resilient theory of natural resources* has four major distinct features.

Firstly, my theory makes good use of all theoretical approaches to the study of natural resources in relation to diversity and discontinuity, be they about the mental argument, the natural argument, the societal argument, or the cultural argument—especially from the most comprehensive combined perspectives of the mind, nature, society, and culture (as will be dealt with in Chapter Two and Chapter Three).

Secondly, just like many other theories of mine in my previous books, my theory here does not heavily favor any specific theory over others in the literature, nor trying to integrate them (as they are not necessarily compatible with each other).

Thirdly, my theory treats the issue of natural resources in the distinctive dialectic context of no diversity without discontinuity, and vice versa.

And fourthly, it contains thirty major theses, namely, (a) the first thesis: the formalness-informalness principle, (b) the second thesis: the absoluteness-relativeness principle, (c) the third thesis: the partiality-totality principle, (d) the fourth thesis: the predictability-unpredictability principle, (e) the fifth thesis: the explicability-inexplicability principle, (f) the sixth thesis: the fiction-reality principle, (g) the seventh thesis: the cognitiveness-noncognitiveness principle, (h) the eighth thesis: the finiteness-transfiniteness principle, (i) the ninth thesis: the preciseness-vagueness principle, (j) the tenth thesis: the simpleness-complicatedness principle, (k) the eleventh thesis: the openness-hiddenness principle, (l) the twelfth thesis: the denseness-emptiness principle, (m) the thirteenth thesis: the change-constancy principle, (n) the fourteenth thesis: the order-chaos principle, (o) the fifteenth thesis: the slowness-quickness principle, (p) the sixteenth thesis: the expansion-contraction principle, (q) the seventeenth thesis: the theory-praxis principle, (r) the eighteenth thesis: the convention-novelty principle, (s) the nineteenth thesis: the evolution-transformation principle, (t) the twentieth thesis: the symmetry-asymmetry principle, (u) the twenty-first thesis: the softness-hardness principle, (v) the twenty-second thesis: the seriousness-playfulness principle, (w) the twenty-third thesis: the activeness-inactiveness principle, (x) the twenty-fourth thesis: the selfness-otherness principle, (y) the twenty-fifth thesis: the regression-progression principle, (z) the twenty-sixth thesis: the sameness-difference principle, (aa) the twenty-seventh thesis: the stability-reaction principle, (bb) the twenty-eighth thesis: the functionality-nonfunctionality principle, (cc) the twenty-ninth thesis: the intentionality-nonintentionality principle, and (dd) the thirtieth thesis: the post-human resilience—to be elaborated in the rest of the book, with a summary in the concluding chapter.

Because of the last thesis on "the post-human resilience," the word "resilient" in the title of my theory is suggestive, because it refers to the adaptive uses of different resources in the face of changed circumstances over time—which then require a multi-faceted exploration of the dialectic relations between diversity and discontinuity, as well as an understanding of the different views regarding these dialectic relations in the study of natural resources (together with other debates as will be discussed in the rest of the book)—and in the end it calls for a need to transcend them by future advanced humans and later post-humans in the context of the great transformations of natural resources in the future world (as will be discussed in the concluding chapter, Chapter Four).

In addition, this book on natural resources is a sequel to my previous books on diversity and discontinuity. The difference, however, is that this book focuses more on diversity and discontinuity in the natural world, whereas those previous books of mine (like *FHC*, *FCD*, *FPHC*, *FPHU*, *FC*, *FPHMIG*, and *BDPD*) which addressed, among many other things, diversity and discontinuity focus more on the human world instead.

With this clarification in mind—the seminal project here, if successful, will fundamentally change the way that we think about diveristy and discontinuity in the study of natural resources together with other debates as will be discussed in the rest of the book), from the combined perspectives of the mind, nature, society, and culture, with enormous implications for the human future and what I originally called its "post-human" fate.

Theory and Meta-Theory

The *resilient* argument, or *the resilient theory of natural resources*, presupposes some kind of meta-theory, in special relation to methodology and ontology—just as all theories, either explicitly or implicitly, require their own versions of meta-theory.

Meta-Theory

As so often clarified in my previous books—theory and meta-theory, broadly speaking, can be further classified into different categories, as shown below:

- On Theory
  - —Ex: in relation to the mind
  - —Ex: in relation to nature
  - —Ex: in relation to society
  - —Ex: in relation to culture
- On Meta-Theory
  - —Ex: in relation to methodology
  - —Ex: in relation to ontology

Each category, in its turn, can then be further sub-divided into different levels of analysis—which will be elaborated in the next two sections (and summarized in *Table 4.43*).

With this reminder in mind, I already proposed (in my previous books) my distinctive contribution to meta-theory, namely, methodology (like "theory of methodological holism," "complex theory of data analysis," and "critical-dialectic theory of formal science," for instance) and ontology (like "theory of existential dialectics" and "contrastive theory of rationality," for instance).

The next two sections are to introduce my meta-theory—especially, existential dialectics and sophisticated methodological holism, respectively—to be repeated, *often verbatim* (with only a few updated revisions), from my previous works, as this is something that I often do in each new book of mine as an introduction to some background information for the convenience of the readers.

In the process, I shall also introduce my numerous new theories (or to be exact, 62 of them in 54 books so far, as of July 2012) in relation to the four categories of the mind, nature, society, and culture in the context of theory (as also summarized in *Table 4.43*).

A Unified Theory of Everything

In the end, my distinctive contribution to both meta-theory and theory serves as a foundation to unify all domains of knowledge (i.e., the social sciences, the formal sciences, the natural sciences, and the humanities) for an overarching understanding of all forms of reality (by way of some ontological principles at the meta-theoretical level and the comprehensive perspectives of the mind, nature, society, and culture at the theoretical level—as will be explained in the next two sections).

This is so, without commiting the sins of reductionism and reverse-reductionism (as often seen in many holistic approaches, with the fad of systems approach as a most recent notorious example, as already critically explained in *The Future of Post-Human Formal Science*, or *FPHFS*), since my distinctive contribution makes good use of different schools of thought without heavily favoring any of them nor trying to integrate them (as they are not necessarily compatible with each other), so as to include (or adjust for) heterogeneity, conflict, subjectivity, and complexity, for example.

In this sense, the word "unified" does not have to mean an integrative approach, in a narrow sense, nor a systems approach, in a broad one, both of which I have critically rebuked (in my previous books), while learning from them.

This overarching project (consisting of numerous theories of mine, as will be summarized in the next section), thus, constitutes my grand unifed theory of everything in the world and beyond.

The Logic of Existential Dialectics

To start, the ontology of existential dialectics can be summarized in this section—*often verbatim* (with only a few updated revisions), from my previous works, as this is something that I regularly do in each new book of mine as an introduction to some background information for the convenience of the readers.

After all, this book, like all others of mine, is in conversation with all previous ones of mine, for the final aim to converge all of them into a larger grand project on the future of intelligent life, both here on earth and there in deep space unto multiverses.

The summary can be organized in relation to (a) the conception of existential dialectics (or its ontics), (b) the syntax of existential dialectics (or its ontomethodology and ontologic), (c) the semantics of existential dialectics (or its ontosemantics), and (d) the pragmatics of existential dialectics (or its ontopragmatics), respectively hereafter—again *often verbatim*.

This means that, for those readers who had read some of my previous books, the summary below is really not much new (*almost verbatim*, with only some updated revisions).

But for those who have never read my previous books, it is a good review of them.

The Conception of Existential Dialectics (or Its Ontics)

This ontology was originally designated as "existential dialectics" in *Beyond Capitalism to Post-Capitalism* (herein abbreviated as *BCPC*), although it was already analyzed in other books of mine like the 2 volumes of *The Future of Human Civilization* (*FHC*), *The Future of Capitalism and Democracy* (*FCD*), *The Future of Post-Human Consciousness* (*FPHC*), and the 2 volumes of *Beyond Democracy to Post-Democracy* (*BDPD*).

Lest any confusion occurs, it is important to stress at the outset that the word 'existential' in "existential dialectics" has nothing to do with Existentialism, which I rebuked in *FHC*, *FCD*, and also *FPHC*. Instead, the word here merely refers to the existence of intelligent life (both primitive and advanced) in a broad sense.

Some Basic Concepts

The conception of existential dialectics (or its ontics) makes use of different concepts (like "sets," "elements," "relations," "operations," "functions," "truth values," "axioms," "postulates," and "principles"—as shown in *Table 4.3*), which are important for the understanding of any logic of ontology.

That clarified—existential dialectics, as a language of ontology, can be analyzed in different ways that a language, as an analogy, is often analyzed, namely, in relation to phonology ("the study of patterns of a language's sounds"), phonetics ("the study of the physical aspects of sounds of human language"), morphonology ("the study of the internal structure of words"), syntax ("the study of how words combine to form grammatical sentences"), semantics ("the study of the meaning of words [lexical semantics] and fixed word combinations [phraseology]"), and pragmatics ("the study of how utterances are used…literally, figuratively, or otherwise…in communicative acts"). (WK 2007)

But since my theory of existential dialectics makes use of the English language for communication here, it does not propose a new way to make the patterns of a language's sounds (as in phonology), to study the physical aspects of sounds (as in phonetics), or even to strive for a new organization of the internal structure of words for the English language (as in morphonology).

For this reason, the language of existential dialectics to be analyzed hereafter is not concerned with the phonological, phonetic, and morphonological aspects of the English language. Instead, the analysis will explore the syntax, semantics, and pragmatics of existential dialectics as a language of ontology to improve our understanding of reality.

The Ontic Dispute

But first, there is an ontic dispute to be addressed, which is related to what I called "the dilemma of ontology" as introduced earlier in *Sec. 1.4* of *The Future of Post-Human Knowledge* (*FPHK*).

A good point of departure concerns what I want to call *the ontic dispute*, on how to understand the very nature of ontology.

Traditionally, scholars in the field used to study the conception of ontology in terms of its entities (e.g., material objects, abstract numbers), properties (e.g., duration, plurality), and relationships (e.g., causation).

This is a good starting point, but it suffers from different problems. Consider three examples of bias below, which are quite common in the literature, and the recent version (of treating concepts and principles in ontology on the basis of "plurality," "dynamism," "duration," "interaction," "life," "consciousness," and "volition") by Reginald Firehammer (2005) only constitutes a latest addition to this old (impoverishing) tradition.

Firstly, the classification of entities, properties, and relationships is too material-centric, in often focusing more on the existence of material entities than on non-material ones.

Secondly, the classification is too anthropo-centric, in often favoring the factor of human agency in depicting reality.

And lastly, the classification is too confused among the three categories relating to each other, in often lumping them together without adequate consideration of their intricated interrelationships (both in relation to each and in relation to all).

A good alternative is precisely what I want to offer in this debate, that is, a new classification of ontological entities, properties, and their relationships in terms of *method*, *structure*, *process*, *agency*, and *outcome* instead. Let me explicate what these words mean below.

In relation to "method"—it concerns how ontology is to be studied. In this sense, the word 'method' here should not be confused with the different usage of the same term 'method' as a technique of investigation in the context of methodology (as already introduced in *Sec. 1.2* of *FPHK*).

In relation to "structure"—it concerns what the general patterns of entities, properties, and relationships are.

In relation to "process"—it concerns how these general patterns change over time.

In relation to "agency"—it concerns how agency affects any change of these general patterns over time (without putting, however, too much emphasis on the role of human agency).

And in relation to "outcome"—it concerns what the outputs of this change over time are.

The Syntax of Existential Dialectics (or Its Ontomethodology and Ontologic)

The syntax of existential dialectics, analogously speaking, refers here to the structure of an ontology which can be studied by way of combining ideas into complex relationships like ontological principles to depict reality in the world—and can be analyzed in seven sub-sections, namely, (a) the selection criteria, (b) further clarifications on classification, (c) ontomethodology, (d) ontologic, (e) the principles as short cuts, (f) the principles as family resemblances, and (g) the dialectic constraints imposed by the principles, respectively hereafter.

The Selection Criteria

To start, not any pair of relationship can be chosen as an ontological principle in existential dialectics; otherwise, there could be billions of them under the sky.

Six illustrations below suffice to clarify this point.

(a) Firstly, an ontological principle cannot *overlap* with other principles, in that it is too closely related with other ones.

Of course, one cannot totally rule out any relationship between two entities, especially in a complex world of everything being linked to everything else—but the comparison here is relative, not absolute.

In this sense, the flexibility-inflexibility pair cannot be an ontological principle, since it has something closely in common with other ones (like the simpleness-complicatedness principle, for instance).

As an illustration, something which is flexible in interpretation is more likely to allow multiple viewpoints and thus makes the overall picture more complicated than otherwise (simple).

(b) Secondly, an ontological principle cannot be *redundant* in relation to other principles, in that it somehow duplicates other principles.

In this sense, for instance, the directness-indirectness pair cannot be an ontological principle, because it somehow duplicates other ontological principles (like the preciseness-vagueness principle, for instance).

As an illustration, something which is direct means to be right to the point and does not need to go to an unnecessarily long loop—so it is less vague, in being clearer about the thing in question.

(c) Thirdly, an ontological principle cannot be *derived* from other ones—as if it is a child-parent relationship.

In this sense, the convenience-inconvenience pair cannot be an ontological principle, because it can be derived from other ones (like the slowness-quickness principle, for instance).

For instance, something which is convenient already implies that it can be obtained quickly (rather than slowly)—all things considered.

(d) Fourthly, an ontological principle cannot be *trivial* (or *parochial*)—as if it is only one component competing with other ones for inclusion in a set.

In this sense, the consistency-inconsistency pair cannot be an ontological principle, because its validity is limited (or parochial), as it can be easily challenged and replaced by opposing major (not minor) theories (e.g., the correspondence theory of truth, the pragmatic theory of truth, the semantic theory of truth, the disclosedness theory of truth, and whatnot), not just in accordance to the consistency (or coherence) theory of truth (for consistency).

Thus, the consistency argument is only one among different opposing arguments in relation to the logic of what constitutes "truth."

(e) Fifthly, an ontological principle cannot be *one-sided*, that is, part of one side (category) of a dialectic pair.

In this sense, the extroversion-introversion pair cannot be an ontological principle, since it is part of the side (category) of "normality" in the normality-abnormaltiy pair (because both extroversion and introversion are two different "normal" personality traits in psychology).

In addition, the normality-abnormality pair is already related to the symmetry-asymmetry principle, the density-emptiness principle, the convention-novelty principle, and the same-difference principle (as will be explained in the section on "the principles as family resemblances" later).

(f) And sixthly, an ontological principle cannot be *misleading*, in that the two apparent opposites are not really opposites at all, upon closer examination.

In this sense, the east-west pair cannot be an ontological principle, since east and west are not really opposites, as there are also north, south, and center to form a set of five opposing elements.

By the same logic, the left-right pair cannot be an ontological principle, since left and right (width) are not really opposites, as there are also top and bottom (height), back and front (depth), and center to form a set of seven elements in four categories. In the field of visual design, this set of seven elements is known as "form" (as opposed to "function"), as addressed in *FPHARCH*. (WK 2011nn)

Although these six examples (criteria) are not exhaustive, they illustrative the selectiveness of any ontological principle to be existentially dialectic.

Further Clarifications on Classification

Lest any misunderstanding carelessly occurs, some further clarifications on classification should be made here (and summarized in *Table 4.8*).

(a) Firstly, the total number of ontological principles is unknown, to be discovered later, as our knowledge of the world becomes more advanced.

Consequently, the principles as introduced in my books are not exhaustive, with new ones being added, whenever more of them are discovered in later research.

At least, future generations can pick up where I leave off and continue the discovery.

In this sense, my work should be treated as a pioneering effort for the development of a systematic, comprehensive analysis of a new general ontology for the future of knowledge.

(b) Secondly, the principles are not rigidly classified, as they can be reclassified in a different way.

For instance, the preciseness-vagueness principle is classified under the category of "structure" but can be reclassified under the category of "method," although in so doing, it has a different meaning in the context of method.

The same logic applies to the same-difference principle under the category of "outcome," which can be reclassified under the category of "structure," although in so doing, once more, it has a different meaning in the context of structure.

(c) Thirdly, the principles are to be understood together, not that each prinicple is to be treated independently of others. After all, the principles are mutually constraining, in that they work together as a whole.

(d) Fourthly, the principles are relevant to all subject matters, but some principles are, relatively speaking of course, more useful to some subject matters than others under certain conditions—as implied in the symmetry-asymmetry principle.

(e) And fifthly, the principles constitute only two levels of analysis, this time, at the ontological and methodological levels—while other levels of analysis (from the perspectives of the mind, nature, society, and culture) are also needed, in order to understand reality in its totality (as explained in the section on "sophisticated methodological holism," so as to avoid the dual dangers of reductionism and reverse-reductionism).

Ontomethodology

With these selection criteria and clarifications in mind—the syntax of existential dialectics can be divided into two main smaller sections for analysis, since they are related to each other in the context of some fundamental principles, that is, (a) ontomethodology and (b) ontologic, to be addressed hereafter, respectively.

In the context of ontomethodology, some ontological principles are proposed here in relation to the nature of methodology, which however is tied up with the general issue of ontology in the context of knowledge.

In this sense, ontomethodology (in the context of these ontological principles in the category of method) are related with ontologic in general (in the context of ontological principles in general).

Some good instances of ontological principles involving ontomethodology include the formalness-informalness principle, the absoluteness-relativeness principle, the partiality-totality principle, the predictability-unpredictability principle, the explicability-inexplicability principle, the fiction-reality principle, and the cognitiveness-noncognitiveness principle, which will be introduced in conjunction with ontological principles in the ontologic of existential dialectics in general hereafter.

Ontologic

With this clarification of the relationship between ontomethodology and ontologic in mind (which is comparable to the relationship between methodology and ontology as described in *Sec. 1.3* of *FPHK*)—the first three principles in existential dialectics were introduced in *BCPC*. And more principles were later added to the logic in my subsequent books (like *FC*, *FAE*, and others), as explained below.

More specifically, in *BCPC*, I proposed three principles (based on previous works of mine, not just something out of the blue) for the ontological logic, namely, (a) the regression-progression principle on the direction of history, (b) the symmetry-asymmetry principle on the relationships among existents, and (c) the change-constancy principle on the alteration of things—or in a more elegant term, the dynamics of space-time (in *FC*).

In *The Future of Complexity* (*FC*), I added three more principles, on top of the three aforementioned, namely, (d) the partiality-totality principle on the relationships between parts and whole, (e) the order-chaos principle on the pattern of things, and (f) the predictability-unpredictability principle on the occurrence of things—as part of the ontological logic of existential dialectics.

In *The Future of Aesthetic Experience* (or simply *FAE*), another principle is added, namely, (g) the evolution-transformation principle (on the multiple kinds of agency).

Also in *FAE*, I then reclassified the 7 principles into four categories, namely, (i) in relation to method, for the partiality-totality principle and the predictability-unpredictability principle, (ii) in relation to process, for the change-constancy principle and the order-chaos principle, (iii) in relation to agency, for the symmetry-asymmetry principle and the evolution-transformation principle, and (iv) in relation to outcome, for the regression-progression principle.

In *The Rise of Authoritarian Liberal Democracy* (*ALD*), I added two more principles, namely, (h) the softness-hardness principle on the force of agency (which is to be classified under the category about agency) and (i) the same-difference principle on the metamorphosis of change (which is to be classified under the category about outcome).

In *The Future of Information Architecture* (*FIA*), I introduce three additional principles, that is, (j) the simpleness-complicadness principle on the interconnection among things, (k) the preciseness-vagueness principle on the refinement of things, and (l) the slowness-quickness principle on the speed of change—with the first two in relation to structure and the third in relation to process.

In *The Future of Post-Human Unconsciousness* (*FPHU*), I added three more principles, that is, (m) the openness-hidden principle on the detection of things, in relation to structure, (n) the explicability-inexplicability principle on the underlying mechanisms of things, in relation to method, and (o) the expansion-contraction principle on the growth of things, in relation to process.

In *The Future of Post-Human Knowledge* (*FPHK*), I added one more principle, that is, (p) the absoluteness-relativeness principle on the multiplicity of entities, in relation to method.

In *The Future of Post-Human Mathematical Logic* (*FPHML*), I added one more principle, that is, (q) the formalness-informalness principle on the formal requirements of systems, in relation to method.

In *The Future of Post-Human Engineering* (*FPHE*), I added one more principle, that is, (r) the theory-praxis principle on the duality of knowledge, in relation to agency.

In *The Future of Post-Human Creative Thinking* (*FPHCT*), I added one more principle, that is, (s) the convention-novelty principle on the nature of creative thinking, in relation to agency.

In *The Future of Post-Human Geometry* (*FPHG*) here, I added one more principle, that is, (t) the finiteness-transfiniteness principle on the nature of numbers, in relation to structure.

In *The Future of Post-Human Urban Planning* (*FPHUP*), I added one more principle, that is, (u) the denseness-emptiness principle on the distribution of entities in space, in relation to structure.

In *The Future of Post-Human Humor* (*FPHH*), I added one more principle, that is, (v) the seriousness-playfulness principle on the extent of seriousness, in relation to agency.

In *The Future of Post-Human Literature* (*FPHLIT*), I added one more principle, that is, (w) the fiction-reality principle on the different faces of reality, in relation to method.

In *The Future of Post-Human Chemistry* (*FPHCHEM*), I added one more principle, that is, (x) the stability-reaction principle on the feedback mechanism for further changes, in relation to outcome.

In *The Future of Post-Human Transportation* (*FPHTRAN*), I added one more principle, that is, (y) the activeness-inactiveness principle on the extent of activeness, in relation to agency.

In *The Future of Post-Human Culinary Art* (*FPHCA*), I added two more principles, that is, (z) the selfness-otherness principle on the magnitude of self-centeredness in relation to agency and (aa) the functionality-nonfunctionality principle on the presence of function in relation to outcome.

In *The Future of Post-Human Waste* (*FPHW*), I added one more principle, that is, (bb) the intentionality-nonintentionality principle on the planning for outcomes in relation to outcome.

And in *Beyond Natural Resources to Post-Human Resources* (*BNR*), I add one more principle, that is, (cc) the cognitiveness-noncognitiveness principle on the understanding of meaning in relation to method.

With this update, there are twenty-nine principles so far in existential dialectics, that is:

- seven principles in the category of "method"—viz., the formalness-informalness principle, the absoluteness-relativeness principle, the partiality-totality principle, the predictability-unpredictability

principle, the explicability-inexplicability principle, the fiction-reality principle, and the cognitiveness-noncognitiveness principle
- five principles in the category of "structure"—viz., the finiteness-transfiniteness principle, the simpleness-complicatedness principle, the preciseness-vagueness principle, the openness-hiddenness principle, and the denseness-emptiness principle
- four principles in the category of "process"—viz., the change-constancy principle, the slowness-quickness principle, the order-chaos principle, and the expansion-contraction principle
- eight principles in the category of "agency"—viz., the theory-praxis principle, the convention-novelty principle, the symmetry-asymmetry principle, the evolution-transformation principle, the softness-hardness principle, the seriousness-playfulness principle, the activeness-inactiveness principle, and the selfness-otherness principle
- five principles in the category of "outcome"—viz., the regression-progression principle, the same-difference principle, the stability-reaction principle, the functionality-nonfunctionality principle, and the intentionality-nonintentionality principle

There can be more than twenty-nine principles in existential dialectics, but they are what I have discovered so far (as summarized in *Table 4.4*).

The Principles as Short Cuts

Yet, these principles should be treated with caution, lest misunderstanding occurs, since they do not constitute rigid dualities (or dichotomies).

The reason is that each pair in an ontological principle consists of two opposites, which are, however, merely short cuts both for multiple variations and degrees, as well as for different interactions with multiple other entities unlike them.

In this light, each pair can end up having hundred (or even thousand, if not more) different versions, which interact with hundred (or even thousand, if not more) other entities—as summarized in *Table 4.5*.

There are two clarifications here.

Firstly, this conception of shortcuts is not mutually exclusive nor absolute, in that the opposites can come in all shapes and sizes, with different degrees. For instance, by analogy, just as there are different degrees of the two colors "white" and "black"—there are likewise different degrees of the opposites in each ontological pair, to the extent that there can be multiple entities (not only two) interacting with each other in each pair.

And secondly, this conception of shortcuts do not ignore other possible entities in interacting with the two opposites (with their different versions) in each pair. For instance, by analogy, there are not only the two colors "white" and "black" as opposites, since there are other colors too besides them like "yellow," "green," "purple," or else. The same logic can be applied to each ontological pair, in that they also interact with other entities, not with only two of them (with their different versions).

Therefore, with these two clarifications in mind—each pair in an ontological principle serves only as an abbreviation for something more complicated and, therefore, although it contains two opposites, it should not be confused as a duality (dualism).

Existential dialectics rejects any dualism (or dichotomy) as too rigid and instead allows the multiplicity of entities, to the extent that between the two opposites in each pair exist many other alternatives to choose from. They are named in that short form for aesthetic elegance only, instead of listing all possible entities between the two opposites in the title (which would be too cumbersome and sound extremely awkward).

To be dialectic is to go beyond any rigid dichotomy and transcend into something different altogether in the long haul. One may be tempted to call the dialectic logic here with a different name like existential "multilectics" (instead of "dialectics"), but this naming is incorrect (or even misleading), for the two reasons aforestated.

After all, all the possible entities between any two opposites in an ontological pair are still about the two opposites, although each ontological pair come in all shapes and sizes (with different degrees) and do not exist by themselves but also interact with other entities unlike them. The virtue of revealing an ontological pair is to show how they relate within themselves (in multiple versions) and also interact with others unlike them (also in multiple versions).

This will be clear in a later sub-section titled "the dialectic constraints imposed by the principles."

The Principles as Family Resemblances

In addition, each ontological principle is generic, with some other comparable ontological pairs to be put in the same family (like a *family resemblance*)—as summarized in *Table 4.6*.

Ludwig Wittgenstein (1953) once suggested the idea of "family resemblance" in explaining different games classified under the same family called *games*.

Why should, for instance, playing football and chess as playing "games," when it is well understood that football is not the same as chess?

The answer is that, although each game is different and has different rules, many of them (though not all) share, more or less, some commonalites (e.g., scoring as necessary for winning).

And this is so, even though some games share more than some others in any given selection of critieria, and no two games are exactly identical. So, Wittgenstein's point here is that there is no essential core which is common to all games, and the best that one can look for is some characteristics which are common to many (but not all) games. (A. Biletzki 2006)

(a) Appropriate Family Resemblances

In the context of existential dialectics, I want to distinguish two types of family resemblances, and, in the absence of better words, let me call them (a) *appropriate* family resemblances and (b) *inappropriate* family resemblances.

In regard to appropriate family resemblances—good examples include some of the pairs as already explained in an earlier section on selection criteria, like the flexibility-inflexibility pair, which can be put in the family resemblance of the simpleness-complicatedness principle, although the two pairs are not exactly identical. Likewise, the directness-indirectness pair can be put in the family resemblance of the preciseness-vagueness principle, although, again, the two pairs are not exactly identical.

In this sense, which specific pair in a family should be used to designate the name of the family can be at times a bit arbitrary, but with good reason.

For illustration, in the context of method, the partiality-totality principle can take the different form like individualisticness-holisticness, just as the explicability-inexplicability principle can take the different form like underlyingness-regularness—although each two pairs are not exactly identical and have slightly different meanings and usages.

A more comprehensive listing of this family resemblance for all other principles is shown in *Table 4.6* on the syntax of existential dialectics.

In fact, this section on family resemblances overlaps with the semantics of existential dialectics, which will be introduced later in the next section (and summarized in *Table 4.11*).

(b) Inappropriate Family Resemblances

However, there are some family resemblances which are not appropriate, and let me distinugish them by calling them, in the absence of better words, (b1) *inadequate* family resemblances, (b2) *compound* family resemblances, (b3) *one-sided* family resemblances, and (b4) *vague* family resemblances (as summarized in *Table 4.10*).

(b1) In regard to "inadequate" family resemblances—there are some pairs which are not adequate, because they fail the test of selection criteria (as already pointed out in an earlier section), especially in relation to the criterion of not being *trivial* (or *parochial*)—as if it is only one component competing with other ones for inclusion in a set.

For instance, the *consistency-inconsistency* pair is inadequate, because it is only one component competing with other ones for inclusion in the larger set of the formalness-informalenss principle (like the *completeness-incompleteness* pair and the *soundness-unsoundness* pair, as discussed in my book titled *The Future of Post-Human Mathematical Logic*, or *FPHML* in short).

The same is true for the *cognition-noncognition* pair, which is related with both the completeness-incompleteness pair and the soundness-unsoundness pair in the larger set of the formalness-informalenss principle.

(b2) In regard to "compound" (or non-atomic) family resemblances—there are some pairs which are not appropriate for a different reason, because they fail the test of selection criteria (as already pointed out in an earlier section), especially in relation to the criterion of not *overlapping* with other principles, in that it is too closely related with other ones.

For instance, the *normality-abnormality* pair is compound (and thus inappropriate), because it overlaps with several other principles, like the symmetry-asymmetry principle, the density-emptiness principle, the convention-novelty principle, and the same-difference principle, for instance (as already discussed in *FPHP*).

The normality-abnormality pair overlaps with the symmetry-asymmetry principle, because being "abnormal" (say, for those with mental illnesses) is often treated not equally by others who are "normal." So, the relationship reveals the symmetry-asymmetry principle.

The normality-abnormality pair also overlaps with the density-emptiness principle, because being "abnormal" (say, for those with mental illnesses) is often not as dense in the overall (not local) distribution of a population as those who are "normal." So, the relationship reveals the density-emptiness principle.

The normality-abnormality pair also overlaps with the convention-novelty principle, because being "abnormal" (say, for those with mental illnesses) often exhibits non-conventional (non-conformist) ideas and behaviors when contrasted with those who are "normal." So, the relationship reveals the convention-novelty principle.

The normality-abnormality pair also overlaps with the same-difference principle, because being "abnormal" (say, for those with mental illnesses) shares some similarities in outcome when compared with those who are "normal" (in that both contribute to the control and regulation of human thoughts and behaviors in society, albeit in different ways). Yet, being "abnormal" also has something different from being "normal," in that abnormal thoughts and behaviors follow a different pattern when contrasted with those which are normal. So, the relationship reveals the same-difference principle.

(b2.1) There is a special case of compound family resemblances (as summarized in *Table 4.10*), which can be called, in the absence of better words, *one-sided and compound* family resemblances.

For instance, the extroversion-introversion pair is a good example of *one-sided and compound* family resemblances, because, firstly, it is "one-sided," in being part of the side (category) of "normality" in the normality-abnormality pair—and, secondly, the normality-abnormality pair is in turn "compounded," because it is related to the symmetry-asymmetry principle, the density-emptiness principle, the convention-novelty principle, and the same-difference principle, as already explained above.

(b3) In regard to "one-sided" family resemblances—there are some pairs which are not appropriate for another different reason, because they fail the test of selection criteria (as already pointed out in an earlier section), especially in relation to the criterion of not being *one-sided*, that is, part of one side (category) of a dialectic pair.

For instance, the extroversion-introversion pair cannot be an ontological principle, since it is part of the side (category) of "normality" in the normality-abnormaltiy pair (because both extroversion and introversion are two different "normal" personality traits in psychology).

By the same logic, the vision-blindness pair and the sound-silence pair cannot be ontological principles, because the pairs are part of the side (category) of "asymmetry" in the symmetry-asymmetry principle (since being able to see and hear for some individuals and being unable to see or hear for others are two contrastive abilities in the world of animals and humans, for instance).

These illustrations of indequate, compound, and one-sided family resemblances show how difficult it is to come up with an ontological principle in existential dialectics out of the blue, because many of them violate the selection criteria in one way or the other. It is all too easy to join two opposing words and simply make them into a pair, but it is all too difficult to pass the test of the selection criteria to be qualified as an ontological principle in existential dialectics.

(b4) And in regard to "vague" family resemblances—there are some pairs which are not appropriate for still another different reason, because the terms in the pairs are vague in meaning, so that they are inappropriate family resemblances if defined in some particular ways, although they can be appropriate family resemblances if defined in other particular ways.

For instance, such pairs like shortness-longness, narrowness-broadness, smallness-largeness, smallness-greatness, microness-macroness, and littleness-muchness can be understood in at least three different meanings and thus have three different possibilities. The first possibility is that they are one-sided as part of finiteness (in the finiteness-transfiniteness principle), if both opposites are defined on a finite scale. The second possibility is that they are one-sided as part of transfiniteness (in the finiteness-transfiniteness principle) if both opposites are defined on a transfinite scale. And the third possibility is that they are appropriate family resemblances in relation to the finiteness-transfiniteness principle if one opposite is defined on a finite scale and the other opposite is defined on a transfinite scale.

Other examples of inappropriate family resemblances are shown in *Table 4.10* on "The Syntax of Existential Dialectics VII: Types of Inappropriate Family Resemblances."

The Dialectic Constraints Imposed by the Principles

The principles, as they constitute the syntax of existential dialectics, are dialectic in character, such that, when they are applied, they impose dialectic constraints on how reality is to be understood.

Consider, say, the symmetry-asymmetry principle as an illustration here, in order to summarize two main characters of the dialectic constraints in question (as summarized in *Table 4.7*).

Firstly, to be dialectic here is to go beyond the narrow dichotomies (and, for that matter, any rigid multi-dimensional classificatory scheme), be they about "self" vs. "world," "freedom" vs. "unfreedom," "barbarity" vs. "civilization," "individuality" vs. "communality," and so on.

One way to do so (to go beyond) is to consider them all in terms of co-existence (without heavily favoring one over the rest). For instance, my theory of "post-civilization" (to be summarized later in the section on the pragmatics of existential dialectics) is to go beyond barbarity and civilization in terms of understanding barbarity and civilization as being co-existent.

And the same logic can be said in relation to my theories of "post-democracy," "post-capitalsm," and others (also to be introduced later in the section on the pragmatics of existential dialectics), in regard to freedom vs. unfreedom, equality vs. inequality, communality vs. individuality, spirituality vs. materiality, and so on.

But to consider them all (in the dichotomies—and, for that matter, in any rigid multi-dimensional classifactory scheme) as co-existent is not the same as to imply that the opposing categories in any classificatory scheme are all equal, since, in accordance to the symmetry-asymmetry principle (as an illustration here), if they are equal in terms of being considered as co-existent, they are asymmetric in terms of being unequal in dialectic interaction (e.g., X can be more dominant than Y in case A, or Y is more dominant than X in case B), although in some special cases, they can be relatively equal (e.g., X and Y are relatively equal in case C).

For this reason, there are different versions of "post-democracy" and "post-capitalism" in my theories. As an illustration, in version I of the theory of post-democracy, freedom is more dominant than equality, whereas in version II of the theory of post-democracy, equality is more so than freedom.

But this "X more than Y" has to be understood in the context of dialectic logic (not in symbolic logic, as conventionally understood), in that both "X" and "Y" are important in post-democracy (in the context of dialectic logic), but in an asymmetry way.

By contrast, in symbolic logic, it often favors one over the other—be it in regard to privileging freedom over equality (in Fascism), favoring freedom relatively more than equality (in Liberal Democracy), or favoring equality relatively more than freedom (in Socialist Democracy). In regard to the latter two cases (about Liberal Democracy and Socialist Democracy), the difference between dialectic logic and symbolic logic can be one in degree, not in kind—in this sense, albeit not in other senses.

The same reasoning can be said about the relatonships between individuality and communality, between spirituality and materiality, and between formal legalism and informal legalism in the different versions of my theory of "post-capitalism."

Secondly, to be dialectic is to go beyond the narrow dichotomies (and, for that matter, any rigid multi-dimensional classificatory scheme) in another way, this time, in a transcendent way, that is, in exploring other possibilities or even other issues not considered within the narrow confines of narrow dichotomies (and, for that matter, any rigid multi-dimensional classificatory scheme).

As an analogy, in this second meaning, to go beyond the narrow dichotomy of "black" vs. "white" is not just to choose both "black" and "white" (as in the first meaning) but also to explore other options (e.g., "green," "purple," "blue," etc.—and, alternatively, "shade," "line," "curve," etc.).

By the same reasoning, to go beyond "democracy" is to transcend democracy (as in version III of my theory of "post-democracy") and to explore other possibilties of lifeforms (e.g., "floating consciousness," "hyper-spatial consciousness," etc., to live beyond the narrow obsession with freedom and equality).

This dialectic character of the principles in existential dialectics has important implications for the pragmatics of existential dialectics (as will be clear shortly, in the section on the pragmatics of existential dialectics).

The Semantics of Existential Dialectics (or Its Ontosemantics)

The syntax of existential dialectics so understood in terms of ontological principles only gives us the structure of ontology in the world, in an abstract (general) sense. These principles by themselves do not tell us the specific meanings in a given context.

In order to grasp the specific meanings of the principles in a given context, it is necessary to study the semantics of existential dialectics (as summarized in *Table 4.11*).

And the analysis of the ontological principles as family resemblances in the previous section on syntax is only a starting point (and thus overlaps a bit with the semantics of existential dialectics here).

For this reason, I have often gone to great lengths in my previous books on different subjects to explain the specific meanings of the principles when applied in different contexts.

Consider two examples below for illustration, which are not exhaustive.

In *FPHST*, for instance, I used the first three principles (i.e., the change-constancy principle, the regression-progression principle, and the symmetry-asymmetry principle) to propose "the perspectival theory of space-time," for a better way to understand space and time—especially, though not exclusively, in relation to future post-human history (as summarized in *Table 1.3*, *Table 1.4*, *Table 1.5*, and *Table 1.6*).

In so doing, I had to introduce concepts and theories specific to the field of physics and other related fields (e.g., "absolute space" and "absolute time" in "classical mechanics" and "relative space-time" in "the theory of relativity").

And in *BNN*, I also exploited the three principles to propose the "transcendent" approach to the study of genes and memes as a new way to understand the interaction between nature and nurture.

In so doing, I had to explore concepts and theories in the world of evolutionary biology (e.g., "mutation," "variation," "adaptation," "selection," and "inheritance" in classical evolutionary theory) and neural biology (e.g., "chromosome," "gene," "DNA," "RNA," "protein," "neuron," "neural network," and "behavior").

Other examples, besides these two here, are given in other books of mine on different topics.

Therefore, this semantic dimension of existential dialectics is important, since, in each of my books when the ontological principles are used, each principle becomes relevant in a different meaning when contextualized within a given case study in question.

In this book, for instance, the principles are relevant in relation to a different subject matter (as summarized in the concluding chapter), and they can be shown in all shapes and sizes (since they are also to be treated as "short cuts," as indicated before), without, however, being reduced to "specific ontology" (as opposed to "general ontology") as explained in *Table 4.9*.

In the technical jargons of semantics, these exercises perform two essential functions of translation (or interpretation), namely, (a) "intensive" interpretation (about the properties or relations of the principles in existential dialectics), and (b) "extensive" interpretation (about the classes or cases to which the properties and relations of the principles in existential dialectics are applied). (WK 2011oo)

But there are limits of translation (or interpretation), and an excellent example is that "there is a problem with the 'objectivity' of translation, as it is relative to different cultural (or sub-cultural) contexts. For instance, one of the still unresolved problems in linguistics is: 'Is there an objective gauge for the quality of translation?'" (*FPHCOM* & WK 2011pp) And the problems of "radical translation" by W. V. Quine and "radical interpretation" by Donald Davidson are another well-known examples, as already discussed in *FPHSEM*.

This is one serious problem about the logic of translation and interpretation that I already addressed in *Sec. 2.5.2* of *FPHSEM*, *Sec. 3.5.1* of *FPHCOM* (and also, for that matter, in *FHC*, in the context of hermeneutics).

The Pragmatics of Existential Dialectics (or Its Ontopragmatics)

Even the semantics of existential dialectics is not enough to understand reality, since semantics by itself does not tell us the nature of intentionality and interests in relation to the matrix of power formations and conflicting interests when existential dialectics as an ontology is applied.

Surely, one cannot completely separate the pragmatics of existential dialectics from its semantics (and for that matter, its syntax), but the point here is to give each a more elaborated analysis.

For this reason, the pragmatics of existential dialectics needs to be studied too, in special relation to the inclusion of power formation and conflicting interests in the application of the principles of existential dialectics (as summarized in *Table 4.12*).

It is thus no wonder that all my previous books have shown in different ways how and why these ontological principles and their theoretical applications can reveal the future world to come in a way that humans have never known, in a totally different battleground for competing human (and later post-human) interests by myriad groups fighting for their own versions of hegemony.

Perhaps a summary of my previous works in relation to existential dialectics and their contributions to my new theories as proposed over the years is deemed revealing here, for the further understanding of the pragmatics of existential dialectics (in close relation to the semantics of existential dialectics too).

That stressed—the following summary in four sub-sections is something that I often do in my previous books too, as a way to introduce the intellectual background of my ideas.

So, for those readers who had read my previous works, the following sub-sections serve as a reminder, as they are repeated hereafter *often verbatim* (with only some updated changes, with the rest kept intact).

These sub-sections are, namely, (a) the two-way street connecting theory with meta-theory, (b) the theoretical application of existential dialectics, (c) direct and indirect applications of existential dialectics, and (d) multiple applications of existential dialectics—to be summaried herefafter, respectively.

The Two-Way Street Connecting Theory and Meta-Theory

However, lest any misunderstanding occurs, it is important to remember that the pragmatics of existential dialectics is not a one-way street (that is, using the ontological principles for theoretical insights in praxis) but a two-way one, namely, (a) from meta-theory to theory, and (b) from theory to meta-theory.

(a) On one side of the street, the ontological principles can inspire some theoretical insights in praxis, that is, in relation to some specific fields.

(b) On the other (opposing) side of the street, however, the study of a subject matter in the specific fields in turn reveals some more hitherto unknown ontological principles to be discovered and identified.

For this reason, three new principles were added in *FC*, one in *FAE*, two in *ALD*, and three in *FIA*, on top of the original three in *BCPC*, for example (as there are more)—after some research on the specific subject matters.

Both theory and meta-theory enrich each other in all my works over the years.

The Theoretical Application of Existential Dialectics

This two-way street connecting theory and meta-theory has contributed to my original construction of many theories in numerous books of mine on numerous fields in all domains of knowledge, which can be classified in terms of specific "meta-theories" (about epistemology, methodology, logic, ontology, etc.) and specific "theories" (about the mind, nature, culture, society, and the rest—as will be further explained later in the sub-section on "the holistic organization of an inquiry" in the section about "sophisticated methodological holism").

The final aim of my numerous new specific theories (on both theory and meta-theory) is to converge all of them (as proposed in my numerous books) into a grand project about the future of intelligent life, both here on earth and in deep space.

In this sense, all these theories of mine serve as a foundation to unify all domains of knowledge for an overarching understanding of all forms of reality (by way of some ontological principles and the comprehensive perspectives of the mind, nature, society, and culture), while allowing for heterogeneity, conflict, subjectivity, and complexity, for example.

With this grand project in mind—the summary of my specific theories on meta-theory are given in this section (on "existential dialectics") and the next (on "sophisticated methodological holism"), whereas the specific theories on theory will be summarized in the following categories of (a) society, (b) culture, (c) the mind, (d) nature, and (e) the rest, as shown below, in that order.

(a) Society

The category of "society" here can be divided into four sub-categories, i.e., (a1) social organizations, (a2) social institutions, (a3) social structure, and (a4) social systems (or trends).

(a1) Social Organizations

In relation to social organizations, I proposed a new thesis on the ambivalent nature of virtual organizations in the future, known as "the theory of e-civic alienation" in *FCD* (and summarized in *Table 4.19*).

And in *FPHO*, another theory called "the combinational theory of organization" was worked out, although it was first proposed in Ch.6 of *ALD*.

(a2) Social Institutions

There are different social institutions, be they about (a2.1) politics, (a2.2) economics, (a2.3) military affairs, (a2.4) mass media, (a2.5) law, (a2.6) education, and (a.7) gambling, for instance.

(a2.1) In relation to politics, I offered different visions on (a2.1.1) political theory, (a2.1.2) comparative politics, and (a2.1.3) international relations, to cite three main ones.

(a2.1.1) In the context of political theory, my vision of a path-breaking political system in future times concerns the different forms of "post-democracy" to supersede democracy unto the post-human age.

My vision of post-democracy was called "the theory of post-democracy" (as summarized in *Table 4.22*, *Table 4.23*, and *Table 4.24*), whereas *Table 4.25* explains the distinctions among democracy, non-democracy, and post-democracy.

And *Table 4.26* stresses the multiple causes of the emergence of post-democracy, while *Table 4.27* clarifies some possible misunderstandings in regard to post-capitalism and post-democracy.

(a2.1.2) In the context of comparative politics, I also showed, in *ALD*, how and why authoritarianism can be an advanced form of liberal democracy (just as it has its own counterpart in non-liberal democracy), or what I originally called "authoritarian liberal democracy" (as summarized in *Table 4.21*), before it is to be superceded by "post-democracy."

(a2.1.3) In the context of international relations (both at the institutional and structural levels), I provided, in *BWT*, "the theory of cyclical progression of empire-building," for a better way to understand the logic of empire-building on earth and beyond—with the rise of what I originally envisioned as "hyper-empires" (like "the Chinese Union" and "the Indian Union"), "meso-empires" (like "the European Union" and "the North American Union") and "micro-empires" (like "the Middle Eastern Union" and "the Latin American Union") in the coming "post-post-Cold War era" as a visionary illustration, as summarized in *Table 4.20*.

This is also related to the issues of "social structure" in (a3), "migration" in (a4,6), "identity" in (b11), and "geography" in (d4), for instance.

In addition, in *FPHWP*, I proposed "the multifaceted theory of war and peace" to understand war and peace, in a small sense—or aggression and pacificity, in a broad sense.

(a2.2) In relation to economics, I also offered another vision, this time to understand the future forms of economic systems, that is, the different forms of "post-capitalism" to supersede capitalism unto the post-human age.

My vision of post-capitalism was referred to (in *FCD*, *BDPD* and *BCPC*) as "the theory of post-capitalism" (as summarized in *Table 4.28*, *Table 4.29*, *Table 4.30*, *Table 4.31*, and *Table 4.32*). And *Table 4.33* clarifies the differences among capitalism, non-capitalism, and post-capitalism, whereas *Table 4.34* shows multiple causes of the emergence of post-capitalism.

This vision on "post capitalism" constitutes the third theoretical application of existential dialectics, this time, in that there is no wealth without poverty (or the wealth/poverty dialectics).

(a2.3) In relation to military affairs, I also proposed "the multifaceted theory of war and peace" (in *FCD*, although it was first so labeled in *FPHK*) for a new way to understand the causes of war—and, for that matter, the rise and fall of different forms of warfare over time.

This theory was further expanded in *FPHWP*, where I suggested a new way to understand war and peace, in a small sense—or aggression and pacificity, in a broad sense.

(a2.4) In relation to mass media, I also proposed "the totalistic theory of communication" in *The Future of Post-Human Mass Media* (or in short, *FPHMM*).

(a2.5) In relation to law, I also proposed "the reconstruction theory of law" in *The Future of Post-Human Law* (or in short, *FPHLAW*). And in relation to criminality, I proposed "the reflective theory of criminality" in *FPHCRIM* for a different way to understand heroes and villains.

(a2.6) In relation to education, I proposed "the heterodox theory of education" in *The Future of Post-Human Education* (or in short, *FPHEDU*).

(a2.7) And in relation to gambling, I proposed "the detached theory of gambling" in *The Future of Post-Human Gambling* (or in short, *FPHGAM*).

(a3) Social Structure

In relation to social structure, I first proposed two additional theories, that is, "the theory of the cyclical progression of hegemony" in *FCD* (and also in *FPHC* and *BDPD*), and "the theory of the cyclical progression of empire-building" (in *BWT*, as already indicated above)—with important implications for the symmetry-asymmetry princple in existential dialectics, just to cite one example.

In addition, this is also related to the issues of "migration" in (a4,6), "identity" in (b11), and "geography" in (d4), for instance.

Then, in *BNN*, I proposed another theory, namely, "the theory of contrastive advantages" to understand the persistence of social structure by gender, race, ethnicity, or any other groupings.

(a4) Social Systems (or Trends)

In relation to social systems, there are several sub-categories for illustration, which can be about (a4.1) urbanization, (a4.2) technology, (a4.3) demography, (a4.4) the environment, (a4.5) social trends as a whole (which also include the current phenomenon of globalization), (a4.6) migration, and (a4.7) resources.

(a4.1) In relation to urbanization, I proposed "the contingent theory of urban planning" in *FPHUP* for a new way to understand density and void—which is also related to the issues of "social trends" (especially in relation to globalization) in (a4.5), "migration" in (a4.6), and "identity" in (b11).

(a4.2) In relation to technology, I worked out "the ambivalent theory of technology" in *FPHE* for a new way to understand the nature of technology.

Some other books (like *FPHMM*, *FAE*, *FCD*, and *FHC*) also deal with technology, and a good example concerns "the synthetic theory of information architecture" in *FIA* on taxonomy, network, and intelligence, in conjunction with the rise of technology.

(a4.3) In relation to demography, I proposed, first in *FHC* (and later in other books like *FCD*, *FPHC*, *FPHU*, etc.), "the theory of post-humanity," in relation to the "post-human" age at some distant point of "after-postmodernity," long after human extinction, to be eventually superseded by "post-humans" of various forms.

Some good candidates of the post-human lifeforms include, for instance, "thinking robots," "thinking machines," "cyborgs," "genetically altered superior beings," "floating consciousness," "hyper-spatial consciousness," "unfolding unconsciousness," "the hyper-sexual body," and "the hyper-martial body." This post-human vision of mine was first originally worked out in *FHC* and further elaborated in both *FCD*, *FPHC*, *FPHU*, *FPHS*, and *FPHMA*, for instance.

Lest any confusion occurs, I need to emphasize two clarifications here about the term "post-human" as a neologism in my works.

Firstly, the word "post-human" here should *not* be confused with another term which looks similar but has a totally different meaning in the literature of postmodernism, namely, "post-humanism"—which constitutes a critique of "humanism" as tradionally understood (especially, though not exclusively, in relation to the idea of progress in science and reason in the Enlightenment project). (WK 2008)

And secondly, the word "post-human" here should also *not* be confused with a similar term which is used to champion the ideology of technology for the future co-existence between humans and cyborgs in "trans-humanism." (WK 2008a)

Instead, my term "post-human" in relation to "posthuman-ism" also rejects "transhumanism" (especially, though not exclusively, in relation to the promises of technology) and refers to something else altogether, that is, the future extinction of humans and its post-human successors in deep space and beyond unto multiverses.

Later, in *BEPF*, I went further and provided a more comprehensive critique of "transhumanism," together with other issues.

With these two clarifications in mind (as summarized in *Table 1.31*)—in the end, a most fundamental question about intelligent life now has an answer, in that, if asked, "What is the future of human civilization?"—my answer in *FCD* (89) is thus: "As addressed in Ch.7 of *FHC*, a later epoch of the age of after-postmodernity (that is, at some point further away from after-postmodernity) will begin, as what I called the 'post-human' history (with the term 'post-human' originally used in my doctoral dissertation at M.I.T., which was finished in November 1995, under the title *After Postmodernity*, still available at M.I.T. library, and was later revised and published as *FHC*). The post-human history will be such that humans are nothing in the end, other than what culture, society, and nature (with some luck) have shaped them into, to be eventually superseded by post-humans (e.g., cyborgs, thinking machines, genetically altered superior beings, and others), if humans are not destroyed long before then."

(a4.4) In relation to the environment, I also addressed the issue of sustainability, especially in *FHC*, *FCD*, and *FPHUP*. For instance, in *FPHUP*, I specifically proposed a new theory, "the contingent theory of urban planning," on the need to go beyond the contemporoary obsession with sustainability, in the context of not only the environment but also other issues.

I also proposed "the interventive-reshaping theory of geography" in *FPHGEOG*—which is related to the issues of "migration" in (a4.6), "geography" in (d4), and "geology" in (d5).

In addition, I proposed "the transfigurative theory of waste" in *FPHW*.

(a4.5) In relation to social trends as a whole, I proposed in Ch.9 of *FCD* (367-8), that "civilizational history will continue into the following cyclical progression of expansion, before it is to be superseded (solely as a high probability, since humans might be destroyed sooner either by themselves or in a gigantic natural calamity) by posthumans at some distant point in after-postmodernity (which I already discussed in *FHC*)" unto multiverses (different constellations of universes):

Local → Regional → Global →Solar →
Galactic → Clustery...→ Multiversal

The current phenomeon of "globalization" lies in the third category on the line above and was extensively analyzed in *FCD*, in close connection with "globalism" at the cultural level—in the context of "postmodernity" (which will be introduced later in the section on history).

In *BDPD*, this thesis of mine was referred to as "the theory of the cyclical progression of system integration and fragmentation," at the systemic level—in close relation to another thesis analyzed in Ch.9 of *FCD*, which was called, in the absence of better words, "the multifaceted theory of war and peace" in *FPHK* (as already indicated above).

In *FC*, I also proposed "the dialectic theory of complexity" to account for the factor of uncertainty, which can have a major role to play in the world.

All this is also related to the issues of "urbanization" in (a4.1), "the environment" in (a4.4), "migration" in (a4,6), "identity" in (b11), "geography" in (d4), and "geology" in (d5).

(a4.6) In relation to migration, I proposed "the resettlement theory of geology" in *FPHGEOL* and "the theory of the cyclical progression of migration" in *FPHMIG*—which are also related to the issues of "social structure" in (a3), "urbanization" in (a4.1), "the environment" in (a4.4), "social trends" in (a4.5), "identity" in (b11), "geography" in (d4), and "geology" in (d5).

(a4.7) And in relation to resources, I propose "the resilient theory of natural resources" in *BNR* —which is also related to the issue of "natural resources" in (d10).

(b) Culture

The category of "culture" here can also be divided into different sub-categories like (b1) aesthetics, (b2) religion, (b3) popular (mass) culture, (b4) logic, (b5) civilization as a whole, (b6) ethics, (b7) epistemology (methodology), (b8) metaphysics (ontology), (b9) ideology, (b10) language, and (b11) identity.

(b1) Aesthetics

In relation to aesthetics, I proposed "the transformative theory of aesthetic experience" in *FAE* about some great transformations of aesthetic experience in future history.

In *FPHLIT*, I further proposed "the comparative-impartial theory of literature" for a new way to understand fiction and non-fiction.

In *FPHPA*, I further proposed "the transdisciplinary theory of performing arts" for a new way to understand the body and its presence—which is also related to the issues of "popular (mass) culture" in (b3) and of "the human body" in (c1.2). And the aesthetic issue also overlaps with (d8) on architecture.

And in *FPHCA*, I further proposed "the inquisitive theory of culinary art" for a new way to understand ingredients and techniques.

(b2) Religion

In relation to religion, I proposed "the comparative theory of religion" in *FHC* (and also in *FCD*, although it was first so labeled in *FPHK*).

And in *FPHR*, I provided a solid foundation of this theory with a better name, that is, "the comparative-substitutive theory of religion" for a new way to understand spirituality.

(b3) Popular (Mass) Culture

In relation to popular (mass) culture, there are some sub-categories for illustration, which can be, for example, (b3.1) mass culture in general and (b3.2) sports (like chess)—both of which are also related to the issues of "aesthetics" in (b1) and of "the human body" in (c1.2).

(b3.1) In relation to mass culture in general, I proposed, in *FHC*, a new way to understand mass culture in terms of some fundamental dilemmas, which is called, in the absence of better words, "the dualistic theory of mass culture" in *FPHO*.

(b3.2) In relation to sports (like chess), I proposed, in *FPHCHESS*, "the mediative-variative theory of chess" for a new way to understand tactics and strategy (in different traditions).

(b4) Logic

In relation to logic, I also worked out, in *FPHML*, "the contrastive theory of rationality" for a new way to understand rationality and knowledge (in the context of mathematical logic).

(b5) Civilization as a Whole

In relation to civilization as a whole, I also argued, in *FHC*, no freedom without unfreedom (as shown in *Table 4.13*) and no equality without inequality (as shown in *Table 4.14*), especially in relation to the seven dimensions of life existence first worked out in *FHC* (i.e., the technological, the everyday, the true, the holy, the sublime/beautiful, the good, and the just).

Even more interestingly, in *BDPD* and later in *BCIV*, I further revealed a theoretical refinement of existential dialectics, this time, in arguing that there is no civilization without barbarity, with human civilization to be eventually superseded by what I originally analyzed as "post-human post-civilization" (which should not be confused with "post-human civilization"), in the context of the freedom/unfreedom and equality/inequality dialectics.

In the final analysis, civilization cannot live without barbarity and has to learn to co-exist with it in ever new ways. Preposteorus as this may seem to many contemporaries—it is no more imperative to preserve civilization than necessary to destroy barbarity, and the ideal of civilization is essentially bankrupt, to be eventually superseded by "post-civilization" (as summarized in *Tables 4.35-4.43*).

This freedom/unfreedom and equality/inequality dialectics holds in pre-moderny, modernity, postmodernity, and what I orignally called "after-postmodernity" in *FHC* and *FCD*—especially, though not exclusively, in the context of the emergence of "post-humans."

There are two specific theories about "after-postmodernity" here. The first theory about "after-postmodernity" is about the trinity of pre-modernity, modernity, postmodernity, and after-postmodernity at the cultural level—and was called, in *BDPD*, "the theory of the trinity of modernity to its after-postmodern counterpart," although it was first analyzed in *FHC* and later in *FCD* and also *BCIV* (as summarized in *Table 1.34*, *Table 1.35*, *Table 1.36*, and *Table 1.37* and will be elaborated further in the later sub-section on "the rest" about future history).

The second theory about "after-postmodernity" is about pre-modernity, modernity, postmodernity, and after-postmodernity at the historical level. For instance, a conclusion in *FHC*, which is shocking indeed for many contemporaries, is none other than that "[t]he post-human history will therefore mark the end of human history as we know it and, for that matter, the end of human dominance and, practically speaking, the end of humans as well. The entire history of human civilization, from its beginning to the end, can be summarized by four words, linked by three arrows (as already discussed in *FHC*)":

Pre-Modernity → Modernity →
Post-Modernity → After-Postmodernity

In *BDPD*, this thesis of mine was specifically called "the theory of the evolution from pre-modernity to after-postmodernity," at the historical level—as shown in *Table 1.34*, *Table 1.35*, *Table 1.36*, and *Table 1.37* (and will be elaborated further in the later sub-section on "the rest" about future history).

Therefore, "[t]he end of humanity in the coming human extinction is the beginning of post-humanity. To say an untimely farewell to humanity is to foretell the future welcome of post-humanity." (P. Baofu 2002: 89) This thesis of mine was known in *BDPD* as "the theory of post-humanity," at the systemic level—as already indicated in (a4.3) above.

(b6) Ethics

In relation to ethics, I proposed, in *BEPE*, "the theory of post-ethics" for a new way to understand morality and immorality

(b7) Epistemology (Methodology)

In relation to epistemology (methodology), I proposed "the complex theory of data analysis" in *FPHDA* for a different way to understand quantitative and qualitative research.

In addition, I already proposed "the critical-dialectic theory of formal science" in *FPHFS* and "the theory of methodological holism" in several books (like Ch.1 *FCD*, Ch.1of *FPHC*, *Sec.2.1* and *Sec.2.5* of *BCPC*, *FC*, *FPHK*, etc.) for a different way to do methodology.

(b8) Metaphysics (Ontology)

I also proposed "the theory of existential dialectics" in several books (like *FHC*, *FCD*, *FPHC*, *BDPD*, *FC*, *FAE*, *ALD, FIA*, *FPHK*, etc,) and "the contrastive theory of rationality" in *FPHML* for a different way to understand ontology.

(b9) Ideology

The issue about ideology was already discussed under the categories of "social institutions" and of "civilization as a whole," especially in relation to my theories of "post-democracy," "post-capitalism," "post-civilization," and so on.

(b10) Language

I already proposed "the multilogical theory of learning" in *FPHL* for a new way to understand the nature of language structure and context .

And I also worked out, in *FPHSEM*, "the interactive theory of semantics" for a new way to understand internality and externality (in the context of semantics).

(b11) Identity

I propose here "the theory of the cyclical progression of migration" in *FPHMIG* for a new way to understand the nature of sameness, otherness, and identity—which is also related to the issues of "social structure" in (a3), "urbanization" in (a4.1), "social trends" (especially in relation to globalization) in (a4.5), "migration" in (a4.6), religion in (b2), and ideology in (b9).

(c) The Mind

The important vision about "post-humans" (as summarized earlier) brings us to the category of "the mind" here, which, as is often understood in everyday's usage, refers to the "brain" and the "body" (together with their interactions) and has different levels of consciousness which can be both "spiritual" and non-spiritual.

Therefore, the "mind" already presupposes the "brain," "body," and "spirit" (or "soul"), so that one does not fall into the intellectual trap of the "mind-body dualism" as debunked in modern (and especially, postmodern) philosophy (which I already went to great lengths to explain in *The Future of Human Civilization*).

That said—the category of "the mind" can be divided into two sub-categories, namely, in relation to (c1) biology and (c2) psychology

(c1) Biology

In relation to biology, there are two concerns here, namely, (c1.1) genes and intelligence and (c1.2) the human body.

(c1.1) In relation to genes and intelligence, I worked out "the theory of contrastive advantages" (which was originally proposed in *FCD* and further worked out in *BNN*, as summarized in *Table 1.24*), to show the nature-nurture interactions of multiple levels in action (e.g., the biological, the psychological, the structural, the systemic, the cultural, and others) for humans and post-humans on earth and beyond, to the extent that different groups, be they on the basis of race, gender, ethnicity, class, age, or else, are not equal, on average, and have not been, nor will they be, even if there can be exceptions.

(c1.2) In relation to the human body, for instance, in *FPHMA*, I suggested "the expansive-contractive theory of martial arts," in which two great future transformations of the martial body were envisioned, that is, both "with the hyper-martial body" and "without the body" (as summarized in *Table 1.32*)—although this issue can also be classified under the category of "culture."

In *FPHS*, I proposed "the theory of virtual sexuality," in which three great future transformations of the sexual body were envisioned, that is, "virtual pleasure," "the hyper-sexual body," and "the post-sexual floating mind" (as summarized in *Table 1.33*).

And in *FPHPA*, I further proposed "the transdisciplinary theory of performing arts" for a new way to understand the body and its presence, with seven great future transformations of the body in performing arts—

which is also related to the issues of "aesthetics" in (b1) and "popular (mass) culture" in (b3).

(c2) Psychology

In relation to psypchology (which also overlaps with biology too, however), I also proposed some greater transformations of the mind to come in the future, especially though not exclusively in the post-human age.

These greater transformations of the mind in psychology can be further sub-divided into different categories, namely, (c2.1) consciousness, (c2.2) unconsciousness (together with subconsciousness), (c2.3) creativity, (c2.4) learning, (c2.5) aggression and pacificity, (c2.6) personality, (c2.7) humor, and (c2.8) artificial intelligence.

(c2.1) In relation to consciousness, I analyzed, in *FPHC*, the nature of consciousness in relation to the conceptual dimensions of consciousness (as shown in *Table 1.12*), the theoretical levels of consciousness (as shown in *Table 1.13*), the thematic issues of consciousness (as shown in *Table 1.14*), and the different dimensions of human existence (viz., having-ness, belonging-ness, and being-ness) in relation to consciousness (as shown in *Table 1.15*, *Table 1.16*, *Table 1.17*, and *Table 1.18*).

In *BCPC*, the nature of consciousness was further explored in relation to cognition, emotion, and behavior, which can be in consciousness and other mental states (e.g., subconsciousness and unconsciousness), in the context of existential dialectics, as shown in *Table 1.19*, *Table 1.20*, *Table 1.21*, *Table 1.22,* and *Table 1.23*.

In fact, in *FCD*, I already proposed three different theories in relation to the limits of cognition, emotion, and behavior, namely, "the theory of cognitive partiality," "the theory of emotional non-neutrality," and "the theory of behavioral alteration."

Then, I argued that the existence of human consciousness will be superseded one day with "floating consciousness" (as first proposed in *FCD*) and "hyper-spatial consciousness" (as first proposed in *FPHC*) as a climax of evolution in consciousness, after the future extinction of human consciousness (as shown below):

Primordial consciousness → Human consciousness →
Post-human consciousness (with floating consciousness and hyper-spatial consciousness as a climax in the evolution of consciousness)

In *BDPD*, these latest theses of mine were called "the theory of floating consciousness" and "the theory of hyper-spatial consciousness," both at the cosmological and psychological levels (as summarized in *Table 1.26* and *Table 1.25*).

(c2.2) In relation to unconsciousness, for instance, in *FPHU*, the unconsious mind will undergo tremendous change, to the rise of what I originally called "unfolding unconsciousess" in relation to anomalous experience, as a climax of evolution in unconsciousness, in conjunction with the conscious mind (both human now and post-human later), especially at the psychological level, as shown below:

Primordial unconsciousness → Human unconsciousness →
Post-human unconsciousness (with unfolding unconsciousness
as a climax in the evolution of unconsciousness)

In *FPHU*, this thesis of mine is called "the theory of unfolding unconsciousness" in a small sense, or "the unfolding theory of anomalous experience," in a larger one, as shown in *Table 1.27* and *Table 1.28*.

(c2.3) In relation to creativity, for instance, in *FPHCT*, I also proposed "the comprehensive theory of creative thinking" to explore the role of creative thinking, in relation to technology and other factors, together with my original proposal of important creative techniques and traits (as shown in *Table 1.29*) and its double nature of desirability and undesirability (as shown in *Table 1.30*).

(c2.4) In relation to learning, for instance, in *FPHL*, I worked out "the multilogical theory of learning" for a new way to understand the nature of structure and context in language, in relation to learning and its future development in the context of the mind.

(c2.5) In relation to aggression (and pacificity), I proposed "the multifaceted theory of war and peace" in *FPHWP* (as mentioned above).

(c2.6) In relation to personality, I worked out "the contrarian theory of personality" in *FPHP* for a new way to understand normality and abnormality.

(c2.7) In relation to humor, I worked out "the metamorphic theory of humor" in *FPHH* for a new way to understand joking and laughing.

(c2.8) And in relation to artificial intelligence, I proposed "the supersession theory of computing" in *FPHCOM* for a new way to understand hardware, software, and the mind.

In addition, I also proposed "the complex theory of data analysis" in *FPHDA* for a different way to understand quantitative and qualitative research in the context of the evolution of new minds.

(d) Nature

The category of "nature" here refers to the state of nature, which can be divided into some sub-categories like (d1) physics, (d2) cosmology, (d3) biology, (d4) geography, (d5) geology, (d6) acoustics, (d7) chemistry, (d8) architecture, (d9) transportation, and (d10) natural resources.

But since biology in (d3) overlaps with biology in (c1)—only the rest of the categories are addressed hereafter.

(d1) Physics

In relation to physics, I also proposed "the perspectival theory of space-time" in *FPHST* (based on *FPHC*) for a new way to understand space and time, as shown in *Table 1.3*, *Table 1.4*, *Table 1.5*, and *Table 1.7.*

In *FPHG* (based on *FPHC* and *FPHST*), I proposed "the selective theory of geometry" for a new way to understand geometry, with implications for the understanding of space-time in the context of infinity, symmetry, and dimensionality for future lifeforms that our world has never known (as shown in *Table 1.6*, *Table 1.7*, *Table 1.8; Table 1.9*, *Table 1.10,* and *Table 1.11*).

And in *FC*, I proposed "the dialectic theory of complexity" for a new way to understand order and chaos in the state of nature and beyond.

(d2) Cosmology

In relation to cosmology, my theories on physics (as indicated above) already have important implications for understanding the nature and the future of the universe (which was also addressed in *FHC*, among other works aforcited).

I also proposed "the theory of hyper-spatial consciousness" in *FPHC* (and also in *FPHG*) for a new way to understand the future evolution of the mind to enter into different dimensions of space-time that our current world has never known.

In addition, in *BCOS*, I proposed "the theory of post-cosmology" for a new way to understand the beginnings and ends of the cosmos and beyond.

(d4) Geography

In relation to geography, I proposed "the theory of the geopower of nature" in *ALD* for a different way to understand the relationships between geography and other entities (like the mind, culture, and society).

I further refined this issue of geography by proposing "the interventive-reshaping theory of geography" in *FPHGEOG*—which is related to the issues of "the environment" in (a4.4), "migration" in (a4.6), and "geology" in (d5).

(d5) Geology

In relation to geology, I proposed "the resettlement theory of geology" in *FPHGEOL* for a different way to understand the relationships between statics and dynamics—which is related to the issues of "the environment" in (a4.4), "migration" in (a4.6), and "geography" in (d4).

(d6) Acoustics

In relation to acoustics, I proposed "the multilateral theory of acoustics" in *FPHA* for a different way to understand the relationships between sound and silence.

(d7) Chemistry

In relation to chemistry, I proposed "the creational theory of chemistry" in *FPHCHEM* for a different way to understand substances and their changes.

(d8) Architecture

In relation to architecture, I proposed "the transcendent theory of architecture" in *FPHARCH* for a different way to understand form and function.

(d9) Transportation

In relation to transportation, I proposed "the panoramic theory of transportation" in *FPHTRAN* for a different way to understand networks and operations.

(d10) Natural Resources

And in relation to natural resources, I propose "the resilient theory of natural resources" in *BNR* —which is also related to the issue of "resources" in (a4.7).

(e) The Rest

The category of "the rest" here refers to what cannot be strictly classified in any of the previous categories (i.e., society, culture, the mind, and nature). A good illustration of "the rest" refers to (e1) history.

(e1) History

The category of history can be divided into some sub-categories like (e1.1) empirical history and (e.1.2) meta-history.

(e1.1) In relation to empirical history, I proposed "the theory of the evolution from pre-modernity to after-postmodernity" (as first worked out in *FHC* and later in *FCD* and *FPHC*), which overlaps—as a reminder from (b5)—with the issue concerning "civilization as a whole" in (b5).

For instance, in both *FCD* and *FPHC*, I worked out the structure of "post-human civilization" in terms of the trinity of after-postmodernity (i.e., "free-spirited after-postmodernity," "post-capitalist after-postmodernity," and "hegemonic after-postmodernity").

Both conceptually and theoretically, the trinity of after-postmodernity is a sequential extension of the trinity of modernity (i.e., "free-spirited modernity," "capitalist modernity,"and "hegemonic modernity") and the trinity of postmodernity (i.e., "free-spirited postmodernity," "capitalist postmodernity," and "hegemonic postmodernity") as first proposed in *FHC*.

And the trinity of pre-modernity (i.e., "pre-free-spirited pre-modernity," "pre-capitalist pre-modernity" and "hegemonic pre-modernity") was later conceived in *BCIV* to complete the historical set from pre-modernity to after-postmodernity.

In *BDPD*, this thesis about the trinity of pre-modernity, modernity, postmodernity, and after-postmodernity was collectively known as "the theory of the trinity of modernity to its after-postmodern counterpart," at the cultural level—as already so mentioned in (b5) earlier, and summarized in *Table 1.34*, *Table 1.35*, *Table 1.36*, and *Table 1.37*.

At the structural level, all these trinities are subject to the existential constraints (e.g., the freedom/unfreedom and equality/inequality dialectics in the context of "the cyclical progression of hegemony"), be the historical epoch in pre-modernity, modernity, postmodernity, or after-postmodernity in future times. Each of the historical epochs has its ever new ways of coming to terms with the ever new (different) mixtures of freedom/unfreedom and equality/inequality.

This is importantly so, not because, as is falsely assumed in conventional wisdom, one certain way is superior (or better) than another in terms of achieving more freedom and less unfreedom, or more equality with less inequality.

On the contrary, indeed, in each of the historical epochs, each increase of unfreedom greets each freedom achieved, just as each increase of inequality welcomes each equality achieved, albeit in ever new (different) ways. In *BDPD*, this thesis of mine was labeled as "the theory of the cyclical progression of hegemony," at the structural level, though it was first analyzed in *FCD*.

In *BDPD*, more theoretical applications of existential dialectics were further examined, in relation to five main features, in the context of the duality of oppression, namely, (a) that each freedom/equality achieved is also each unfreedom/inequality created, (b) that the subsequent oppressiveness is dualistic, both by the Same against the Others and itself and by the Others against the Same and themselves, (c) that both oppression and self-oppression can be achieved by way of downgrading differences (between the Same and the Others) and of accentuating them, (d) that the relationships are relatively asymmetric among them but relatively symmetric within them, even when the Same can be relatively asymmetric towards itself in self-oppression, and the Others can be likewise towards themselves, and (e) that symmetry and asymmetry change over time, with ever new players, new causes, and new forms, be the locality here on Earth or in deep space unto multiverses—as summarized in *Table.4.13*, *Table 4.14*, *Table 4.15*, and *Table 4.16*.

The same logic also holds both in relation to wealth and poverty (as addressed in *BCPC* and summarized in *Table 4.17* on the wealth/poverty dialectics) and in relation to civilization and barbarity (as addressed in *BCIV* and summarized in *Table 4.18* on the civilization/barbarity dialectics).

In *BDPD*, this thesis on existential dialectics was labeled as "the theory of existential dialectics."

(e.1.2) And in relation to meta-history, I proposed "the multifold theory of history" in *FPHHIST* for a different way to understand universality and relativity (together with other debates).

Direct and Indirect Applications of Existential Dialectics

A different way to appreciate the usefulness of existential dialectics is by way of the analysis of its (a) direct and (b) indirect applications.

(a) In direct applications, on the one hand, the logic of existential dialectics can shed some theoretical insights on diverse phenomena in the world, and good instances are the usage of the principles of existential dialectics for the theoretical insights on the freedom/unfreedom dialectics, the equality/inequality dialectics, and the wealth/poverty dialectics—as introduced earlier in (b5) and (e) in the previous sub-section.

For instance, my books like *FPHST*, *BNN*, *FAE*, *FC*, *FIA*, *FPHU*, *FPHK*, *FPHML*, *FPHE*, *FPHMM*, *FPHCT*, *FPHG*, *FPHUP*, *FPHL*, *FPHO*, *FPHMA*, *FPHS*, *FPHLAW*, *FPHFS*, *FPHWP*, *FPHP*, *BEPE*, *BCOS*, *FPHGEOL*, *FPHCHESS*, *FPHR*, *FPHEDU*, *FPHH*, *FPHA*, *FPHGAM*, *FPHCOM*, *FPHLIT*, *FPHCHEM*, *FPHDA*, *FPHGEOG*, *FPHPA*, *FPHSEM*, *FPHARCH*, *FPHCRIM*, *FPHHIST*, *FPHMIG*, *FPHTRAN*, *FPHCA*, *FPHW*, and *BNR* use the principles to reveal some theoretical insights on the perspectives of space and time (as in *FPHST*), of nature and nurture (as in *BNN*), of beauty and ugliness (as in *FAE*), of order and chaos (as in *FC*), of taxonomy and network (as in *FIA*), of the unconscious mind (as in *FPHU*), of the nature of knowledge (as in *FPHK*), of mathematical logic (as in *FPHML*), of engineering (as in *FPHE*), of mass media (as in *FPHMM*), of creative thinking (as in *FPHCT*), of geometry (as in *FPHG*), of urban planning (as in *FPHUP*), of language (as in *FPHL*), of organization (as in *FPHO*), of martial arts (as in

FPHMA), of sexuality (as in *FPHS*), of law (as in *FPHLAW*), of formal science (as in *FPHFS*), of aggression and pacificity (as in *FPHWP*), of normality and abnormality (as in *FPHP*). of morality and immorality (as in *BEPE*), of the beginnings and ends of the universe (as in *BOCS*), of statics and dynamics (as in *FPHGEOL*), of tactics and strategy (as in *FPHCHESS*), of secularness and sacredness (as in *FPHR*), of teaching and learning (as in *FPHEDU*), of joking and laughing (as in *FPHH*), of sound and silence (as in *FPHA*), of risk and caution (as in *FPHGAM*), of hardware and software (as in *FPHCOM*), of fiction and non-fiction (as in *FPHLIT*), of substances and their changes (as in *FPHCHEM*), of quantitative and qualitative research (as in *FPHDA*), of environments and their interactions (as in *FPHGEOG*), of the body and its presence (as in *FPHPA*), of internality and externality (as in *FPHSEM*). of form and function (as in *FPHARCH*), of heroes and villains (as in *FPHCRIM*), of universality and relativity (as in *FPHHIST*), of Sameness and Otherness (as in *FPHMIG*), of networks and operations (as in *FPHTRAN*), of ingredients and techniques (as in *FPHCA*), of uselessness and usefulness (as in *FPHW*), and of diversity and discontinuity (as in *BNR*).

(b) In indirect applications, on the other hand, the theoretical insights can further be used to reveal other phenomena directly from them (viz., the theoretical insights) and therefore indirectly from the principles themselves.

A good illustration is of course the use of the theoretical insights on the freedom/unfreedom and equality/inequality dialectics for the understanding of the civilization/barbarity dialectics.

This distinction between direct and indirect applications may sound a bit academic, since even in indirect applications, the phenomena under study can still be directly related back to the principles themselves.

In the previous example, as an illustration, the civilization/barbarity dialectics can be directly related to the principles of existential dialectics without the intermediate role of the freedom/unfreedom and equality/inequality dialectics.

Multiple Levels of Application

There is another issue to be clarified, however. In other words, the theoretical insights can be applied to multiple levels of analysis—even though, in a given example, it may refer to one level or a few only.

For instance, in the example concerning the freedom/unfreedom dialectics, it can be used at the structural level (e.g., in relation to the theory of the cyclical progression of hegemony), but it can also be exploited for other levels (e.g., the theory of post-capitalism at the institutional level).

All these levels of application should not be misleadingly construed, as I stress this before, as a one-way street (that is, to use the ontological principles for theoretical application at multiple levels of specific analysis) but a two-way one, in which theoretical insights in praxis, when studied in more specific contexts, can in turn refine the nature of existential dialectics (for example, with the addition of new principles).

Sophisticated Methodological Holism

The summary of my original meta-theory on methodology is provided below, again, to be repeated *often verbatim* from my previous works—as this is something I regularly do in each new book to introduce my past works, which are in conversation with this current one.

This means that, for those readers who have not read any of my previous books, the summary below is helpful.

That clarified—this original contribution of mine to the study of methodology takes the form of an approach known as "the theory of methodological holism" or "methodological holism" in short, as already worked out in *FPHC* (2004).

I have made good use of this methodological approach of mine for all of my previous works.

The Confusion of Holism with Systems Approach

Yet, it is imperative to stress at the outset, as I often emphasized this in all my previous books, that my approach of "methodological holism" does not oppose or exclude "methodological individualism" (as some readers may be tempted to assume, as is conventionally understood) but actually includes it.

The confusion here has to do with equating holism with systems approach in general (be it about systems theory, chaos theory, complexity theory, or else), which I debunked, while learning something from it, as already elaborated in *The Future of Complexity* (2007)—and *The Future of Post-Human Formal Science* (2010) is another addition to this critique of the intellectual fad of systems approach.

In other words, my meta-theory of holism, or sophisticated methodological holism in short, makes a distinction between doing holistic work and doing systems work, for two reasons.

The first reason is that holism, in my theory, does not imply systems approach, since it rejects systems approach because of the myriad problems that the literature on systems approach has been unable to resolve, while learning something from it.

And the second reason is that my theory is related to the other theory of mine, that is, existential dialectics, which does not exclude methodological individualism (but includes it, because it simply goes beyond both reductionism and reverse-reductionism in a dialectic way).

For these reasons (and others too, as summarized in *Table 4.1*), my version of methodological holism is *sophisticated*—not *vulgar* as sometimes used by inapt scholars employing the same term.

The Ontological Constraints

Sophisticated methodological holism is subject to some ontological constraints, and some good examples include "the formalness-informalness principle," "the absoluteness-relativeness principle," "the partiality-totality principle," "the predictability-unpredictability principle," "the explicability-inexplicability principle," and "the fiction-reallity principle" in existential dialectics.

Of course, there can be more than six ontological principles in relation to method, but the examples here are sufficient to illustrate the point in question.

With this caveat in mind—the first two were already analyzed in previous books of mine like *FC* and *FAE*, whereas the third one was addressed in *FPHU*, the fourth and fifth ones were accounted for in *FPHK* and *FPHML*, respectively, and the last one was addressed in *FPHLIT*.

For now, it suffices to show that sophisticted methodological holism, because of these ontological constraints on methodology, targets two major sins of methodology, namely, what I call, in the absence of better words, (a) *reductionism* and (b) *reverse-reductionism*—both of which come in all shapes and sizes, to be summarized below (and also shown in *Table 4.2*).

Against the Varieties of Reductionism

There are four versions of reductionism to be summarized here, which sophisticated methodological holism rejects, namely, (a1) conceptual, (a2) theoretical, (a3) methodological, and (a4) ontological forms of reductionism.

(a1) In conceptual reductionism, a good case in point concerns myriad dualities like mind vs. body, self vs. world, democracy vs non-democracy, and the like (as already addressed in *FHC*, *FPHC*, and *BDPD*, for instance).

(a2) In theoretical reductionism, an illuminating case study is best exemplified by what I originally called "the foundation fallacy" in *FPHST*, in any attempt to naively understand space-time from the physical perspective as the foundation and, consequently, to dangerously dismiss (or belittle) other perspectives.

In *FAE*, I elaborated further these versions of reductionism in the literature on aesthetics (e.g., form vs. content, representation vs. expression, critics vs. artists, and externalism vs. internalism).

In *FIA*, I revealed other forms of reductionism in the literature on information architecture (e.g., the constructivist argument and the representational argument).

In *FPHU*, I showed the persistent legacy of reductionism, this time, in the literature on anomalous experience (e.g., the obsession with physics, chemistry, and biology for explaining anomalous experience).

In *FPHE*, I examined another case of reductionism in action, in the context of engineering (e.g., technical constraints vs. normative constraints).

In *FPHMM*, I elaborated one more version of reductionism, in relation to the three domains of communication (e.g., the competing views on sending, connecting, and receiving).

In *FPHCT*, I explored another version of reductionism, in relation to invention and innovation (e.g., the bio-psychological argument vs. the socio-cultural arguments).

In *FPHG*, I identified another version of reductionism, in relation to infinity, symmetry, and dimensionality (e.g., the Euclidean argument vs. the non-Euclidean arguments).

In *FPHUP*, I analyzed another version of reductionism, in relation to density and void (e.g., the engineering argument and the ecology argument).

In *PFHL*, I examined another version of reductionism, in relation to structure and context (e.g., the structuralist argument and the contextualist argument).

In *PFHO*, I explored another version of reductionism, in relation to communcation, decision-making, and leadership (e.g., the rational-system argument vs. the natural-system argument vs. the open-system argument).

In *PFHMA*, I revealed another version of reductionism, in relation to the martial body and spirit (e.g., the spiritual argument vs. the materialist argument vs. the defensive argument).

In *PFHS*, I analyzed another version of reductionism, in relation to the sexual body and spirit (e.g., the naturalist argument vs. the constructivist argument).

In *PFHLAW*, I examined another version of reductionism, in relation to law (e.g., the necessity argument vs. the contengency argument).

In *FPHWP*, I show another version of reductionism, in relation to war and peace (e.g., the aggressivist argument vs. the pacifist argument).

In *BEPE*, I analyzed another version of reductionism, in relation to morality and immorality (e.g., the objectivist argument vs. the non-objectivist argument vs. the skeptical argument).

In *BCOS*, I examined another version of reductionism, in relation to the contested beginnings and speculative ends of the universe (e.g., the scientific argument vs. the religious argument vs. the esoteric argument vs. the metaphysical argument).

In *FPHP*, I showed another version of reductionism, in relation to normality and abnormality (e.g., the natural argument vs. the social argument vs. the cultural argument vs. the mental argument).

In *FPHGEOL*, I scrutinized another version of reductionism, in relation to statics and dynamics (e.g., the catastrophe argument vs. the uniformity argument vs. the revision argument).

In *FPHCHESS*, I showed another version of reductionism, in relation to tactics and strategy (e.g., the natural argument vs. the social argument vs. the cultural argument vs. the mental argument).

In *FPHR*, I revealed another version of reductionism, in relation to secularness and sacredness (e.g., the critical argument vs. the skeptical argument vs. the theist argument).

In *FPHEDU*, I examined another version of reductionism, in relation to teaching and learning (e.g., the teacher-centered argument vs. the student-centered argument vs. the balanced argument).

In *FPHH*, I explored another version of reductionism, in relation to joking and laughing (e.g., the natural argument vs. the social argument vs. the cultural argument vs. the mental argument).

In *FPHA*, I revealed another version of reductionism, in relation to sound and silence (e.g., the natural argument vs. the social argument vs. the cultural argument vs. the mental argument).

In *FPHGAM*, I revealed another version of reductionism, in relation to risk and caution (e.g., the natural argument vs. the social argument vs. the cultural argument vs. the mental argument).

In *FPHCOM*, I showed another version of reductionism, in relation to hardware and software (e.g., the natural argument vs. the social argument vs. the cultural argument vs. the mental argument).

In *FPHLIT*, I proposed another version of reductionism, in relation to fiction and non-fiction (e.g., the natural argument vs. the social argument vs. the cultural argument vs. the mental argument).

In *FPHCHEM*, I proposed another version of reductionism, in relation to substances and their changes (e.g., the natural argument vs. the social argument vs. the cultural argument vs. the mental argument).

In *FPHDA*, I proposed another version of reductionism, in relation to quantitative and qualitative research (e.g., the natural argument vs. the social argument vs. the cultural argument vs. the mental argument).

In *FPHGEOG*, I proposed another version of reductionism, in relation to environments and their interactions (e.g., the natural argument vs. the social argument vs. the cultural argument vs. the mental argument).

In *FPHPA*, I proposed another version of reductionism, in relation to the body and its presence (e.g., the natural argument vs. the social argument vs. the cultural argument vs. the mental argument).

In *FPHSEM*, I proposed another version of reductionism, in relation to internality and externality (e.g., the natural argument vs. the social argument vs. the cultural argument vs. the mental argument).

In *FPHARCH*, I proposed another version of reductionism, in relation to form and function (e.g., the natural argument vs. the social argument vs. the cultural argument vs. the mental argument).

In *FPHCRIM*, I proposed another version of reductionism, in relation to heroes and villains (e.g., the natural argument vs. the social argument vs. the cultural argument vs. the mental argument).

In *FPHHIST*, I proposed another version of reductionism, in relation to universality and relativity (e.g., the natural argument vs. the social argument vs. the cultural argument vs. the mental argument).

In *FPHMIG*, I proposed another version of reductionism, in relation to Sameness and Otherness (e.g., the natural argument vs. the social argument vs. the cultural argument vs. the mental argument).

In *FPHTRAN*, I proposed another version of reductionism, in relation to networks and operations (e.g., the natural argument vs. the social argument vs. the cultural argument vs. the mental argument).

In *FPHCA*, I proposed another version of reductionism, in relation to ingredients and techniques (e.g., the natural argument vs. the social argument vs. the cultural argument vs. the mental argument).

In *FPHW*, I proposed another version of reductionism, in relation to uselessness and usefulness (e.g., the natural argument vs. the social argument vs. the cultural argument vs. the mental argument).

And in *BNR*, I propose another version of reductionism, in relation to diversity and discontinuity (e.g., the natural argument vs. the social argument vs. the cultural argument vs. the mental argument).

(a3) In methodological reductionism, a good illustration can be the debate between different versions of qualitative and quantitative methods (as already analyzed in *FC* and also *FHC*).

In *FPHML*, I examined similar reductionism, this time, in the literature on mathematical logic (e.g., the obsession with consistency, soundness, and completeness).

And in *FPHFS*, I explored the problems of reductionism in the context of formal science (e.g., the analytical argument, as opposed to the synthetic argument).

(a4) In ontological reductionism, an excellent instance involves another debate, this time between emergentism and reductionism in complexity theory (as addressed in *FC*) and also in psychology (as elaborated in *FPHC*, in the context of Being and Belonging).

But to target against the varieties of reductionism constitutes only a side of the same coin, and the other side concerns the varieties of reverse-reductionism.

Against the Varieties of Reverse-Reductionism

The other side of the same coin is the reverse version of reductionism, which is what I want to call, in the absence of better words, *reverse-reductionism*.

My sophisticated version of methodological holism targets against the varieties of reverse-reductionism (just as it also rejects the ones of reductionism).

Perhaps nothing expresses better the popularity of reverse-reductionim than the "anything-goes" mentality in postmodernism of our time, as shown in the following four versions.

(b1) In conceptual reverse-reductionism, any concept of "art" (e.g., fine art, applied art, outsider art, junk art) is welcome in postmodernism (as already analyzed in Ch.4 of *FHC*—and also in *FAE*).

(b2) In theoretical reverse-reductionism, a variety of art and literary theories co-exist. Take the case of literary studies, as there are now Literary Structuralism, Marxist Literary Criticism, New Criticism, Phenomenology, Hermeneutics, Language-Game Literary Criticism, Feminist Literary Criticism, Reception Theory, Reader Response Criticism, Poststructuralism, Semiotics, Pyschoanalytic Literary Criticism, just to cite some well-known ones, with no one being said to be better than any others (as detailedly analyzed in Ch.4 of *FHC*). (S. Raman 1997)

In *BNN*, I even introduced "the compromise fallacy" as another good illustration of theoretical reverse-reductionism, in misleadingly treating both genetic and environmental approaches as equally valid.

(b3) In methodological reverse-reductionism, multiple methodologies are deemed as acceptable in postmodernism (e.g., doing art without praxis, doing art with praxis, and doing art by sublation), as analyzed in Ch.4 of *FHC*.

And in *FPHFS*, I exposed the problems of reverse-reductionism in the context of formal science in relation to systems theory (e.g., the synthetic argument).

(b4) In ontological reverse-reductionism, no privileged ontology is allowed, and the door is open practically for anything in postmodernism (e.g., the equal status of the ontology of Being vs. that of Becoming, as already addressed in Ch.4 of *FHC*—and also in *FPHC*).

In *FAE*, I also introduced another version of reverse-reductionism, that is, "the pluralist fallacy," in the context of understanding aesthetic experience, for instance—although this fallacy has been committed not exclusively in relation to the ontological level (but also at the conceptual, theoretical, and methodological ones).

These dual dangers against reverse-reductionism (in this sub-section) and reductionism (in the previous sub-section) are something that sophisticated methodological holism rejects. And in *FPHFS*, I addressed these two problems further.

This point about the dual dangers is important enough, since many scholars often favor one against the other, but it is vital to target both.

The Holistic Organization of an Inquiry

With these dual dangers against reductionism and reverse-reductionism in mind—sophisticated methodological holism suggests that an inquiry of any given phenomenon is more complete, if treated in the context of a comprehensive analysis at all relevant levels, which challengingly encompass all the fields of human knowledge, ranging from the natural sciences and formal sciences through the social sciences to the humanities.

This section is something that I had already stressed in all my previous books and repeat (*often verbatim*) hereafter. But, for those readers who had read my previous books, this serves as a reminder.

With this reminder in mind—there are multiple ways to engage in a holistic inquiry with all relevent levels of analysis. Over the years, I have proposed different ways to fulfill this holistic methodological requirement, as repeated in my previous books.

Hereafter is a summary of four major ways, namely, (a) by discipline, (b) by domain, (c) by subject, and (d) by meta-analysis—all of which fulfill the holistic requirement of sophisticated methodological holism on a given issue.

(a) In a holistic organization by discipline, a good illustration is the multiple levels of analysis, namely, (a1) the micro-physical, (a2) the chemical, (a3) the biological, (a4) the psychological, (a5) the organizational, (a6) the institutional, (a7) the structural, (a8) the systemic, (a9) the cultural, (a10) the macro-physical (cosmological), and (a11) other relevant levels, which are either a combination of all these levels or the practical applications with a combination of them.

This holistic organization by discipline is used in *FCD* and *BCPC*, for instance.

(b) In a holistic organization by domain, some good candidates include the classification of different perspectives of analysis in relation to (b1) nature, (b2) the mind, (b3) society, and (b4) culture, as already worked out in *BCIV*, *FPHST*, *FC*, *FAE*, *FPHU*, *FPHK*, *FPHE*, and *FPHFS*, just to cite some examples.

Culture in (b4), in this re-classification, is the same as culture in (a9) and can be further divided into different sub-categories, with good examples like (a9i) epistemology (methodology), (a9ii) aesthetics, (a9iii) ethics, (a9iv) metaphysics (ontology), and (a9v) religion. For clarification, epistemology (methodology) in (a9i) and metaphysics (ontology) in (a9iv) overlap with the category of "meta-theory."

Society in (b3) includes the organizational in (a5), the institutional in (a6), the structural in (a7), and the systemic in (a8).

The mind (b2) has more to do with the chemical in (a2), the biological in (a3), and the psychological in (a4), although it can overlap with (a1), (b3) and (b4), for instance.

It should be stressed that the "mind" here also includes the categories like the "brain," "body," and the "spirit" (or the "soul")—whenever needed in a given context, so that one does not fall into the intellectual trap of the "mind-body dualism" in the history of modern (and especially, postmodern) philosophy, which I already went to great lengths to explain in *The Future of Human Civilization*. After all, in everyday's usage, the "mind" already implies the "brain" and the "body" (together with their interactions) and has different levels of consciousness which can be both "spiritual" and non-spiritual.

And nature in (b1) refers to the micro-physical in (a1) and the macro-physical (cosmological) in (a10), although it can also overlap with (a2), (a3) and (a4).

(c) In a holistic organization by subject, some excellent examples concern the analysis of consciousness in *FPHC* (i.e., on Having, Belonging, and Being); of civilization in *FHC* (i.e., the True, the Holy, the Everyday, the Technological, the Beautiful/Sublime, the Good, and the Just); of international politics in *BWT* (i.e., Hyper-Empires, Meso-Empires, and Micro-Empires); of nature and nurture in *BNN* (i.e., genes and memes); of information architecture in *FIA* (i.e., taxonomy and network); of mathematical logic in *FPHML* (i.e., consistency, soundness, and completeness); of communication in *FPHMM* (i.e., sending, connecting, and receiving); of creative thinking in *FPHCT* (i.e., invention and innovation); of geometry in *FPHG* (i.e., infinity, symmetry, and dimensionality); of urban planning in *FPHUP* (i.e., density and void), of language in *FPHL* (i.e., structure and context), of organization in *FPHO* (i.e., communication, decision-making, and leadership), of martial arts in *FPHMA* (i.e., the martial body and the martial spirit), of sexuality in *FPHS* (i.e., the sexual body and the sexual spirit), of law in *FPHLAW* (i.e., necessity and contingency), of war and peace in *FPHWP* (i.e., aggression and pacificity), of personality in *FPHP* (i.e., normality and abnormality), of ethics in *BEPE* (i.e., morality and immorality), of cosmology in *BCOS* (i.e., the beginnings and ends of the universe), of geology in *FPHGEOL* (i.e., statics and dynamics), of chess in *FPHCHESS* (i.e., tactics and strategy), of religion in *FPHR* (i.e., secularness and sacredness), of education in *FPHEDU* (i.e., teaching and learning), of humor in *FPHH* (i.e., laughing and joking), of acoustics in *FPHA* (i.e., sound and silence), of gambling in *FPHGAM* (i.e., risk and caution), of computing in *FPHCOM* (i.e., hardware and software), of literature in *FPHLIT* (i.e., fiction and non-fiction), of chemistry in *FPHCHEM* (i.e., substances and their changes), of data analysis in *FPHDA* (i.e., quantitative and qualitative research), of geography in *FPHGEOG* (i.e., environments and their interactions), of performing arts in *FPHPA* (i.e., the body and its presence), of semantics in *FPHSEM* (i.e., internality and externality), of architecture in *FPHARCH* (i.e., form and function), of criminality in *FPHCRIM* (i.e., heroes and villains), of history in *FPHHIST* (i.e., universality and relativity), of migration in *FPHMIG* (i.e., Sameness and Otherness), of transportation in *FPHTRAN* (i.e., networks and operations), of culinary art in *FPHCA* (i.e., ingredients and techniques), of waste in *FPHW* (i.e., uselessness and usefulness), and of natural resources in *BNR* (i.e., diversity and discontinuity)—by incorporating all the relevent levels of analysis as cited above in each.

(d) And in a holistic organization by meta-analysis, two good cases in point involve the works on post-capitalism in *BCPC* and on authoritarian liberal democracy in *ALD*, with the classification of analysis in theory and meta-theory.

The essential point to remember here, as I thus stress in all my previous books, is that the multiple levels of analysis can be reorganized in many different ways, insofar as none of the levels (if relevant to an inquiry in question) is ignored or dismissed, to avoid the danger of reductionism (and for that matter, the one of reverse-reductionism, in the opposite direction).

Three Clarifications

Three clarifications are needed here, however, to avoid any hasty misunderstanding.

Firstly, the entities in each classification are not mutually exclusive. For instance, in the holisitc organization by domain, nature in (b1) can also be linked to the chemical in (a2), the biological in (a3), and the systemic in (a8). And the mind (b3) can alternatively be related to the micro-physical in (a1) and the chemical in (a2), for example.

Secondly, there is also the important factor of luck (or uncertainty in conventional usage), but it is already implied or allowed in each of the organizations of an inquiry—especially in relation to the predictability-unpredictability principle and the order-chaos principle in existential dialectics. This topic about uncertainty was already addressed in *FHC*—and more extensively, in *FC*, in the context of order and chaos (and later, to an extent, in *FPHK*, *FPHFS*, and *FPHGAM*, for example).

And lastly, the comparison in a classification is not absolute, but relative, as there are often some shades of gray, not exactly black or white (figuratively speaking).

Some further clarifications and qualifications of sophisticated methodological holism are shown in *Table 4.2*.

Some Distinctive Usefulness

In the end, sophisticated methodological holism—when applied, especially though not exclusively, as illustrated in my numerous works—can enrich the understanding of reality in some distinctive ways. Here are four examples (as revealed in each of my books).

Firstly, it provides a comprehensive analysis of a subject matter, from which much can be learned about reality, both in relation to the perspectives of the mind, nature, society, and culture.

Secondly, it offers a new classification of the subject matter.

Thirdly, it suggests some visions of the future in relation to the subject matter in question.

And fourthly, it proposes some insights on meta-theory (e.g., methodology and ontology) in general—with the clear understanding, however, of the dilemma of specific vs. general ontology (as shown in *Table 4.9* on the syntax of existential dialectics in the context of the dilemma of ontology).

For this very reason, all of these ways are important, without reducing one into the analysis of another. (*FPHCT*)

Chapter Outline

With this summary of my two original meta-theories (on methodology and ontology, *almost verbatim* from my previous works) in mind—this book is organized in four main parts, namely, (a) Part I on the introduction to this book, (b) Part II on diversity, (c) Part III on discontinuity, and (d) Part IV on the conclusion of the book.

The book is thus organized in four chapters.

The introductory chapter, that is, Chapter One titled "Introduction—The Ambience of Natural Resources," begins with a question to a fashionable ideology, provides a definition and history of natural resources, and then introduces diversity and discontinuity in the study of natural resources, as well as the multiple aspects of natural resources—together with the theoretical debate on natural resources, my resilient theory of natural resources, the relationship between theory and meta-theory, existential dialectics, sophisticated methodological holism, and some further clarifications.

Chapter Two titled "Diversity and its Ambivalent Characters" examines the nature of diversity in natural resources from the four perspectives of the mind, nature, society, and culture, with a constructive assessment of each.

Chapter Three titled "Discontinuity and its Dual Effects" analyzes the nature of discontinuity in natural resources from the four perspectives of the mind, nature, society, and culture, with a deconstructive critique of each.

The last chapter titled "Conclusion—The Future of Natural Resources" summarizes the analysis in the book in the dialectic context of diversity and discontinuity in natural resources with thirty major theses, namely, (a) the first thesis: the formalness-informalness principle, (b) the second thesis: the absoluteness-relativeness principle, (c) the third thesis: the partiality-totality principle, (d) the fourth thesis: the predictability-unpredictability principle, (e) the fifth thesis: the explicability-inexplicability principle, (f) the sixth thesis: the fiction-reality principle, (g) the seventh thesis: the cognitiveness-noncognitiveness principle, (h) the eighth thesis: the finiteness-transfiniteness principle, (i) the ninth thesis: the preciseness-vagueness principle, (j) the tenth thesis: the simpleness-complicatedness principle, (k) the eleventh thesis: the openness-hiddenness principle, (l) the twelfth thesis: the denseness-emptiness principle, (m) the thirteenth thesis: the change-constancy principle, (n) the fourteenth thesis: the order-chaos principle, (o) the fifteenth thesis: the slowness-quickness principle, (p) the sixteenth thesis: the expansion-contraction principle, (q) the seventeenth thesis: the theory-praxis principle, (r) the eighteenth thesis: the convention-novelty principle, (s) the nineteenth thesis: the evolution-transformation principle, (t) the twentieth thesis: the symmetry-asymmetry principle, (u) the twenty-first thesis: the softness-hardness principle, (v) the twenty-second thesis: the seriousness-playfulness principle, (w) the twenty-third thesis: the activeness-inactiveness principle, (x) the twenty-fourth thesis: the selfness-otherness principle, (y) the twenty-fifth thesis: the regression-progression principle, (z) the twenty-sixth thesis: the sameness-difference principle, (aa) the twenty-seventh thesis: the stability-reaction principle, (bb) the twenty-eighth thesis: the functionality-nonfunctionality principle, (cc) the twenty-ninth thesis: the intentionality-nonintentionality principle, and (dd) the thirtieth thesis: the post-human resilience—to be elaborated in the rest of the book, with a summary in the concluding chapter.

This seminal project here, if successful, will fundamentally change the way that we think about diversity and discontinuity in the study of natural resources (together with other debates as will be discussed in the rest of the book), from the combined perspectives of the mind, nature, society, and culture, with enormous implications for the human future and what I originally called its "post-human" fate.

Some Clarifications

However, some additional (last) clarifications are needed.

These clarifications, lest any potential misunderstanding by the readers occurs, refer to (a) the conversation with my previous books, (b) the illustrative selection of case studies and examples, (c) detailed analysis versus overall analysis, (d) three distinctive features of using quotations, (e) two difficulties with understanding, (f) repetition, (g) chapter organization to counter conventional wisdom, and (h) the use of neologisms.

These clarifications are something that I also regularly repeat in my previous books too, so these points are repeated hereafter, *often verbatim*—albeit in a different context this time.

In Conversation with My Previous Books

Firstly, my *resilient theory of natural resources* is constructed on the theoretical foundation of my previous books (i.e., *FHC*, *FCD*, *FPHC*, *BDPD*, *BCPC*, *BCIV*, *FPHST*, *BNN*, *BWT*, *FC*, *FAE*, *ALD*, *FIA*, *FPHU*, *FPHK*, *FPHML*, *FPHE*, *FPHMM*, *FPHCT*, *FPHG*, *FPHUP*, *FPHL*, *FPHO*, *FPHMA*, *FPHS*, *FPHLAW*, *FPHFS*, *FPHWP*, *FPHP*, *BCOS*, *BEPE*, *FPHGEOL*, *FPHCHESS*, *FPHR*, *FPHEDU*, *FPHH*, *FPHA*, *FPHGAM*, *FPHCOM*, *FPHLIT*, *FPHCHEM*, *FPHDA*, *FPHGEOG*, *FPHPA*, *FPHSEM*, *FPHARCH*, *FPHCRIM*, *FPHHIST*, *FPHMIG*, *FPHTRAN*, *FPHCA*, and *FPHW*—as explicated in "The List of Abbreviations").

For this reason, many of my books have some familiar words in their titles like "The Future of Post-Human ____" or "Beyond ____to ____," as well as some tables for background information—as they constitute a series of books with the final aim to converge all of my original theories on numerous fields into a unified theory of everything (as already summarized in the section on theory and meta-theory in Chapter One).

Just consider the following two illustrations, namely, (a) in relation to other related theoretical debates and (b) in relation to different visions as worked out in my previous books.

(a) The first illustration is that the theoretical debate here also involves a few other related theoretical issues.

These other related theoretical issues have to do with some deeply contested theoretical debates, which, however, I had already gone to great lengths to analyze in my numerous (previous) books, as summarized below. It should also be stressed, lest misunderstanding occurs, that each of the theoretical debates (as listed below) does not exhaust the issues covered in the books, as each book covers more issues than what is listed below (for illustration only).

- On Extinction
  –Ex: 2 volumes, *The Future of Human Civilization*
  –Ex: *The Future of Capitalism and Democracy*
  –Ex: *The Future of Post-Human Unconsciousness*
  –Ex: *The Future of Post-Human Consciousness*

- On Fragmentation
  - –Ex: *The Future of Complexity*
  - –Ex: *The Future of Capitalism and Democracy*
  - –Ex: 2 volumes, *The Future of Human Civilization*
  - –Ex: *The Future of Post-Human Migration*
- On Diversity
  - –Ex: *The Future of Post-Human Migration*
  - –Ex: 2 volumes, *Beyond Democracy to Post-Democracy*
- On the Environment
  - –Ex: *The Future of Post-Human Waste*
  - –Ex: *The Future of Post-Human Geography*
  - –Ex: *The Future of Post-Human Geology*
  - –Ex: *The Future of Post-Human Urban Planning*
  - –Ex: *The Future of Post-Human Migration*
  - –Ex: *The Future of Capitalism and Democracy*
  - –Ex: 2 volumes, *The Future of Human Civilization*
- On Food
  - –Ex: *The Future of Post-Human Culinary Art*
- On Consumption
  - –Ex: *Beyond Capitalism to Post-Capitalism*
  - –Ex: *The Future of Capitalism and Democracy*
  - –Ex: 2 volumes, *The Future of Human Civilization*
- On Drugs
  - –Ex: *The Future of Post-Human Chemistry*
  - –Ex: *The Future of Post-Human Sexuality*
  - –Ex: *The Future of Post-Human Consciousness*
- On Technology
  - –Ex: *The Future of Post-Human Engineering*
  - –Ex: 2 volumes, *The Future of Human Civilization*
  - –Ex: *The Future of Post-Human Computing*
- On Engineering
  - –Ex: *The Future of Post-Human Engineering*
- On Space Colonization
  - –Ex: *The Future of Post-Human Geography*
  - –Ex: *The Future of Post-Human Migration*
  - –Ex: *The Future of Capitalism and Democracy*
  - –Ex: 2 volumes, *The Future of Human Civilization*
- On Economics
  - –Ex: *The Future of Capitalism and Democracy*
  - –Ex: 2 volumes, *The Future of Human Civilization*
  - –Ex: *Beyond Capitalism to Post-Capitalism*

- On Ethics
  - –Ex: *Beyond Ethics to Post-Ethics*
  - –Ex: 2 volumes, *The Future of Human Civilization*
- On Spirituality
  - –Ex: 2 volumes, *The Future of Human Civilization*
  - –Ex: *The Future of Capitalism and Democracy*
  - –Ex: *The Future of Post-Human Religion*
  - –Ex: *Beyond Civilization to Post-Civilization*
  - –Ex: *The Future of Post-Human Sexuality*
  - –Ex: *The Future of Post-Human Martial Arts*
  - –Ex: *The Future of Post-Human Unconsciousness*
  - –Ex: *The Future of Post-Human Consciousness*
  - –Ex: *Beyond Ethics to Post-Ethics*
- On Aesthetics
  - –Ex: *The Future of Aesthetic Experience*
  - –Ex: *The Future of Post-Human Culinary Art*
  - –Ex: *The Future of Post-Human Architecture*
  - –Ex: *The Future of Post-Human Martial Arts*
  - –Ex: *The Future of Post-Human Performing Art*
  - –Ex: *The Future of Post-Human Literature*
- On Globalization
  - –Ex: *The Future of Capitalism and Democracy*
  - –Ex: 2 volumes, *The Future of Human Civilization*
- On Civilization
  - –Ex: *Beyond Civilization to Post-Civilization*
  - –Ex: 2 volumes, *The Future of Human Civilization*
- On Epistemology and Ontology
  - –Ex: *The Future of Post-Human Knowledge*
  - –Ex: *The Future of Post-Human Consciousness*
  - –Ex: 2 volumes, *The Future of Human Civilization*
- On Rationality
  - –Ex: *The Future of Post-Human Mathematical Logic*
  - –Ex: *The Future of Post-Human Humor*
- On Freedom
  - –Ex: 2 volumes, *The Future of Human Civilization*
  - –Ex: *The Future of Capitalism and Democracy*
  - –Ex: *Beyond Capitalism to Post-Capitalism*
  - –Ex: 2 volumes, *Beyond Democracy to Post-Democracy*
  - –Ex: *Beyond Civilization to Post-Civilization*
- On Communication
  - –Ex: *The Future of Post-Human Mass Media*

–Ex: *The Future of Post-Human Organization*
- On Learning
  –Ex: *The Future of Post-Human Language*
  –Ex: *The Future of Post-Human Education*
- On Psychology
  –Ex: *The Future of Post-Human Unconsciousness*
  –Ex: *The Future of Post-Human Consciousness*
  –Ex: *The Future of Capitalism and Democracy*
  –Ex: *The Future of Post-Human Creative Thinking*
  –Ex: *The Future of Post-Human Personality*
- On Cognition
  –Ex: *The Future of Post-Human Unconsciousness*
  –Ex: *The Future of Post-Human Consciousness*
  –Ex: *The Future of Capitalism and Democracy*
  –Ex: *The Future of Post-Human Creative Thinking*
  –Ex: *The Future of Post-Human Personality*
- On Intuition
  –Ex: *The Future of Post-Human Unconsciousness*
  –Ex: *The Future of Post-Human Consciousness*
  –Ex: *Beyond Capitalism to Post-Capitalism*
  –Ex: *The Future of Capitalism and Democracy*
- On Emotion
  –Ex: *The Future of Post-Human Unconsciousness*
  –Ex: *The Future of Post-Human Consciousness*
  –Ex: *The Future of Aesthetic Experience*
  –Ex: *Beyond Capitalism to Post-Capitalism*
  –Ex: *The Future of Capitalism and Democracy*
  –Ex: *The Future of Post-Human Personality*
- On Behavior
  –Ex: *The Future of Post-Human Language*
  –Ex: *The Future of Post-Human Unconsciousness*
  –Ex: *The Future of Post-Human Consciousness*
  –Ex: *Beyond Capitalism to Post-Capitalism*
  –Ex: *The Future of Capitalism and Democracy*
  –Ex: *The Future of Post-Human Sexuality*
  –Ex: *The Future of Post-Human Personality*
  –Ex: *The Future of Post-Human Martial Arts*
- On Being and Belonging
  –Ex: *The Future of Post-Human Consciousness*
  –Ex: 2 volumes, *The Future of Human Civilization*
- On Different Dimensions of Human Existence

  –Ex: 2 volumes, *The Future of Human Civilization*
- On Modernity, Post-Modernity, and After-Postmodernity
  –Ex: 2 volumes, *The Future of Human Civilization*
  –Ex: *Beyond Civilization to Post-Civilization*
- On Pre-Modernity
  –Ex: *Beyond Civilization to Post-Civilization*

Therefore, those who are interested in these other theoretical issues (and debates) which are related to the current context can consult these other books of mine for more details.

(b) And the second illustration is that one of the major theses of this book concerns the future, be it human and later post-human, which is closely linked with the different visions that I had worked out in my previous books, in relation to the mind, nature, society, and culture.

Of course, for the convenience of the readers, a summary of some of these visions will be provided whenever necessary, as shown in the two sections on meta-theory above (*often verbatim*) and in different tables at the end of this chapter and the concluding chapter, for instance.

Yet, I still expect the readers to read my previous books directly for more details—as this book is written in conversation with my previous ones.

Case Studies and Examples

Secondly, the case studies and examples in the chapters are not exhaustive but solely illustrative. Nor are they necessarily mutually exclusive, as they can be reclassified in a different way. And exceptions are allowed.

One is tempted to assume, however, that, since the case studies and their examples are not exhaustive, more case studies and examples are needed. But the problem here is that just adding more of them does not necessarily change the conclusions to be drawn and may even lead to redundancy.

After all, the case studies and examples used in this book are chosen with care and deemed sufficient for the purpose at hand, even though they are solely illustrative.

Detailed Analysis versus Overall Synthesis

Thirdly, there is a distinction between detailed analysis and overall synthesis, in that the former is, especially though not exclusively, more for the specialized viewpoint of a specific discipline, whereas the latter is, especially though not exclusively, more for the broad horizon of an interdisciplinary approach—relatively speaking, of course.

Or, by analogy, there is a distinction between depicting a "tree" (with tiny details) in a "forest" and viewing the entire "forest" itself (with the broader horizon).

This book, like all previous ones of mine, tries to strike a balance between the two forms of understanding, although this does not imply that the two forms should be exactly equally offered, in an exact 50-50 balance.

Instead, the book, more often than not, does not allow the tiny details to obscure the larger picture in a given case study, because the entire book is inter-disciplinary in nature and tries to look at the entire "forest," not being bogged down with the detailed analysis of an individual "tree" in the forest, by analogy.

In this sense, although detailed analysis can be provided whenever needed for a given case study, the important point to remember is the overall horizon with its broadness of scope.

Three Distinctive Features of Using Quotations

Fourthly, the way that quotes are used in this book (and others of mine) has three distinctive features, as explained below.

The first distinctive feature is that, at the end of each quotation, sometimes there are more than one reference—in which case the first reference is for the original citation, and the rest of the references are for the convenience of the reader for more information about the issue(s) under quotation.

The second distinctive feature is that quotations are used as often as possible, so as to let others speak for themselves, without the nuisance of putting my own words into their mouths. Even more importantly, this allows me to use the original text as an evidence for critical analysis at the end of each section.

And the third distinctive feature is that all these quotations are used for my own purposes, in that I often creatively (or originally) re-organize them for a new way of analyzing the issue(s) in question.

Two Difficulties with Understanding

Fifthly, to understand this book (or any other book of mine) is difficult for two main reasons.

In regard to the first difficulty, this book, just like many other books of mine, is in conceptual and theoretical conversation with other books of mine (55 books, as of August 2012), so this requires a vast background information on numerous subject matters.

And in regard to the second difficulty, this book also analyzes the subject matter in a way very different from the expectation of a conventional specialist reader, since it examines reality outside the specialist's interest (or outside his or her "comfort zone"), from the four broad perspectives of the mind, nature, society, and culture, such that it covers all domains of knowledge (in the social sciences, the natural sciences, the formal sciences, and the humanities). In other words, those who are truly interested in my project for a unified theory of everything would find many books of mine highly stimulating, but those who are specialists in a subject matter (like physics, geography, chess, and so on, in a strictly specialist sense) will not find them suitable to their specialist taste (or expectation).

Not surprisingly, a conventional specialist reader would find these two difficulties (with understanding) almost practically impossible to overcome, as they almost require the mind of a polymath. Even then, the best advice is that one should still try one's best to understand it.

Repetition

Sixthly, many of my books have something in common, and the reason is to show "the unified theory of everything" in all domains of knowledge.

For this reason, there is certain repetition in all books (especially in regard to the background information, together with the tables, in the introductory chapter and the concluding chapter). Therefore, these books are all titled "The Future of Post-Human ____" (or at times "Beyond ____ to ____") and constitute the "post-human" series.

Chapter Organization to Counter Conventional Wisdom

Seventhly, all of my books aim to deconstruct conventional wisdom and to replace it with new thinking, and the chapters in my books are often organized in two distinctive ways for this purpose

The first distinctive way is to show the polarized thinking in conventional wisdom by devoting two main chapters on opposites and then deconstructing them later. For example, in *Beyond Ethics to Post-Ethics* (2011), Chapter Two first introduces the conventional wisdom for morality but then shows that morality has its negative side often unsaid—and Chapter Three does just the opposite, that is, first introduces the conventional wisdom against immorality but then shows that immorality has its bright side often unspoken of.

And the second distinctive way is to show the opposites (embedded in conventional wisdom) in a dialectic way. For instance, in the above example, the task here is to show how and why the analysis of morality (in Chapter Two) is inadequate without the inclusion of its immorality, just as the examination of immorality (in Chapter Three) is insufficient without the consideration of its morality—to the extent that there is no morality without immorality, and vice versa.

But there are two qualifications.

The first qualification is that the opposites do not have to be always mutually exclusive in the conventional meaning (like morality vs. immorality in *BEPE* and heroes vs. villains in *FPHCRIM*)—but are simply disproportional in focus in other cases (in that some thinkers focus more on one than on the other), and good examples are networks vs. operations (in *FPHTRAN*), ingredients vs. techniques (in *FPHCA*), and form vs. function (in *FPHARCH*). In other words, there are multiple meanings of the opposites.

And the second qualification is that some books (like the 2 volumes of *The Future of Human Civilization*, the 2 volumes of *Beyond Democracy to Post-Democracy*, and *The Future of Capitalism and Democracy*) are very lengthy and analyze many issues in more than two main chapters. Yet, there is a general pattern of introducing and then deconstructing conventional wisdom in my works.

The Use of Neologisms

And finally, I use different neologisms in my books, mostly to introduce my original concepts and theories, and good instances include those here or elsewhere in my previous books (e.g., "the resilient theory of natural resources," "post-democracy," "hyper-spatial consciousness," "post-capitalism," and whatnot). Needless to say, they are used here solely for our current intellectual convenience, as they will be renamed differently in different ways in future history.

As I thus wrote in *FCD* (508-9), "all these terms 'post-capitalism,' 'post-democracy'...and other ones as introduced in...[the] project (e.g.,...'posthuman elitists,' and 'posthuman counter-elitists,' just to cite a few of them) are more for our current intellectual convenience than to the liking of future humans and post-humans, who will surely invent more tasteful neologisms to call their own eras, entities, and everything else, for that matter. But the didactic point here is to use the terms to foretell what the future might be like, not that its eras and entities must be called so exactly and permanently. After all, William Shakespeare (1995: Act II, Scene II, Line 47) well said long ago: 'What is in a name? That which we call a rose by any other name would smell as sweet.'"

As I had also stressed time and again before, each of the neologisms can be re-written as a different "X," only to be re-named differently by the powers that be in different eras of future history.

With these clarifications for thought in mind—I now proceed to Chapter Two on diversity and its ambivalent characters.

Table 1.1. Multiple Aspects of Natural Resources

- **Diversity and Discontinuity**
  - —Diversity
    - Ex: "conservation"
    - Ex: "management"
    - Ex: "equilibrium"
  - —Discontinuity
    - Ex: "extinction"
    - Ex: "fragmentation"
    - Ex: "reduction (in range)"

- **Source of Origin**
  - —Ex: "biotic" resources
  - —Ex: "abiotic" resources

- **Stage of Development**
  - —Ex: "potential resources"
  - —Ex: "actual resources"
  - —Ex: "reserve resources"
  - —Ex: "stock resources"

- **Renewability**
  - —Ex: "renewable" resources
  - —Ex: "non-renewable" resources

- **Earth's Spheres**
  - —Ex: the "atmosphere" (corresponding to air)
  - —Ex: the "cryosphere" (corresponding to ice)
  - —Ex: the "hydrosphere" (corresponding to water)
  - —Ex: the "biosphere" (corresponding to life)
  - —Ex: the "lithosphere" (corresponding to rocks)
  - —Ex: the "pedosphere" (corresponding to soil)

Notes: The categories are solely illustrative (not exhaustive), and the comparison is relative (not absolute), nor are they necessarily mutually exclusive. And some can be easily re-classified elsewhere. As generalities, they allow exceptions.
Source: From *Sec. 1.3* & *Sec. 1.4* of *BNR*. See book for citations.

Table 1.2. Nine Great Future Transformations of Natural Resources (Part I)

- **In Relation to the Mind**
  —*New Goals*
  - This search search for "new goals" will give future advanced humans and later their successors, namely, the post-humans, different motivations to find natural resources in a way that current humans cannot—in relation to my previous visions on the future "post-human" supersession of the human mind like "the supersession computing," "metamorphic humor," "the contrarian personality," "the hyper-martial body," "the hyper-sexual body," "floating consciousness," "hyper-spatial consciousness," "thinking machines," "unfolding unconsciousness," "thinking robots," "genetically altered superior beings," "cyborgs," etc.

  —*New Means*
  - This quest for "new means" will give future advanced humans and later their successors, namely, the post-humans, the capacity to find novel options to achieve the goals in regard to natural resources in a way that current humans cannot.

- **In Relation to Nature**
  —*New Comparisons*
  - This search for "new comparisons" will give future advanced humans and later their successors, namely, the post-humans, the ability to learn from animals, plants, and the universe, through comparison and contrast, about the study of natural resources in a way that current humans cannot.

(continued on next page)

Table 1.2. Nine Great Future Transformations of Natural Resources (Part II)

• **In Relation to Nature (*cont'd*)**

—*New Integrations*

• This search for "new integrations" will give future advanced humans and later their successors, namely, the post-humans, the ability to live and interact with nature for natural resources in a way that current humans cannot. This is especially so, in the context of the great future transformations of geology (in *FPHGEOL*), of geography (in *FPHGEOG*), of migration (in *FPHMIG*), and of transportation (in *FPHTRAN*)—in relation to my previous visions on the future "post-human" adventures into different dimensions of space-time unto multiverses like "the alteration of space-time" (*Table 1.3* & *Table 1.4*), "multiverses" (*Table 1.7* & *Table 1.8*), "hyperspace" (*Table 1.9* & *Table 1.25*), "time travel" (*Table 1.10* & *Table 1.11*), and others.

This extension is consistent with my other theory, namely, "the theory of the cyclical progression of system integration and fragmentation" in *FCD* (2002)—at the social-systemic level (as already clarified in *Sec. 1.8* and shown below):

Local → Regional → Global →Solar →
Galactic → Clustery…→ Multiversal

For instance, most advanced forms of future intelligent life in deep space until multiverses (e.g., "floating consciousness," "hyper-spatial consciousness," "unfolding unconsciousness," etc.) will need to adapt to different environments by seeking different natural resources (both renewable and non-renewable) through space colonization (as already discussed in *FPHMIG*, *FPHST*, *FPHC*, *FCD*, and *FHC*) in the distant future.

(continued on next page)

Table 1.2. Nine Great Future Transformations of Natural Resources (Part III)

- **In Relation to Society**
  - —*New Praxes*
    - This search for "new praxes" will give future advanced humans and later their successors, namely, the post-humans, the ability to engage in practice (like activism, politics, personal experience, etc.) for the future of natural resources in a way that current humans cannot.
  - —*New Technologies*
    - This search for "new technologies" will give future advanced humans and later their successors, namely, the post-humans, the ability to come up with different techniques, materials, and experiments for natural resources in a way that current humans cannot.

      For instance, most advanced forms of future intelligent life in deep space until multiverses will make good use of their technological prowess (a) to restore life through cloning (as already discussed in *Sec. 2.4.1*—and also in *FCD* and *FHC*), (b) to extract different energy sources in outerspace through space colonization (as already discussed in *FPHMIG*, *FPHST*, *FPHC*, *FCD*, and *FHC*), (c) to create new forms of matter and energy in the distant future (as already discussed in *FPHST* and other works—and summarized in *Table 1.4*), and (d) to explore the technological frontiers of the micro-world (as already discussed in *FPHC*—and summarized in *Table 1.5*).
  - —*New Stratifications*
    - This emergence of "new stratifications" will lead future advanced humans and later their successors, namely, the post-humans, to face new inequalities for natural resources in a way that current humans cannot.

(continued on next page)

Table 1.2. Nine Great Future Transformations of Natural Resources (Part IV)

- **In Relation to Culture**
  - —*New Sub-Fields*
    - This search for "new sub-fields" will give future advanced humans and later their successors, namely, the post-humans, the ability to explore new specialties for the understanding of natural resources in a way that current humans cannot.
  - —*New Interpretations*
    - This search for "new interpretations" will give future advanced humans and later their successors, namely, the post-humans, the ability to come up with different evaluations and views about the study of natural resources in a way that current humans cannot.

Notes: The categories and examples are solely illustrative (not exhaustive), and the comparison is relative (not absolute), nor are they necessarily mutually exclusive. And some can be easily re-classified elsewhere. As generalities, they allow exceptions.

Source: From *Ch.4* of *BNR*. See text for citations.

Table 1.3. The Theoretical Debate on Space-Time (Part I)

- **Isaac Newton's Absolutist (Substantivist) Theory of Space-Time**
  —space and time are independent from each other. The structure of space-time is $E3 \times E1$ (with the structure of space, *P*, as the set of spatial locations in a three-dimensional Euclidean space, *E3*, and the structure of time as the set of temporal moments, *T*, in the one-dimensional real time, *E1*).
  —space and time are also independent from the effects of matter and events. The existence of space and time is possible even in a world absent of matter (and, for that matter, even in a world absent of events), as if they were material objects but with their total unchangingness thorough time.

- **Albert Einstein's Relativist Theory of Space-Time**
  —space and time are interchangeable (not absolute), just as matter and energy are equivalent (not independent) with the famous equation, $E = mc^2$ (as in the special theory of relativity in 1905).
  —space-time and matter-energy are also relative in a grand union (as in the general theory of relativity in 1915). Thus, each pair affects the other pair, as "matter 'tells' spacetime how to curve [in a non-Euclidean geometry] and curved spacetime 'tells' matter how to behave….Space contracts near mass and dilates away from it. Time dilates near mass and contracts away from it. Clocks positioned farther away from the mass of the earth run faster than clocks closer to the earth." (L. Shlain 1991: 328-330)

(continued on next page)

Table 1.3. The Theoretical Debate on Space-Time (Part II)

- **Peter Baofu's Perspectival Theory of Space-Time**

  —space and time can be understood from multiple perspectives, be they in relation to culture, society, nature, and the mind, with each perspective revealing something about the nature of space-time and simultaneously delimiting its view. This is subject to "the regression-progression principle" in existential dialectics.

  —each perspective of space and time exists in society and culture with good reasons, with some being more successful and hegemonic (dominant) than others. This is subject to "the symmetry-asymmetry principle" in existential dialectics.

  —space and time will not last, to be eventually superseded (altered) by post-humans in different forms (e.g.,stretching/shrinking space-time, engineering more dimensions of space-time, and manipulating multiverses), be they here in this universe or in multiverses. Thus, even the physical existence of space-time cannot last forever, with ever more transformations in the process. This is subject to "the change-constancy principle" in existential dialectics.

  —the conventional wisdom (especially by physicists) of treating the physical perspective of space and time as the foundation of all other perspectives (of space and time) and of regarding them as much less important is a form of reductionism,committing what I call *the foundation fallacy*, in misleadingly dismissing the multiple perspectives of space and time in relation to culture, society, nature, and the mind.

Notes: The examples in each category are solely illustrative (not exhaustive), and the comparison is relative (not absolute), nor are they necessarily mutually exclusive. Some can be easily re-classified elsewhere. As generalities, they allow exceptions.

Source: A summary of *Sec.1.1*, *Sec.1.2*, and *Sec.1.3* of *FPHST*—and, for that matter, the rest of *FPHST*

Table 1.4. Main Reasons for Altering Space-Time

- **The Need to Make New Energy-Matter**
  —Ex: manipulating molecular bonds for new materials
  —Ex: creating nanotechnologies on the atomic scale
  —Ex: engineering the atomic nucleus
  —Ex: restructuring most elementary particles
  —Ex: inventing new forms of matter and energy

- **The Need to Create New Space-Time**
  —Ex: creating "warp drive" (as in science fiction) for space travel
  —Ex: creating "pocket universes"

- **The Need to Conquer the Cosmos unto Multiverses**
  —Ex: spreading floating consciousness and hyper-spatial consciousness, besides other forms that humans have never known, in the cosmos and beyond unto multiverses for ultimate conquest

Notes: The examples in each category are solely illustrative (not exhaustive) nor necessarily mutually exclusive, and the comparison is relative (not absolute). As generalities, they allow exceptions. Also, it should be stressed that the three reasons are all related, in that they all contribute to the evolution of intelligent life in the cosmos unto multiverses in the most distant future beyond our current knowledge.
Sources: A summary of *Sec.6.2* of *FPHST*. See also *FHC*, *FCD*, *FPHC*, *FC*, *FPHU*, and *FPHG*, for example.

Table 1.5. The Technological Frontiers of the Micro-World

- **Type I-Minus**
  —Ex: building structures and mining

- **Type II-Minus**
  —Ex: playing with the genetic makeups of living things

- **Type III-Minus**
  —Ex: manipulating molecular bonds for new materials

- **Type IV-Minus**
  —Ex: creating nanotechnologies on the atomic scale

- **Type V-Minus**
  —Ex: engineering the atomic nucleus

- **Type VI-Minus**
  —Ex: restructuring most elementary particles

- **Type Ω-Minus**
  —Ex: altering the structure of space-time

Notes: As already indicated in *Sec.4.4.2.2* of *FPHC*, the problem with this micro-classification (from Barrow's work) is that the civilization types (with the exception of Type Ω-Minus, for example) are not quite distinct, since many of them can be achieved more or less in a civilization, to the extent that Type II-minus and Type III-minus, just to cite two plausible types, can be historically contemporaneous, relatively speaking, unlike the vast historical distance between, say, Type O and Type I (or Type I and Type II) civilizations. In other words, the micro-classification here is not very useful to understand civilization types but is revealing to see the technological frontiers of the micro-world.
Sources: A reconstruction from J.Barrow (1998:133), as originally shown in *Table 4.7* of *FPHC*. See *FPHC* for more info.

Table 1.6. Finity, Transfinity, and Infinity (Part I)

- **Infinity**
  —The term "infinity," which is "symbolically represented with ∞," derives from the Latin *infinitas* (or "unboundedness"), to refer to "several distinct concepts—usually linked to the idea of 'without end'—which arise in philosophy, mathematics, and theology." (WK 2008c)
  —In the context of mathematics, "infinity" is often "treated as if it were a number (i.e., it counts or measures things: 'an infinite number of terms') but it is a different type of 'number' from the real numbers. Infinity is related to limits,...large cardinals,...projective geometry, extended real numbers and the absolute Infinite," for instance. (WK 2008c)

- **Transfinity**
  —To avoid confusion, my usage of the word "transfinite" (as used here in the ontological principle) is not the same as the one which was originally coined by Georg Cantor to refer to "cardinal numbers or ordinal numbers that are larger than all finite numbers, yet not necessarily absolutely infinite" (in the mind of God), as opposed to "relative" infinity (in the mind of Man). (WK 2008d) In other words, for Cantor, there are two kinds of infinity, "absolute" and "relative"—with "relative" infinity or "transfinity" (in the mind of Man) being dependent on "absolute" infinity (in the mind of God) for its existence.
  —By contrast, my usage of the word "transfinite" here differs radically from the one used by Cantor (and other mathematicians) and is more limited, in light of the problems confronting any attempt to understand the idea of infinity, be it by intuition, imagination, and conception (as already shown in the prevous three sub-sections).
  —With this clarification in mind, I allow numbers which can be many times larger—or smaller, for that matter—than the finite things that we encounter in daily life, but they do not have to be related to the idea of infinity at all (which may not exist).

(continued on next page)

Table 1.6. Finity, Transfinity, and Infinity (Part II)

• **Transfinity** ***(cont'd)***

—Of course, there may be some *borderline* cases, in which it is not clear whether the number in question is transfinite (in my usage) or simply a mathematical convenience. A good example of a borderline case is the Planck unit of length for "the smallest space possibly measured in nature," which is "less than billionths of trillionths of trillionths of an inch" (or something like 1.6×10^{-35} meters). (P. Baofu 2006a; N. McAleer 1987: 219; WKV 2008; D. Corbett 2008)

—The ideas concerning symmetry and dimensions in the examples above were addressed in Chapter Three and Chapter Four of *FPHG*. For now, it suffices to further add that it is mind-boggling to even conceive or imagine extreme numbers like the Planck unit of length (or others). Does it really exist at all? Or is it merely a mathematical convenience to describe a physical situation that the human mind does not understand properly?

—In any event, "transfinity" (in my usage) can be used as a further extension (of number) to the finite numbers in daily life—albeit with some borderline cases in mind.

• **Finity**

—It refers to numbers which are bounded (that is, with an end), especially (though not exclusively) in relation to things in everyday life.

Source: From *Sec. 1.4* and *Sec. 2.2.3* of *FPHG*

Table 1.7. Theoretical Speculations of Multiverses

- **"Baby Universes" (Ex: Andre Linde and others)**
  —Ex: In a flat universe theory, "even if our part of it eventually collapses,...some spots in the cosmos would suddenly start inflating on their own, creating brand-new 'baby universes.'" (P. Baofu 2000: 623)

- **"Parallel Universes" (Ex: Stephen Hawking and others)**
  —Ex: In quantum cosmology, there allows the existence of infinite numbers of parallel universes, with tunneling among them. (M. Kaku 1994: 256) Hawking later revised his views on this.

- **"Pocket Universes" (Ex: Alan Guth)**
  —Ex: "As the pocket universes live out their lives and recollapse or dwindle away, new universes are generated to take their place....While life in our pocket universe will presumably die out, life in the universe as a whole will thrive for eternity." (A. Guth 1997: 248; P. Baofu 2002: 482)

- **"Brane Worlds" (Ex: Warren Siegel, Lisa Randall, and others)**
  —Ex: Our universe is stuck on a membrane of space-time embedded in a larger cosmos, with different brane worlds connecting and/or colliding with each other.

Notes: These examples are solely illustrative (not exhaustive), and some of the items can be reclassified somewhere else. Nor are they always mutually exclusive. Since they are generalities, exceptions are expected.
Source: From *Table 4.8* of *FPHC*

Table 1.8. The Confusion between "Many Worlds" and "Multiverse" (Part I)

• **First Problem**

—The first problem in this second confusion concerns a debate between the "many-worlds" interpretation of quantum mechanics by Hugh Everett (1957) and others, and the Copenhagen interpretation by Niels Bohr and Werner Heisenberg on the effects of quantum states after interacting with an external environment (e.g., by a measuring observer).

—On the one hand, the Copenhagen interpretation of quantum mechanics is more indeterministic in its concern with the probabilities of the outcomes of a wave function after interacting with an external environment (e.g., by a measuring observer using light to detect, for example, "an electron in a particular region around the nucleus at a particular time." (WK 2008e) Each wave function for the state of a particle refers to "a mathematical representation used to calculate the probability for it to be found in a location, or state of motion." (WK 2008f) And the "wavefunction collapse" means that "the act of measurement causes the calculated set of probabilities to 'collapse' to the value defined by the measurement." (WK 2008f)

—On the other hand, the Everett's interrpetation is more deterministic, in rejecting "the objective reality of wavefunction collapse" and "instead explaining the subjective appearance of wavefunction collapse with the mechanism of quantum decoherence." (WK 2008g) The words "quantum decoherence" here refers to "the mechanism by which quantum systems interact with their environments to exhibit probabilistically additive behavior," in such a way that "the quantum nature of the system is simply 'leaked' into the environment." (WK 2008g) For example, "[a]s a result of an interaction, the wave functions of the system and the measuring device become entangled with each other. Decoherence happens when different portions of the system's wavefunction become entangled *in different ways* [italic added] with the measuring device...." (WK 2008g) Thus, for Everett, the appearance of "wavefunction collapse" can be explained by "quantum decoherence" for quantum states to branch out in different independent worlds after interacting with an external environment.

(continued on next page)

Table 1.8. The Confusion between "Many Worlds" and "Multiverse" (Part II)

- **First Problem** ***(cont'd)***
  —Yet, a major criticism against Everett's interpretation is that these different independent worlds in Everett's sense "will never be accessible to us," because of the interference by the measuring observer in an external environment. (WK 2008g)
  —For illustration, "once a measurement is done, the measured system becomes entangled with both the physicist who measured it and a huge number of other particles, some of which are photons flying away towards the other end of the universe; in order to prove that the wave function did not collapse one would have to bring all these particles back and measure them again, together with the system that was measured originally. This is completely impractical, but even if one can theoretically do this, it would destroy any evidence that the original measurement took place (including the physicist's memory)." (WK 2008g)
  —But then, the critics ask, If these independent worlds in Everett's sense are inaccessible to us, how do we know much of anything really about them?

- **Second Problem**
  —What is relevant at the quantum mechanical level may not necessarily apply to the world at the physicsl level.
  —In fact, the persistent inability among contemporoary physicists to unifiy the theory of relativity for large objects at the physical level and quantum mechanics for tiny objects at the quantum mechanical level should warn us about the danger of reductionism, by reducing the understanding of different universes ("parallel universes") at the physical level from a generalization of "many worlds" at the quantum mechanical level.

- **Third Problem**
  —There is no empirical evidence to suggest that there are infinite (or extremely many) possible worlds in the cosmos as the many-worlds model by Everett would have us to believe.

Source: From *Sec.4.4.1* of *FPHG*

Table 1.9. Hyperspace and Its Challenge (Part I)

• **First Problem**

—Hyperspace for intra- and inter-universal travel is assumed to be possible, on the basis of four main assumptions; although there may be others, these are the often cited ones.

—The first assumption concerns the existence of "entry" and "exit" points in hyperspace, which is easier said than done to enter, go through, and then exit hyperspace.

—For instance, using hyperspace "requires complex calculations of single or multiple hyperspace jumps and the control of the jump.... Peturbations such as those experienced by ship in space from the gravitional field around an object such as a planet or even a star are exacerbated in hyperspatial travel, since mass in real space distorts hyperspace in an equal measure. 'Jumping' near to a gravitational mass is likely to make resulting exit from hyperspace to be highly uncertain, with the level of improbability *i* increasing with the square of the distance to the nearest gravitional 'well.'" (WK 2008h)

—Besides, "[o]nly spaceships equipped with a special force field can enter hyperspace, because exposure to hyperspace even for short period of time is hazardous to unprotected humans." (WK 2008h)

• **Second Problem**

—The second assumption concerns the faster-than-light speed. It is well known that Einstein argued for the speed of light as the ultimate limit, but nowadays phyiscists do allow the violation of this limit under certain conditions (like the hypothetical existence of tachyons, as will be explained further in a later sub-section on time travel).

—Even then, there is a challenge to explain in science fiction "why ships can travel faster than light in hyperspace," and two good arguments are that "hyperspace may be smaller than real space and therefore a star ship's propulsion seems to be greatly multiplied, or else the speed of light in hyperspace is not a barrier as it is in real space. Whatever the reasoning, the general effect is that ships traveling in hyperspace seem to have broken the speed of light, appearing at their destinations much more quickly and without the shift in time that the Theory of Relativity would suggest." (WK 2008h)

(continued on next page)

Table 1.9. Hyperspace and Its Challenge (Part II)

- **Third Problem**
  —The third assumption concerns the speed of time (to be measured, say, by a clock), but this is elaborated in a different section on the related issue of "time travel" (as shown in a different table on time travel).

- **Fourth Problem**
  —The fourth assumption concerns spatial shortcut, as it is possible, because of the curvature of space, in that "the manifold of ordinary three-dimensional space is curved in four or more 'higher' spacial dimensions (a 'hyperspace' in the geometric sense…). This curvature causes certain widely separated points in three-dimensional space to nonetheless be 'adjacent' to each other four-dimensionally. Creating an aperture in 4D space (a wormhole) between these locations can allow instantaneous transit between the two locations; a common comparison is that of a folded piece of paper, where a hole punched through two folded sections is more direct than a line drawn between them on the sheet. This idea probably arose out of certain popular descriptions of General Relativity and/or Riemannian manifolds, and may be the original form from which later concepts of hyperspace arose." (WK 2008h)
  —But this does not mean that hyperpsace is always safe. In fact, "in some science fiction, the danger of hyperspace travel is due to the chance that the route through hyperspace may take a ship too close to a celestial body with a large gravitational field, such as a star. In such scenarios, if a starship passes too close to a large gravitational field while in hyperspace, the ship is forcibly pulled out of hyperspace and reverts to normal space. Therefore, certain hyperspace 'routes' may be mapped out that are safe, not passing too close to stars or other dangers." (WJ 2008h)

Source: From *Sec.4.4.1* of *FPHG*

Table 1.10. The Problems of Time Travel into the Future (Part I)

- **First Problem**

—The first problem concerns "time dilation under the Theory of Special Relativity" (which separates space-time from mass-energy as special cases). Suppose you are on a moving train and an observer outside the train is watching you along the tracks. As your train moves along at the speed of light (or something close to it), "time, as measured by your watch, ticked along at a slower pace than time measured by the observer. Not only that, distance changed, too. For the observer, a one-foot ruler whizzing by on the train would have measured less than a foot." (NO 2000)

—In fact, you may not even notice that this slower clicking of the clock; in fact, everything looks normal to you inside the train as it was before: "The weird thing is that, for you on the train, time wouldn't seem to be moving slower and your ruler wouldn't be shorter—all would appear normal." (NO 2008) By contrast, you instead think that "time on the rest of the Earth would appear to be ticking along slower and its rulers would be shorter." (NO 2000) So, both of you insist that the other clock (not his) is clicking more slowly—assuming, of course, an inertial frame of reference.

—Now, change the story a bit, in that, this time, instead of an "inertial reference frame," acceleration is allowed, and you are on your way to board a spaceship. As an illustration, suppose "[y]ou board a spaceship and take off for deep space. The ship approaches the speed of light. Time for you seems to pass as it always has. It takes you about five seconds to tie your shoe. But to an observer on Earth (assuming he or she could watch you), you are moving at a snail's pace. It takes hours to tie your shoe." (NO 2000) In any event, "you continue on your journey. You slow down, stop, and accelerate back to Earth. You arrive home. You have aged two years during your flight. Two hundred years have passed on Earth. You have successfully travelled forward through time." (NO 2000) In this case, both the observer (if he is still alive) and you agree that you have aged less.

(continued on next page)

Table 1.10. The Problems of Time Travel into the Future (Part II)

- **First Problem *(cont'd)***
  —In other words, when you are back to Earth, a lot of things have changed since, and you may find rather strange to be in a new environment which you hardly recognize, when compared with what you used to think as your city, or the Earth (two years ago, well, in accordance to your time of measurement). This means that you are in the future (two hundred years later), even though you are still relatively young (only two years older). But, can you go back? Well, no. The reason is that "[a]ccording to relativity, you can only move through time in one direction." (NO 2000)

- **Second Problem**
  —The second problem concerns "time dilation under the Theory of General Relativity" (which unifies space-time with mass-energy in a general theory), the same result of time dilation occurs.
  —For instance, when "one clock is deeper in a gravity well than the other," a result is that "the clock deeper in the well" will "tick…more slowly; this effect must be taken into account when calibrating the clocks on the satellites of the Global Positioning System, and it could lead to significant differences in rates of aging for observers at different distances from a black hole." (WK 2008i)
  —More precisely, by way of another example, "it has been calculated that, under general relativity, a person could travel forward in time at a rate four times that of distant observers by residing inside a spherical shell with a diameter of 5 meters and the mass of Jupiter. For such a person, every one second of their 'personal' time would correspond to four seconds for distant observers. Of course, squeezing the mass of a large planet into such a structure is not expected to be within our technological capabilities in the near future." (WK 2008i; R. Gott 2002)
  —The point here, however, is that time travel into the future, in this sense, is allowed, theoretically speaking, but in a very limited way, from the theory of relativity. But one can also ask, What if the theory of relativity were wrong? Well, in that case, there would be a need to find a better theory to explain time travel into the future.

Source: From *Sec.4.4.2* of *FPHG*

Table 1.11. The Problems of Time Travel into the Past (Part I)

• **First Problem**

—The first problem concerns the faster-than-light-speed, since, in accordance to the theory of special relativity, when an object goes faster than the speed of light, something weird can happen, in that, instead of having the clock clicking more slowly, it is moving backward, in relation to some inertial frame of reference—which then raisies the issue of whether time travel into the past is possible at all.

—But the problem here is about the violation of the law of casuality. In everyday language, an excellent illustration of causality violation is that, suppose you travel back into the past and then kill your parents, this then means that you could not have been born. But this is not true, since you are alive.

—Yet, "in the case of a hypothetical signal moving faster than light, there would always be some frames in which the signal was received before it was sent, so that the signal could be said to have moved backwards in time." The violation of casuality exists in this case.

—The first part of the problem is that, in accordance to the theory of special relativity, "it would take an infinite amount of energy to accelerate a slower-than-light object to the speed of light"—let alone the energy needed to propel an object to go faster than the speed of light. (WK 2008i) Although some suggest the possibility of "negative energy," it remains to be seen to what extent this is true.

—And the second part of the problem is that, "although relativity does not forbid the theoretical possibility of tachyons which move faster than light at all times, when analyzed using quantum field theory it seems that it would not actually be possible to use them to transmit information faster than light and there is no evidence for their existence." (WK 2008i; S. Chase 1993)

(continued on next page)

Table 1.11. The Problems of Time Travel into the Past (Part II)

- **Second Problem**
  - —The second problem concerns cosmic string and black holes, since it is the theory of general relativity which "extends the special theory to cover gravity, illustrating it in terms of curvature in spacetime caused by mass-energy and the flow of momentum." (WK 2008i)
  - —On the other hand, "[g]eneral relativity describes the universe under a system of field equations…and there exist solutions to these equations that permit what are called 'closed time-like curves,' and hence time travel into the past….The first of these was proposed by Kurt Gödel, a solution known as the Gödel metric, but his (and many others') example requires the universe to have physical characteristics that it does not appear to have." (WK 2008i; K. Thorne 1994)
  - —But then, the problem now becomes: "Whether general relativity forbids closed time-like curves for all realistic conditions is unknown." (WK 2008i)

- **Third Problem**
  - —The third problem concerns wormholes and related ones, and "[w]ormholes are a hypothetical warped spacetime which are also permitted by the Einstein field equations of general relativity, although it would be impossible to travel through a wormhole unless it was what is known as a traversable wormhole." (WK 2008i; M. Visser 1995) In other words, "a wormhole is a hypothetical topological feature of spacetime that is fundamentally a 'shortcut' through space and time. Spacetime can be viewed as a 2D surface, and when 'folded' over, a wormhole bridge can be formed. A wormhole has at least two mouths which are connected to a single throat or tube. If the wormhole is traversable, matter can 'travel' from one mouth to the other by passing through the throat." (WK 2008j)
  - —But the problem here is three-fold, as shown in what follows.

(continued on next page)

Table 1.11. The Problems of Time Travel into the Past (Part III)

- **Third Problem *(cont'd)***
  - —The first part of the problem is that "there is no observational evidence for wormholes." (WK 2008j)
  - —The second part of the problem is that it is not clear how exactly wormholes (and black holes, for that matter) could allow time travel into the past at all.
  - —And the third part of the problem is that if using wormholes (and black holes, for that matter) to travel into the past requires something to go faster than light too, then this raises another issue concerning the two problems as previously indicated in the sub-section on the faster-than-light-speed.

- **Fourth Problem**
  - —The fourth problem is that some physicists have argued that, even should time travel into the past be possible, by way of these three methods, the person still would not return to the same past history that he exactly experienced before, but in a different one. (WK 2008i)
  - —In other words, "strictly speaking," time travel into the past is not "really" possible, in light of these criticisms, so there is no need to worry that one could travel into the past and kill one's parent, as a way to violate the law of casuality.

Source: From *Sec.4.4.2* of *FPHG*

Table 1.12. The Conceptual Dimensions of Consciousness (and Other Mental States)

- **On Heredity and Time**
  —Heredity and the Environment
  —The Past and the Present

- **On Layers of Mental States and Abnormality**
  —Consciousness, Unconsciousness, and Preconsciousness
  —Normality and Paranormality

- **On Organicity and Motivation**
  —Mechanicity and Organicity
  —Primary Motivations and Multiple Motivations

- **On Other (Mostly Epistemic) Considerations**
  —Synthesis and Analysis
  —Situation and the Subject
  —Process and Outcome
  —Reasoning and Other Modes of Thinking
  —Meta-Conceptual Nominalism and Realism

Notes:: These examples are solely illustrative (not exhaustive), and some of the items can be reclassified somewhere else. Nor are they always mutually exclusive. Since they are generalities, exceptions are expected.
Sources: From *FPHC*. A re-construction, but with my own contribution, originally from G.Lindzey & C.Hall, *Introduction to Theories of Personality* (NY: John Wiley & Sons, 1985).

Table 1.13. The Theoretical Levels of Consciousness (and Other Mental States) (Part I)

- **At the Micro-Physical Theoretical Level**
  - —Quantum-Mechanics
  - —Electromagnetism
- **At the Chemical Theoretical Level**
  - —Biochemistry
- **At the Biological Theoretical Level**
  - —Evolutionary Biology
  - —Neuroscience
  - —Artificial Intelligence
- **At the Psychological Theoretical Level**
  - —Psychodynamic Psychology
    - Psychoanalytical Psychology
    - Analytical Psychology
    - Socially Oriented Psychology
  - —Experimental (Behavioral) Psychology
    - Operant Reinforcement Theory
    - Stimulus-Response Theory
    - Social Learning Theory
  - —Cognitive (Gestalt) Psychology
    - Humanist Psychology
    - Existential Psychology
    - Field Theory
  - —Psychometric Psychology
    - Idiographics
    - Constitutional Psychology
  - —Social Psychology
    - Symbolic Interactive Theory
    - Social Exchange Theory

(continued on next page)

Table 1.13. The Theoretical Levels of Consciousness (and Other Mental States) (Part II)

- **At the Organizational Theoretical Level**
  —Managerial-Bureaucratic Theory
  —Oligarchic Theory
  —Network Theory

- **At the Institutional Theoretical Level**
  —Functionalist Theory
  —Anomic Theory

- **At the Structural Theoretical Level**
  —Conflict Theory
    - Marxian Theory
    - Critical Theory
    - Weberian Theory

  —Games Theory (in Formal Theory)
  —Feminist Theory
    - Feminist Concerns
    - Feminist Hopes

- **At the Systemic Theoretical Level**
  —Equilibrium Theory
  —System Theory
  —Chaos Theory

- **At the Cosmological Theoretical Level**
  —Superluminal Model
  —The Theory of Floating Consciousness

(continued on next page)

Table 1.13. The Theoretical Levels of Consciousness (and Other Mental States) (Part III)

- **At the Cultural Theoretical Level**
  - —Substantive Theories
    - Structuralist Theory
    - Post-Structuralist Theory (in Postmodernism)
  - —Meta-Theories
    - Epistemic Objectivism vs. Epistemic Historicism
    - Epistemic Subjectivism vs. Epistemic Non-Subjectivism
      - –Phenomenology
      - –Ethnomethodology
      - –Hermeneutics
    - Epistemic Relativism vs. Epistemic Absolutism
    - Epistemic Reductionism vs. Epistemic Emergencism

- **At Other Levels**
  - —Structuration Theory
  - —Reflexive Socioanalysis

Notes: These examples are solely illustrative (not exhaustive), and some of the items can be reclassified somewhere else. Nor are they always mutually exclusive. Since they are generalities, exceptions are expected.
Source: From many different sources as indicated in *FPHC*

Table 1.14. The Thematic Issues of Consciousness (and Other Mental States)

- **The Factor of History**
  —Pre-Modernity
  —Modernity
  —Postmodernity
  —After-Postmodernity

- **The Importance of Needs**
  —Having (e.g., the everyday, the technological)
  —Belonging (e.g., the just and the good)
  —Being (e.g., the true, the holy, the sublime/beautiful, and the good)

Notes: These examples are solely illustrative (not exhaustive). Nor are they always mutually exclusive. Since they are generalities, exceptions are expected. Also, some categories overlap with others; for instance, the good is also related with being-ness, just as the everyday can be connected with belong-ness.
Sources: From *FPHC*, based on *FHC* and *FCD*

Table 1.15. Having, Belonging, and Being in Consciousness

- **Having (e.g., About the Everyday and Technological)**
  —Ex: Physiological needs (e.g., thirst, hunger, sex)
  —Ex: Safety (e.g., stability, freedom from fear and anxiety)

- **Belonging (e.g., About the Good and Just)**
  —Ex: Social interaction
  —Ex: Friendship, acquaintance
  —Ex: Love, family
  —Ex: Self-respect, respect from others

- **Being (e.g., About the True, Holy, Beautiful/Sublime, and Good)**
  —Ex: Understanding of reality about self and world
  —Ex: Spiritual quest for holiness, beauty, sublimity, and goodness

Notes: These examples are solely illustrative (not exhaustive), and some of the items can be reclassified somewhere else. Nor are they always mutually exclusive. Since they are generalities, exceptions are expected.
Sources: From *FPHC*, based on *FHC* and *FCD*

Table 1.16. The Having-Ness of Consciousness (and Other Mental States)

- **At the Micro-Physical Level**
  —Ex: The finer physics of nutrition

- **At the Chemical Level**
  —Ex: Food guide pyramid

- **At the Bio-Psychological Level**
  —Ex: The hierarchy of needs

- **At the Institutional Level**
  —Ex: The production of commodity fetishism

- **At the Organizational Level**
  —Ex: Mass standardization (and customization); bureaucratic rationality

- **At the Structural Level**
  —Ex: The divide between the haves and the have-nots

- **At the Cultural Level**
  —Ex: The bondage of greater expectations

- **At the Systemic Level**
  —Ex: The technological transformation of poverty with transvaluations

- **At the Cosmological Level**
  —Ex: Floating consciousness in deep space without the human physiological needs

Notes: These examples are solely illustrative (not exhaustive), and some of the items can be reclassified somewhere else. Nor are they always mutually exclusive. Since they are generalities, exceptions are expected.
Sources: From *FPHC*. See also *FCD* and *FHC*.

Table 1.17. The Belonging-Ness of Consciousness (and Other Mental States)

- **At the Micro-Physical Level**
  —Ex: Statistical physics and social networks

- **At the Chemical Level**
  —Ex: Olfaction and social relations

- **At the Bio-Psychological Level**
  —Ex: Dual human nature on relationships

- **At the Institutional Level**
  —Ex: Private property and social cleavage

- **At the Organizational Level**
  —Ex: Legal formalism and iron cage

- **At the Structural Level**
  —Ex: Class conflicts and social alienation

- **At the Cultural Level**
  —Ex: Ruthless society and the culture of law

- **At the Systemic Level**
  —Ex: Human extinction and post-human relations

- **At the Cosmological Level**
  —Ex: The transcendence of equality in the cosmos and beyond

Notes: These examples are solely illustrative (not exhaustive), and some of the items can be reclassified somewhere else. Nor are they always mutually exclusive. Since they are generalities, exceptions are expected.
Sources: From *FPHC*. See also *FCD* and *FHC*.

Table 1.18. The Being-Ness of Consciousness (and Other Mental States) (Part I)

- **At the Micro-Physical Level**
  —Ex: Space-time and matter-energy in classical mechanics, quantum mechanics, and the theory of relativity

- **At the Chemical Level**
  —Ex: Drugs and well being

- **At the Biological Level**
  —Ex: Neuroscience and mystical experiences

- **At the Psychological Level**
  —Ex: Reductionism and emergencism

- **At the Institutional Level**
  —Ex: The proliferation of cults and sects, and the corporate-art industry

- **At the Organizational Level**
  —Ex: The organization of ideas in the scheme of things

- **At the Structural Level**
  —Ex: Disciplinary power and control in the order of things

- **At the Systemic Level**
  —Ex: Mutualism and ecological psychology

- **At the Cultural Level**
  —Ex: The recurrence of competing spirits, and floating consciousness as a climax of evolution

(continued on next page)

Table 1.18. The Being-Ness of Consciousness (and Other Mental States) (Part II)

- **At the Cosmological Level**
  —Ex: The search for elsewherewhen, and the evolution of other consciousnesses (e.g., the hyper-spatial consciousness)

Notes: These examples are solely illustrative (not exhaustive), and some of the items can be reclassified somewhere else. Nor are they always mutually exclusive. Since they are generalities, exceptions are expected.
Sources: From *FPHC*. See also *FCD* and *FHC*.

Table 1.19. Cognitive Partiality in Different Mental States

- **The Conscious**
  - —*Biased*
    - Ex: Self-overratingness
    - Ex: Self-modesty
    - Ex: Self-fulfilling prophesy
  - —*Shallow*
    - Ex: Attribution error
    - Ex: Heuristics
    - Ex: Illusory thinking
  - —*Short-term*
    - Ex: Deindividuation
  - —*Materialistic*
    - Ex: Hierarchy of needs

- **The Subconscious**
    - Ex: mental reconstruction
    - Ex: stereotyping

- **The Unconscious**
    - Ex: automatic processing
    - Ex: cognitive intuition

Notes: These examples are solely illustrative (not exhaustive), and some of the items can be reclassified somewhere else. Nor are they always mutually exclusive. Since they are generalities, exceptions are expected. And the comparison is relative, not absolute.
Source: From *BCPC*

Table 1.20. Emotional Non-Neutrality and Behavioral Alteration in Different Mental States

- **The Emotional**
  - —*Conscious*
    - Ex: Feelings
  - —*Subconscious*
    - Ex: Moods
  - —*Unconscious*
    - Ex: Empathy
    - Ex: Instincts

- **The Behavioral**
  - —*Conscious*
    - Ex: Foot-in-the-door
    - Ex: Role playing
    - Ex: Saying-is-believing
  - —*Subconscious*
    - Ex: Discrimination
  - —*Unconscious*
    - Ex: Non-Verbal Communication

Notes: These examples are solely illustrative (not exhaustive), and some of the items can be reclassified somewhere else. Nor are they always mutually exclusive. Since they are generalities, exceptions are expected. And the comparison is relative, not absolute.
Source: From *BCPC*

Table 1.21. The Limits of Intuition in Unconsciousness

- **Cognitive Intuition (or Intuitive Thinking)**
  —*Features*
  Ex: mental schemas
  Ex: expertise
  Ex: blindsight
  —*Problems*
  Ex: inaccurate
  Ex: non-explanatory

- **Emotional Intuition**
  —*Features*
  Ex: empathy
  Ex: instincts
  —*Problems*
  Ex: over-generalized
  Ex: non-explanatory

- **Behavioral Intuition**
  —*Features*
  Ex: prosemics
  Ex: kinesics
  —*Problems*
  Ex: unreliable
  Ex: non-explanatory

Notes: These examples are solely illustrative (not exhaustive), and some of the items can be reclassified somewhere else. Nor are they always mutually exclusive. Since they are generalities, exceptions are expected. And the comparison is relative, not absolute.
Source: From *BCPC*

Table 1.22. The Wealth/Poverty Dialectics in Different Mental States: The Case of Cognition

- **The Conscious**
  - —*Biased*
    - Ex: If richer in self-serving bias, then poorer in understanding others
  - —*Shallow*
    - Ex: If richer in heuristic thinking, then poorer in comprehending reality
  - —*Short-term*
    - Ex: If richer in short-term tendency, then poorer in long-term planning
  - —*Materialistic*
    - Ex: If richer in physiological obsession, then poorer in spiritual enlightenment
- **The Subconscious**
  - Ex: If richer in mental reconstruction, then poorer in accuracy of recall
  - Ex: If richer in stereotyping, then poorer in understanding groups
- **The Unconscious**
  - Ex: If richer in automatic processing, then poorer in depth of analysis
  - Ex: If richer in cognitive intuition, then poorer in reliability of judgments

Notes: These examples are solely illustrative (not exhaustive), and some of the items can be reclassified somewhere else. Nor are they always mutually exclusive. Since they are generalities, exceptions are expected. And the comparison is relative, not absolute.
Source: From *BCPC*

Table 1.23. The Wealth/Poverty Dialectics in Different Mental States: The Case of Emotion and Behavior

- **Emotion**
  - —*Conscious*
    - Ex: If richer in relying on familiarity and similarity for social attraction, then poorer in having diverse social relations
  - —*Subconscious*
    - Ex: If richer in moodiness, then poorer in sober thinking
  - —*Unconscious*
    - Ex: If richer in empathy, then poorer in task-oriented efficiency
    - Ex: If richer in instincts, then poorer in critical thinking

- **Behavior**
  - —*Conscious*
    - Ex: If richer in foot-in-the-door technique, then poorer in instant Gratification
    - Ex: If richer in role playing, then poorer in identity stability
  - —*Subconscious*
    - Ex: If richer in discriminating, then poorer in social harmony
  - —*Unconscious*
    - Ex: If richer in kinesics and proxemics, then poorer in verbal skills, relatively speaking.

Notes: These examples are solely illustrative (not exhaustive), and some of the items can be reclassified somewhere else. Nor are they always mutually exclusive. Since they are generalities, exceptions are expected. And the comparison is relative, not absolute.
Source: From *BCPC*

Table 1.24. The Theoretical Debate on Nature vs. Nurture (Part I)

- **The Environmental Approach**
  - —*Thesis*: It focuses, relatively speaking, more on the environment (culture and society) in explaining the achievement gap among individuals and for that matter, countries or regions, when contrasted with the natural factors.
  - —*Discourse*: Examples include Jose Ortega y Gasset ("Man has no nature; what he has is history"), Ashley Montagu ("Man is man because he has no instincts, because everything he is and has become he has learned from his culture, from the man-made part of the environment, from other human beings"), Stephen Jay Gould ("[The] brain [is] capable of a full range of behaviors and predisposed to none"), and Jesse Jackson (who blames white racism for the failure of blacks to close the achievement gap between whites and blacks in America). The works on dependency theory in international political economy (with a Marxian influence) and on the Protestant work ethic (by Max Weber) also point to this environmental direction.

- **The Genetic Approach**
  - —*Thesis*: It focuses instead, relatively speaking again, on hereditory factors (e.g., genes and evolution) in explaining the achievement gap among individuals and for that matter, countries or regions, when contrasted with the envrionmental factors.
  - —*Discourse*: Examples are Hans Eysenck and William Sheldon (in constitutional psychology), Konrad Lorenz (in his work on innate aggressive human nature), Gary Marcus (on the complexities of human thought by a tiny number of genes) and Robert Plomin, together with Michael Owen and Peter McGuffin (on the genetic basis of complex human behaviors).

(continued on next page)

Table 1.24. The Theoretical Debate on Nature vs. Nurture (Part II)

- **The Compromise Approach**
  —*Thesis*: It seeks the middle-of-the-road argument in regard to nature and nurture and regards all differences among individuals and groups as the result of the mixture of both nature and nurture, more or less equally.
  —*Discourse*: Examples include C. Murray and R. Herrnstein ("It seems highly likely to us that both genes and environment have something to do with this issue") and Dan Dennett ("Surely 'everyone knows' that the nature-nurture debate was resolved long ago, and neither side wins since everything – is – a – mixture – of – both – and - it's – all – very -complicated....")

- **The Transcendent Approach**
  —*Thesis*: It goes beyond both nature and nurture (without, however, committing the compromise fallacy) in showing their closely intertwined interactions in producing the behavioral differences as often seen in individual human endeavors on the micro scale, and for that matter, in country (or regional) endeavors on the macro one—in the context of my five theses, namely, (a) the compromise fallacy, (b) no oppression without self-oppression, (c) no success without failure, (d) the factor of randomness, and (e) the post-human vision, to be elaborated in Chapter Six.
  —*Discourse*: Peter Baofu proposed this approach on the basis of his "theory of contrastive advantages" (as an original theoretical contribution to the debate, which was first proposed and analyzed in *The Future of Capitalism and Democracy*). In the end, the human genes will not last, to be eventually superseded by post-human life forms, so the debate between genes and memes has obscured something profoundly important about the future that the world has never known. And the debate is also misleading and faulty in its dichotomy.

Source: From *BNN*. See the book for citations and details.

Table 1.25. Physical Challenges to Hyper-Spatial Consciousness

- **The Understanding of a Higher-Dimensional World of Space-Time**
  —Ex: 4 for traditional aspects of space-time (e.g., length, width, breadth and time) plus 6 more new dimensions in theory of hyper-space, with profound implications for practical applications to new forms of consciousness.

- **The Mastering of Dark Matter and Dark Energy**
  —Ex: "ordinary matter" (e.g., atoms, molecules) as a mere 4.4% of the universe, with 23% made of "cold dark matter" and the rest (about 73%) of mysterious "dark energy," with fundamental significance to questions about the limit of the speed of energy (or info), the availability of energy for use, and the nature of space-time, just to cite some examples.

- **The Exploration of Multiverses**
  —Ex: theoretical speculation of other universes (e.g., "baby universes," "gateways" in black holes, "wave function of the universe," "many worlds," "brane worlds"), with potentially seminary discoveries of different physical laws in relation to matter-energy and space-time, and vital differences to the future of post-human conquest of other universes (for the emergence of new forms of consciousness).

Notes: These examples are solely illustrative (not exhaustive), and some of the items can be reclassified somewhere else. Nor are they always mutually exclusive. Since they are generalities, exceptions are expected. The point here is to give a rough picture of the evolution of consciousness to the hyper-spatial consciousness and others totally unknown to current earthlings. As a note of clarification, it makes no difference to my argument as to whether or not the hyper-spatial consciousness may emerge before, during, and after floating consciousness.
Source: From *Table 4.5* of *FPHC*

Table 1.26. The Theory of Floating Consciousness (Part I)

- **At the Micro-Physical Level**
  —Ex: intelligent life without the human physical-chemical system

- **At the Chemical Level**
  —Ex: space radiation and toxins

- **At the Bio-Psychological Level**
  —Ex: exo-biological evolution in deep space
  —Ex: genetic engineering of new beings

- **At the Institutional Level**
  —Ex: post-capitalism
  —Ex: post-democracy

- **At the Organizational Level**
  —Ex: less legal-formalistic routines

- **At the Structural Level**
  —Ex: alien forms of violence

- **At the Cultural Level**
  —Ex: transcending freedom
  —Ex: transcending equality

(continued on next page)

Table 1.26. The Theory of Floating Consciousness (Part II)

- **At the Cosmological Level**
  —Ex: parallel universes
  —Ex: pocket universes

- **At the Systemic Level**
  —Ex: space habitats (in zero-gravity)

Notes: Each example draws from the works of different scholars in the field. For instance, at the cosmological level, the idea of parallel universes is from the theoretical speculation in quantum cosmology by Stephen Hawking and others, while the one of pocket universes comes from the theoretical work of Allan Guth at MIT. And at the institutional level, I proposed post-capitalism and post-democracy in *FCD* (and later, from *BDPD* and *BCPC*). In addition, the examples are solely illustrative (not exhaustive), and some of the items can be reclassified somewhere else. Nor are they always mutually exclusive. Since they are generalities, exceptions are expected.
Sources: From *FPHC*—and, originally, from *FCD*

Table 1.27. The Potential of Unfolding Unconsciousness

- **Superior Senses**
  - —*Through space*
    - Ex: clair-sentience (in feeling)
    - Ex: clair-voyance (in seeing)
    - Ex: clair-audience (in hearing)
    - Ex: clair-austance (in tasting)
    - Ex: clair-alience (in smelling)
    - Ex: clair-cognizance (in knowing)
  - —*Through time*
    - Ex: pre-cognition (in knowing), pre-sentiment (in feeling)
    - Ex: retro-cognition (in knowing), retro-sentiment (in feeling)

- **Superior Contacts**
  - —*With living minds*
    - Ex: telepathy
  - —*With the dead*
    - Ex: mediumship
  - —*With other-living entities*
    - Ex: telekinesis

Notes: The categories and examples are solely illustrative (not exhaustive), and the comparison is relative (not absolute), nor are they necessarily mutually exclusive. And some can be easily re-classified elsewhere. As generalities, they allow exceptions.
Source: A summary of Ch.5 of *FPHU*

Table 1.28. The Future Exploration of Unfolding Unconsciousness

- **In the Context of Nature**
  —Ex: the higher-dimensional space-time in hyperspace theory
  —Ex: the particle-wave interactions in quantum mechanics
  —Ex: the study of brain waves in electromagnetic theory

- **In the Context of the Mind**
  —Ex: the unconscious fantasies in psychoanalysis
  —Ex: the mind-altering drugs in biochemistry
  —Ex: the manipulation of neural activity in neurobiology

- **In the Context of Culture**
  —Ex: the correlation between anomalous experience and moral/religious/aesthetic interests in culture studies
  —Ex: the role of myths and language in epistemology

- **In the Context of Society**
  —Ex: different organizational agendas in anomalous research
  —Ex: the scholarly divide on anomalous study
  —Ex: interests and propaganda in institutioanl research
  —Ex: the transformative power of technology in research on social systems
  —Ex: the natural and non-natural selection of other worlds in exo-biology
  —Ex: the role of the post-human conscious mind in qualitative demography

Notes: The examples in each category are solely illustrative (not exhaustive), and the comparison is relative (not absolute), nor are they necessarily mutually exclusive. And some can be easily re-classified elsewhere. As generalities, they allow exceptions.
Source: A summary of Chs.2-5 of *FPHU*, in conjunction with the rest of the book

Table 1.29. Creative Techniques and Traits

- **Creative Techniques**
  - —*Comprehensiveness*
    - Ex: Broadness of Scope
    - Ex: In-Depth Analysis
    - Ex: Multifaceted Taxonomy
    - Ex: Numerous Theories
  - —*Arrangement*
    - Ex: Separation (or Subtraction)
    - Ex: Combination (or Addition)
    - Ex: Replacement
    - Ex: Accommodation
    - Ex: Reversal
    - Ex: Stretching
  - —*Serendipity*
    - Ex: Play
    - Ex: Chance

- **Creative Traits**
  - —*Openness / Risk*
    - Ex: Ability to welcome new ideas or to do new things, with risk
  - —*Discipline*
    - Ex: Ability to sacrifice whatever it takes to succeed
  - —*Resilience / Confidence*
    - Ex: Ability to take failures to mind, not to heart
  - —*Trust*
    - Ex: Ability to cooperate with others for team work, if necessary

Notes: The examples in the categories are solely illustrative (not exhaustive), and the comparison is relative (not absolute), nor are they necessarily mutually exclusive. And some can be easily re-classified elsewhere. As generalities, they allow exceptions.
Source: From *Sec.4.2* of *FPHCT*. For more details, also read the rest of the book.

Table 1.30. The Desirability of Creativity, and Its Dark Sides (Part I)

- **In Relation to Instrumental Rationality**
  —*Desirability*
  - Ex: For the sake of efficiency and effectiveness in this capitalist era of our time.

  —*Undesirability*
  - Ex: What Max Weber (1930) called the "iron cage" of capitalism, where live the "sensualists without heart, specialists without spirit"—or something which the Frankfurt School has forcefully asked us to be freed from.

- **In Relation to Substantive Rationality**
  —*Desirability*
  - Ex: For the sake of God, the King, Motherland, or other comparable ideals.

  —*Undesirability*
  - Ex: Domination and oppression of various forms—something which the moderns had struggled to free themselves from in the first place, since the dawn of modernity.

- **In Relation to Autonomous Rationality**
  —*Desirability*
  - Ex: For the sake of itself, the autonomy of creative endeavor.

  —*Undesirability*
  - Ex: The decadence of its own degeneration (e.g., the donwsides of postmodernism in our postmodern times, or what Nietzsche once called the unbearable "unreality" and "falseness" of the autonomous artist's "innermost existence"—and, in other cases, the painful suffering from different mental illnesses for those highly creative individuals).

(continued on next page)

Table 1.30. The Desirability of Creativity, and Its Dark Sides (Part II)

Notes: The examples in the categories are solely illustrative (not exhaustive), and the comparison is relative (not absolute), nor are they necessarily mutually exclusive. And some can be easily re-classified elsewhere. As generalities, they allow exceptions.
Also, I already went to great lenghts to explain the desirability and its dark sides of all these three rationalities (although there may be others too) in the two volumes of *FHC*—and, for that matter, *FCD*.
Source: From *Sec.4.9* of *FPHCT*. For more details, also read the rest of the book.

Table 1.31. Posthuman-Ism, Post-Humanism, and Trans-Humanism

- **Post-Humanism**
  —The neologism "post-human" used in my books should *not* be confused with another term which looks similar but has a totally different meaning in the literature of postmodernism, namely, "post-humanism"—which constitutes a critique of "humanism" as tradionally understood (especially, though not exclusively, in relation to the idea of progress in science and reason in the Enlightenment project). (WK 2008)
  —My works reject the project of "postmodernism" and propose the future world of what I orignally called "after-postmodernity" in *FHC* and *FCD*, for instance.

- **Trans-Humanism**
  —Also, the neologism "post-human" used in my books should *not* be confused with a similar term which is used to champion the ideology of technology for the fturue co-existence between humans and cyborgs in "trans-humanism." (WK 2008a)
  —Instead, my term "post-human" in relation to "posthuman-ism" also rejects "trans-humanism" (especially, though not exclusively, in relation to the promises of technology) and refers to something else altogether, that is, the future extinction of humans and its post-human successors in deep space and beyond unto multiverses.
  —Chapter Two of my book on ethics (*BEPE*) provides a more comprehensive critique of trans-humanism.

- **Posthuman-Ism**
  —With these two clarifications in mind, the "post-humans" (as envisioned in my books) can take different forms, and I proposed, in my different books, some of them, such as "unfolding unconsciousness," "floating consciousness," "hyper-spatial consciousness," "thinking machines," "thinking robots," "genetically altered superior beings," "cyborgs," and others (as already analyzed in my previous works).
  —They will evolve in conjunction with other visions of mine in relation to nature, society, and culture.

Sources: From *Sec.1.6* of *FPHE* (and also *BEPE*). For more details, also read the rest of the books (and other books of mine).

Table 1.32. Three Great Future Transformations of the Martial Body (Part I)

- **Virtual Battle**
  —Ex: virtual fight
  —Ex: virtual memory

- **The Hyper-Martial Body**
  —*In relation to the physical dimension*
  - Ex: striking (e.g., punching, kicking, trapping, acupressure-striking, and so on in an amazing way)
  - Ex: grapping (e.g., throwing, pinning, joint-locking, and so on in an amazing way)
  - Ex: running and walking (e.g., on different surfaces)
  - Ex: jumping (e.g., through walls, trees, etc.)
  - Ex: flying (e.g., through space)
  - Ex: enduring pain (e.g., absorbing punches, kicks, and so on in an amazing way)
  - Ex: using extremely advanced hi-tech weapons
  - Ex: healing (e.g., with special energy, etc.)
  - Ex: entering (e.g., into different dimensions of space)
  - Ex: sensing (e.g., in extraordinary ways through space and time, etc.)
  - Ex: wearing hi-tech suits for defensive and offensive purposes (e.g., like a hi-tech armor)

  —*In relation to the mental dimension*
  - Ex: the ability to relax the martial body in a way that current humans cannot.
  - Ex: the ability to concentrate the martial mind in a way that current humans are not accustomed to.
  - Ex: the ability to breathe and be timely in preparing the martial body for fighting in a way that current humans are not good at.

- **The Post-Martial Floating Mind**
  —Ex: the mind without the human body
  —Ex: the quest for spirituality beyond martial arts

(continued on next page)

Table 1.32. Three Great Future Transformations of the Martial Body (Part II)

Notes: The examples in the categories are solely illustrative (not exhaustive), and the comparison is relative (not absolute), nor are they necessarily mutually exclusive. And some can be easily re-classified elsewhere. As generalities, they allow exceptions.
Source: From *Sec. 2.5.1* of *FPHMA*. See text for citations.

Table 1.33. Three Great Future Transformations of the Sexual Body

- **Virtual Pleasure**
  —Ex: virtual Sex
  —Ex: virtual Recall

- **The Hyper-Sexual Body**
  —*In relation to the physical dimension*
  - Ex: the ability to perform differenet sexual positions without the various physical limitations that humans currently have.
  - Ex: the ability to produce unusual amount of sexual energy to endure in sexual acts in a way that humans currently cannot.
  - Ex: the ability to engage in multiple sexual experiences comparable to (but different from) non-human sexuality (e.g., hermaphrodite, monoecious, sex-switching, and many other types) that current humans cannot.

  —*In relation to the mental dimension*
  - Ex: the ability to relax the sexual body in a way that current humans are not accustomed to.
  - Ex: the ability to concentrate the sexual mind in a way that current humans are not good at.
  - Ex: the ability to breathe and be patient in preparing the sexual body for sexuality in a way that current humans are not able to.

- **The Post-Sexual Floating Mind**
  —Ex: the mind without the human body
  —Ex: the quest for spirituality beyond sexuality

Notes: The examples in the categories are solely illustrative (not exhaustive), and the comparison is relative (not absolute), nor are they necessarily mutually exclusive. And some can be easily re-classified elsewhere. As generalities, they allow exceptions.
Source: From *Sec. 3.4.2* of *FPHS*. See text for more info.

Table 1.34. The Trinity of Pre-Modernity

- **Pre-Free-Spirited Pre-Modernity (Pre-Modernism) and Its Internal Split**
  —Competing worldviews and values both within and between linear centric (e.g., Islamic, Christian, Judaic, Imperial Roman) and cyclical-centric (e.g., Confucian, Taoist, Hindu, and Buddhist) orientations
  —Compare modernism with pre-modernism here in relation to the seven dimensions of human existence like the true and the holy (e.g., different versions of epistemic dogmas and religious superstitions), the everyday and the technological (e.g.,different versions of non technophilism and non-consumerism), the beautiful/sublime (e.g., different versions of aesthetic non-autonomy), and the good and the just (e.g., different versions of moral particularism).

- **Pre-Capitalist Pre-Modernity (Pre-Modernization) and Its Own Discontents**
  —Competing versions of societal arrangements (e.g., feudalism, monarchism, and the holy order)

- **Hegemonic Pre-Modernity and Its Countervailing Forces**
  —Different power centers and their enemies (e.g., the Roman Empire and the "barbarian hordes," the "Holy Crusades" and the Muslims, the Middle Kingdom and the invading tribes, different social castes in India, and warring Greek city-states)

Notes: The examples in each category are solely illustrative (not exhaustive) nor necessarily mutually exclusive, and the comparison is relative (not absolute). As generalities, they allow exceptions. Also, it does not matter what the "base" era is in the analysis of any trinity. And in the present context, the "base" era is modernity (for instance, with its "free-spirited modernity" and the other two parts). So, for pre-modernity, the trinity takes the form of, say, "pre-free-spirited pre-modernity," together with the other two parts.
Sources: From Ch.2 of *BCIV* and also the 2 volumes of *FHC*

Table 1.35. The Trinity of Modernity (Part I)

- **Free-Spirited Modernity (Modernism) and Its Internal Split**
  —*On the True and the Holy*
    - The freedom from the dogmas of the past to the better understanding of, and union with, the world and self (Ch.3 of *FHC*)
    - Alternative discourses: about the true (e.g., anti-science discourses) and the holy (non-mainstream theologies) (Ch.3 of *FHC*)

  —*On the Technological and the Everyday*
    - The freedom from life harshness to the higher technophilic, consumeristic lifeform (Ch.2 of *FHC*)
    - Alternative discourses: about the everyday (e.g., transcendental mindsets) and the technological (e.g., Arcadianism) (Ch.2 of *FHC*)

  —*On the Good and the Just*
    - The freedom from the theo-aristocratic tyranny to the moral universality for a just society (Ch.5 of *FHC*)
    - Alternative discourses: about the just (e.g., Communism, Anarchism) and the good (e.g., Nazism/Fascism, and Zarathustrianism) (Chs.5-6 of *FHC*)

  —*On the Beautiful and the Sublime*
    - The freedom from the external distortion of aesthetic pleasure to the boundless infinity of totality in artistic autonomy (Ch.4 of *FHC*)
    - Alternative discourses: about the beautiful/sublime (e.g., kitsch and historical avant-gardism) (Ch.4 of *FHC*)

- **Capitalist Modernity (Modernization) and Its Own Discontents**
  —*During the Industrial Revolution*
    - Ex: Marx on the institution of inequality (Ch.1 of *FHC*)

  —*During the Modern Rational-Instrumental Epoch*
    - Ex: Weber on the politics of soft liberal institutions (Ch.5 of *FHC*)

  —*During the Great Depression*
    - Ex: Keynes on the myth of the free market (Chs.1-3 of *FHC*)

  —*During the Cold War*
    - Ex: Lasch on the narcissistic culture industry (Chs.2-3 of *FHC*)

(continued on next page)

Table 1.35. The Trinity of Modernity (Part II)

- **Hegemonic Modernity and Its Countervailing Forces**

—*The Legacies of Colonialism and Imperialism*

- Ex: European colonization of most of the modern world (Ch.1 of *FHC*)

—*The Struggle for Decolonialization*

- Ex: The countervailing forces of resentment, rechantment, and regionalism (Chs.1 & 6 of *FHC*)

Notes: The examples in each category are solely illustrative (not exhaustive) nor necessarily mutually exclusive, and the comparison is relative (not absolute). As generalities, they allow exceptions.

Sources: From the 2 volumes of *FHC*—and also from *FCD*

Table 1.36. The Trinity of Post-Modernity

- **Free-Spirited Postmodernity (Postmodernism) and Its Internal Split**
  —*On the True and the Holy*
  - Postmodern performative turn for knowing and its enemies (Ch.3 of *FHC*)
  - Postmodern comparative theology and its opponents (Ch.3 of *FHC*)

  —*On the Technological and the Everyday*
  - Postmodern corporate technological mindset and its adversaries (Ch.2 of *FHC*)
  - Postmodern postmaterialism and its critics (Ch.2 of *FHC*)

  —*On the Good and the Just*
  - Postmodern politics of difference and its foes (Ch.5 of *FHC*)

  —*On the Beautiful and the Sublime*
  - Postmodern deconstruction and its dissenters (Ch.4 of *FHC*)

- **Capitalist Postmodernity (Postmodernization) and Its Own Discontents**
  —*During the Post-Cold War and Beyond*
  - Ex: post-Fordism and its shortcomings (Ch.6 of *FHC*; Chs.6-7 of *FCD*)

- **Hegemonic Postmodernity and Its Countervailing Forces**
  —*The Debate on the Global Village*
  - Ex: uni-civilizationalism vs. multi-civilizationalism (Ch.6 of *FHC*)

  —*The Resistance Movement*
  - Ex: rechantment and the politics of civilizational claims (e.g., Islamic, Confucian and other ethos in relation to the Same) (Ch.6 of *FHC*; Ch.10 of *FCD*)
  - Ex: resentment and the politics of resurgence (e.g., the rising Chinese superpower, the growing EU, and other players in relation to the U.S. and her allies) (Ch.6 of *FHC*; Ch.8 of *FCD*)
  - Ex: regionalism and the politics of inequality (e.g., trans- or international blocs, the North-South divide, NGO's) (Ch.6 of *FHC*; Ch.5 of *FCD*)

Notes: The examples in each category are solely illustrative (not exhaustive) nor necessarily mutually exclusive, and the comparison is relative (not absolute). As generalities, they allow exceptions.
Sources: From *FCD* and the 2 volumes of *FHC*

Table 1.37. The Trinity of After-Postmodernity

- **Free-Spirited After-Postmodernity (After-Postmodernism) and Its Internal Split**
  —The discourse of naked contingency (Ch.10 of *FCD*; Ch.4 of *FPHC*)

- **Post-Capitalist After-Postmodernity (After-Postmodernization) and Its Own Discontents**
  —Different versions of post-capitalism and post-democracy, and their enemies (Ch.10 of *FCD*; Chs.3-4 of *FPHC*)

- **Hegemonic After-Postmodernity and Its Countervailing Forces**
  —The Cyclical Progression of Hegemony in Multiverses (Chs.9-10 of *FCD*; Ch.4 of *FPHC*)

Notes: The examples in each category are solely illustrative (not exhaustive) nor necessarily mutually exclusive, and the comparison is relative (not absolute). As generalities, they allow exceptions.
Sources: From *FCD* and also *FHC*

• Part Two •

Diversity

CHAPTER 2

DIVERSITY AND ITS AMBIVALENT CHARACTERS

Mankind can find happiness only in unity, not in diversity."
—Sri Sathya Sai Baba (TE 2012a)

The Desirability of Diversity

The study of natural resources involves both diversity and discontinuity—as already discussed in *Sec. 1.3*.

The relationship between diversity and discontinuity is dialectic, since there is no diversity without discontinuity—and vice versa.

Yet, a scholarly danger here is to reduce the analysis of one to that of the other; thus, this book is to devote a whole chapter on each, with the understanding, of course, that the two are related.

This dialectic relationship between diversity and discontinuity can be revealed by showing how and why the analysis of diversity (in Chapter Two) is inadequate without the inclusion of discontinuity, just as the examination of discontinuity (in Chapter Three) is insufficient without the consideration of diversity.

The issues concerning diversity in the study of natural resources are to be addressed in this chapter (Chapter Two), whereas the issues concerning discontinuity in the study of natural resources will be analyzed in the next chapter (Chapter Three) instead.

With this dialectic treatment in mind—a good way to examine diversity and its ambivalent characters in the study of natural resources is by way of the evaluation of the extent to which it is in fact both possible (or impossible) and desirable (or undesirable).

This can be done by way of a comprehensive analysis of diversity in the context of natural resources from the four perspectives of the mind, nature, society, and culture (in accordance to my sophisticated methodological holism, as explained in *Sec. 1.9*).

In other words, this chapter is organized in four main sections, namely, in relation to (2.2) diversity and the mind, (2.3) diversity and nature, (2.4) diversity and society, and (2.5) diversity and culture—to be addressed hereafter, in that order (and summarized in *Table 2.1*).

Diversity and the Mind

Diversity, when examined from the perspective of the mind, can illuminate its possibility (or impossibility) and desirability (or undesirability), and this can be shown by way of two case studies, namely, (2.2.1) diversity, psychological benefits, and biology, and (2.2.2) diversity, dietary health, and agriculture—to be addressed in what follows, respectively.

Diversity, Psychological Benefits, and Biology

A good way to evaluate the possibility (or impossibility) and desirability (or undesirability) of diversity from the perspective of the mind is to explore diversity, psychological benefits, and biology.

An argument here is that there is a relationship between human health and biodiversity; after all, "biodiversity is the degree of variation of life forms within a given species, ecosystem, biome, or an entire planet" and is therefore "a measure of the health of ecosystems," which in turn has an impact on human health. (WK 2012i)

Types of Biodiversity

But "the term *biological diversity* or *biodiversity* can have many interpretations. It is most commonly used to replace the more clearly defined and long established terms, species diversity and species richness. Biologists most often define biodiversity as the 'totality of genes, species, and ecosystems of a region.' An advantage of this definition is that it seems to describe most circumstances and presents a unified view of the traditional three levels at which biological variety has been identified," namely, at the levels of genes, species, and ecosystems. (WK 2012i; T. Larsson 2001)

Yet, "in 2003 Professor Anthony Campbell at Cardiff University, UK and the Darwin Centre, Pembrokeshire, defined a fourth level: Molecular Diversity." (WK 2012i) So now, there are four types of biodiversity as shown below (and summarized in *Table 2.2*):

- "molecular diversity"
- "genetic diversity"
- "species diversity"
- "ecosystem diversity"

This multilevel interpretation was first generally given "in a paper by Bruce A. Wilcox commissioned by the International Union for the Conservation of Nature and Natural Resources (IUCN) for the 1982 World National Parks Conference. Wilcox's definition was 'Biological diversity is the variety of life forms...at all levels of biological systems (i.e., molecular, organismic, population, species and ecosystem)....' The 1992 United Nations Earth Summit defined 'biological diversity' as 'the variability among living organisms from all sources, including... terrestrial, marine, and other aquatic ecosystems, and the ecological complexes of which they are part: this includes diversity within species, between species and of ecosystems.' This definition is used in the United Nations Convention on Biological Diversity." (WK 2012i; D. Hawksworth 1996; B. Wilcox 1984)

Distribution of Biodiversity

Besides existing at different levels of "biological organization" (as indicated above)—biodiversity is also not evenly distributed, because "the diversity of all living things (biota) depends on temperature, precipitation, altitude, soils, geography and the presence of other species. The study of the spatial distribution of organisms, species, and ecosystems is the science of biogeography." (WK 2012i; K. Gaston 2004)

In general, it is known that "diversity consistently measures higher in the tropics and in other localized regions such as Cape Floristic Province and lower in polar regions," and "terrestrial biodiversity is up to 25 times greater than ocean biodiversity." (WK 2012i; M. Benton 2001)

In any event, two variations in biodiversity are often mentioned, namely, (a) latitudinal gradients and (b) hotspots, as explained below (and summarized in *Table 2.2*).

(a) Latitudinal Gradients

The first variation in biodiversity has to do with "latitudinal gradients," in that "there is an increase in biodiversity from the poles to the tropics. Thus localities at lower latitudes have more species than localities at higher latitudes. This is often referred to as the latitudinal gradient in species diversity. Several ecological mechanisms may contribute to the gradient, but the ultimate factor behind many of them is the greater mean temperature at the equator compared to that of the poles." (WK 2012i; D. Currie 2004; A. Allen J 2006)

(b) Hotspots

And the second variation in biodiversity has to do with "hotspots," and a hotspot is "a region with a high level of endemic [uniquely local] species," and the term was "first named in 1988 by Dr. Sabina Virk. Many hotspots have large nearby human populations. While hotspots are spread all over the world, the majority are forest areas and most are located in the tropics." (WK 2012i; N. Myers 1988 & 1990; J. McKee 2004)

For example, "Brazil's Atlantic Forest is considered one such hotspot, containing roughly 20,000 plant species, 1,350 vertebrates, and millions of insects, about half of which occur nowhere else. The island of Madagascar, particularly the unique Madagascar dry deciduous forests and lowland rainforests, possess a high ratio of endemism. Since the island separated from mainland Africa 65 million years ago, many species and ecosystems have evolved independently. Indonesia's 17,000 islands cover 735,355 square miles (1,904,560 km2) and contain 10% of the world's flowering plants, 12% of mammals and 17% of reptiles, amphibians and birds—along with nearly 240 million people. Many regions of high biodiversity and/or endemism arise from specialized habitats which require unusual adaptations, for example alpine environments in high mountains, or Northern European peat bogs." (WK 2012i; D. Normile 2010)

Evolution of Biodiversity

Biodiversity has gone through a long history on Earth, as it is "the result of 3.5 billion years of evolution. The origin of life has not been definitely established by science, however some evidence suggests that life may already have been well-established only a few hundred million years after the formation of the Earth. Until approximately 600 million years ago, all life consisted of archaea, bacteria, protozoans and similar single-celled organisms." (WK 2012i)

For instance, "the history of biodiversity during the Phanerozoic (the last 540 million years) starts with rapid growth during the Cambrian explosion—a period during which nearly every phylum of multicellular organisms first appeared. Over the next 400 million years or so, invertebrate diversity showed little overall trend, and vertebrate diversity shows an overall exponential trend." (WK 2012i; S. Sahney 2010)

There are "estimates of the present global macroscopic species diversity," which "vary from 2 million to 100 million, with a best estimate of somewhere near 13–14 million, the vast majority arthropods. Diversity appears to increase continually in the absence of natural selection." (WK 2012i; S. Okasha 2010)

Psychological Benefits of Biodiversity

The causal relationship between biodiversity and its psychological effects on human health "is becoming an international political issue, as scientific evidence builds on the global health implications of biodiversity loss." (WK 2012i; E. Chivian 2008; C. Corvalan 2005)

One major reason has to do with "the issue of climate change, as many of the anticipated health risks of climate change are associated with changes in biodiversity (e.g. changes in populations and distribution of disease vectors, scarcity of fresh water, impacts on agricultural biodiversity and food resources etc.) This is because the species most likely to disappear are those that buffer against infectious disease transmission, while surviving species tend to be the ones that increase disease transmission, such as that of West Nile Virus, Lyme disease and Hantavirus, according to a study done and co-authored by Felicia Keesing, ecologist at Bard College, and Drew Harvell, associate director for Environment of the Atkinson Center for a Sustainable Future (ACSF) at Cornell University." (WK 2012i; K. Ramanujan 2010; CBD 2009)

In addition, "the growing demand and lack of drinkable water on the planet presents an additional challenge to the future of human health. Partly, the problem lies in the success of water suppliers to increase supplies, and failure of groups promoting preservation of water resources. While the distribution of clean water increases, in some parts of the world it remains unequal. According to 2008 World Population Data Sheet, only 62% of least developed countries are able to access clean water." (WK 2012i)

In general, "some of the health issues influenced by biodiversity include dietary health and nutrition security, infectious disease, medical science and medicinal resources, social and psychological health. Biodiversity is also known to have an important role in reducing disaster risk, and in post-disaster relief and recovery efforts." (WK 2012i; K. Gaston 2007)

A good illustration is how "bodiversity provides critical support for drug discovery and the availability of medicinal resources. A significant proportion of drugs are derived, directly or indirectly, from biological sources: at least 50% of the pharmaceutical compounds on the US market are derived from plants, animals, and micro-organisms, while about 80% of the world population depends on medicines from nature (used in either modern or traditional medical practice) for primary healthcare." (WK 2012i; E. Chivian 2008; J. Roopesh 2008)

Problems with the Study of Psychological Benefits and Biodiversity

However, for the critics, this study of psychological benefits and biodiversity has some major problems. Consider, for illustration, a few examples below (as summarized in *Table 2.2*).

Firstly, one main criticism of the study of psychological benefits and biodiversity is that biodiversity is not stable and subject to different threats.

For example, for the critics, biodiversity has been subject to different threats like "habitat destruction," "overexploitation," "human overpopulation," "introduced and invasive species," "climate change," and so on. (WK 2012i)

Secondly, another main criticism of the study of psychological benefits and biodiversity is that there are major extinctions in the history of biodiveristy.

For instance, for the critics, history does not solely favor biodiversity, as there are periodic mass extinctions which constitute different checks on biodiveristy, and the following five mass extinctions in the history of the Earth are excellent examples: (WK 2012s; J. Baez 2006; C. Labandeira 1993)

- "The Cretaceous-Paleogene (or K–Pg) Extinction Event"— which is "the most recent" and "occurred approximately 65.5 million years ago....About 17% of all families, 50% of all genera and 75% of species became extinct."
- "The Triassic-Jurassic Extinction Event (End Triassic)"—which occurred "205 millioin years ago at the Triassic-Jurassic transition. About 23% of all families and 48% of all genera (20% of marine families and 55% of marine genera) went extinct."
- "The Permian-Triassic Extinction Event (End Permian)"—which occurred "251 million years ago at the Permian-Triassic transition. Earth's largest extinction killed 57% of all families and 83% of all genera (53% of marine families, 84% of marine genera, about 96% of all marine species and an estimated 70% of land species) including insects."
- "The Late Devonian Extinction"—which occurred "375-360 million years ago near the Devonian-Carboniferous transition. At the end of the Frasnian Age in the later part(s) of the Devonian Period, a

prolonged series of extinctions eliminated about 19% of all families, 50% of all genera and 70% of all species."

- "The Ordovician–Silurian Extinction Event (End Ordovician or O-S)"—which occurred "450-440 million years ago at the Ordovician-Silurian transition. Two events occurred that killed off 27% of all families and 57% of all genera."

Thus, these extinction events had caused "periodic, massive losses of diversity." (WK 2012s)

Thirdly, still another main criticism of the study of psychological benefits and biodiversity is that the actual evidences are not sufficient.

For instance, "even though terrestrial biodiversity declines from the equator to the poles, some studies claim that this characteristic is unverified in aquatic ecosystems, especially in marine ecosystems. The latitudinal distribution of parasites does not follow this rule. Other instances of great diversity in higher latitudes have also been recorded." (WK 2012i; H. Hillebrand 2004; A. Moustakas 2005; S. Morand 2010) After all, "accurately measuring differences in biodiversity can be difficult. Selection bias amongst researchers may contribute to biased empirical research for modern estimates of biodiversity. In 1768, Rev. Gilbert White [1909] succinctly observed of his Selborne, Hampshire [that] 'all nature is so full, that that district produces the most variety which is the most examined.'" (WK 2012i) Furthermore, although "the fossil record suggests that the last few million years featured the greatest biodiversity in history…not all scientists support this view, since there is uncertainty as to how strongly the fossil record is biased by the greater availability and preservation of recent geologic sections. Some scientists believe that corrected for sampling artifacts, modern biodiversity may not be much different from biodiversity 300 million years ago, whereas others consider the fossil record reasonably reflective of the diversification of life." (WK 2012i; S. Sahney 2010; J. Alroy 2001)

Fourthly, an additional main criticism of the study of psychological benefits and biodiversity is that the causal relationship is not so clear-cut as the advocates would like us to believe.

For instance, in a study by R. Fuller (2007) and his colleagues, it was found that, "although these associations [between psychological benefits and biodversity] emerge quite clearly, untangling causality is not easy. First, it is unlikely that people were randomly distributed among parks, perhaps preferring those that enhance their psychological well-being. Selection by individuals of environments that enhance personal well-being makes interpretation of the causal structure of these relationships difficult....Second, the increase in psychological well-being with species richness and the accurate assessment of richness levels presumably operate through some proxy mechanism. Positive relationships between the number of habitats in the greenspace and (i) psychological benefits gained by users and (ii) perceived levels of plant diversity hint that gross structural habitat heterogeneity might cue the perceptions and benefits of biodiversity."

Fifthly, another additional main criticism of the study of psychological benefits and biodiversity is that there are many different causes of human health problems.

For example, for the critics, an individual may suffer from human health problems which are not necessarily related to biodiversity, and good examples include overconsumption, junk diet, personality disorders, etc.

Sixthly, still another additional main criticism of the study of psychological benefits and biodiversity is that there can be alternative resources for health improvement, even when biodiversity occurs.

For instance, for the critics, although it is true that "biodiversity provides critical support for drug discovery and the availability of medicinal resources" to improve human health—it is also important to point out that "the decline in output from the pharmaceutical sector since the mid-1980s can be attributed to a move away from natural product exploration ('bioprospecting') in favor of genomics and synthetic chemistry" for health improvement instead. (WK 2012i; E. Hawkins 1992)

And seventhly, one last main criticism of the study of psychological benefits and biodiversity is that different levels of diversity may not correspond.

For instance, for the critics, "measuring diversity at one level in a group of organisms may not precisely correspond to diversity at other levels," even though some cases like "tetrapod (terrestrial vertebrates) taxonomic and ecological diversity show...a very close correlation." (WK 2012i; S. Sahney 2010)

Beyond Biodiversity and Psychological Benefits

These criticisms of the relationship between psychological benefits and biodiversity should not be misunderstood as a total rejection of its validity but serve to show us the two opposing sides of the debate, such that the possibility (or impossibility) and desirability (or undesirability) of diversity (from the perspective of the mind with diversity, psychological benefits, and biology as a case study here) are not to the extent that the respective defenders would like us to believe.

Furthermore, the analysis of diversity, psychological benefits, and biology can teach us something important about the ontological principles in existential dialectics, and good examples include the formalness-informalness principle, the absoluteness-relativeness principle, the partiality-totality principle, the predictability-unpredictability principle, the explicability-inexplicability principle, the fiction-reality principle, the cognitiveness-noncognitiveness principle, the finiteness-transfiniteness principle, the preciseness-vagueness principle, the simpleness-complicatedness principle, the openness-hiddenness principle, the denseness-emptiness principle, the change-constancy principle, the order-chaos principle, the slowness-quickness principle, the expansion-contraction principle, the theory-praxis principle, the convention-novelty principle, the evolution-transformation principle, the symmetry-asymmetry principle, the softness-hardness principle, the seriousness-playfulness principle, the activeness-inactivess principle, the selfness-otherness principle, the regression-progression principle, the same-difference principle, the stability-reaction principle, the functionality-nonfunctionality principle, and the intentionality-nonintentionality principle.

For instance, in relation to the formalness-informalness principle, if there is formalness (e.g., the formal logical requirement of soundness, consistency, and completeness in a system of ideas, as in the study of psychological benefits and biodiversity), there is informalness (e.g., the non-formal existence of unsoundness, inconsistency, and incompleteness in the study of psychological benefits and biodiversity, as shown in the criticisms that biodiversity is not stable and subject to different threats, that there are major extinctions in the history of biodiveristy, that the actual evidences are not sufficient, that the causal relationship is not so clear-cut as the advocates would like us to believe, that there are many different causes of human health problems, that there can be alternative resources for health improvement even when biodiversity occurs, and that different levels of diversity may not correspond). And the reverse direction also holds true.

In relation to the absoluteness-relativeness principle, if there is absoluteness (e.g., the absolute view of those advocates of the "latitudinal gradients" in biodiversity, in that "there is an increase in biodiversity from the poles to the tropics"), there is relativeness (e.g., what is true for those advocates of the "latitudinal gradients" in biodiversity is not necessarily so for the critics, who "claim that this characteristic is unverified in aquatic ecosystems, especially in marine ecosystems," that "the latitudinal distribution of parasites does not follow this rule," and that "other instances of great diversity in higher latitudes have also been recorded"). And the reverse direction also holds true.

In relation to the partiality-totality principle, if there is partiality (e.g., the partial view of the advocates of biodiversity on the basis of fossil records), there is totality (e.g., the more holistic view of biodiversity, such that the whole is not the sum of its parts, that is, it cannot be reduced to the sum of opposing views of the advocates and their critics, since there will emerge new views in the future not yet known today, just as there are already some skeptical views nowadays which suggest that "there is uncertainty as to how strongly the fossil record is biased by the greater availability and preservation of recent geologic sections," whereas other views are more in-between). And the reverse direction also holds true.

In relation to the predictability-unpredictability principle, if there is predictability (e.g., the predictable tendency of many scientists to focus on biodiversity at the macro level, since "contemporary biodiversity physics is 'firmly fixated on the visible [macroscopic] world'"), there is unpredictability (e.g., the more difficult task to predict exactly how biodiversity works in the nonvisible, microscopic world, since "less than 1% of all species that have been described have been studied beyond simply noting their existence," while "the vast majority of Earth's species are microbial"). (WK 2012i; E. Wilson 2002; S. Nee 2004) And the reverse direction also holds true.

In relation to the explicability-inexplicability principle, if there is explicability (e.g., the explanation by the advocates of the "latitudinal gradients" in biodiversity on the basis of "the greater mean temperature at the equator compared to that of the poles"), there is inexplicability (e.g., the lack of sufficient explanation by the advocates of the "latitudinal gradients" in biodiversity, since the critics had pointed out some studies which instead showed that "this characteristic [of latitudinal gradients] is unverified in aquatic ecosystems, especially in marine ecosystems," that "the latitudinal distribution of parasites does not follow this rule," and "other instances of great diversity in higher latitudes have also been recorded"). And the reverse direction also holds true.

In relation to the fiction-reality principle, if there is fiction (e.g., the fictional aspect of the "latitudinal gradients" in biodiversity, insofar as "the latitudinal distribution of parasites does not follow this rule," and "other instances of great diversity in higher latitudes have also been recorded"), there is reality (e.g., the realistic aspect of the "latitudinal gradients" in biodiversity, insofar as many terrestrial ecosystems follow the rule of latitudinal gradients). And the reverse direction also holds true.

In relation to the cognitiveness-noncognitiveness principle, if there is cognitiveness (e.g., a view of things on the basis of reason and evidences, as shown in the empirical study by Fuller and his colleagues on "the…associations [between psychological benefits and biodversity]"), there is noncognitiveness (e.g., a view of things on the basis of non-cognitive factors like envy, jealousy, power, nationality, race, gender, age, class, greed, lust, status, anger, sadness, joy, fear, wish, etc., as shown in the study by R. Fuller and his colleagues, when they found that, "it is unlikely that people were randomly distributed among parks, perhaps preferring those that enhance their psychological well-being," so "selection by individuals of environments that enhance personal well-being makes interpretation of the causal structure of these relationships difficult," because of this factor of personal preference). And the reverse direction also holds true.

In relation to the finiteness-transfiniteness principle, if there is finiteness (e.g., the finite number of the levels of biodiversity, namely, "molecular," "genetic," "species," and "ecosystem"), there is transfiniteness (e.g., the transfinite number of all the individual organisms, species, and ecosystems that have ever existed at the four levels of biodiversity in the history of the Earth). And the reverse direction also holds true.

In relation to the preciseness-vagueness principle, if there is preciseness (e.g., the precise identification of four levels of biodiversity in the article aforecited), there is vagueness (e.g., the vagueness in the identification of the four levels of biodiversity in the article aforecited, since it is not clear why there must be only 4, not 5, 6, 7, and so on). And the reverse direction also holds true.

In relation to the simpleness-complicatedness principle, if there is simpleness (e.g., the relatively simple analysis of biodiversity, when "less than 1% of all species that have been described have been studied beyond simply noting their existence," and "contemporary biodiversity physics is 'firmly fixated on the visible [macroscopic] world,'" but "the vast majority of Earth's species are microbial" and non-visible), there is complicatedness (e.g., the relatively more complicated analysis of biodiversity, if one wants to study in depth not only "the visible [macroscopic] world'" but also the non-visible "microbial" world). (WK 2012i; E. Wilson 2002; S. Nee 2004) And the reverse direction also holds true.

In relation to the openness-hiddenness principle, if there is openness (e.g., the open exploration by some scientists of biodiversity in nature), there is hiddenness (e.g., the hidden bias by some scientists in the exploration of biodiversity in nature, because "selection bias amongst researchers may contribute to biased empirical research for modern estimates of biodiversity," as "Rev. Gilbert White [1909] succinctly observed of his Selborne, Hampshire [that] 'all nature is so full, that that district produces the most variety which is the most examined'"). (WK 2012i) And the reverse direction also holds true.

In relation to the denseness-emptiness principle, if there is denseness (e.g., the relatively denser concentration of species in the tropical regions), there is emptiness (e.g., the relatively less dense, or more empty, concentration of species in the polar regions). And the reverse direction also holds true.

In relation to the change-constancy principle, if there is change (e.g., the ever new ways to study diversity, as shown in the study of psychological benefits and biodiversity), there is constancy (e.g., the ever constant existence of different problems in the ever new ways to study diversity, as shown in the problems with the study of psychological benefits and biodiversity pointed out by the critics). And the reverse direction also holds true.

In relation to the order-chaos principle, if there is order (e.g., the more or less orderly appearance of reality, if looked from the vantage point of the period before "the Cambrian explosion," when "most organisms were simple, composed of individual cells occasionally organized into colonies"), there is chaos (e.g., the more or less chaotic appearance of reality, if looked from the vantage point of the period known as "the Cambrian explosion," especially in regard to "the long-running puzzlement about the appearance of the Cambrian fauna, seemingly abruptly and from nowhere"). (WK 2012t) And the reverse direction also holds true.

In relation to the slowness-quickness principle, if there is slowness (e.g., the relatively slower speed of the emergence of new life forms before the period known as "the Cambrian explosion"), there is quickness (e.g., the relatively quicker speed of the emergence of new life forms, as shown in "the relatively rapid appearance…of most major animal phyla" in the period known as "the Cambrian explosion"). (WK 2012t) And the reverse direction also holds true.

In relation to the expansion-contraction principle, if there is expansion (e.g., the relative expansion of life forms, as shown in "the Phanerozoic eon [the last 540 million years]," which "marked a rapid growth in biodiversity via the Cambrian explosion"), there is contraction (e.g., the relative decrease of life forms, as shown in the period after the Cambrian explosion, when "the next 400 million years included repeated, massive biodiversity losses classified as mass extinction events"). (WK 2012i; UCMP 2012) And the reverse direction also holds true.

In relation to the theory-praxis principle, if there is theory (e.g., the theoretical construction of the relationship between psychological benefits and biodiversity), there is praxis (e.g., the practical application of the understanding of the relationship between psychological benefits and biodiversity to the field of pharmaceutical industry, as shown in the statistics that "a significant proportion of drugs are derived, directly or indirectly, from biological sources: at least 50% of the pharmaceutical compounds on the US market are derived from plants, animals, and micro-organisms"). (WK 2012i) And the reverse direction also holds true.

In relation to the convention-novelty principle, if there is convention (e.g., the conventional wisdom about the "latitudinal gradients" in biodiversity), there is novelty (e.g., the alternative novel challenge to the conventional wisdom about the "latitudinal gradients" in biodiversity, by the new idea that "this characteristic [of latitudinal gradients] is unverified in aquatic ecosystems, especially in marine ecosystems," that "the latitudinal distribution of parasites does not follow this rule," and "other instances of great diversity in higher latitudes have also been recorded"). And the reverse direction also holds true.

In relation to the evolution-transformation principle, if there is evolution (e.g., the natural evolution of humans to benefit from biodiversity, like the impact on human health), there is transformation (e.g., the technical transformation of human ability to benefit from biodiversity, by the invention of the pharmaceutical sector for "natural product exploration [bioprospecting']," "genomics and synthetic chemistry" for health improvement). (WK 2012i) And the reverse direction also holds true.

In relation to the symmetry-asymmetry principle, if there is symmetry (e.g., the existence of different life forms, like microorganisms, animals, and humans in the world of biodiversity), there is asymmetry (e.g., humans can have a greater threat to biodiversity on Earth than animals, as shown in the problems like human overpopulation, overexploitation, etc.). And the reverse direction also holds true.

In relation to the softness-hardness principle, if there is softness (e.g.,the appreciation of the idea of the "latitudinal gradients" in biodiversity by the advocates), there is hardness (e.g., the critique of the idea of the "latitudinal gradients" in biodiversity by the critics). And the reverse direction also holds true.

In relation to the seriousness-playfulness principle, if there is seriousness (e.g., the serious business of those who study the nature of biodiversity), there is playfulness (e.g., the playful part of those who study the nature of biodiversity, when they play around with different ideas of biodiveristy over time, like "molecular diversity," "genetic diversity," "speices diversity," "ecosystem diversity," etc.). And the reverse direction also holds true.

In relation to the activeness-inactivess principle, if there is activeness (e.g., the relative activeness of those species that can survive in changing natural circumstances to "increase disease transmission," such as "that of West Nile Virus, Lyme disease and Hantavirus, according to a study" by Felicia Keesing and her colleagues), there is inactiveness (e.g., the relative inactiveness of those species that cannot survive in changing natural circumstances, so that they cannot "increase disease transmission").

In relation to the selfness-otherness principle, if there is selfness (e.g., the tendency to be mostly centered on one's self or group, as shown in "the latitudinal distribution of parasites"), there is otherness (e.g., the tendency to be mostly centered on others, as shown in those species that "most likely to disappear" but that can "buffer against infectious disease transmission" to benefit others like humans, according to a study" by Felicia Keesing and her colleagues). And the reverse direction also holds true.

In relation to the regression-progression principle, if there is regression (e.g., the regression made by the study of biodiversity and psychological benefits, as shown in the problems pointed out by the critics), there is progression (e.g., the progress made by the study of biodiversity and psychological benefits, as shown in its contribution to a better understanding of the impact of biodiversity on human health aforecited). And the reverse direction also holds true.

In relation to the same-difference principle, if there is similarity in outcome (e.g., the contribution to the molding and control of beliefs, values, and behaviors, regardless of whether this be done by way of the evolutinoary theory to explain biodiversity, or by way of the creation theory to explain biodiversity), there is difference in outcome (e.g., the contribution to the molding and control of beliefs, values, and behaviors by way of the evolutinoary theory to explain biodiversity for a relatively more scientific worldview—but the contribution to the molding and control of beliefs, values, and behaviors by way of the creation theory to explain biodiversity for a relatively more religious worldview). And the reverse direction also holds true.

In relation to the stability-reaction principle, if there is stability (e.g., the relatively more stable condition of the belief in the "latitudinal gradients" in biodiversity among the advocates), there is reaction (e.g., the belief in the "latitudinal gradients" in biodiversity among the advocates led to further changes later on, when some critics nowadays suggest instead that "this characteristic [of latitudinal gradients] is unverified in aquatic ecosystems, especially in marine ecosystems," that "the latitudinal distribution of parasites does not follow this rule," and "other instances of great diversity in higher latitudes have also been recorded"). And the reverse direction also holds true.

In relation to the functionality-nonfunctionality principle, if there is functionality (e.g., the presence of action for which something is fitted, as shown in the trend that "contemporary biodiversity physics is 'firmly fixated on the visible [macroscopic] world,'" which serves the function of understanding biodiversity in mainstream society), there is nonfunctionality (e.g., the relative lack of action for which something is fitted, as shown in the call by some critical scientists for an analysis of biodiversity not only in "the visible [macroscopic] world'" but also in the non-visible "microbial" world, because "the vast majority of Earth's species are microbial" and non-visible, such that the older way of being "fixated on the visible [macroscopic] world" is becoming more and more untenable and no longer functional in the coming future of research on biodiversity) And the reverse direction also holds true.

And in relation to the intentionality-nonintentionality principle, if there is intentionality (e.g., the planning or design of something for a certain outcome, as shown in the empirical study by Fuller and his colleagues for a confirmation of the causal relationship between psychological benefits and biodiversity), there is nonintentionality (e.g., the relative absence of planning or design of something for a certain outcome, such that the outcome can even be counter-intentional in being the opposite of what was originally planned, as shown in the findings by Fuller and his colleagues, when they came to realize that the relationship between psychological benefits and biodiversity may not be causal but can be illusionary, because people may simply "prefer…those that enhance their psychological well-being," so "selection by individuals of environments that enhance personal well-being makes interpretation of the causal structure of these relationships difficult"). And the reverse direction also holds true.

Diversity, Dietary Health, and Agriculture

Another way to evaluate the possibility (or impossibility) and desirability (or undesirability) of diversity from the perspective of the mind is to explore diversity, dietary health, and agriculture.

An argument here is that there is a relationship between dietary health and agricultural biodiversity.

The term "agricultural biodiversity" (also known as "agrobiodiversity") refers to "the result of the careful selection and inventive developments of farmers, herders and fishers over millennia. Agricultural biodiversity is a vital sub-set of biodiversity. It is a use of life, i.e. ancillary biotechnologies, by Mankind whose food and livelihood security depend on the sustained management of those diverse biological resources that are important for food and agriculture." (WK 2012u; UN 1996)

In fact, agricultural biodiversity "includes all forms of life directly relevant to agriculture: rare seed varieties and animal breeds (farm biodiversity), but also many other organisms such as soil fauna, weeds, pests, predators, and all of the native plants and animals (wild biodiversity) existing on and flowing through the farm. However, most attention in this field is given to crop varieties and to crop wild relatives." (WK 2012u)

Scope of Agrobiodiversity

It is thus no wonder that the scope of agricultural biodiversity is quite broad, as it can include the following: (WK 2012u)

- "Domesticated Crop and 'Wild' Plants" (or "Crop Wild Relatives)"—which include "woody perennials" [like "forest genetic resources"] and "aquatic plants (used for food and other natural resources based products), domestic and wild animals (used for food, fibre, milk, hides, furs, power, organic fertilizer), fish and other aquatic animals, within field, forest, rangeland and aquatic ecosystems"
- "Non-Harvested Species within Production Agroecosystems"—which "support food provision, including soil micro-biota, pollinators and so on"
- "Non-Harvested Species in the Wider Environment"—which "support food production agroecosystems (agricultural, pastoral, forest and aquatic ecosystems)"

Thus, the scope of agricultural biodiversity is broad indeed.

Dietary Health and Agrobiodiversity

Agricultural biodiversity is important for the main reason that humans are dependent on it for dietary health, *inter alia*.

In general, "agricultural biodiversity is not only the result of human activity but human life is dependent on it not just for the immediate provision of food and other natural resources based goods, but for the maintenance of areas of land and waters that will sustain production and maintain agroecosystems and the wider biological and environmental services (biosphere)." (WK 2012u)

More specifically, agricultural biodiversity provides the following kinds of support: (WK 2012u)

- "Sustainable production of food and other agricultural products emphasising both strengthening sustainability in production systems at all levels of intensity and improving the conservation, sustainable use and enhancement of the diversity of all genetic resources for food and agriculture, especially plant and animal genetic resources, in all types of production systems"
- "Biological or life support to production emphasising conservation, sustainable use and enhancement of the biological resources that support sustainable production systems, particularly soil biota, pollinators and predators"
- "Ecological and social services provided by agro-ecosystems such as landscape and wildlife protection, soil protection and health (fertility, structure and function), water cycle and water quality, air quality, CO2 sequestration, etc."

These supports are important for human (and non-human) life.

Threats to Agrobiodiversity

But agricultural biodiversity is not stable and can be subject to different threats, and a good example is "genetic erosion in agricultural and livestock biodiversity," which is "the loss of genetic diversity, including the loss of individual genes, and the loss of particular combinants of genes (or gene complexes) such as those manifested in locally adapted landraces. The term genetic erosion is sometimes used in a narrow sense, such as for the loss of alleles or genes, as well as more broadly, referring to the loss of varieties or even species." (WK 2012u)

Normally, "the major driving forces behind genetic erosion in crops are: variety replacement, land clearing, overexploitation of species, population pressure, environmental degradation, overgrazing, policy and changing agricultural systems." (WK 2012u; UN 1997)

However, "the main factor...is the replacement of local varieties by high yielding or exotic varieties or species. A large number of varieties can also often be dramatically reduced when commercial varieties (including GMOs) are introduced into traditional farming systems. Many researchers believe that the main problem related to agro-ecosystem management is the general tendency towards genetic and ecological uniformity imposed by the development of modern agriculture. Pressures for that ecological uniformity on farmers and breeders is caused by the food industry demand for more and more raw materials consistency." (WK 2012u; L. Gao 2003; L. Guarino 1995)

The Case of the Irish Potato Famine

An excellent case study of the human dependency on agricultural biodiversity is "the Irish Potato Famine" (1845-1852), which "was a period of mass starvation, disease and emigration....During the famine approximately 1 million people died and a million more emigrated from Ireland, causing the island's population to fall by between 20% and 25%." (WK 2012v; C. Kinealy 1995; D. Ross 2002)

Historians found that "the proximate cause of famine was a potato disease commonly known as potato blight. Although blight ravaged potato crops throughout Europe during the 1840s, the impact and human cost in Ireland—where one-third of the population was entirely dependent on the potato for food—was exacerbated by a host of political, social and economic factors which remain the subject of historical debate." (WK 2012v; C. Ó Gráda 2006; C. Woodham-Smith 1991)

More than that, "the famine was a watershed in the history of Ireland. Its effects permanently changed the island's demographic, political and cultural landscape. For both the native Irish and those in the resulting diaspora, the famine entered folk memory and became a rallying point for various nationalist movements, as the whole island was then part of the United Kingdom of Great Britain and Ireland. Modern historians regard it as a dividing line in the Irish historical narrative, referring to the preceding period of Irish history as 'pre-Famine.'" (WK 2012v; C. Kinealy 1995)

Problems with the Study of Dietary Health and Agrobiodiversity

However, for the critics, this study of dietary health and agrobiodiversity has some major problems. Consider, for illustration, a few examples below.

Firstly, one main criticism of the study of dietary health and agrobiodiversity is that agrodiversity is not assured for human dependency in regard to dietary health.

For example, good reasons for the instability of agrobiodiversity are the main driving forces (as discussed above), namely, "variety replacement, land clearing, overexploitation of species, population pressure, environmental degradation, overgrazing, policy and changing agricultural systems."

Secondly, another main criticism of the study of dietary health and agrobiodiversity is that the relationship between agrobiodiversity and the Irish Potato Famine is much exaggerated.

For instance, for the critics, there are other main causes of the Irsh Potato Famine, and important contributing factors leading to the eventual famine include "laws that restricted the rights of Irish Catholics," the "middleman system," "tenants, subdivisions, and bankruptcy," "potato dependency," and so on.(WK 2012v)

Thirdly, still another main criticism of the study of dietary health and agrobiodiversity is that there are alternative causes of human dietary health.

For instance, a study by Paul Johnson (2010) and Paul Kenny at the Scripps Research Institute suggested that junk food consumption had contributed to "unhealthy" dietary problems in our time. (WK 2012w)

Fourthly, an additional main criticism of the study of dietary health and agrobiodiversity is that the nature of agrodiversity is still not well known.

For instance, for the critics, different studies offer different insights, and a good example is "a meta-analysis of studies comparing biodiversity, which "noted that, when compared to organic cropping systems, conventional systems had significantly lower species richness and abundance"—but other studies ("16% of studies,") "did find a greater level of species richness in conventional systems" instead. (WK 2012u; J. Bengtsoon 2005) Moreover, "other cropping systems like permaculture have undergone little study to determine relative levels of biodiversity compared to other cropping systems," so the overall nature of agrobiodiversity remains to be investigated. (WK 2012u)

And fifthly, another additional main criticism of the study of dietary health and agrobiodiversity is that agrobiodiveristy can contribute to dietary problems (via crop diseases).

For instance, when "coffee rust attacked coffee plantations in Sri Lanka, Brazil, and Central America in 1970," scientists discovered that "the diseases are themselves a form of biodiversity." (WK 2012i)

Beyond Dietary Health and Agrobiodiversity

These criticisms of the study on dietary health and agrobiodiversity do not mean, however, a total rejection of its validity but serve to show us the two opposing sides of the debate, such that the possibility (or impossibility) and desirability (or undesirability) of diversity (from the perspective of the mind with diversity, dietary health and agriculture as a case study here) are not to the extent that the respective defenders would like us to believe.

In fact, the analysis of diversity, dietary health and agriculture shows us something valuable about the ontological principles in existential dialectics, and good examples include the formalness-informalness principle, the absoluteness-relativeness principle, the partiality-totality principle, the predictability-unpredictability principle, the explicability-inexplicability principle, the fiction-reality principle, the cognitiveness-noncognitiveness principle, the finiteness-transfiniteness principle, the preciseness-vagueness principle, the simpleness-complicatedness principle, the openness-hiddenness principle, the denseness-emptiness principle, the change-constancy principle, the order-chaos principle, the slowness-quickness principle, the expansion-contraction principle, the theory-praxis principle, the convention-novelty principle, the evolution-transformation principle, the symmetry-asymmetry principle, the softness-hardness principle, the seriousness-playfulness principle, the activeness-inactivess principle, the selfness-otherness principle, the regression-progression principle, the same-difference principle, the stability-reaction principle, the functionality-nonfunctionality principle, and the intentionality-nonintentionality principle.

For instance, in relation to the formalness-informalness principle, if there is formalness (e.g., the formal logical requirement of soundness, consistency, and completeness in a system of ideas, as shown in the study on dietary health and agrobiodiversity), there is informalness (e.g., the non-formal existence of unsoundness, inconsistency, and incompleteness in the study on dietary health and agrobiodiversity, as shown in the criticisms that agrodiversity is not assured for human dependency in regard to dietary health, that the relationship between agrobiodiversity and the Irish Potato Famine is much exaggerated, that there are alternative causes of human dietary health, that the nature of agrodiversity is still not well known, and that agrobiodiveristy can contribute to dietary problems [via crop diseases]). And the reverse direction also holds true.

In relation to the absoluteness-relativeness principle, if there is absoluteness (e.g., the absolute view of "a meta-analysis of studies comparing biodiversity, which "noted that, when compared to organic cropping systems, conventional systems had significantly lower species richness and abundance"), there is relativeness (e.g., what is right for "a meta-analysis of studies comparing biodiversity" is not necessarily so for for the critics, since different studies offer different insights, as shown in other studies ["16% of studies"] which "did find a greater level of species richness in conventional systems" instead). And the reverse direction also holds true.

In relation to the partiality-totality principle, if there is partiality (e.g., the partial view of "a meta-analysis of studies comparing biodiversity," which "noted that, when compared to organic cropping systems, conventional systems had significantly lower species richness and abundance"), there is totality (e.g., the more holistic view of agrobiodiversity, such that the whole is not the sum of its parts, that is, it cannot be reduced to the sum of opposing views about differrent cropping systems, since there will emerge new views in the future not yet known today, just as there are already some alternative views nowadays which warn us that "other cropping systems like permaculture have undergone little study to determine relative levels of biodiversity compared to other cropping systems," so the overall nature of agrobiodiversity remains uncertain). And the reverse direction also holds true.

In relation to the predictability-unpredictability principle, if there is predictability (e.g., the predictable tendency of the advocates of agrobiodiversity to stress the positive impact of biodiversity on dietary health), there is unpredictability (e.g., the more difficult task to predict exactly the extent to which agrobiodiversity is contritutive to better dietary health on a particular future occasion, since scientists also discovered that agrobiodiveristy can contribute to dietary problems [via crop diseases] like the case of "coffee rust"). And the reverse direction also holds true.

In relation to the explicability-inexplicability principle, if there is explicability (e.g., the explanation by those who advocated the causal relationship between agrobiodiversity and the Irish Potato Famine on the basis of the potato disease or blight), there is inexplicability (e.g., the lack of sufficient explanation, by those who advocated the causal relationship between agrobiodiversity and the Irish Potato Famine, of why it is necessarily true, since, for the critics, there are other main causes of the Irsh Potato Famine, and important factors include "laws that restricted the rights of Irish Catholics," the "middleman system," "tenants, subdivisions, and bankruptcy," "potato dependency," and so on). And the reverse direction also holds true.

In relation to the fiction-reality principle, if there is fiction (e.g., the fictional aspect of the argument for agrobiodiversity as contritutive to better dietary health, insofar as scientists also discovered that agrobiodiveristy can contribute to dietary problems [via crop diseases] like the case of "coffee rust"), there is reality (e.g., the realistic aspect of the argument for agrobiodiversity as contritutive to better dietary health, insofar as, in some estimates, "about 80 percent of humans' food supply comes from just 20 kinds of plants" and "humans use at least 40,000 species"). (WK 2012i) And the reverse direction also holds true.

In relation to the cognitiveness-noncognitiveness principle, if there is cognitiveness (e.g., a view of things on the basis of reason and evidences, as shown in the study of the Irish Potato Famine, which found that "the proximate cause of famine was a potato disease commonly known as potato blight"), there is noncognitiveness (e.g., a view of things on the basis of non-cognitive factors like envy, jealousy, power, nationality, race, gender, age, class, greed, lust, status, anger, sadness, joy, fear, wish, etc., as shown in the study of the Irish Potato Famine, when "the impact and human cost in Ireland…was exacerbated by a host of political, social and economic factors which remain the subject of historical debate" among different groups today). And the reverse direction also holds true.

In relation to the finiteness-transfiniteness principle, if there is finiteness (e.g., the finite number of the kinds of support by agrobiodiversity, like the 3 kinds in the article aforecited), there is transfiniteness (e.g., the transfinite number of all the individual microorganisms, species, and ecosystems that have ever existed in the history of agrobiodiversity on Earth). And the reverse direction also holds true.

In relation to the preciseness-vagueness principle, if there is preciseness (e.g., the precise identification of three major scopes of agrobiodiversity in the article aforecited), there is vagueness (e.g., the vagueness in the identification of the three major scopes of agrobiodiversity in the article aforecited, since it is not clear why there must be only 3, not 4, 5, 6, and so on). And the reverse direction also holds true.

In relation to the simpleness-complicatedness principle, if there is simpleness (e.g., the relatively simple analysis of the Irish Potato Famine on the basis of the potato disease or blight only), there is complicatedness (e.g., the relatively more complicated analysis of the Irish Potato Famine, by challenging the claims and assumptions, as shown in the more complicated analysis of other important contributing factors like "laws that restricted the rights of Irish Catholics," the "middleman system," "tenants, subdivisions, and bankruptcy," "potato dependency," and so on). And the reverse direction also holds true.

In relation to the openness-hiddenness principle, if there is openness (e.g., the open exploration of different cropping systems in agrobiodiversity, as shown by "a meta-analysis of studies comparing biodiversity," which "noted that, when compared to organic cropping systems, conventional systems had significantly lower species richness and abundance"), there is hiddenness (e.g., the hidden bias in the "meta-analysis of studies comparing biodiversity," since different studies offer different insights, as shown in other studies ["16% of studies"] which "did find a greater level of species richness in conventional systems" instead). And the reverse direction also holds true.

In relation to the denseness-emptiness principle, if there is denseness (e.g., the relatively denser concentration of potatoes in Irish diet during the early 19th century), there is emptiness (e.g., the relatively less dense, or more empty, concentration of potatoes in Chinese diet during the early 19th century). And the reverse direction also holds true.

In relation to the change-constancy principle, if there is change (e.g., the ever new ways to understand biodiversity, as shown in the study of dietary health and agrobiodiversity), there is constancy (e.g., the ever constant existence of different problems in the ever new ways to understand biodiversity, as shown in the problems with the study of dietary health and agrobiodiversity pointed out by the critics). And the reverse direction also holds true.

In relation to the order-chaos principle, if there is order (e.g., the more or less orderly appearance of the Irish dietary situation in the early 19th century, when it was dependent on potatoes), there is chaos (e.g., the more or less chaotic appearance of the Irsh dietary situation during the Irish Potato Famine, when "blight ravaged potato crops throughout Europe during the 1840s" and "the impact and human cost in Ireland…was exacerbated by a host of political, social and economic factors which remain the subject of historical debate"). (WK 2012v) And the reverse direction also holds true.

In relation to the slowness-quickness principle, if there is slowness (e.g., the relatively slower speed of the fall of the Irish population before the Irish Potato Famine, when the demographic situation was more normal), there is quickness (e.g., the relatively quicker speed of the fall of the Irish population during the Irish Potato Famine and its aftermath, when "approximately 1 million people died and a million more emigrated from Ireland, causing the island's population to fall by between 20% and 25%"). (WK 2012v; D. Ross 2002; C. Kinealy 1995) And the reverse direction also holds true.

In relation to the expansion-contraction principle, if there is expansion (e.g., the relative expansion of the American population in the 19th century, when it gained from the arrival of Irish immigrants and others), there is contraction (e.g., the relative decline of the Irish population, when it fell "by between 20% and 25%" during the Irish Potato Famine and its aftermath). And the reverse direction also holds true.

In relation to the theory-praxis principle, if there is theory (e.g., the theoretical understanding of the history of the Irish Potato Famine, like its causes and impacts, as shown in the study aforecited), there is praxis (e.g., the practical application of the understanding of the history of the Irish Potato Famine like its causes and impacts to the field of politics, when nowadays, there are different memorials throughout Ireland to memorialise the tragedy, as shown by the memorial at "Custom House Quays, Dublin," with "the thin sculptural figures, by artist Rowan Gillespie, who stand as if walking towards the emigration ships on the Dublin Quayside"). (WK 2012v) And the reverse direction also holds true.

In relation to the convention-novelty principle, if there is convention (e.g., the conventional wisdom about the positive impact of agrobiodiversity on dietary health), there is novelty (e.g., the alternative novel challenge to the conventional wisdom about the positive impact of agrobiodiversity on dietary health, by the new idea of "coffee rust" which "attacked coffee plantations in Sri Lanka, Brazil, and Central America in 1970," when scientists discovered that "the diseases are themselves a form of biodiversity"). And the reverse direction also holds true.

In relation to the evolution-transformation principle, if there is evolution (e.g., the natural evolution of humans to hunt for food in the state of nature in a primitive way), there is transformation (e.g., the technical transformation of human ability to get food by the invention of different agricultural systems, like organic cropping, conventional, and permacultural systems). And the reverse direction also holds true.

In relation to the symmetry-asymmetry principle, if there is symmetry (e.g., the potato "blight ravaged potato crops throughout Europe" in the middle 19th century), there is asymmetry (e.g., "although blight ravaged potato crops throughout Europe during the 1840s, the impact and human cost in Ireland…was exacerbated by a host of political, social and economic factors," so that Ireland suffered more than its neighbors). (WK 2012v) And the reverse direction also holds true.

In relation to the softness-hardness principle, if there is softness (e.g., the defense of the "middleman system" by its advocates in early 19th century Ireland), there is hardness (e.g., the critique of the "middleman system" by some critical historians in the article aforecited). And the reverse direction also holds true.

In relation to the seriousness-playfulness principle, if there is seriousness (e.g., the serious business of those who try to find out the causes of the Irish Potato Famine), there is playfulness (e.g., the playful part of those who try to find out the causes of the Irish Potato Famine, when they play around with different possible contributing factors over time, like "laws that restricted the rights of Irish Catholics," the "middleman system," "tenants, subdivisions, and bankruptcy," "potato dependency," "potato blight," and so on). And the reverse direction also holds true.

In relation to the activeness-inactivess principle, if there is activeness (e.g., the relative activeness of those who were lucky enugh to not suffer from the Irish Potato Famine in the 1848-52 period and were able to work), there is inactiveness (e.g., the relative inactiveness of those who were unfortunate enugh to get sick or die due to the Irish Potato Famine in the 1848-52 period, so they could not engage in productive work).

In relation to the selfness-otherness principle, if there is selfness (e.g., the tendency to be mostly centered on one's self or group, as shown by those who advocated the "middleman system" in 19th century Ireland, because they benefited from it at the expense of others), there is otherness (e.g., the tendency to be mostly centered on others, as shown by those critical historians who tried to provide a more objective analysis of the Irish Potato Famine for the advancement of historical knowledge). And the reverse direction also holds true.

In relation to the regression-progression principle, if there is regression (e.g., agrobiodiversity can contribute to crop diseases resulting in human suffering), there is progression (e.g., agrobiodiversity can contribute to food supply for human dietary health). And the reverse direction also holds true.

In relation to the same-difference principle, if there is similarity in outcome (e.g., the contribution to the molding and control of beliefs, values, and behaviors, regardless of whether this be done by way of focusing on agrobiodiversity in understanding the Irish Potato Famine, or by way of focusing on "political, social and economic factors" in understanding the Irish Potato Famine), there is difference in outcome (e.g., the contribution to the molding and control of beliefs, values, and behaviors by way of focusing on agrobiodiversity for a relatively more naturalistic worldview—but the contribution to the molding and control of beliefs, values, and behaviors by way of focusing on "political, social and economic factors" for a relatively more societal worldview). And the reverse direction also holds true.

In relation to the stability-reaction principle, if there is stability (e.g., the relatively more stable condition of the potato dependency in 19th century Ireland, which had contribtued to the Irish Potato Famine), there is reaction (e.g., the potato dependency in 19th century Ireland which had contribtued to the Irish Potato Famine then led to further changes later on, when the Irish became less dependent on potatoes over time by consuming other resources, even though this took many decades for them to overcome it). And the reverse direction also holds true.

In relation to the functionality-nonfunctionality principle, if there is functionality (e.g., the presence of action for which something is fitted, as shown in the potato dependency in 19th century Ireland, as it served the essential function of providing food supply for the Irish at the time), there is nonfunctionality (e.g., the relative lack of action for which something is fitted, as shown in the less potato dependency in 21st century Ireland, when higher living standards allow better nutrition, so that the potato dependency is no longer functional for the Irish nowadays, because of the current higher consumption level). And the reverse direction also holds true.

And in relation to the intentionality-nonintentionality principle, if there is intentionality (e.g., the planning or design of something for a certain outcome, as shown in the investigation by some scientists of the outbreak of "coffee rust" which "attacked coffee plantations in Sri Lanka, Brazil, and Central America in 1970"), there is nonintentionality (e.g., the relative absence of planning or design of something for a certain outcome, such that the outcome can even be counter-intentional in being the opposite of what was originally planned, as shown in the investigation by some scientists of the outbreak of "coffee rust" in Sri Lanka, Brazil, and Central America in 1970," when they discovered, to their surprise. that "the diseases are themselves a form of biodiversity," so agrobiodiversity has its dark side too). And the reverse direction also holds true.

Diversity and Nature

Diversity, when examined from the perspective of nature, can show us its possibility (or impossibility) and desirability (or undesirability), and this can be shown by way of two case studies, namely, (2.3.1) diversity, conservation biology, and environmental protection, and (2.3.2) diversity, ecology, and adaptive management—to be addressed in what follows, respectively.

Diversity, Conservation Biology, and Environmental Protection

A good way to evaluate the possibility (or impossibility) and desirability (or undesirability) of diversity from the perspective of nature is to explore diversity, conservation biology, and envionmental protection.

An argument here is that there is a relationship between conservation biology and environmental protection, in that conservation biology can contribute to the protection of "species, their habitats, and ecosystems from excessive rates of extinction and the erosion of biotic interactions," with the help of "sciences, economics, and the practice of natural resource management." (WK 2012c; S. Sahney 2008; B. Wilcox 1980; M. Soule 1986 & 198a; M. Hunter 1996; G. Meffe 2006; F. van Dyke 2008)

History of the Protection of Natural Resources

Human efforts to protect natural resources are more a modern phenomenon and can be described in the following periods (and summarized in *Table 2.3*). (WK 2012c; F. van Dyke 2008)

(a) Natural Resource Conservation

In the older days, "resource ethics grew out of necessity through direct relations with nature. Regulation or communal restraint became necessary to prevent selfish motives from taking more than could be locally sustained, therefore compromising the long-term supply for the rest of the community. This social dilemma with respect to natural resource management is often called the 'Tragedy of the Commons.' From this principal, conservation biologists can trace communal resource based ethics throughout cultures as a solution to communal resource conflict." (WK 2012c; F. van Dyke 2008; G. Hardin 1968; C. Kay 1997; E. Wilson 2007)

For example, "the Alaskan Tlingit peoples and the Haida of the Pacific Northwest had resource boundaries, rules, and restrictions among clans with respect to the fishing of Sockeye Salmon. These rules were guided by clan elders who knew lifelong details of each river and stream they managed. There are numerous examples in history where cultures have followed rules, rituals, and organized practice with respect to communal natural resource management." (WK 2012c; F. van Dyke 2008; R. Mason 2004; D. Wilson 2002)

In addition, "conservation ethics are also found in early religious and philosophical writings. There are examples in the Tao, Shinto, Hindu, Islamic and Buddhist traditions. In Greek philosophy, Plato lamented about pasture land degradation: 'What is left now is, so to say, the skeleton of a body wasted by disease; the rich, soft soil has been carried off and only the bare framework of the district left.' In the bible, through Moses, God commanded to let the land rest from cultivation every seventh year. Before the 18th century, however, much of European culture considered it a pagan view to admire nature. Wilderness was denigrated while agricultural development was praised. However, as early as AD 680 a wildlife sanctuary was founded on the Farne Islands by St Cuthbert in response to his religious beliefs." (WK 2012c; F. van Dyke 2008; R. Primack 2004; E. Hamilton 1961; D. Evans 1997)

(b) Early Naturalists

By the 18th century, "natural history was a major preoccupation..., with grand expeditions and the opening of popular public displays in Europe and North America. By 1900 there were 150 natural history museums in Germany, 250 in Great Britain, 250 in the United States, and 300 in France." (WK 2012c; P. Farber 2000)

Later on, "preservationist or conservationist sentiments are a development in the late 18th to early 20th century. The 19th century fascination with natural history engendered a fervor to be the first to collect rare specimens with the goal of doing so before they became extinct by other such collectors. Although his artistic works and romantic depiction of avian life inspired many bird enthusiasts and conservation organizations, the writings of John James Audubon, by modern standards, show insensitivity toward bird conservation as he shot and collected hundreds of specimens. Inspired by him, however, the first chapter of the Audubon Society started in 1905 for the purpose of protecting birds." (WK 2012c; P. Farber 2000; D. Evans 1997)

(c) 19th Century Conservation

But it was in the 19th century which witnessed the coming of the age of conservation, since "the modern concept of ecosystem services can be found in the late 19th century." (WK 2012c)

In the early years of the 19th century, "biogeography was ignited through efforts of Alexander von Humboldt, Lyell and Darwin; their efforts, while important in relating species to their environments, were part of the naturalist tradition and fell short of conservation biology proper. Darwin, for example, hunted and shot birds and kept natural history cabinets in line with Victorian tradition." (WK 2012c)

And "modern roots of conservation biology can be found in the late 19th century Enlightenment period particularly in England and Scotland. A number of thinkers, among them notably Lord Monboddo, described the importance of 'preserving nature'; much of this early emphasis had its origins in Christian theology." (WK 2012c; E. Cloyd 1972; D. Evans 1997)

(d) 20th Century Conservation

By the 20th century, "actions in the United Kingdom, United States, and Canada emphasized the protection of habitat areas pursuant to visions of such people as John Muir, Theodore Roosevelt, and Aldo Leopold." (WK 2012c)

In fact, "the term *conservation* came into use in the late 19th century and referred to the management, mainly for economic reasons, of such natural resources as timber, fish, game, topsoil, pastureland, and minerals. In addition it referred to the preservation of forests (forestry), wildlife (wildlife refuge), parkland, wilderness, and watersheds. Western Europe was the source of much 19th century progress for conservation biology, particularly the British Empire with the Sea Birds Preservation Act 1869. However, the United States made contributions to this field starting with thinking of Thoreau and taking form with the Forest Act of 1891, John Muir's founding of the Sierra Club in 1892, the founding of the New York Zoological Society in 1895 and establishment of a series of national forests and preserves by Theodore Roosevelt from 1901 to 1909." (WK 2012c)

Later, in the mid-20th century, there were "efforts…to target individual species for conservation, notably efforts in big cat conservation in South America led by the New York Zoological Society. In the early 20th century the New York Zoological Society was instrumental in developing concepts of establishing preserves for particular species and conducting the necessary conservation studies to determine the suitability of locations that are most appropriate as conservation priorities; the work of Henry Fairfield Osborn Jr., Carl E. Akeley, Archie Carr and Archie Carr III is notable in this era." (WK 2012c) Rabinowitz, A. R. 1986; M. Carr 1994)

In the 1970s, "the United Nations acted to conserve sites of outstanding cultural or natural importance to the common heritage of mankind. The programme was adopted by the General Conference of UNESCO in 1972. As of 2006, a total of 830 sites are listed: 644 cultural, 162 natural. The first country to pursue aggressive biological conservation through national legislation was the United States, which passed back to back legislation in the Endangered Species Act (1966) and National Environmental Policy Act (1970), which together injected major funding and protection measures to large scale habitat protection and threatened species research." (WK 2012c; US 1973)

In 1980, "a significant development was the emergence of the urban conservation movement. A local organization was established in Birmingham, UK, a development followed in rapid succession in cities across the UK, then overseas. Although perceived as a grassroots movement, its early development was driven by academic research into urban wildlife. Initially perceived as radical, the movement's view of conservation being inextricably linked with other human activity has now become mainstream in conservation thought. Considerable research effort is now directed at urban conservation biology. The Society for Conservation Biology originated in 1985." (WK 2012c)

And "by 1992, most of the countries of the world had become committed to the principles of conservation of biological diversity with the Convention on Biological Diversity; subsequently many countries began programmes of Biodiversity Action Plans to identify and conserve threatened species within their borders, as well as protect associated habitats." (WK 2012c)

Approaches to to the Protection of Natural Resources

With this history in mind—there are different approaches to the protection of natural resources, and hereafter are three major ones (as summarized in *Table 2.3*).

(a) Voluntary Environmental Agreements

The first approach is more voluntary. For instance, "in industrialized countries, voluntary environmental agreements often provide a platform for companies to be recognized for moving beyond the minimum regulatory standards and, thus, support the development of best environmental practice." (WK 2012k; P. Karamanos 2001)

However, "in developing countries, such as throughout Latin America, these agreements are more commonly used to remedy significant levels of non-compliance with mandatory regulation." (WK 2012k; A. Blackman 2008)

(b) Ecosystems Approach

The second approach is more ecosystem-oriended. In other words, "an ecosystems approach to resource management and environmental protection aims to consider the complex interrelationships of an entire ecosystem in decision making rather than simply responding to specific issues and challenges." (WK 2012k)

This then means that "the decision-making processes under such an approach would be a collaborative approach to planning and decision making that involves a broad range of stakeholders across all relevant governmental departments, as well as representatives of industry, environmental groups and community. This approach ideally supports a better exchange of information, development of conflict-resolution strategies and improved regional conservation." (WK 2012k; CIPA 2001)

(c) International Environmental Agreements

And the third approach is more international in orientation. This approach is motivated by the understanding that "many of the earth's resources are especially vulnerable because they are influenced by human impacts across many countries." (WK 2012k)

Thus, "many attempts are made by countries to develop agreements that are signed by multiple governments to prevent damage or manage the impacts of human activity on natural resources. This can include agreements that impact factors such as climate, oceans, rivers and air pollution. These international environmental agreements are sometimes legally binding documents that have legal implications when they are not followed and, at other times, are more agreements in principle or are for use as codes of conduct. These agreements have a long history with some multinational agreements being in place from as early as 1910 in Europe, America and Africa. Some of the most well-known multinational agreements include: the Kyoto Protocol, Vienna Convention on the Protection of the Ozone Layer and Rio Declaration on Development and Environment." (WK 2012k; R. Mitchell 2003)

Planning of the Protection of Natural Resources

These different approaches have something in common, in that there are some stages in any systemtic planning of the protection of natural resources.

For instance, C. Margules (2000) and R. Pressey identified "six interlinked stages in the systematic planning," as shown below (and summarized in *Table 2.3*): (WK 2012k).

- "Compile data on the biodiversity of the planning region"
- "Identify conservation goals for the planning region"
- "Review existing conservation areas"
- "Select additional conservation areas"
- "Implement conservation actions"
- "Maintain the required values of conservation areas"

Of course, there can be other stages, but Margules and Pressey proposed only these six.

Problems with the Protection of Natural Resources

However, for the critics, this study of conservation biology and environmental protection has some major problems. Consider, for illustration, a few examples below (as summarized in *Table 2.3*).

Firstly, one main criticism of the study of conservation biology and environmental protection is that it does not give sufficient attention to the important problem of poor management.

For instance, for the critics, "the main issues for developing countries like Brazil and Mexico are that protected areas suffer from encroachment and poor management. In Brazil, protected areas are increasing, but there are significant challenges caused by human impacts. Logging and mining are potentially huge threats to protected areas." (WK 2012k)

Secondly, another main criticism of the study of conservation biology and environmental protection is that it often ignores the factors of power abuse and financial shortage.

For instance, for the critics, "African governments face several challenges in implementing environmental protection mechanisms. In Tanzania for example these include lack of financial resources to manage protected areas, poor governance and corruption…." (WK 2012k; D. Pallangyo 2007)

Thirdly, still another main criticism of the study of conservation biology and environmental protection is that different approaches have different weaknesses.

For instance, the problem with the voluntary approach to the protection of natural resources is that, "due to the difficulties inherent in evaluating effectiveness, their use is often questioned and, indeed, the environment may well be adversely affected as a result." (WK 2012k)

Fourthly, an additional main criticism of the study of conservation biology and environmental protection is that the priority to focus on hotspots for biodiveristy in the protection of natural resources is misleading and dangerous.

For instance, "those arguing in favor of setting priorities for coldspots [not hotspots] point out that there are other measures to consider beyond biodiversity. They point out that emphasizing hotspots downplays the importance of the social and ecological connections to vast areas of the Earth's ecosystems where biomass, not biodiversity, reigns supreme. It is estimated that 36% of the Earth's surface, encompassing 38.9% of the worlds vertebrates, lacks the endemic species to qualify as biodiversity hotspot. Moreover, measures show that maximizing protections for biodiversity does not capture ecosystem services any better than targeting randomly chosen regions. Population level biodiversity (i.e. coldspots) are disappearing at a rate that is ten times that at the species level. The level of importance in addressing biomass versus endemism as a concern for conservation biology is highlighted in literature measuring the level of threat to global ecosystem carbon stocks that do not necessarily reside in areas of endemism. A hotspot priority approach would not invest so heavily in places such as steppes, the Serengeti, the Arctic, or taiga. These areas contribute a great abundance of population (not species) level biodiversity and ecosystem services, including cultural value and planetary nutrient cycling." (WK 2012k; E. Underwood 2008; S. Leroux 2007; R. Naidoo 2008; C. Wood 2008; S. Running 2008; W. Kurz 2008; P. Kareiva 2003)

And fifthly, another additional main criticism of the study of conservation biology and environmental protection is that it is too utilitarian in ethical orientation.

For instance, the critics questioned the utilitarian position by K. Chan (2008) who argued that "conservationists must advocate for biodiversity and can do so" for "the greatest good for the greatest number of people for the longest time," because the critics argue instead that "nature has an intrinsic value that is independent of anthropocentric usefulness or utilitarianism. Intrinsic value advocates that a gene, or species, be valued because they have a utility for the ecosystems they sustain. Aldo Leopold was a classical thinker and writer on such conservation ethics" along the deontological line. (WK 2012c; P. Farber 2000)

Beyond the Protection of Natural Resources

These criticisms of the study of conservation biology and environmental protection do not entail a total rejection of its usefulness but serve to show us the two opposing sides of the debate, such that the possibility (or impossibility) and desirability (or undesirability) of diversity (from the perspective of nature with diveristy, conservation biology, and environmental protection as a case study here) are not to the extent that the respective defenders would like us to believe.

Furthermore, the analysis of diveristy, conservation biology, and environmental protection reveals some fascinating lessons about the ontological principles in existential dialectics, and good examples include the formalness-informalness principle, the absoluteness-relativeness principle, the partiality-totality principle, the predictability-unpredictability principle, the explicability-inexplicability principle, the fiction-reality principle, the cognitiveness-noncognitiveness principle, the finiteness-transfiniteness principle, the preciseness-vagueness principle, the simpleness-complicatedness principle, the openness-hiddenness principle, the denseness-emptiness principle, the change-constancy principle, the order-chaos principle, the slowness-quickness principle, the expansion-contraction principle, the theory-praxis principle, the convention-novelty principle, the evolution-transformation principle, the symmetry-asymmetry principle, the softness-hardness principle, the seriousness-playfulness principle, the activeness-inactivess principle, the selfness-otherness principle, the regression-progression principle, the same-difference principle, the stability-reaction principle, the functionality-nonfunctionality principle, and the intentionality-nonintentionality principle.

For instance, in relation to the formalness-informalness principle, if there is formalness (e.g., the formal logical requirement of soundness and completeness in a system of ideas, as in the study of conservation biology and environmental protection), there is informalness (e.g., the non-formal existence of unsoundness and incompleteness in the study of conservation biology and environmental protection, as shown in the criticisms that it does not give sufficient attention to the important problem of poor management, that it often ignores the factors of power abuse and financial shortage, that different approaches have different weaknesses, that the priority to focus on hotspots for biodiveristy in the protection of natural resources is misleading and dangerous, and that it is too utilitarian in ethical orientation). And the reverse direction also holds true.

In relation to the absoluteness-relativeness principle, if there is absoluteness (e.g., the absolute view of K. Chan, who argued that "conservationists must advocate for biodiversity and can do so" for "the greatest good for the greatest number of people for the longest time"), there is relativeness (e.g., what is right for K. Chan in regard to his utilitarian ethics for conservation is not necessarily so for the critics, who argue instead that "nature has an intrinsic value that is independent of anthropocentric usefulness or utilitarianism"). And the reverse direction also holds true.

In relation to the partiality-totality principle, if there is partiality (e.g., the partial view of the volunary approach to the protection of natural resources, because there are alternative ones too, as cited earlier), there is totality (e.g., the more holistic view of the protection of natural resources, such that the whole is not the sum of its parts, that is, it cannot be reduced to the sum of the opposing views in the three different approaches as discussed earlier, since there will emerge new views in the future not yet known today, just as there are others which do not identify with any of them). And the reverse direction also holds true.

In relation to the predictability-unpredictability principle, if there is predictability (e.g., the predictable tendency of those for the voluntary approach to target developing countries with these agreements so as "to remedy significant levels of non-compliance with mandatory regulation"), there is unpredictability (e.g., the more difficult task to predict exactly the extent to which with the voluntary approach will be successful on a particular future occasion, "due to the difficulties inherent in evaluating effectiveness," such that "their use is often questioned and, indeed, the environment may well be adversely affected as a result"). (WK 2012k) And the reverse direction also holds true.

In relation to the explicability-inexplicability principle, if there is explicability (e.g., the explanation by K. Chan of the need for conservation on the basis of "the greatest good for the greatest number of people for the longest time"), there is inexplicability (e.g., the lack of sufficient explanation by K. Chan of why utilitarian ethics is necessarily valid for conservation, since, for the critics, "nature has an intrinsic value that is independent of anthropocentric usefulness or utilitarianism"). And the reverse direction also holds true.

In relation to the fiction-reality principle, if there is fiction (e.g., the fictional aspect of the ecosystems approach, insofar as it is "idealistic" in calling for "a broad range of stakeholders across all relevant governmental departments, as well as representatives of industry, environmental groups and community"), there is reality (e.g., the realistic aspect of the ecosystems approach, insofar as it stresses the need to address "the complex interrelationships of an entire ecosystem in decision making rather than simply responding to specific issues and challenges"). And the reverse direction also holds true.

In relation to the cognitiveness-noncognitiveness principle, if there is cognitiveness (e.g., a view of things on the basis of reason and evidences, as shown in the study of environmental protection of natural resources in Tanzania in the article aforecited), there is noncognitiveness (e.g., a view of things on the basis of non-cognitive factors like envy, jealousy, power, nationality, race, gender, age, greed, lust, status, class, anger, sadness, joy, fear, wish, etc., as shown in the study of environmental protection of natural resources in Tanzania, when researchers discovered that "African governments face several challenges in implementing environmental protection mechanisms," and "in Tanzania for example these include…corruption" because of greed). And the reverse direction also holds true.

In relation to the finiteness-transfiniteness principle, if there is finiteness (e.g., the finite number of the categories of concerns with biodiversity in conservation efforts, like "genes," "species," and "ecosystems" in the article aforecited), there is transfiniteness (e.g., the transfinite number of all the individual "genes," "species," and "ecosystems" that have ever existed in the history of biodiversity on Earth). And the reverse direction also holds true.

In relation to the preciseness-vagueness principle, if there is preciseness (e.g., the precise identification of six stages in the systematic planning" of the protection of natural resources in the article aforecited), there is vagueness (e.g., the vagueness in the identification of the six stages in the systematic planning" of the protection of natural resources in the article aforecited, since it is not clear why there must be only 6, not 7, 8, 9, and so on). And the reverse direction also holds true.

In relation to the simpleness-complicatedness principle, if there is simpleness (e.g., the relatively simple analysis of utilitarian ethics for conservation efforts by K. Chan), there is complicatedness (e.g., the relatively more complicated analysis of utilitarian ethics for conservation efforts, by challenging the claims and assumptions, as shown in the problems pointed out by the critics as cited). And the reverse direction also holds true.

In relation to the openness-hiddenness principle, if there is openness (e.g., the open exploration of the priority to emphasize hotspots for the understanding of biodiversity), there is hiddenness (e.g., the hidden bias in the priority to emphasize hotspots for the understanding of biodiversity, because the critics point out that "emphasizing hotspots downplays the importance of the social and ecological connections to vast areas of the Earth's ecosystems where biomass, not biodiversity, reigns supreme," so "a hotspot priority approach would not invest so heavily in places such as steppes, the Serengeti, the Arctic, or taiga," but "these areas contribute a great abundance of population [not species] level biodiversity and ecosystem services, including cultural value and planetary nutrient cycling"). And the reverse direction also holds true.

In relation to the denseness-emptiness principle, if there is denseness (e.g., the relatively denser concentration of concerns with "endemism" in a "hotspot priority approach"), there is emptiness (e.g., the relatively less dense, or more empty, concentration of concerns with "endemism" in a "coldspot priority approach"). And the reverse direction also holds true.

In relation to the change-constancy principle, if there is change (e.g., the ever new ways to understand ecosystems, as shown in the study of conservation biology and environmental protection), there is constancy (e.g., the ever constant existence of different problems in the ever new ways to understand ecosystems, as shown in the problems with the study of conservation biology and environmental protection pointed out by the critics). And the reverse direction also holds true.

In relation to the order-chaos principle, if there is order (e.g., the more or less orderly appearance of conservation efforts, if looked from the sole vantage point of ecological issues like biodiversity), there is chaos (e.g., the more or less chaotic appearance of conservation efforts, if looked from the multifaceted vantage point of myriad linkages, including "encroachment and poor management" as in the case studies of Brazil and Mexico, "lack of financial resources" as well as "poor governance and corruption" as in the case studies of African countries). And the reverse direction also holds true.

In relation to the slowness-quickness principle, if there is slowness (e.g., the relatively slower speed of a decision making process in conservation if the approach is to require "the complex interrelationships of an entire ecosystem in decision making…that involves a broad range of stakeholders across all relevant governmental departments, as well as representatives of industry, environmental groups and community"), there is quickness (e.g., the relatively quicker speed of a decision making process in conservation if the approach is to "simply respond…to specific issues and challenges" locally at a time). And the reverse direction also holds true.

In relation to the expansion-contraction principle, if there is expansion (e.g., the relative expansion of the commitment to "the principles of conservation of biological diversity" in the 1990s, as shown in "the Convention on Biological Diversity" in 1992 accepted by "most of the countries of the world"), there is contraction (e.g., the relative decline of the opponents of "the principles of conservation of biological diversity" in the 1990s). And the reverse direction also holds true.

In relation to the theory-praxis principle, if there is theory (e.g., the theoretical construction of the "ecosystems approach," which requires, at least on paper, "the complex interrelationships of an entire ecosystem in decision making…that involves a broad range of stakeholders across all relevant governmental departments, as well as representatives of industry, environmental groups and community"), there is praxis (e.g., the practical application of the "ecosystems approach to the real world of international relations, such that the approach turns out to be "idealistic" in practice, as shown by the problems confronted by many countries in implementing conservation efforts aforecited). And the reverse direction also holds true.

In relation to the convention-novelty principle, if there is convention (e.g., the conventional wisdom about the "hotspot priority approach), there is novelty (e.g., the alternative novel challenge to the conventional wisdom about the "hotspot" priority approach, by the new idea of the "coldspot" priority approach). And the reverse direction also holds true.

In relation to the evolution-transformation principle, if there is evolution (e.g., the natural evolution of humans to use natural resources in the state of nature in a primitive way), there is transformation (e.g., the technical transformation of human ability to use natural resoources by the invention of conservation biology for a more systematic planning which allows the use of natural resources to be sustainable in the long run). And the reverse direction also holds true.

In relation to the symmetry-asymmetry principle, if there is symmetry (e.g., the existence of voluntary international agreements for conservation in developed and developing countries), there is asymmetry (e.g., "in industrialized countries, voluntary environmental agreements often provide a platform for companies to be recognized for moving beyond the minimum regulatory standards and, thus, support the development of best environmental practice"—but "in developing countries, such as throughout Latin America, these agreements are more commonly used to remedy significant levels of non-compliance with mandatory regulation") And the reverse direction also holds true.

In relation to the softness-hardness principle, if there is softness (e.g., "companies to be recognized for moving beyond the minimum regulatory standards" in developed countries), there is hardness (e.g., "significant levels of non-compliance with mandatory regulation" in developing countries). And the reverse direction also holds true.

In relation to the seriousness-playfulness principle, if there is seriousness (e.g., the serious business of those trying to deal with conservation efforts), there is playfulness (e.g., the playful part of those trying to deal with conservation efforts, when they play around with different possible approaches over time, such as "voluntary international agreements," "ecosystems approach," "international environmental agreements," etc.). And the reverse direction also holds true.

In relation to the activeness-inactivess principle, if there is activeness (e.g., the activeness of those who are willing to "mov[e]…beyond the minimum regulatory standards," so as to do more for conservation efforts), there is inactiveness (e.g., the relative inactiveness of those who show "non-compliance with mandatory regulation" because of "poor governance," "corruption," etc.).

In relation to the selfness-otherness principle, if there is selfness (e.g., the tendency to be mostly centered on one's self or group, as shown by those who show "non-compliance with mandatory regulation" because of greed and corruption), there is otherness (e.g., the tendency to be mostly centered on others, as shown by those who care a lot about "the tragedy of the commons" by participating in conservation efforts for the future benefits of humanity). And the reverse direction also holds true.

In relation to the regression-progression principle, if there is regression (e.g., the regression made by the study on conservation biology and environmental protection , as shown in the problems pointed out by the critics aforecited), there is progression (e.g., the progress made by the study of conservation biology and environmental protection, as shown in its contribution to a better protection of natural resources aforecited). And the reverse direction also holds true.

In relation to the same-difference principle, if there is similarity in outcome (e.g., the contribution to the molding and control of beliefs, values, and behaviors, regardless of whether this be done by way of an extrinsic [utilitarian] conservation effort, or by way of intrinsic conservation effort), there is difference in outcome (e.g., the contribution to the molding and control of beliefs, values, and behaviors by way of an extrinsic [utilitarian] conservation effort for a relatively more consequentialist lifestyle—but the contribution to the molding and control of beliefs, values, and behaviors by way of intrinsic conservation effort for a relatively more deontological lifestyle). And the reverse direction also holds true.

In relation to the stability-reaction principle, if there is stability (e.g., the relatively more stable condition of the "hotspot" priority approach, as shown by the fact that "the predominant approach to date has been to focus efforts on endangered species by conserving biodiversity hotspots"), there is reaction (e.g., the "hotspot" priority approach then led to further changes later on, when nowadays, "some scientists" argue that "it is more cost effective, logical, and socially relevant to invest in *biodiversity coldspots*" instead). (WK 2012c; W. Turner 2007) And the reverse direction also holds true.

In relation to the functionality-nonfunctionality principle, if there is functionality (e.g., the presence of action for which something is fitted, as shown in the study that "Alaskan Tlingit peoples and the Haida of the Pacific Northwest had resource boundaries, rules, and restrictions among clans with respect to the fishing of Sockeye Salmon," as these rules served the function of resolving "communal resource conflict" at the time), there is nonfunctionality (e.g., the relative lack of action for which something is fitted, as shown in different "international environmental agreements" nowadays, like "the Kyoto Protocol, Vienna Convention on the Protection of the Ozone Layer and Rio Declaration on Development and Environment," so the "resource boundaries, rules, and restrictions" among the Alaskan Tlingit peoples and the Haida of the Pacific Northwest are no longer adequately functional in a globalized world at our time). And the reverse direction also holds true.

And in relation to the intentionality-nonintentionality principle, if there is intentionality (e.g., the planning or design of something for a certain outcome, as shown in the "voluntary international agreements" in developing countries, which "are more commonly used to remedy significant levels of non-compliance with mandatory regulation"), there is nonintentionality (e.g., the relative absence of planning or design of something for a certain outcome, such that the outcome can even be counter-intentional in being the opposite of what was originally planned, as shown in the "voluntary international agreements" in developing countries, where, "due to the difficulties inherent in evaluating effectiveness, their use is often questioned and, indeed, the environment may well be adversely affected as a result" instead). (WK 2012k) And the reverse direction also holds true.

Diversity, Ecology, and Adaptive Management

A good way to evaluate the possibility (or impossibility) and desirability (or undesirability) of diversity from the perspective of nature is to explore diversity, ecology, and adaptive management.

An argument here is that there is a relationship between adaptive management and better decision making to deal with ecological resources in the face of uncertainty.

The term "adaptive management" or "AM" (also known as "adaptive resource management" or "ARM") is "a structured, iterative process of robust decision making in the face of uncertainty, with an aim to reducing uncertainty over time via system monitoring. In this way, decision making simultaneously meets one or more resource management objectives and, either passively or actively, accrues information needed to improve future management." (WK 2012z)

History of Adaptive Management

Historically, "the use of adaptive management techniques can be traced back to peoples from ancient civilizations. For example, the Yap people of Micronesia have been using adaptive management techniques to sustain high population densities in the face of resource scarcity for thousands of years. In using these techniques, the Yap people have altered their environment creating, for example, coastal mangrove depressions and seagrass meadows to support fishing and termite resistant wood." (WK 2012z; Stankey 1997; M. Falanrue 1984)

But "the origin of the adaptive management concept can be traced back to ideas of scientific management pioneered by Frederick Taylor in the early 1900s. While the term 'adaptive management' evolved in natural resource management workshops through decision makers, managers and scientists focusing on building simulation models to uncover key assumptions and uncertainties." (WK 2012z; B. Bormann 1999)

Some suggested that "two ecologists at The University of British Columbia, C. S. Holling (1978) and C. J. Walters (1986) further developed the adaptive management approach as they distinguished between passive and adaptive management practice. Kai Lee, notable Princeton physicist, expanded upon the approach in the late 1970s and early 1980s while pursuing a post-doctorate degree at UC Berkeley. The approach was further developed at the International Institute for Applied Systems Analysis (IIASA) in Vienna, Austria, while C. S. Holling was director of the Institute. In 1992, Hilbourne described three learning models for federal land managers, around which adaptive management approaches could be developed," and "these are reactive, passive and active." (WK 2012z)

In practice, "adaptive management has probably been most frequently applied in Australia and North America, initially applied in fishery management, but received more broad application in the 1990s and 2000s. One of the most successful applications of adaptive management has been in the area of waterfowl harvest management in North America, most notably for the mallard." (WK 2012z; F. Johnson 1993; J. Nichols 2007).

Especially in the area of conservation, "adaptive management…can trace its roots back to at last the early 1990s, with the establishment of the Biodiversity Support Program (BSP) in 1989. BSP was a USAID-funded consortium of WWF, The Nature Conservancy (TNC), and World Resources Institute (WRI). Its Analysis and Adaptive Management Program sought to understand the conditions under which certain conservation strategies were most effective and to identify lessons learned across conservation projects. When BSP ended in 2001, TNC and Foundations of Success (FOS, a non-profit which grew out of BSP) continued to actively work in promoting adaptive management for conservation projects and programs. The approaches used included Conservation by Design (TNC) and Measures of Success (FOS)." (WK 2012z)

And "in 2004, the Conservation Measures Partnership (CMP)—which includes several former BSP members—developed a common set of standards and guidelines for applying adaptive management to conservation projects and programs." (WK 2012z)

Features of Adaptive Management

With this history in mind—"adaptive management can proceed as either passive adaptive management or active adaptive management, depending on how learning takes place. Passive adaptive management values learning only insofar as it improves decision outcomes (i.e. passively), as measured by the specified utility function. In contrast, active adaptive management explicitly incorporates learning as part of the objective function, and hence, decisions which improve learning are valued over those which do not." (WK 2012z; C. Holling 1978; C. Walters 1986)

In both cases, "passive and active adaptive management" share the following main features (as summarized in *Table 2.4*): (WK 2012z)

- "Iterative decision-making (evaluating results and adjusting actions on the basis of what has been learned)"
- "Feedback between monitoring and decisions (learning)"
- "Explicit characterization of system uncertainty through multi-model inference"
- "Bayesian [subjective] inference"
- "Embracing risk and uncertainty as a way of building understanding"

This then means that "adaptive management is particularly applicable for systems in which learning via experimentation is impractical." (WK 2012z)

Processes of Adaptive Management

Because of this tendency to learn in the face of uncertainty—adaptive management borrows "a number of scientific and social processes," and some good examples include the following (as summarized in *Table 2.4*): (WK 2012z)

- "Management is linked to appropriate temporal and spatial scales."
- "Management retains a focus on statistical power and controls."
- "Use of computer models to build synthesis and an embodied ecological consensus."

- "Use of embodied ecological consensus to evaluate strategic alternatives."
- "Communication of alternatives to political arena for negotiation of a selection."

In other words, "the achievement of these objectives requires an open management process which seeks to include past, present and future stakeholders. Adaptive management needs to at least maintain political openness, but usually aims to create it. Adaptive management must therefore be a scientific and social process. It must focus on the development of new institutions and institutional strategies in balance with scientific hypothesis and experimental frameworks." (WK 2012z)

Applications of Adaptive Management

In any event, adaptive management has been applied in different projects over the years. For instance, "some recent write-ups of adaptive management in conservation include: wildlife protection (SWAP, 2008), forests ecosystem protection (CMER, 2010), coastal protection and restoration (LACPR, 2009), natural resource management (water, land and soil), species conservation especially, fish conservation from overfishing (FOS, 2007) and climate change (DFG, 2010)." (WK 2012z)

Some other important examples include the following: (WK 2012z)

- "The Department of Fish and Game (DFG) is developing and implementing adaptation strategies to help protect, restore and manage fish and wildlife, with the understanding that some level of climate change will occur and that it will have profound effects on Massachusetts habitats."
- "The Adaptive Management program was created by CMR to provide science-based recommendations and technical information to assist the Forest Practices Board. In April 2010, the Forest Practices Adaptive Management Annual Science Conference was held in Washington."
- "In 2009, The Louisiana Coastal Protection and Restoration (LACPR) Technical Report has been developed by the United States Army Corps of Engineers (USACE) according to adaptive management process."

Of course, there can be other examples too, so the ones above are illustrative, not exhaustive.

Problems with the Study of Adaptive Management

However, for the critics, this study of adaptive management has some major problems. Consider, for illustration, a few examples below (as summarized in *Table 2.4*).

Firstly, one main criticism of the study of adaptive management is that it is risky enough.

For example, for the critics, the scientific and social processes that adaptive management borrows also have their main drawback, in that, if any of these processes do not work out as planned, then the consequences are not pretty at all, as the failures "can seriously compromise effective adaptive management decision making," in the following ways: (WK 2012z; C. Elzinga 1998; A. Moore 2009; B. Bormann 1999):

- "The monitoring is never completed."
- "The monitoring data are not analyzed and cannot ensure it's definitely accurate."
- "The analyzed results are not conclusive."

Thus, adaptive management is quite risky indeed.

Secondly, another main criticism of the study of adaptive management is that it suffers from "a common lament" about "significant uncertainty". (WK 2012n; J. Arvai 2006)

In other words, for the critics, "the pairing of significant uncertainty about the behaviour and response of ecological systems with urgent calls for near-term action constitutes a difficult reality, and a common lament" for those practitioners of adaptive management," because "scientific analysis of the environment deals with several dimensions of ecological uncertainty," as shown below: (WK 2012n; J. Arvai 2006)

- "Structural Uncertainty"—"resulting from the misidentification, or lack of information pertaining to the relationships between ecological variables"
- "Parameter Uncertainty"—"referring to 'uncertainty associated with parameter values that are not known precisely but can be assessed and reported in terms of the likelihood…of experiencing a defined range of outcomes'"
- "Stochastic Uncertainty"—"stemming from chance or unrelated factors."

Therefore, adaptive management has to constantly deal "with situations of high levels of uncertainty." (WK 2012n; C. Walters 1986; C. Holling 1978; L. Gunderson 2002)

Thirdly, still another main criticism of the study of adaptive management is that adaptive management has not yet been widely accepted in the corporate world.

For instance, a recent study by the Boston Consulting Group showed that environmental management (of ecological systems) "has yet to reach the majority of global companies operating in the markets," and one major reason is that it is so difficult to "model…an economically viable case for the switch," especially in the face of uncertainty (so embedded in adaptive management). (WK 2012z; UN 2010; M. Berns 2009) In addition, "deriving actively adaptive policies is technically very difficult, which prevents it being more commonly applied." (WK 2012z)

And fourthly, an additional main criticism of the study of adaptive management is that scientists still disagree about the features of adaptive management.

For instance, while "two ecologists at The University of British Columbia, C. S Holling (1978) and C. J Walters (1986)…developed the adaptive management approach as they distinguished between passive and adaptive management practice"—others beg to differ, as, "in 1992, [R.] Hilbourne described three learning models for federal land managers, around which adaptive management approaches could be developed," and "these are reactive, passive and active." (WK 2012z)

Beyond Adaptive Management

However, these criticisms of adaptive management should not be misconstrued as a total rejection of its usefulness but serve to show us the two opposing sides of the debate, such that the possibility (or impossibility) and desirability (or undesirability) of diversity (from the perspective of nature with diversity, ecology, and adaptive management as a case study here) are not to the extent that the respective defenders would like us to believe.

In fact, the analysis of diversity, ecology, and adaptive management can teach us a valuable lesson about the ontological principles in existential dialectics, and good examples include the formalness-informalness principle, the absoluteness-relativeness principle, the partiality-totality principle, the predictability-unpredictability principle, the explicability-inexplicability principle, the fiction-reality principle, the cognitiveness-noncognitiveness principle, the finiteness-transfiniteness principle, the preciseness-vagueness principle, the simpleness-complicatedness principle, the openness-hiddenness principle, the denseness-emptiness principle, the change-constancy principle, the order-chaos principle, the slowness-quickness principle, the expansion-contraction principle, the theory-praxis principle, the convention-novelty principle, the evolution-transformation principle, the symmetry-asymmetry principle, the softness-hardness principle, the seriousness-playfulness principle, the activeness-inactivess principle, the selfness-otherness principle, the regression-progression principle, the same-difference principle, the stability-reaction principle, the functionality-nonfunctionality principle, and the intentionality-nonintentionality principle.

For instance, in relation to the formalness-informalness principle, if there is formalness (e.g., the formal logical requirement of soundness and completeness in a system of ideas, as in the study of adaptive management), there is informalness (e.g., the non-formal existence of unsoundness and incompleteness in the study of adaptive management, as shown in the criticisms that it is risky enough, that it suffers from "a common lament" about "significant uncertainty," that adaptive management has not yet been widely accepted in the corporate world, that scientists still disagree about the nature of adaptive management, etc.). And the reverse direction also holds true.

In relation to the absoluteness-relativeness principle, if there is absoluteness (e.g., the absolute view of C. Holling and C. Walters in regard to "passive and adaptive management practice"), there is relativeness (e.g., what is right for C. Holling and C. Walters in regard to "passive and adaptive management practice" is not necessarily so for the critics like R. Hilbourne, who proposed instead "three learning models..., around which adaptive management approaches could be developed," and "these are reactive, passive and active"). And the reverse direction also holds true.

In relation to the partiality-totality principle, if there is partiality (e.g., the partial view of "the Yap people of Micronesia" in regard to their "adaptive management techniques to sustain high population densities in the face of resource scarcity" at the local level), there is totality (e.g., the more holistic view of adaptive management techniques, such that the whole is not the sum of its parts, that is, they cannot be reduced to the sum of the opposing views of the Yap people, modern scientists like C. Holling and C. Walters, etc., since there will emerge new views in the future not yet known today, just as there are already nowadays alternative views like the ones by Hilbourne and the critics). And the reverse direction also holds true.

In relation to the predictability-unpredictability principle, if there is predictability (e.g., the predictable tendency of those practitioners of adaptive management systems to call "for near-term action" with "a structured, iterative process of robust decision making in the face of uncertainty"), there is unpredictability (e.g., the more difficult task to predict exactly the extent to which the practitioners of adaptive management systems will be successful on a particular future occasion, because of the three kinds of "ecological uncertainty," namely, "structural uncertainty," "parameter uncertainty," and "stochastic uncertainty"). And the reverse direction also holds true.

In relation to the explicability-inexplicability principle, if there is explicability (e.g., the explanation by the advocates of adaptive management that it is desirable on the basis of being able to carry out "near-term action" with "a structured, iterative process of robust decision making in the face of uncertainty"), there is inexplicability (e.g., the lack of sufficient explanation by the advocates of adaptive management of why it is necessarily desirable, since, for the critics, adaptive management has not yet been widely accepted in the corporate world, because "it is so difficult to "model…an economically viable case for the switch," and, in addition, "deriving actively adaptive policies is technically very difficult, which prevents it being more commonly applied"). And the reverse direction also holds true.

In relation to the fiction-reality principle, if there is fiction (e.g., the fictional aspect of adaptive management, insofar as, if any of its processes do not work out as planned, the consequences are not pretty at all, as the failures "can seriously compromise effective adaptive management decision making"), there is reality (e.g., the realistic aspect of adaptive management, insofar as it is willing to deal with "decision making" in regard to ecological systems, while acknowledging the existence of high "uncertainty"). And the reverse direction also holds true.

In relation to the cognitiveness-noncognitiveness principle, if there is cognitiveness (e.g., a view of things on the basis of reason and evidences, as shown in the use of adaptive management for dealing with ecological isues in the face of uncertainty), there is noncognitiveness (e.g., a view of things on the basis of non-cognitive factors like envy, jealousy, power, nationality, race, gender, age, class, greed, lust, status, anger, sadness, joy, fear, wish, etc., as shown in the use of adaptive management in environmental management by some companies, which accept the switch, not because of the soundness of the techniques, but because it can promote its business reputation as environmentally friendly, that is, as a "strategic advantage" for its profit-seeking purpose). (WK 2012n; B. Daily 2001; M. Berns 2009) And the reverse direction also holds true.

In relation to the finiteness-transfiniteness principle, if there is finiteness (e.g., the finite number of the features and processes of adpative management, as shown in the 5 features and 5 processes aforecited), there is transfiniteness (e.g., the transfinite number of all the individual genes, microorganisms, species, habitats, ecosystems, etc. that have ever existed in the history of ecology on Earth and which can be studied for adaptive management). And the reverse direction also holds true.

In relation to the preciseness-vagueness principle, if there is preciseness (e.g., the precise identification of 5 features of adaptive management in the article aforecited), there is vagueness (e.g., the vagueness in the identification of the 5 features of adpative management in the article aforecited, since it is not clear why there must be only 5, not 6, 7, 8, and so on). And the reverse direction also holds true.

In relation to the simpleness-complicatedness principle, if there is simpleness (e.g., the relatively simple analysis of adaptive management by its advocates), there is complicatedness (e.g., the relatively more complicated analysis of adaptive management, by challenging the claims and assumptions, as shown in the problems pointed out by the critics as cited). And the reverse direction also holds true.

In relation to the openness-hiddenness principle, if there is openness (e.g., the open exploration of different kinds of adaptive management, as shown by the work of C. Holling and C. Walters on "passive and adaptive management practice"), there is hiddenness (e.g., the hidden bias in C. Holling and C. Walters on "passive and adaptive management practice," since the critics like R. Hilbourne proposed instead "three learning models..., around which adaptive management approaches could be developed," and "these are reactive, passive and active"). And the reverse direction also holds true.

In relation to the denseness-emptiness principle, if there is denseness (e.g., the relatively denser concentration of concerns with decision making about ecological issues in the face of uncertainty among the practitioners of adaptive management), there is emptiness (e.g., the relatively less dense, or more empty, concentration of concerns with decision making about ecological issues in the face of uncertainty among those in the corporate world, because business people, unlike the practitioners of adaptive management, are more interested in models which are "economically viable" and "technically" feasible instead, for their profit-seeking purpose). And the reverse direction also holds true.

In relation to the change-constancy principle, if there is change (e.g., the ever new ways to study ecology, as shown in the work on adaptive management), there is constancy (e.g., the ever constant existence of different problems in the ever new ways to study ecology, as shown in the problems with the work on adaptive management pointed out by the critics). And the reverse direction also holds true.

In relation to the order-chaos principle, if there is order (e.g., the more or less orderly view of reality, if looked from the perspective of those in the corporate world who are more interested in models which are "economically viable" and "technically" feasible), there is chaos (e.g., the more or less chaotic view of reality, if looked from the practitioners of adaptive management who are more interested in models which deal with several dimensions of ecological uncertainty," like "structural uncertainty," "parameter uncertainty," and "stochastic uncertainty"). And the reverse direction also holds true.

In relation to the slowness-quickness principle, if there is slowness (e.g., the relatively slower speed of the techniques of adaptive management to be accepted by "the majority of global companies operating in the markets"), there is quickness (e.g., the relatively quicker speed of the techniques of adaptive management to be accepted by the Nature Conservancy, the World Resources Institute, etc.). And the reverse direction also holds true.

In relation to the expansion-contraction principle, if there is expansion (e.g., the relative expansion of "the demand for lateral decision making that supports affective participation" in environmental management compatible with adaptive management nowadays), there is contraction (e.g., the relative decline of the "traditional…hierarchal decision making" in environmental management compatible with adaptive management nowadays). (WK 2012n; S. Selin 1995) And the reverse direction also holds true.

In relation to the theory-praxis principle, if there is theory (e.g., the theoretical construction of the features and processes of adaptive management), there is praxis (e.g., the practical application of the features and processes of adaptive management to the real world, like its application by such organizations as DFG, CMR, LACPR, and so on, as cited earlier). And the reverse direction also holds true.

In relation to the convention-novelty principle, if there is convention (e.g., the conventional wisdom about the two types of adaptive management by C. Holling and C. Walters, namely, "passive and adaptive management practice"), there is novelty (e.g., the alternative novel challenge to the conventional wisdom about the two types of adaptive management, by the new idea of R. Hilbourne, who proposed instead "three learning models," and "these are reactive, passive and active"). And the reverse direction also holds true.

In relation to the evolution-transformation principle, if there is evolution (e.g., the natural evolution of humans to use natural resources in the state of nature in a primitive way and on a relatively much smaller scale), there is transformation (e.g., the technical transformation of human ability to manage natural resources by the invention of the techniques of adaptive management nowadays). And the reverse direction also holds true.

In relation to the symmetry-asymmetry principle, if there is symmetry (e.g., the existence of different models of decision making), there is asymmetry (e.g., the popularity of the decision making modeling with the features and processes of adaptive management among those in such organizations as DFG, CMR, LACPR, and so on—but the popularity of the decision making modeling which must be "economically viable" and "technically" feasible among "the majority of global companies operating in the markets"). And the reverse direction also holds true.

In relation to the softness-hardness principle, if there is softness (e.g., the appreciation by Kai Lee of the classification of adaptive management into two types by Holling and Walters), there is hardness (e.g., the critique by Hilbourne of the classification of adaptive management into two types by Holling and Walters, as Hilbourne proposed three types instead). And the reverse direction also holds true.

In relation to the seriousness-playfulness principle, if there is seriousness (e.g., the serious business of those who try to make adaptive management successful), there is playfulness (e.g., the playful part of those who try to make adaptive management successful, as they play around with different possibilities in regard to its features and processes over time, in order to make them better). And the reverse direction also holds true.

In relation to the activeness-inactivess principle, if there is activeness (e.g., the relative activeness of those ordinary employees in "lateral decision making," where they are encouraged to engage in "affective participation"), there is inactiveness (e.g., the relative inactiveness of those ordinary employees in "traditional…hierarchal decision making," where they are to follow what is decided by those at the top instead). And the reverse direction also holds true.

In relation to the selfness-otherness principle, if there is selfness (e.g., the tendency to be mostly centered on one's self or group, as shown by "the majority of global companies operating in the markets," as they are reluctant to adopt adaptive management, because it is not "economically viable" and "technically" feasible, as their major concern is profit-making), there is otherness (e.g., the tendency to be mostly centered on others, as shown in those non-profit organizations like FOS and others, which "continued to actively work in promoting adaptive management for conservation projects and program," such that humanity can benefit from a more sustainable way to use natural resources). And the reverse direction also holds true.

In relation to the regression-progression principle, if there is regression (e.g., the regression made by adaptive management, as shown in the problems pointed out by the critics aforecited), there is progression (e.g., the progress made by adaptive management, as shown in its contribution to a better way to make decision in the face of uncertainty, while addressing ecological concerns aforecited). And the reverse direction also holds true.

In relation to the same-difference principle, if there is similarity in outcome (e.g., the contribution to the molding and control of beliefs, values, and behaviors, regardless of whether this be done by way of "lateral decision making" or by way of "hierarchal decision making"), there is difference in outcome (e.g., the contribution to the molding and control of beliefs, values, and behaviors by way of "lateral decision making" for a relatively more egalitarian lifeworld—but the contribution to the molding and control of beliefs, values, and behaviors by way of "hierarchal decision making" for a relatively more inegalitarian lifeworld). And the reverse direction also holds true.

In relation to the stability-reaction principle, if there is stability (e.g., the relatively more stable condition of the "hierarchal decision making" in the older days), there is reaction (e.g., the "hierarchal decision making" in the older days then led to further changes later on, when nowadays, there are alternative models like "lateral decision making" in adaptive management for learning). And the reverse direction also holds true.

In relation to the functionality-nonfunctionality principle, if there is functionality (e.g., the presence of action for which something is fitted, as shown in the "hierarchal decision making" in the older days, when it served the function of the inegalitarian social world at the time), there is nonfunctionality (e.g., the relative lack of action for which something is fitted, as shown in the "lateral decision making" in adaptive management nowadays, so that the "hierarchal decision making" in the older days is no longer functional in our relatively more democratic time). And the reverse direction also holds true.

And in relation to the intentionality-nonintentionality principle, if there is intentionality (e.g., the planning or design of something for a certain outcome, as shown in the processes of adaptive management, which are designed to deal with decision making in the face of uncertainty), there is nonintentionality (e.g., the relative absence of planning or design of something for a certain outcome, such that the outcome can even be counter-intentional in being the opposite of what was originally planned, as shown in the processes of adaptive management, which can backfire at times, so the failures "can seriously compromise effective adaptive management decision making" instead). And the reverse direction also holds true.

Diversity and Society

Diversity, when examined from the perspective of society, can reveal some powerful features in relation to its possibility (or impossibility) and desirability (or undesirability), and this can be shown by way of two case studies, namely, (2.4.1) diversity, cloning, and the role of technology, and (2.4.2) diversity, natural capital, and ecological economics—to be addressed hereafter.

Diversity, Cloning, and the Role of Technology

A good way to evaluate the possibility (or impossibility) and desirability (or undesirability) of diversity from the perspective of society is to explore diversity, cloning, and the role of technology.

An argument here is that there is a relationship between the restoration of life for biodiversity and the technology of cloning. (WK 2012e)

Extinction and the Restoration of Life

To restore the life of an extinct species for biodiversity is to assume that it is extinct and needs to be restored.

By definition, "a species becomes extinct when the last existing member dies. Extinction therefore becomes a certainty when there are no surviving individuals that are able to reproduce and create a new generation. A species may become functionally extinct when only a handful of individuals survive, which cannot reproduce due to poor health, age, sparse distribution over a large range, a lack of individuals of both sexes (in sexually reproducing species), or other reasons." (WK 2012e)

Thus, "pinpointing the extinction…of a species requires a clear definition of that species. If it is to be declared extinct, the species in question must be uniquely identifiable from any ancestor or daughter species, or from other closely related species. Extinction of a species (or replacement by a daughter species) plays a key role in the punctuated equilibrium hypothesis of Stephen Jay Gould and Niles Eldredge [1985]." (WK 2012e) Gould and Eldredge proposed that "the degree of gradualism commonly attributed to Charles Darwin is virtually nonexistent in the fossil record," and evolution can occur abruptly and rapidly, as shown by the "rare and geologically rapid events of branching speciation called cladogenesis." (WK 2012x)

But "extinction" should not be confused with "local extinction," since, "in ecology, extinction is often used informally to refer to local extinction, in which a species ceases to exist in the chosen area of study, but still exists elsewhere. This phenomenon is also known as extirpation. Local extinctions may be followed by a replacement of the species taken from other locations; wolf reintroduction is an example of this. Species which are not extinct are termed extant. Those that are extant but threatened by extinction are referred to as threatened or endangered species." (WK 2012e)

In our time, "an important aspect of extinction…is human attempts to preserve critically endangered species, which is reflected by the creation of the conservation status 'Extinct in the Wild' (EW). Species listed under this status by the International Union for Conservation of Nature (IUCN) are not known to have any living specimens in the wild, and are maintained only in zoos or other artificial environments. Some of these species are functionally extinct, as they are no longer part of their natural habitat and it is unlikely the species will ever be restored to the wild. When possible, modern zoological institutions attempt to maintain a viable population for species preservation and possible future reintroduction to the wild through use of carefully planned breeding programs." (WK 2012e; P. Maas 2007)

And it is important to note that "the extinction of one species' wild population can have knock-on effects, causing further extinctions. These are also called 'chains of extinction.' This is especially common with extinction of keystone species." (WK 2012e; C. Quince 2006)

Yet, with the technology of cloning, it is possible to restore the life of an extinct species for biodiversity.

Types of Cloning

In biology, the term "cloning" was originally used to refer to "the process of producing similar populations of genetically identical individuals that occurs in nature when organisms such as bacteria, insects or plants reproduce asexually"—but, nowadays, "in biotechnology," cloning "refers to processes used to create copies of DNA fragments (molecular cloning), cells (cell cloning), or organisms." (WK 2012m)

In other words, there are 3 types of cloning in biotechnology, to be discussed below (and summarized in *Table 2.5*).

(a) Molecular Cloning

In molecular cloning, the task here is to make "multiple molecules. Cloning is commonly used to amplify DNA fragments containing whole genes, but it can also be used to amplify any DNA sequence such as promoters, non-coding sequences and randomly fragmented DNA. It is used in a wide array of biological experiments and practical applications ranging from genetic fingerprinting to large scale protein production." (WK 2012m)

Generally, "cloning of any DNA fragment essentially involves four steps," as shown below: (WK 2012m)

- "Fragmentation"—"breaking apart a strand of DNA"
- "Ligation"— "gluing together pieces of DNA in a desired sequence"
- "Transfection"—"inserting the newly formed pieces of DNA into cells"
- "Screening/Selection"—"selecting out the cells that were successfully transfected with the new DNA"

Of course, "although these steps are invariable among cloning procedures, a number of alternative routes can be selected," so the above ones are illustrative. (WK 2012m; P. Russel 2005)

(b) Cell Cloning

In cell cloning, there are two sub-types, namely, (i) cloning of unicellular organisms and (ii) research cloning (or cloning in a stem-cell research).

(i) In regard to cloning of "unicellular organisms"—the task here is "to derive a population of cells from a single cell. In the case of unicellular organisms such as bacteria and yeast, this process is remarkably simple and essentially only requires the inoculation of the appropriate medium. However, in the case of cell cultures from multi-cellular organisms, cell cloning is an arduous task as these cells will not readily grow in standard media." (WK 2012m)

(ii) And in regard to "research cloning" (also known as "therapeutic cloning" or "cloning in a stem cell research")—the task here is "to create embryos for research or therapeutic purposes. The most likely purpose for this is to produce embryos for use in stem cell research....The goal is not to create cloned human beings (called 'reproductive cloning'), but rather to harvest stem cells that can be used to study human development and to potentially treat disease." (WK 2012m)

In general, "therapeutic cloning is achieved by creating embryonic stem cells in the hopes of treating diseases such as diabetes and Alzheimer's. The process begins by taking out the nucleus that contains the DNA from an egg and putting it in a nucleus from an adult. In the case of someone with Alzheimer's disease, the nucleus from a skin cell of that patient is placed into an empty egg. The reprogrammed cell begins to develop into an embryo because the egg reacts with the transferred nucleus. The embryo will become genetically identical to the patient. The embryo will then form a blastocyst which has the potential to form/become any cell in the body." (WK 2012m; N. Halim 2002; MED 2011)

(c) Organism Cloning

And in "organism cloning" (also known as "reproductive cloning"), the task here is to create "a new multicellular organism, genetically identical to another. In essence this form of cloning is an asexual method of reproduction, where fertilization or inter-gamete contact does not take place. Asexual reproduction is a naturally occurring phenomenon in many species, including most plants...and some insects. Scientists have made some major achievements with cloning, including the asexual reproduction of sheep and cows....Cloning, or asexual propagation, has been common practice in the horticultural world for hundreds of years." (WK 2012m; AGG 2012)

There are three types of organism cloning, namely, (i) horticultural cloning, (ii) parthenogenesis, and (iii) artificial cloning of organisms. (WK 2012m)

(i) In regard to horticultural cloning—a clone belongs "descendants of a single plant which were produced by vegetative reproduction or apomixis. Many horticultural plant cultivars are clones, having been derived from a single individual, multiplied by some process other than sexual reproduction. As an example, some European cultivars of grapes represent clones that have been propagated for over two millennia....Many trees, shrubs, vines, ferns and other herbaceous perennials form clonal colonies." (WK 2012m)

(ii) In regard to parthenogenesis—"clonal derivation exists in nature in some animal species and is referred to as…reproduction of an organism by itself without a mate….This is an asexual form of reproduction that is only found in females of some insects, crustaceans and lizards. The growth and development occurs without fertilization by a male. In plants, parthenogenesis means the development of an embryo from an unfertilized egg cell, and is a component process of apomixis. In species that use the XY sex-determination system, the offspring will always be female. An example is the 'Little Fire Ant' (*Wasmannia auropunctata*)." (WK 2012m)

(iii) In regard to the artificial cloning of organisms—the process here is known as "somatic cell nuclear transfer" (SCNT), which is "to create animals that are genetically identical. This process entails the transfer of a nucleus from a donor adult cell (somatic cell) to an egg that has no nucleus. If the egg begins to divide normally it is transferred into the uterus of the surrogate mother." (WK 2012m)

In addition, "artificial embryo splitting or embryo twinning may also be used as a method of cloning, where an embryo is split in the maturation before embryo transfer. It is optimally performed at the 6- to 8-cell stage, where it can be used as an expansion of IVF to increase the number of available embryos. If both embryos are successful, it gives rise to monozygotic (identical) twins." (WK 2012m; K. Illmensee 2009)

Well-known examples of the artificial cloning of organisms include "carp" in China in 1963, "Dolly the sheep" in Scotland in 1996, "water buffalo" in the Philippines in 2007, and so on. (WK 2012m; *PBS* 2003)

Cloning of Extinct and Endangered Species

But a controversial case of cloning has to do with the cloning of extinct and endangered species, which involves "the reconstruction of functional DNA from extinct species" and "has, for decades, been a dream of some scientists." (WK 2012m)

After all, this hope was "dramatized in the best-selling novel by Michael Crichton and high budget Hollywood thriller *Jurassic Park*. In real life, one of the most anticipated targets for cloning was once the Woolly Mammoth, but attempts to extract DNA from frozen mammoths have been unsuccessful, though a joint Russo-Japanese team is currently working toward this goal. And in January 2011, it was reported by Yomiuri Shimbun that a team of scientists headed by Akira Iritani of Kyoto University had built upon research by Dr. Wakayama, saying that they will extract DNA from a mammoth carcass that had been preserved in a Russian laboratory and insert it into the egg cells of an African elephant in hopes of producing a mammoth embryo. The researchers said they hoped to produce a baby mammoth within six years." (WK 2012m; BBC 2003)

Other cases are a bit closer to reality,; for example, "in 2001, a cow named Bessie gave birth to a cloned Asian gaur, an endangered species, but the calf died after two days. In 2003, a banteng was successfully cloned, followed by three African wildcats from a thawed frozen embryo. These successes provided hope that similar techniques (using surrogate mothers of another species) might be used to clone extinct species. Anticipating this possibility, tissue samples from the last *bucardo* (Pyrenean Ibex) were frozen in liquid nitrogen immediately after it died in 2000. Researchers are also considering cloning endangered species such as the giant panda and cheetah. The 'Frozen Zoo' at the San Diego Zoo now stores frozen tissue from the world's rarest and most endangered species." (WK 2012m; H. Perlman 2000; G. Pence 2005)

More interesting, "the first hybrid human clone was created in November 1998, by Advanced Cell Technologies. It was created from a man's leg cell, and a cow's egg whose DNA was removed. It was destroyed after 12 days. Since a normal embryo implants at 14 days, Dr Robert Lanza, ACT's director of tissue engineering, told the Daily Mail newspaper that the embryo could not be seen as a person before 14 days," but "making an embryo…may have resulted in a complete human had it been allowed to come to term, according to ACT." (WK 2012m; BBC 1999)

Problems with the Technology of Cloning

However, for the critics, the study of the restoration of life and the technology of cloning has some major problems. Consider, for illustration, a few examples below (as summarized in *Table 2.5*).

Firstly, one main criticism of this study of the restoration of life and the technology of cloning is that it is still technological infeasible in our time.

For instance, for the critics, "in 2003, scientists attempted to clone the extinct Pyrenean Ibex (*C. p. pyrenaica*). This initial attempt failed; of the 285 embryos reconstructed, 54 were transferred to 12 mountain goats and mountain goat-domestic goat hybrids, but only two survived the initial two months of gestation before they too died. In 2009, a second attempt was made to clone the Pyrenean Ibex; one clone was born alive, but died seven minutes later, due to physical defects in the lungs." (WK 2012e; S. Connor 2009; R. Gray 2009) And worse, "researchers have found several abnormalities in cloned organisms....The cloned organism may be born normal and resemble its non-cloned counterpart, but majority of the time will express changes in its genome later on in life." (WK 2012m; HEA 2001)

Secondly, another main criticism of this study of the restoration of life and the technology of cloning is that there remain some deep ethical issues

For instance, for the critics, there are deep "bioethical and philosophical objections" to "the cloning of extinct creatures," which have yet to be satisfactorily answered. (WK 2012e; A. Zitner 2000) For illustration, "many religious organizations oppose all forms of cloning, on the grounds that life begins at conception....From the standpoint of classical liberalism, concerns also exist regarding the protection of the identity of the individual and the right to protect one's genetic identity." (WK 2012m; M. Farley 2001)

Thirdly, still another main criticism of the study of the restoration of life and the technology of cloning is that what counts as extinct species may turn out to be pseudoextinct ones, so the restoration of life through cloning may be misleading.

For instance, for the critics, when "daughter species that evolve from a parent species carry on most of the parent species' genetic information, and even though the parent species may become extinct, the daughter species lives on....Extinction of a parent species where daughter species or subspecies are still alive is also called *pseudoextinction*." (WK 2012e)

Fourthly, an additional main criticism of the study of the restoration of life and the technology of cloning is that cloning does not restore an exactly identical life.

For instance, for the critics, in the artificial reproduction of organisms, "such clones are not strictly identical since the somatic cells may contain mutations in their nuclear DNA. Additionally, the mitochondria in the cytoplasm also contains DNA and during SCNT this mitochondrial DNA is wholly from the cytoplasmic donor's egg, thus the mitochondrial genome is not the same as that of the nucleus donor cell from which it was produced. This may have important implications for cross-species nuclear transfer in which nuclear-mitochondrial incompatibilities may lead to death." (WK 2012m)

Fifthly, another additional main criticism of the study of the restoration of life and the technology of cloning is that the more difficult question of how to raise the restored species (in the absence of parents) remains unresolved, so it may not be viable in the long run.

For instance, for the critics, "even if males and females were to be cloned, the question would remain open whether they would be viable at all in the absence of parents that could teach or show them their natural behavior." (WK 2012m)

And sixthly, still another additional main criticism of the study of the restoration of life and the technology of cloning is that cloning distracts the attention and takes away funding from solving the serious issue of animal extinction.

For instance, "in a 2006 review, David Ehrenfeld concluded that cloning in animal conservation is an experimental technology that, at its state in 2006, could not be expected to work except by pure chance and utterly failed a cost-benefit analysis. Furthermore, he said, it is likely to siphon funds from established and working projects and does not address any of the issues underlying animal extinction (such as habitat destruction, hunting or other overexploitation, and an impoverished gene pool)." (WK 2012m)

Beyond the Technology of Cloning

These criticisms of the restoration of life and the technology of cloning, however, do not imply a total rejection of its usefulness but serve to show us the two opposing sides of the debate, such that the possibility (or impossibility) and desirability (or undesirability) of diversity (from the perspective of society with diversity, cloning, and the role of technology as a case study here) are not to the extent that the respective defenders would like us to believe.

Besides, the analysis of diversity, cloning, and the role of technology reveals important insights about the ontological principles in existential dialectics, and good examples include the formalness-informalness principle, the absoluteness-relativeness principle, the partiality-totality principle, the predictability-unpredictability principle, the explicability-inexplicability principle, the fiction-reality principle, the cognitiveness-noncognitiveness principle, the finiteness-transfiniteness principle, the preciseness-vagueness principle, the simpleness-complicatedness principle, the openness-hiddenness principle, the denseness-emptiness principle, the change-constancy principle, the order-chaos principle, the slowness-quickness principle, the expansion-contraction principle, the theory-praxis principle, the convention-novelty principle, the evolution-transformation principle, the symmetry-asymmetry principle, the softness-hardness principle, the seriousness-playfulness principle, the activeness-inactivess principle, the selfness-otherness principle, the regression-progression principle, the same-difference principle, the stability-reaction principle, the functionality-nonfunctionality principle, and the intentionality-nonintentionality principle.

For instance, in relation to the formalness-informalness principle, if there is formalness (e.g., the formal logical requirement of soundness and completeness in a system of ideas, as in the study on the restoration of life and the technology of cloning), there is informalness (e.g., the non-formal existence of unsoundness and incompleteness in the study on the restoration of life and the technology of cloning, as shown in the criticisms that it is still technological infeasible in our time, that there remain some deep ethical issues to be resolved, that what counts as extinct species may turn out to be pseudoextinct ones so the restoration of life through cloning may be misleading, that cloning does not restore an exactly identical life, that the more difficult question of how to raise the restored species remains unresolved so it may not be viable in the long run, that cloning distracts the attention and takes away funding from solving the serious issue of animal extinction, etc.). And the reverse direction also holds true.

In relation to the absoluteness-relativeness principle, if there is absoluteness (e.g., the absolute view of the advocates about cloning for biodivesity), there is relativeness (e.g., what is right for the advocates of cloning for biodivesity is not necessarily so for the critics, since there are deep "bioethical and philosophical objections" to "the cloning of extinct creatures," which have yet to be satisfactorily answered, as indicated earlier). And the reverse direction also holds true.

In relation to the partiality-totality principle, if there is partiality (e.g., the partial view of cloning, if looked from the vantage point of cloning an individual animal in question, like the cloning of the "a cow named Bessie" in 2001), there is totality (e.g., the more holistic view of cloning, such that the whole is not the sum of its parts, that is, it cannot be reduced to the sum of different cases of individual cloning, since, as the critics pointed out, "for such a program to succeed, a sufficient number of individuals would have to be cloned, from the DNA of different individuals [in the case of sexually reproducing organisms] to create a viable population," which also means that there will emerge new problems and viewpoints in the future not yet known today). And the reverse direction also holds true.

In relation to the predictability-unpredictability principle, if there is predictability (e.g., the predictable tendency of some scientists to continue experimenting with cloning in our time), there is unpredictability (e.g., the more difficult task to predict exactly the extent to which cloning will be successful on a particular future occasion, since, for the critics, there are already quite a number of failed attempts in the past, as David Ehrenfeld once warned that "cloning in animal conservation is an experimental technology that, at its state in 2006, could not be expected to work except by pure chance"). And the reverse direction also holds true.

In relation to the explicability-inexplicability principle, if there is explicability (e.g., the explanation by the advocates of cloning that it is desirable on the basis of biodiversity), there is inexplicability (e.g., the lack of sufficient explanation by the advocates of cloning of why it is necessarily desirable, because, for the critics, there remain some deep ethical and philosophical issues to be resolved). And the reverse direction also holds true.

In relation to the fiction-reality principle, if there is fiction (e.g., the fictional aspect of cloning, insofar as there were already various failed attempts in the past, especially in regard to the artificial cloning of organisms, including human cloning), there is reality (e.g., the realistic aspect of cloning, insofar as there are successful cases of cloning, especially in regard to horticultural cloning). And the reverse direction also holds true.

In relation to the cognitiveness-noncognitiveness principle, if there is cognitiveness (e.g., a view of things on the basis of reason and evidences, as shown in some well known cases of cloning like the ones about "carp" in China in 1963, "Dolly the sheep" in Scotland in 1996, "water buffalo" in the Philippines in 2007, and so on), there is noncognitiveness (e.g., a view of things on the basis of non-cognitive factors like envy, jealousy, power, nationality, race, gender, age, class, greed, lust, status, anger, sadness, joy, fear, wish, etc., as shown in the fraudulent 2005 research on "the world's first cloned dog" by South Korea's top human cloning scientist Hwang Woo-suk, which was later found to be "fabricated" instead). (WK 2012m; BBC 2005) And the reverse direction also holds true.

In relation to the finiteness-transfiniteness principle, if there is finiteness (e.g., the finite number of the categories of cloning, like "molecular cloning," "cell cloning," and "organism cloning"), there is transfiniteness (e.g., the transfinite number of all the individual molecules, cells, and organisms that have ever been cloned in history, both naturally and artifically). And the reverse direction also holds true.

In relation to the preciseness-vagueness principle, if there is preciseness (e.g., the precise identification of 4 steps in molecular cloning in the article aforecited, like "fragmentation," "ligation," "transfection" and "screening/selection"), there is vagueness (e.g., the vagueness in the identification of the 4 steps in molecular cloning in the article aforecited, since it is not clear why there must be only 4, not 5, 6, and so on, since, for the critics, "a number of alternative routes can be selected"). (WK 2012m) And the reverse direction also holds true.

In relation to the simpleness-complicatedness principle, if there is simpleness (e.g., the relatively simple analysis of cloning by the advocates), there is complicatedness (e.g., the relatively more complicated analysis of cloning, by challenging the claims and assumptions, as shown in the problems pointed out by the critics as cited). And the reverse direction also holds true.

In relation to the openness-hiddenness principle, if there is openness (e.g., the open exploration of the artificial cloning of organisms), there is hiddenness (e.g., the hidden bias in the artificial cloning of organisms, since "researchers have found several abnormalities in cloned organisms," because "the cloned organism may be born normal and resemble its non-cloned counterpart, but majority of the time will express changes in its genome later on in life"). (WK 2012m; R. Jaenisch 2001) And the reverse direction also holds true.

In relation to the denseness-emptiness principle, if there is denseness (e.g., the relatively denser concentration of successful cases of cloning, as shown in the finding that the "European cultivars of grapes represent clones that have been propagated for over two millennia"), there is emptiness (e.g., the relatively less dense, or more empty, concentration of successful cases of human cloning, as of 2012). And the reverse direction also holds true.

In relation to the change-constancy principle, if there is change (e.g., the ever new ways to deal with biodiversity, as shown in the case of cloning with the use of technology), there is constancy (e.g., the ever constant existence of different problems in the ever new ways to deal with biodiversity, as shown in the problems with the technology of cloning pointed out by the critics aforecited). And the reverse direction also holds true.

In relation to the order-chaos principle, if there is order (e.g., the more or less orderly view of cloning, if looked from the vantage point of its proponents who promise successful cloning for biodiversity), there is chaos (e.g., the more or less chaotic view of cloning, if looked from the vantage point of its critics like David Ehrenfeld, who "concluded that cloning in animal conservation is an experimental technology that, at its state in 2006, could not be expected to work except by pure chance"). And the reverse direction also holds true.

In relation to the slowness-quickness principle, if there is slowness (e.g., the relatively slower speed of human cloning to be achieved at all in reality nowadays, although it was successful in science fiction works like

Aldous Huxley's *Brave New World* in 1932), there is quickness (e.g., the relatively quicker speed of horticultural cloing to be achieved, as shown in the case that the "European cultivars of grapes represent clones that have been propagated for over two millennia"). (WK 2012m) And the reverse direction also holds true.

In relation to the expansion-contraction principle, if there is expansion (e.g., the increase of different cases of cloning for biodiversity nowadays), there is contraction (e.g., the decrease of some species nowadays, because they become extinct or endangered). And the reverse direction also holds true.

In relation to the theory-praxis principle, if there is theory (e.g., the theoretical understanding of the process of cloning in biology), there is praxis (e.g., the practical application of the understanding of the process of cloning in biology to the field of food science, as shown in 2006, when "the U.S. Food and Drug Administration [FDA] approved the consumption of meat and other products from cloned animals"). (WK 2012m; MSNBC 2006) And the reverse direction also holds true.

In relation to the convention-novelty principle, if there is convention (e.g., the conventional wisdom about the process of "somatic cell nuclear transfer" [SCNT]), there is novelty (e.g., the alternative novel challenge to the conventional wisdom about the process of "somatic cell nuclear transfer" [SCNT], by the new idea of "artificial embryo splitting or embryo twinning," which "may also be used as a method of cloning"). And the reverse direction also holds true.

In relation to the evolution-transformation principle, if there is evolution (e.g., the natural evolution of humans and animals to die in the state of nature), there is transformation (e.g., the technical transformation of human ability to restore life by the invention of cloning, although there are still a lot of works needed to be done). And the reverse direction also holds true.

In relation to the symmetry-asymmetry principle, if there is symmetry (e.g., the existence of different lifeforms on Earth, like plants and humans), there is asymmetry (e.g., plants use "vegetative reproduction or apomixes" naturally—but humans use "somatic cell nuclear transfer" [SCNT] artificially). And the reverse direction also holds true.

In relation to the softness-hardness principle, if there is softness (e.g., species can be reproduced in nature, as in "apomixes"), there is hardness (e.g., species can be eliminated in nature, as in "extinction"). And the reverse direction also holds true.

In relation to the seriousness-playfulness principle, if there is seriousness (e.g., the serious business of those who try to understand the nature of cloning), there is playfulness (e.g., the playful part of those who try to understand the nature of cloning, when they play around with different ways of molecular cloning over the years, like the traditional "four steps," "a number of alternative routes," and so on). And the reverse direction also holds true.

In relation to the activeness-inactivess principle, if there is activeness (e.g., the activeness of those clones that survive birth, like Dolly the sheep which lived from 1996 to 2006), there is inactiveness (e.g., the relative inactiveness of those clones that did not quite survive birth and died shortly after, like Bessie the cow, which died two days after birth in 2001). (WK 2012m) And the reverse direction also holds true.

In relation to the selfness-otherness principle, if there is selfness (e.g., the tendency to be mostly centered on one's self or group, as shown in the greed of South Korea's top human cloning scientist Hwang Woo-suk, when his 2005 research on "the world's first cloned dog" was later found to be "fabricated"), there is otherness (e.g., the tendency to be mostly centered on others, as shown in the proven cases of cloning like Dolly the sheep, which aim to provide "a viable outcome of the continuing advancements in our science and technology" for the human future). (WK 2012e; N. Wade 2008) And the reverse direction also holds true.

In relation to the regression-progression principle, if there is regression (e.g., the regression made by cloning, as shown in the problems pointed out by the critics), there is progression (e.g., the progress made by cloning, as shown in its contribution to the restoration of life for biodiversity aforecited). And the reverse direction also holds true.

In relation to the same-difference principle, if there is similarity in outcome (e.g., the contribution to the molding and control of beliefs, values, and behaviors, regardless of whether this be done by way of the restoration of life by cloning, or by way of the beginning of life only at conception in the Catholic tradition), there is difference in outcome (e.g., the contribution to the molding and control of beliefs, values, and behaviors by way of the restoration of life by cloning for a relatively more technophilic lifeworld—but the contribution to the molding and control of beliefs, values, and behaviors by way of the beginning of life only at conception in the Catholic tradition for a relatively more religious lifeworld). And the reverse direction also holds true.

In relation to the stability-reaction principle, if there is stability (e.g., the relatively more stable condition of the idea about the beginning of life only at conception in the Catholic tradition in the older days), there is reaction (e.g., the idea about the beginning of life only at conception in the Catholic tradition in the older days then led to further changes later on, when many scientists of cloning nowadays work on the restoration of life through the use of biotechnology). And the reverse direction also holds true.

In relation to the functionality-nonfunctionality principle, if there is functionality (e.g., the presence of action for which something is fitted, as shown in the idea about the beginning of life only at conception in the Catholic tradition in the older days, which served the function of religious beliefs at the time), there is nonfunctionality (e.g., the relative lack of action for which something is fitted, as shown in the modern view of using biotechnology to restore life through cloning, so that the idea about the beginning of life only at conception in the Catholic tradition in the older days can no longer serve the function of dealing with biodiveristy in our time, especially in light of extinct and endangered species, among other life forms). And the reverse direction also holds true.

And in relation to the intentionality-nonintentionality principle, if there is intentionality (e.g., the planning or design of something for a certain outcome, as shown in the advocation of cloning to resolve the threat to biodiversity), there is nonintentionality (e.g., the relative absence of planning or design of something for a certain outcome, such that the outcome can even be counter-intentional in being the opposite of what was originally planned, as shown in the warning in 2006 by David Ehrenfeld who showed that "cloning…is likely to siphon funds from established and working projects and does not address any of the issues underlying animal extinction [such as habitat destruction, hunting or other overexploitation, and an impoverished gene pool]," contrary to the original goal). And the reverse direction also holds true.

Diversity, Natural Capital, and Ecological Economics

Another good way to evaluate the possibility (or impossibility) and desirability (or undesirability) of diversity from the perspective of society is to explore diversity, natural capital, and ecological economics.

An argument here in ecological economics is that there is a relationship between the investment in natural capital and the enhancement of life quality.

And the term "ecological economics" refers to "a transdisciplinary field of academic research that aims to address the interdependence and coevolution of human economies and natural ecosystems over time and space. It is distinguished from environmental economics, which is the mainstream economic analysis of the environment, by its treatment of the economy as a subsystem of the ecosystem and its emphasis upon preserving natural capital." (WK 2012y; A. Xepapadeas 2008; J. van den Bergh 2001)

History of Ecological Economics

Historically, "early interest in ecology and economics dates back to the 1960s and the work by Kenneth Boulding and Herman Daly, but the first meetings occurred in the 1980s. It began with a 1982 symposium in Sweden which was attended by people who would later be instrumental in the field, including Robert Costanza, Herman Daly, Charles Hall, Ann-Mari Jansson, Bruce Hannon, H. T. Odum, and David Pimentel." (WK 2012y)

Many of them "were ecosystem ecologists or mainstream environmental economists, with the exception of Daly. In 1987, Daly and Costanza edited an issue of *Ecological Modeling* to test the waters. A book entitled *Ecological Economics*, by Juan Martinez-Alier, was published later that year. 1989 saw the foundation of the International Society for Ecological Economics and publication of its journal, *Ecological Economics*, by Elsevier. Robert Costanza was the first president of the society and first editor of the journal, currently edited by Richard Howarth." (WK 2012y; R. Costanza 2003)

Some earlier works "can be traced back to the Romantics of the 19th century as well as some Enlightenment political economists of that era. Concerns over population were expressed by Thomas Malthus, while John Stuart Mill hypothesized that the 'stationary state' of an economy was desirable, anticipating later insights of modern ecological economists, without having had their experience of the social and ecological costs of the dramatic post-World War II industrial expansion. As Martinez-Alier explores in his book the debate on energy in economic systems can also be traced into the 19th century e.g. Nobel prize-winning chemist, Frederick Soddy (1877–1956). Soddy criticized the prevailing belief of the economy as a perpetual motion machine, capable of generating infinite wealth—a criticism echoed by his intellectual heirs in the now emergent field of ecological economics." (WK 2012y)

Later on, "the Romanian economist Nicholas Georgescu-Roegen (1906–1994), who was among Daly's teachers at Vanderbilt University, provided ecological economics with a modern conceptual framework based on the material and energy flows of economic production and consumption....Articles by Inge Ropke (2004 & 2005) and Clive Spash (1999) cover the development and modern history of ecological economics and explain its differentiation from resource and environmental economics, as well as some of the controversy between American and European schools of thought." (WK 2012y)

Nature in Ecological Economics

In ecological economics, nature is not only acknowledged but also emphasized.

For instance, "a simple circular flow of income diagram is replaced in ecological economics by a more complex flow diagram reflecting the input of solar energy, which sustains natural inputs and environmental services which are then used as units of production. Once consumed, natural inputs pass out of the economy as pollution and waste. The potential of an environment to provide services and materials is referred to as an 'environment's source function,' and this function is depleted as resources are consumed or pollution contaminates the resources. The 'sink function' describes an environment's ability to absorb and render harmless waste and pollution: when waste output exceeds the limit of the sink function, long-term damage occurs. Some persistent pollutants, such as some organic pollutants and nuclear waste are absorbed very slowly or not at all; ecological economists emphasize minimizing 'cumulative pollutants.' Pollutants affect human health and the health of the climate." (WK 2012y; J. Harris 2006)

A related central issue concerns "the Earth's carrying capacity," as "early economists such as Thomas Malthus pointed out the finite carrying capacity of the earth, which was also central to the MIT study *Limits to Growth*. Diminishing returns suggest that productivity increases will slow if major technological progress is not made. Food production may become a problem, as erosion, an impending water crisis, and soil salinity (from irrigation) reduce the productivity of agriculture. Ecological economists argue that industrial agriculture, which exacerbates these problems, is not sustainable agriculture, and are generally inclined favorably to organic farming, which also reduces the output of carbon." (WK 2012y; J. Harris 2006)

In our time, "global warming is now widely acknowledged as a major issue, with all national scientific academies expressing agreement on the importance of the issue. As the population growth intensifies and energy demand increases, the world faces an energy crisis. Some economists and scientists forecast a global ecological crisis if energy use is not contained—the Stern report is an example. The disagreement has sparked a vigorous debate on issue of discounting and intergenerational equity." (WK 2012y)

Ideology in Ecological Economics

Because of this emphasis on (intergenerational) "equity" and "sustainability," ecological economics is not value-free and thus differs from mainstream economics which "has attempted to become a value-free 'hard science,' but ecological economists argue that value-free economics is generally not realistic." (WK 2012y)

In addition, "ecological economics is more willing to entertain alternative conceptions of utility, efficiency, and cost-benefits such as positional analysis or multi-criteria analysis. Ecological economics is typically viewed as economics for sustainable development, and may have goals similar to green politics." (WK 2012y; P. Soderbaum 2004)

Approach of Ecological Economics

It is thus no surprise that ecological economics, because of its emphasis on nature and its ideology of sustainable development, becomes a distinctive approach to economics.

After all, there are different schools of economic thought.

On the one hand, there are mainstream approaches, and one of them is known as "neoclassical economics," with its focus on "the Economic Man" as "a utility maximizing consumer." (P. Soderbaum 2004)

On the other hand, there are alternative non-mainstream ("more heterodox") approaches, and one of them is "ecological economics," which adds "natural capital…to the typical capital asset analysis of land, labor, and financial capital. Ecological economics uses tools from mathematical economics, but may apply them more closely to the natural world. Whereas mainstream economists tend to be technological optimists, ecological economists are inclined to be technological pessimists. They reason that the natural world has a limited carrying capacity and that its resources may run out. Since destruction of important environmental resources could be practically irreversible and catastrophic, ecological economists are inclined to justify cautionary measures based on the precautionary principle." (WK 2012y; R. Costanza 1989)

One excellent example to contrast the two opposing schools of economic thought concerns "tropical rainforest ecosystems" in "the Yasuni region of Ecuador. While this area has substantial deposits of bitumen it is also one of the most diverse ecosystems on Earth and some estimates establish it has over 200 undiscovered medical substances in its genomes—most of which would be destroyed by logging the forest or mining the bitumen. Effectively, the instructional capital of the genomes is undervalued by both classical and neoclassical means which would view the rainforest primarily as a source of wood, oil/tar and perhaps food. Increasingly the carbon credit for leaving the extremely carbon-intensive ('dirty') bitumen in the ground is also valued—the government of Ecuador set a price of US$350M for an oil lease with the intent of selling it to someone committed to never exercising it at all and instead preserving the rainforest. Bill Clinton, Paul Martin and other former world leaders have become closely involved in this project which includes lobbying for the issue of International Monetary Fund special drawing rights to recognize the rainforest's value directly within the framework of the Bretton Woods institutions." (WK 2012y)

Besides these two opposing schools ("neoclassical" and "ecological") in the context of natural resources, there are also others like "environmental economics," which is more in-between the two.

Problems with Ecological Economics

However, for the critics, the field of ecological economics has some major problems. Consider, for illustration, a few examples below.

Firstly, one main criticism of the field of ecological economics is that its estimates (or data) on ecosystems to back up its claims are often highly speculative (and thus questionable).

For example, the work by Robert Costanza on the "valuation of the global ecosystem in 1997 was criticized by articles in *Ecological Economics* , Volume 25, Issue 1," although "the critics acknowledged the positive potential for economic valuation of the global ecosystem." (WK 2012y)

Secondly, another main criticism of the field of ecological economics is that it has its own share of ideological bias, just as its rival, neoclassical economics, has its own bias too.

For instance, even Peter Söderbaum (2004), himself an advocate of ecological economics, confessed that "many of us who joined the ecological economics movement in the 1980s did it for ethical and ideological reasons." In fact, a distinctive ideological aspect of ecological economics is that its proponents "are inclined to be technological pessimists." (WK 2012y)

Thirdly, still another main criticism of the field of ecological economics is that there are alternative schools of thought in economics, so ecological economics does not have the last word.

For instance, for the critics, "various competing schools of thought exist in the field" of economics, so ecological economics is just one of them and its claims are not privileged. (WK 2012y)

And fourthly, an additional main criticism of the field of ecological economics is that it evolves over time to become more mainstream and thus betrays itself, contrary to its original goal.

For instance, "there has then been a move to regard such things as natural capital and ecosystems functions as goods and services. However, this is far from uncontroversial within ecology or ecological economics due to the potential for narrowing down values to those found in mainstream economics and the danger of merely regarding Nature as a commodity. This has been referred to as ecologists 'selling out on Nature.' There is then a concern that ecological economics has failed to learn from the extensive literature in environmental ethics about how to structure a plural value system." (WK 2012y; G. Daily 1997; MEA 2005; D. McCauley 2006)

Beyond Ecological Economics

These criticisms of ecological economics, however, do not entail a total rejection of its usefulness but serve to show us the two opposing sides of the debate, such that the possibility (or impossibility) and desirability (or undesirability) of diversity (from the perspective of society with diversity, natural capital, and ecological economics as a case study here) are not to the extent that the respective defenders would like us to believe.

In addition, the analysis of diversity, natural capital, and ecological economics can cast a bright light on the relevance of the ontological principles in existential dialectics, and good examples include the formalness-informalness principle, the absoluteness-relativeness principle, the partiality-totality principle, the predictability-unpredictability principle, the explicability-inexplicability principle, the fiction-reality principle, the cognitiveness-noncognitiveness principle, the finiteness-transfiniteness principle, the preciseness-vagueness principle, the simpleness-complicatedness principle, the openness-hiddenness principle, the denseness-emptiness principle, the change-constancy principle, the order-chaos principle, the slowness-quickness principle, the expansion-contraction principle, the theory-praxis principle, the convention-novelty principle, the evolution-transformation principle, the symmetry-asymmetry principle, the softness-hardness principle, the seriousness-playfulness principle, the activeness-inactivess principle, the selfness-otherness principle, the regression-progression principle, the same-difference principle, the stability-reaction principle, the functionality-nonfunctionality principle, and the intentionality-nonintentionality principle.

For instance, in relation to the formalness-informalness principle, if there is formalness (e.g., the formal logical requirement of soundness, completeness, and consistency in a system of ideas, as in ecological economics), there is informalness (e.g., the non-formal existence of unsoundness, incompleteness, and inconsistency in ecological economics, as shown in the criticisms that its estimates are highly speculative and thus questionable, that it is essentially ideological just as neoclassical economics has its own share of ideological bias too, that there are alternative schools of thought so it does not have the last word, and that it evolves over time to become more mainstream and betrays itself, contrary to its original goal").

In relation to the absoluteness-relativeness principle, if there is absoluteness (e.g., the absolute view of ecological economics, in regard to its own ideology), there is relativeness (e.g., what is right for ecological economists in regard to their own ideology is not necessarily so for neoclassical economists, who maintain their own version instead, that is, with their focus on "the Economic Man" as "a utility maximizing consumer" and on technological optimism). And the reverse direction also holds true.

In relation to the partiality-totality principle, if there is partiality (e.g., the partial view of "neoclassical economics," with its narrow focus on "the Economic Man" as "a utility maximizing consumer"), there is totality (e.g., the more holistic view of economic thought, such that the whole is not the sum of its parts, that is, it cannot be reduced to the sum of opposing views like those of neoclassical economics and ecological economics, since there will emerge new approaches in the future not yet known today, just as there are already critics nowadays who question them all, as shown earlier). And the reverse direction also holds true.

In relation to the predictability-unpredictability principle, if there is predictability (e.g., the predictable tendency of ecological economists to focus on investment in natural capital), there is unpredictability (e.g., the more difficult task to predict exactly the extent to which ecological economists will achieve its own objectives on a particular future occasion, since, for the critics, it can evolve over time to become more mainstream and betrays itself, contrary to its original goal). And the reverse direction also holds true.

In relation to the explicability-inexplicability principle, if there is explicability (e.g., the explanation by the advocates of ecological economists of their ideology on the basis of stressing natural capital for sustainable development), there is inexplicability (e.g., the lack of sufficient explanation by the advocates of ecological economists of why their own ideology is necessarily desirable, since, for the critics, there are alternative schools of thought, so it does not have the last word). And the reverse direction also holds true.

In relation to the fiction-reality principle, if there is fiction (e.g., the fictional aspect of the "attempted valuation of the global ecosystem in 1997" by Robert Costanza, insofar as it "concluded on $33 trillion with a range from $16 trillion to $54 trillion," which the critics rejected as speculative in *Ecological Economics*, Volume 25, Issue 1"), there is reality (e.g., the realistic aspect of the "attempted valuation of the global ecosystem in 1997" by Robert Costanza, insofar as "the critics acknowledged the positive potential for economic valuation of the global ecosystem"). And the reverse direction also holds true.

In relation to the cognitiveness-noncognitiveness principle, if there is cognitiveness (e.g., a view of things on the basis of reason and evidences, as shown in the use of evidences in ecological economics, like the consideration of "the Earth's carrying capacity," in that, for example, "food production may become a problem, as erosion, an impending water crisis, and soil salinity [from irrigation] reduce the productivity of agriculture"), there is noncognitiveness (e.g., a view of things on the basis of non-cognitive factors like envy, jealousy, power, nationality, race, gender, age, class, greed, lust, anger, sadness, joy, fear, wish, etc., as shown in the fear embedded in many ecological economists who "are inclined to be technological pessimists"). And the reverse direction also holds true.

In relation to the finiteness-transfiniteness principle, if there is finiteness (e.g., the finite number of the categories of capital in economics, like "natural capital," "human-made capital," etc.), there is transfiniteness (e.g., the transfinite number of all the individual units of natural capital, human-made capital, etc. in history that have ever existed on Earth). And the reverse direction also holds true.

In relation to the preciseness-vagueness principle, if there is preciseness (e.g., the precise identification of 3 opposing schools of economic thought in the context of natural resources in the article aforecited, namely, "neoclassical economics," "environmental economics," and "ecological economics"), there is vagueness (e.g., the vagueness in the identification of the 3 opposing schools of economic thought in the context of natural resources in the article aforecited, namely, "neoclassical economics," "environmental economics," and "ecological economics, since it is not clear why there must be only 3, not 4, 5, 6, and so on). And the reverse direction also holds true.

In relation to the simpleness-complicatedness principle, if there is simpleness (e.g., the relatively simple analysis of ecological economics by its advocates), there is complicatedness (e.g., the relatively more complicated analysis of ecological economics, by challenging the claims and assumptions, as shown in the problems pointed out by the critics as cited). And the reverse direction also holds true.

In relation to the openness-hiddenness principle, if there is openness (e.g., the open exploration of the "limited carrying capacity" of the Earth, so that "its resources may run out," as studied by those in ecological economics), there is hiddenness (e.g., the hidden bias in ecological economics, since, "whereas mainstream economists tend to be technological optimists, ecological economists are inclined to be technological pessimists"). And the reverse direction also holds true.

In relation to the denseness-emptiness principle, if there is denseness (e.g., the relatively denser concentration of concerns with natural capital in ecological economics), there is emptiness (e.g., the relatively less dense, or more empty, concentration of concerns with natural capital in neoclassical economics). And the reverse direction also holds true.

In relation to the change-constancy principle, if there is change (e.g., the ever new ways to study economics, as shown in the rise of ecological economics), there is constancy (e.g., the ever constant existence of different problems in the ever new ways to study economics, as shown in the problems with ecological economics pointed out by the critics). And the reverse direction also holds true.

In relation to the order-chaos principle, if there is order (e.g., the more or less orderly view of economics, if looked from the sole vantage point of ecological economics, with its focus on natural capital), there is chaos (e.g., the more or less chaotic view of economics, if looked from all the conflicting, competing approaches, such as "neoclassical economics," "ecological economics," "resource and environmental economics," "some of the controversy between American and European schools of thought," and so on). (WK 2012y; I. Røpke 2004; C. Spash 1999) And the reverse direction also holds true.

In relation to the slowness-quickness principle, if there is slowness (e.g., the relatively slower speed of natural resources to be consumed, if operated under the guidance of ecological economics, with its focus on sustainaiblity), there is quickness (e.g., the relatively quicker speed of natural resources to be consumed, if operated under the guidance of neoclassical economics, with its focus on technology, in that "every technology can be improved upon or replaced by innovation, and that there is a substitute for any and all scarce materials" in the natural world). (WK 2012y) And the reverse direction also holds true.

In relation to the expansion-contraction principle, if there is expansion (e.g., the relative expansion of the approach to economics with the inclusion of the ecosystems for analysis nowadays), there is contraction (e.g., the relative decline of the approach to economics with the neglect of the ecosystems for analysis nowadays). And the reverse direction also holds true.

In relation to the theory-praxis principle, if there is theory (e.g., the theoretical construction of "natural capital" in ecological economics), there is praxis (e.g., the practical application of "natural capital" in ecological economics to the field of politics, as shown in the policy of "the government of Ecuador" to "preserve the rainforest" of "the Yasuni region," together with the political support of "Bill Clinton, Paul Martin and other former world leaders"). (WK 2012y) And the reverse direction also holds true.

In relation to the convention-novelty principle, if there is convention (e.g., the conventional wisdom about "natural resources,…by considering [capital] as interchangeable with human-made capital—labor and technology"), there is novelty (e.g., the alternative novel challenge to the conventional wisdom about "natural resources" on the basis of "human-made capital—labor and technology," by the new idea of "ecological economics" which focuses on natural capital instead). (WK 2012y) And the reverse direction also holds true.

In relation to the evolution-transformation principle, if there is evolution (e.g., the natural evolution of humans to consume natural resources in the state of nature in a primitive way), there is transformation (e.g., the technical transformation of human ability to consumer natural resources by the invention of ecological economics, with its focus on natural capital for sustainable development). And the reverse direction also holds true.

In relation to the symmetry-asymmetry principle, if there is symmetry (e.g., the existence of different lifeforms in nature, like humans and animals), there is asymmetry (e.g., humans have damaged the environment on a larger scale than animals in the history of the Earth, as shown by the human impact on "global warming"). And the reverse direction also holds true.

In relation to the softness-hardness principle, if there is softness (e.g., the replacement of natural capital by man-made capital), there is hardness (e.g., the depletion of natural resources). And the reverse direction also holds true.

In relation to the seriousness-playfulness principle, if there is seriousness (e.g., the serious business of those who try to understand the nautre of economics), there is playfulness (e.g., the playful part of those who try to understand the nautre of economics, when they play around with different approaches to economics over the centures, like "neoclassical economics," "ecological economics," "environmental economics," etc.). And the reverse direction also holds true.

In relation to the activeness-inactivess principle, if there is activeness (e.g., the relative activeness of those lands for food production, when they are healthy in being benefited from "organic farming" and do not suffer from soil problems like "erosion, an impending water crisis, soil salinity [from irrigation]," etc.), there is inactiveness (e.g., the relative inactiveness of those lands for food production, when they suffer from soil problems like "erosion, an impending water crisis, soil salinity [from irrigation]," etc).

In relation to the selfness-otherness principle, if there is selfness (e.g., the tendency to be mostly centered on one's self or group, as shown in those profit-seeking companies which are interesting in exploiting "the rainforest" of "the Yasuni region"), there is otherness (e.g., the tendency to be mostly centered on others, as shown in the effort of "the government of Ecuador" to "preserve the rainforest" of "the Yasuni region," together with the political support of "Bill Clinton, Paul Martin and other former world leaders," for the benefit of humanity). (WK 2012y) And the reverse direction also holds true.

In relation to the regression-progression principle, if there is regression (e.g., the regression made by ecological economics, as shown in the problems pointed out by the critics aforecited), there is progression (e.g., the progress made by ecological economics, as shown in its contribution to more investment in natural capital for sustainable development aforecited). And the reverse direction also holds true.

In relation to the same-difference principle, if there is similarity in outcome (e.g., the contribution to the molding and control of beliefs, values, and behaviors, regardless of whether this be done by way of ecological economics, or by way of mainstream economics), there is difference in outcome (e.g., the contribution to the molding and control of beliefs, values, and behaviors by way of ecological economics for a relatively more technologically pessimist worldview—but the contribution to the molding and control of beliefs, values, and behaviors by way of mainstream economics for a relatively more technologically optimist worldview). And the reverse direction also holds true.

In relation to the stability-reaction principle, if there is stability (e.g., the relatively more stable condition of neoclassical economics in the older days, with its focus on "the Economic Man" as "a utility maximizing consumer"), there is reaction (e.g., neoclassical economics in the older days then led to further changes later on, when nowadays "ecological economics" adds "natural capital…to the typical capital asset analysis of land, labor, and financial capital," since "ecological economists argue that neoclassical economics has ignored the environment"). And the reverse direction also holds true.

In relation to the functionality-nonfunctionality principle, if there is functionality (e.g., the presence of action for which something is fitted, as shown in the popularity of neoclassical economics in the older days, when it served the function of economic rationality along the utilitarian line at the time), there is nonfunctionality (e.g., the relative lack of action for which something is fitted, as shown in the rise of ecological economics nowadays, with its focus on natural capital for sustainable development, so that the neoclassical utilitarian mindset in the older days becomes increasingly nonfunctional in our time, when the expectations are higher, in switching from "growth [quantitative increase in economic output]" to "development [qualitative improvement of the quality of life]"). (WK 2012y) And the reverse direction also holds true.

And in relation to the intentionality-nonintentionality principle, if there is intentionality (e.g., the planning or design of something for a certain outcome, as shown in the rise of ecological economics to focus on natural capital for sustainable development), there is nonintentionality (e.g., the relative absence of planning or design of something for a certain outcome, such that the outcome can even be counter-intentional in being the opposite of what was originally planned, as shown in the criticism that ecological economics evolves over time to become more mainstream and betrays itself, contrary to its original goal, as "this has been referred to as ecologists 'selling out on Nature'"). And the reverse direction also holds true.

Diversity and Culture

Diversity, when examined from the perspective of culture, can cast a bright light on its possibility (or impossibility) and desirability (or undesirability), and this can be shown by way of two case studies, namely, (2.5.1) diversity, environmental management, and ethics, and (2.5.2) diversity, animal watching, and leisure activities—to be addressed hereafter, in that order.

Diversity, Environmental Management, and Ethics

An excellent case study here to understand diversity from the perspective of culture has to do with diversity, environmental management, and ethics.

An argument here is that there is a relationship between ethics and the man-nature interactions.

And the term "environmental management" (also known as "environmental resource management") refers to "'a purposeful activity with the goal to maintain and improve the state of an environmental resource affected by human activities.' It is not, as the phrase suggests, the management of the environment as such, but rather the management of the interaction and impact of human societies on the environment. Environmental resources management aims to ensure that ecosystem services are protected and maintained for equitable use by future human generations…." (WK 2012n; C. Pahl-Wost 2006)

Ethical Aspects of Environmental Management

Environmental management so understood is related to ethics, in that its "ethical aspects involve the cultural and social issues relating to the environment, and dealing with changes to it. 'All human activities take place in the context of certain types of relationships between society and the bio-physical world (the rest of nature),' and so, there is a great significance in understanding the ethical values of different groups around the world." (WK 2012n; M. Colby 1991)

In general, "two schools of thought exist in environmental ethics: Anthropocentrism [closely linked to social ecology] and Ecocentrism [closely linked to deep ecology], each influencing a broad spectrum of environmental resource management styles along a continuum. These styles perceive 'different evidence, imperatives, and problems, and prescribe different solutions, strategies, technologies, roles for economic sectors, culture, governments, and ethics, etc.'" (WK 2012n; M. Colby 1991)

Let's consider each of the two schools of environmental ethics below, in that order.

Anthropocentrism, and Environmental Management

On the one hand, there is the school of thought known as "anthropocentrism" (closely linked to "social ecology") in environmental ethics, which is "an inclination to evaluate reality exclusively in terms of human values" and is "an ethic reflected in the major interpretations of

Western religions and the dominant economic paradigms of the industrialized world. Anthropocentrism supports an understanding of nature as existing solely for the benefit of man and as a commodity to be used for the good of humanity and improved human quality of life. Anthropocentric environmental resource management is therefore not the conservation of the environment solely for the environment's sake, but rather the conservation of the environment, and ecosystem structure, for human sake." (WK 2012n; M. Colby 1991; L. White 1967; M. Berman 1981; D. Pepper 1984)

Anthropocentrism can come in all shapes and sizes, and hereafter are three major roots, namely, (a) philosophical, (b) religious, and (c) genetic—to be discussed below (and summarized in *Table 2.6*).

(a) Philosophical

The first root of anthropocentrism to be considered here is philosophical.

For instance, "environmental philosophy…is used to draw attention to a systematic bias in traditional Western attitudes to the non-human world. Val Plumwood has argued that anthropocentrism plays an analogous role in green theory to androcentrism in feminist theory and ethnocentrism in anti-racist theory. Plumwood calls human-centredness 'anthrocentrism' to emphasize this parallel." (WK 2012aa; A. Naess 1973; V. Plumwood 1993 & 1996)

In general, "defenders of anthropocentrist views point out that maintenance of a healthy, sustainable environment is necessary for human well-being as opposed for its own sake….It is important to take note that many devoted environmentalists encompass a somewhat anthropocentric-based philosophical view supporting the fact that they will argue in favor of saving the environment for the sake of human populations. 'We should be concerned to promote a rich, diverse, and vibrant biosphere. Human flourishing may certainly be included as a legitimate part of such a flourishing.'" (WK 2012aa)

(b) Religious

The second root of anthropocentrism to be considered here is religious.

For instance, "in the 1985 CBC series 'A Planet For the Taking,' David Suzuki explored the Old Testament roots of anthropocentrism and how it shaped our view of non-human animals. Some Christian proponents of anthropocentrism base their belief on the Bible, such as the verse 1:26 in the Book of Genesis: 'And God said, Let us make man in our image, after our likeness: and let them have dominion over the fish of the sea, and over the fowl of the air, and over the cattle, and over all the earth, and over every creeping thing that creepeth upon the earth.'" (WK 2012aa)

But "the use of the word 'dominion' in the Genesis is controversial. Many Biblical scholars, especially Roman Catholic and other non-Protestant Christians, consider this to be a flawed translation of a word meaning 'stewardship,' which would indicate that mankind should take care of the earth and its various forms of life, but is not inherently better than any other form of life. The current Latin *Vulgate*, the official Bible of the Catholic Christian church, states that God holds man responsible for the care and fate of all earthly creatures." (WK 2012aa)

(c) Evolutionary

And the third root of anthropocentrism to be considered here is genetic (in biology).

For instance, "some secular proponents of human exceptionalism point to evidence of unusual rapid evolution of the brain and the emergence of exceptional aptitudes. As one commentator put it, 'Over the course of human history, we have been successful in cultivating our faculties, shaping our development, and impacting upon the wider world in a deliberate fashion, quite distinct from evolutionary processes.' Jenia Meng uses evidence from an international survey to argue that anthropocentrism is inevitable in all human societies and that it is an result of gene-centrism. She also notes that although anthropocentrism is unavoidable, excessive anthropocentrism should be avoided." (WK 2012aa)

Ecocentrism, and Environmental Management

On the other hand, there is the opposing school of thought known as "ecocentrism" (closely linked to "deep ecology") in environmental ethics, which stresses "the intrinsic value of nature while maintaining an understanding that 'human beings must use and even exploit nature to survive and live.' It is this fine ethical line that ecocentrists navigate between 'fair use and downright abuse.' At an extreme end of the ethical scale, ecocentrism includes philosophies such as ecofeminism and deep ecology which evolved as a reaction to the dominant anthropocentric paradigms. 'In its current form, it is an attempt to synthesize many old and some new philosophical attitudes about the relationship between nature and human activity, with particular emphasis on ethical, social, and spiritual aspects that have been downplayed in the dominant economic worldview.'" (WK 2012n; R. Purser 1996; M. Colby 1991)

This "ecocentric ethic was conceived by Aldo Leopold [1949] and recognizes that all species, including humans, are the product of a long evolutionary process and are inter-related in their life processes." (WK 2012bb; D. Lindenmeyer 2005)

In general, ecocentrism is against three contrastive philosophies, namely, (a) against anthropocentrism, (b) against technocentrism, and (c) against biocentrism—as discussed below, in that order (and summarized in *Table 2.6*).

(a) Against Anthropocentrism

The first contrastive philosophy that ecocentrism is against has to do with "anthropocentrism."

The reason is that "ecocentrism is taken by its proponents to constitute a radical challenge to long-standing and deeply rooted anthropocentric attitudes in Western culture, science, and politics. Anthropocentrism is alleged to leave the case for the protection of non-human nature subject to the demands of human utility, and thus never more than contingent on the demands of human welfare. An ecocentric ethic, by contrast, is believed to be necessary in order to develop a non-contingent basis for protecting the natural world....Deep ecologist Arne Naess has identified anthropocentrism as a root cause of the ecological crisis, human overpopulation, and the extinctions of many non-human species. Others point to the gradual historical realization that humans are not the centre of all things, that 'A few hundred years ago, with some reluctance, Western people admitted that the planets, Sun and stars did not circle around their abode. In short, our thoughts and concepts though irreducibly anthropomorphic need not be anthropocentric.'" (WK 2012bb; S. Rowe 1994; A. Naess 1973)

(b) Against Technocentrism

The second contrastive philosophy that ecocentrism is against has to do with "technocentrism."

The reason is that "ecocentrism is also contrasted with technocentrism (meaning values centered on technology) as two opposing perspectives on attitudes towards human technology and its ability to affect, control and even protect the environment. Ecocentrics, including 'deep green' ecologists, see themselves as being subject to nature, rather than in control of it. They lack faith in modern technology and the bureaucracy attached to it. Ecocentrics will argue that the natural world should be respected for its processes and products, and that low impact technology and self-reliance is more desirable than technological control of nature." (WK 2012bb; SUS 2012)

But "technocentrics…have absolute faith in technology and industry and firmly believe that humans have control over nature. Although technocentrics may accept that environmental problems do exist, they do not see them as problems to be solved by a reduction in industry. Rather, environmental problems are seen as problems to be solved using science. Indeed, technocentrics see that the way forward for developed and developing countries and the solutions to our environmental problems today lie in scientific and technological advancement." (WK 2012bb; SUS 2012)

(c) Against Biocentrism

And the third contrastive philosophy that ecocentrism is against has to do with "biocentrism."

The reason is that, although "the distinction between biocentrism and ecocentrism is ill-defined," "ecocentrism recognizes Earth's interactive living and non-living systems rather than just the Earth's organisms (biocentrism) as central in importance. The term has been used by those advocating 'left biocentrism,' combining deep ecology with an 'anti-industrial and anti-capitalist' position " (WK 2012bb).

Problems with Environmental Ethics

However, for the critics, environmental ethics has some major problems. Consider, for illustration, a few examples below (as summarized in *Table 2.6*).

Firstly, one main criticism of environmental ethics is that it is divided between competing schools of thought.

For example, anthropocentrism and ecocentrism (as discussed earlier) offer opposing approaches to environmental ethics, such that each is in the eye of the beholder.

Secondly, another main criticism of environmental ethics is that its reasoning is simplistic.

For instance, for the critics, "one of the first extended philosophical essays addressing environmental ethics, John Passmore's *Man's Responsibility for Nature*, has been criticised by defenders of deep ecology because of its anthropocentrism, often claimed to be constitutive of traditional Western moral thought….The problem with a 'shallow' viewpoint [in anthropocentrism] is not that it is human-centred but that according to William Grey: 'What's wrong with shallow views is not their concern about the well-being of humans, but that they do not really consider enough in what that well-being consists. According to this view, we need to develop an enriched, fortified anthropocentric notion of human interest to replace the dominant short-term, sectional and self-regarding conception.'" (WK 2012aa; J. Passmore 1974; R. Routley 1980; W. Grey 1993) And a "2012 documentary *The Superior Human?* systematically analyzes anthropocentrism and concludes that value is fundamentally an opinion, and since life forms naturally value their own traits, some humans are misled to believe that they are actually more valuable than other species. This natural bias, combined with a received sense of comfort and an excuse for exploitation of non-humans cause anthropocentrism to remain in society." (WK 2012aa; S. McAnallen 2012)

Thirdly, still another main criticism of environmental ethics is that its implementation can be destructive.

For instance, in regard to the religious root of anthropocentrism, "some evangelical Christians have…been critical, viewing a human-centered worldview, rather than a Christ-centered or God-centered worldview, as a core societal problem. According to this viewpoint, humanity placing its own desires ahead of the teachings of the Bible leads to rampant selfishness and behaviour viewed as sinful." (WK 2012aa)

Fourthly, an additional main criticism of environmental ethics is that it is self-defeating.

For instance, "critics of ecocentrism have argued that it opens the doors to an anti-humanist morality that risks sacrificing human well-being for the sake of an ill-defined 'greater good.'" (WK 2012bb)

And fifthly, another additional main criticism of environmental ethics is that it focuses too much on the whole without sufficient attention to the individuals.

For instance, "Marti Kheel [1990]…argue that self-realization and identification with all nature places too much emphasis on the whole, at the expense of the independent being." (WK 2012cc).

Beyond Environmental Ethics

These criticisms of environmental ethics, however, do not suggest a total rejection of its usefulness but serve to show us the two opposing sides of the debate, such that the possibility (or impossibility) and desirability (or undesirability) of diversity (from the perspective of culture with diversity, enviromental management, and ethics as a case study here) are not to the extent that the respective defenders would like us to believe.

Furthermore, the analysis of diversity, enviromental management, and ethics can illuminate the relevance of the ontological principles in existential dialectics, and good examples include the formalness-informalness principle, the absoluteness-relativeness principle, the partiality-totality principle, the predictability-unpredictability principle, the explicability-inexplicability principle, the fiction-reality principle, the cognitiveness-noncognitiveness principle, the finiteness-transfiniteness principle, the preciseness-vagueness principle, the simpleness-complicatedness principle, the openness-hiddenness principle, the denseness-emptiness principle, the change-constancy principle, the order-chaos principle, the slowness-quickness principle, the expansion-contraction principle, the theory-praxis principle, the convention-novelty principle, the evolution-transformation principle, the symmetry-asymmetry principle, the softness-hardness principle, the seriousness-playfulness principle, the activeness-inactivess principle, the selfness-otherness principle, the regression-progression principle, the same-difference principle, the stability-reaction principle, the functionality-nonfunctionality principle, and the intentionality-nonintentionality principle.

For instance, in relation to the formalness-informalness principle, if there is formalness (e.g., the formal logical requirement of soundness, consistency, and completeness in a system of ideas, as in environmental ethics), there is informalness (e.g., the non-formal existence of unsoundness, inconsistency, and incompleteness in environmental ethics, as shown in the criticisms that it is divided between competing schools of thought, that its reasoning is simplistic, that its implementation can be destructive, that it is self-defeating, that it focuses too much on the whole without sufficient attention to the individuals, etc.). And the reverse direction also holds true.

In relation to the absoluteness-relativeness principle, if there is absoluteness (e.g., the absolute view of ecocentrism in regard to its claim about "the intrinsic value of nature"), there is relativeness (e.g., what is right for ecocentrism is not necessarily so for anthropocentrism, since they constitute two opposing schools of environmental ethics). And the reverse direction also holds true.

In relation to the partiality-totality principle, if there is partiality (e.g., the partial view of anthropocentrism in regard to man-nature interactions), there is totality (e.g., the more holistic view of man-nature interactions, such that the whole is not the sum of its parts, that is, it cannot be reduced to the sum of opposing views like ecocentrism, anthropocentrism, technocentrism, etc., since there will emerge new views in the future not yet known today, just as there are critics nowadays who do not side exclusively with any of them). And the reverse direction also holds true.

In relation to the predictability-unpredictability principle, if there is predictability (e.g., the predictable tendency of anthropocentrists to "support…an understanding of nature as existing solely for the benefit of man and as a commodity to be used for the good of humanity and improved human quality of life"), there is unpredictability (e.g., the more difficult task to predict exactly the extent to which anthropocentrism will succeed in a particular future era, since, for the critics, it would lead "to rampant selfishness and behaviour viewed as sinful," among other problems aforediscussed). And the reverse direction also holds true.

In relation to the explicability-inexplicability principle, if there is explicability (e.g., the explanation by the advocates of environmental ethics of the importance of focusing on "the whole" of man-nature interactions), there is inexplicability (e.g., the lack of sufficient explanation by the advocates of environmental ethics of why focusing on "the whole" of man-nature interactions is necessarily desirable, since, for the critics like Marti Kheel, "self-realization and identification with all nature places too much emphasis on the whole, at the expense of the independent being"). And the reverse direction also holds true.

In relation to the fiction-reality principle, if there is fiction (e.g., the fictional aspect of ecocentrism in deep ecology, insofar as its reasoning is not "deep" enough, because, for the critic like Murray Bookchin [1987], "deep ecology fails to link environmental crises with authoritarianism and hierarchy" and therefore fails to see that "environmental problems are firmly rooted in the manner of human social interaction," so "an ecologically sustainable society could still be socially exploitative"), there is reality (e.g., the realistic aspect of ecocentrism in deep ecology, insofar as it points out the need to consider "the intrinsic value of nature while maintaining an understanding that 'human beings must use and even exploit nature to survive and live"). (WK 2012n & 2012cc) And the reverse direction also holds true.

In relation to the cognitiveness-noncognitiveness principle, if there is cognitiveness (e.g., a view of things on the basis of reason and evidences, as shown in the case of anthropocentrism, when its advocates "point out that maintenance of a healthy, sustainable environment is necessary for human well-being"), there is noncognitiveness (e.g., a view of things on the basis of non-cognitive factors like envy, jealousy, power, nationality, race, gender, age, class, greed, lust, anger, sadness, joy, fear, wish, etc., as shown in the case of anthropocentrism, when "some Christian proponents of anthropocentrism base their belief on the Bible, such as the verse 1:26 in the Book of Genesis," that is, on the basis of religious faith, not of empirical proof). And the reverse direction also holds true.

In relation to the finiteness-transfiniteness principle, if there is finiteness (e.g., the finite number of the categories of lifeforms in the man-nature interactions as understood in ecocentrism, such as microorganisms, plants, animals, humans, etc.), there is transfiniteness (e.g., the transfinite number of all the individual microorganisms, plants, animals, humans, etc. in the history of man-nature interaction on Earth that have ever existed). And the reverse direction also holds true.

In relation to the preciseness-vagueness principle, if there is preciseness (e.g., the precise identification of 3 roots of anthropocentrism in the article aforecited), there is vagueness (e.g., the vagueness in the identification of the 3 roots of anthropocentrism in the article aforecited, since it is not clear why there must be only 3, not 4, 5, 6, and so on). And the reverse direction also holds true.

In relation to the simpleness-complicatedness principle, if there is simpleness (e.g., the relatively simple analysis of each school of environmental ethics by its advocates), there is complicatedness (e.g., the relatively more complicated analysis of each school of environmental ethics, by challenging the claims and assumptions, as shown in the problems pointed out by the critics as cited). And the reverse direction also holds true.

In relation to the openness-hiddenness principle, if there is openness (e.g., the open exploration of the need for a "maintenance of a healthy, sustainable environment…for human well-being" in anthropocentrism), there is hiddenness (e.g., the hidden bias in anthropocentrism, since, for the critics, "anthropocentrism" is "a root cause of the ecological crisis, human overpopulation, and the extinctions of many non-human species"). And the reverse direction also holds true.

In relation to the denseness-emptiness principle, if there is denseness (e.g., the relatively denser concentration of concerns with "the conservation of the environment solely for the environment's sake" in ecocentrism), there is emptiness (e.g., the relatively less dense, or more empty, concentration of concerns with "the conservation of the environment solely for the environment's sake" in anthropocentrism). And the reverse direction also holds true.

In relation to the change-constancy principle, if there is change (e.g., the ever new ways to understand ethics, as shown in environmental ethics), there is constancy (e.g., the ever constant existence of different problems with the ever new ways to understand ethics, as shown in the problems with environmental ethics pointed out by the critics as cited earlier). And the reverse direction also holds true.

In relation to the order-chaos principle, if there is order (e.g., the more or less orderly view of nature, if looked from the Christian vantage point of "stewardship" on the basis of the Book of Genesis, in that "mankind should take care of the earth and its various forms of life"), there is chaos (e.g., the more or less chaotic view of nature, if looked from the vantage point of laissez-faire capitalists, who prefer to compete with each other for resources in the spirit of social Darwinism, where chance or luck also has a role to play). And the reverse direction also holds true.

In relation to the slowness-quickness principle, if there is slowness (e.g., the relatively slower speed of the environment to be affected or controlled by technology under the guidance of ecocentrism), there is quickness (e.g., the relatively quicker speed of the environment to be affected or controlled by technology under the guidance of technocentrism). And the reverse direction also holds true.

In relation to the expansion-contraction principle, if there is expansion (e.g., the relative expanson, in our time, of the influence of anthropocentrism, since it is used by some defenders like Wesley J. Smith, who argued that it "gives rise to human duties to each other, the natural world, and to treat animals humanely"), there is contraction (e.g., the relative decline, in our time, of the elitist idea that "groups of superior men" should "be able to justify their enslavement, exploitation, or even genocide of inferior human groups on factual and moral grounds akin to those we now rely on to justify our treatment of the animals we harness as beasts of burden, that we butcher for food and clothing, or that we destroy as disease-bearing pests or as dangerous predators?"). (WK 2012aa; M. Adler 1993) And the reverse direction also holds true.

In relation to the theory-praxis principle, if there is theory (e.g., the theoretical construction of anthropocentrism in environmental ethics), there is praxis (e.g., the practical application of anthropocentrism in environmental ethics to the field of politics, as shown in the use of anthropocentrism as "the grounding for some naturalistic concepts of human rights," as "defenders of anthropocentrism argue that it is the necessary fundamental premise to defend universal human rights"). (WK 2012aa) And the reverse direction also holds true.

In relation to the convention-novelty principle, if there is convention (e.g., the conventional wisdom of anthropocentrism, since it is "an ethic reflected in the major interpretations of Western religions and the dominant economic paradigms of the industrialised world"), there is novelty (e.g., the alternative novel challenge to the conventional wisdom of anthropocentrism, by the new idea of ecocentrism, which is "a radical challenge to long-standing and deeply rooted anthropocentric attitudes in Western culture, science, and politics"). (WK 2012n & 2012bb) And the reverse direction also holds true.

In relation to the evolution-transformation principle, if there is evolution (e.g., the natural evolution of humans to "butcher" animals "for food and clothing" in the state of nature in a primitive way), there is transformation (e.g., the technical transformation of human treatment of animals by the invention of anthropocentrism, which "gives rise to human duties…to treat animals humanely"). (WK 2012aa; M. Adler 1993) And the reverse direction also holds true.

In relation to the symmetry-asymmetry principle, if there is symmetry (e.g., the existence of different lifeforms like animals and humans), there is asymmetry (e.g., humans can treat animals in a way on a larger scale that animals cannot do to humans, especially in the exploitative "treatment of the animals we harness as beasts of burden, that we butcher for food and clothing, or that we destroy as disease-bearing pests or as dangerous predators?"). (WK 2012aa; M. Adler 1993) And the reverse direction also holds true.

In relation to the softness-hardness principle, if there is softness (e.g., humans can "treat animals humanely"), there is hardness (e.g., humans can kill animals "as beasts of burden," "for food and clothing," as "disease-bearing pests or as dangerous predators"). (WK 2012aa; M. Adler 1993) And the reverse direction also holds true.

In relation to the seriousness-playfulness principle, if there is seriousness (e.g., the serious business of those in anthropocentrism to understand man-nature interactions), there is playfulness (e.g., the playful part of those in anthropocentrism to understand man-nature interactions, when they play around with different justifications or roots over the decades, like the "philosophical," "religious," and "genetic" arguments aforecited). And the reverse direction also holds true.

In relation to the activeness-inactivess principle, if there is activeness (e.g., the relative activeness of those animals which are treated humanely by humans and are able to survive and live healthily), there is inactiveness (e.g., the relative inactiveness of those animals which are caught or killed by humans, because they are treated "as beasts of burden," "for food and clothing," as "disease-bearing pests or as dangerous predators"). And the reverse direction also holds true.

In relation to the selfness-otherness principle, if there is selfness (e.g., the tendency to be mostly centered on one's self or group, as shown by those "technocentrics, including imperialists," who "have absolute faith in technology and industry," so they go around to colonize and exploit nature for their own interests), there is otherness (e.g., the tendency to be mostly centered on others, as shown by those ecocentrists, who "believe in the intrinsic value of nature"). (WK 2012n & 2012bb) And the reverse direction also holds true.

In relation to the regression-progression principle, if there is regression (e.g., the regression made by environmental ethics, as shown in the problems pointed out by the critics), there is progression (e.g., the progress made by environmental ethics, as shown in its contribution to a more humane treatment of nature aforecited). And the reverse direction also holds true.

In relation to the same-difference principle, if there is similarity in outcome (e.g., the contribution to the molding and control of beliefs, values, and behaviors, regardless of whether this be done by way of ecocentricism, or by way of technocentrism), there is difference in outcome (e.g., the contribution to the molding and control of beliefs, values, and behaviors by way of ecocentricism for a relatively more naturalistic lifeworld—but the contribution to the molding and control of beliefs, values, and behaviors by way of technocentrism for a relatively more technophilic lifeworld). And the reverse direction also holds true.

In relation to the stability-reaction principle, if there is stability (e.g., the relatively more stable condition of anthropocentrism in the older days, because of its being "an ethic reflected in the major interpretations of Western religions and the dominant economic paradigms of the industrialised world"), there is reaction (e.g., anthropocentrism in the older days then led to further changes later on, when it has to come to terms with its new rival like ecocentrism nowadays). And the reverse direction also holds true.

In relation to the functionality-nonfunctionality principle, if there is functionality (e.g., the presence of action for which something is fitted, as shown in anthropocentrism in the older days, which served the important function of being "an ethic reflected in the major interpretations of Western religions and the dominant economic paradigms of the industrialised world"), there is nonfunctionality (e.g., the relative lack of action for which something is fitted, as shown in the rise of ecocentrism in our time to challenge anthropocentrism, which is no longer functional as the dominant legitimate ethics nowadays, since a "2012 documentary *The Superior Human?* systematically analyzes anthropocentrism and concludes that value is fundamentally an opinion, and since life forms naturally value their own traits, some humans are misled to believe that they are actually more valuable than other species"). (WK 2012aa; S. McAnallen 2012) And the reverse direction also holds true.

And in relation to the intentionality-nonintentionality principle, if there is intentionality (e.g., the planning or design of something for a certain outcome, as shown in the use of ecocentrism to justify "the intrinsic value of nature while maintaining an understanding that 'human beings must use and even exploit nature to survive and live'"), there is nonintentionality (e.g., the relative absence of planning or design of something for a certain outcome, such that the outcome can even be counter-intentional in being the opposite of what was originally planned, as shown in the critique of ecocentrism, in that it "risks sacrificing human well-being for the sake of an ill-defined 'greater good'" in nature, contrary to its original claim). WK 2012bb) And the reverse direction also holds true.

Diversity, Animal Watching, and Leisure Activities

Another way to evaluate the possibility (or impossibility) and desirability (or undesirability) of diversity from the perspective of culture concerns diversity, animal watching, and leisure activities.

An argument here is that there is a relationship between biodiversity and animal watching as part of leisure activities, in that "biodiversity enriches leisure activities such as hiking, birdwatching or natural history study. Biodiversity inspires musicians, painters, sculptors, writers and other artists. Many cultures view themselves as an integral part of the natural world which requires them to respect other living organisms." (WK 2012i)

Consider, for illustration, the leisure activities of (a) birdwatching and (b) fishkeeping—to be described below, in that order (and summarized in *Table 2.7*).

The Leisure Activity of Birdwatching

The leisure activity of "birdwatching" (aka "birding") is "the observation of birds as a recreational activity. It can be done with the naked eye, through a visual enhancement device like binoculars and telescopes, or by listening for bird sounds. Birding often involves a significant auditory component, as many bird species are more easily detected and identified by ear than by eye. Most birdwatchers pursue this activity mainly for recreational or social reasons, unlike ornithologists, who engage in the study of birds using more formal scientific methods." (WK 2012dd; P. Dunne 2003; B. Oddie 1980)

(a) Birders vs. Birdwatchers

However, there are different groups who enjoy the observation of birds. For instance, "in North America, many birders differentiate themselves from birdwatchers, and the term *birder* is unknown to most lay people. At the most basic level, the distinction is one of dedication or intensity. Generally, self-described birders are more versed in minutiae like identification (aural and visual), molt, distribution, migration timing, and habitat usage. Whereas these dedicated *birders* may often travel specifically in search of birds, *birdwatchers* have been described by some enthusiasts as having a more limited scope, perhaps not venturing far from their own yards or local parks to view birds." (WK 2012dd; Dunne 2003)

In accordance to the "Birding Glossary" appeared in the 1969 *Birding* (Volume 1, No.2), there is the following disctintion: (WK 2012dd)

- "Birder"—"The acceptable term used to describe the person who seriously pursues the hobby or sport of birding" and "may be professional or amateur"
- "Birding"—"A sport and/or hobby in which individuals enjoy the challenge of bird study, listing, or other general activities involving bird life"
- "Bird-Watcher"—"A rather ambiguous term used both to describe the person who watches birds for any reason at all, and, more recently, to refer to a person who watches girls" and "should not be used to refer to the serious birder"

Of course, there can be other terms too (not distinguished by the "Birding Glossary"), like "twitching," which is "a British term used to mean 'the pursuit of a previously-located rare bird.'" (WK 2012dd)

(b) History of Birdwatching

Historically, "the study of birds and natural history became fashionable in Britain during the Victorian Era," but "it was mainly collection oriented with eggs and later skins being the artifacts of interest. Wealthy collectors made use of their contacts in the colonies to obtain specimens from around the world. It was only in the late 19th century that the call for bird protection began leading to the rising popularity of observations on living birds." (WK 2012dd)

For instance, "the Audubon Society was started to protect birds from the growing trade in feathers in the United States while the Royal Society for the Protection of Birds began in Britain. The term 'birdwatching' appeared for the first time as the title of a book 'Bird Watching' by Edmund Selous in 1901." (WK 2012dd; S. Moss 2004)

And "in North America, the identification of birds, once thought possible only by shooting was made possible by the emergence of optics and field identification guides. The earliest field guide in the US was *Birds through an Opera Glass* (1889) by Florence Bailey. Birding in North America was focused in the early and mid-20th century in the eastern seaboard region, and was influenced by the works of Ludlow Griscom and later Roger Tory Peterson. *Bird Neighbors* (1897) by Neltje Blanchan was an early birding book which sold over 250,000 copies. It was illustrated with color photographs of stuffed birds." (WK 2012dd; M. Barrow 1998; G. Shearin 2008; T. Dunlap 2011)

Later, "the organization and networking of those interested in birds began through organizations like the Audubon Society that was against the killing of birds and the American Ornithologists' Union (AOU). The rising popularity of the car increased the mobility of birdwatchers and this made new locations accessible to those interested in birds. Networks of birdwatchers in the UK began to form in the late 1930s under the British Trust for Ornithology (BTO). The BTO saw the potential to produce scientific results through the networks, unlike the Royal Society for the Protection of Birds (RSPB) which like the Audubon Society originated from the bird protection movement. Like the AOU in North America, the BOU had a focus mainly in collection based taxonomy." (WK 2012dd; H. Macdonald 2002; S. Moss 2004)

However, "the BOU changed focus to ecology and behaviour only in the 1940s. The BTO movement towards 'organized birdwatching,' was opposed by the RSPB which claimed that the 'scientification' of the pastime was 'undesirable'. This stand was to change only in 1936 when the RSPB was taken over by Tom Harrisson and others. Harrisson was instrumental in the organization of pioneering surveys of the Great Crested Grebe." (WK 2012dd; K. Johnson 2004; S. Moss 2004)

The advance in transportation technology also contributed to the spread of birdwatching. For instance, "by the 1960s air-travel became feasible and long distance holiday destinations opened up with the result that by 1965, Britain's first birding tour company, *Ornitholidays*, was started by Lawrence Holloway. Travelling far away also led to problems in name usage, British birds like 'Wheatear,' 'Heron,' and 'Swallow' needed adjectives to differentiate them in places where there were several related species. The falling cost of air-travel made flying to remote birding destinations a possibility for a large number of people towards the 1980s. The need for global guides to birds became more relevant and one of the biggest projects that began was the 'Handbook of the Birds of the World' which started in the 1990s with Josep del Hoyo a country doctor in Catalonia, Jordi Sargatal and ornithologist Andy Elliott." (WK 2012dd; S. Moss 2004)

The rest of the world also became more interesed in birdwatching; for instance, "since the second half of the 20th century an increasing number of people in developing countries have engaged in this activity.

Transnational birding has played an important role in this, as citizens from developing countries that engage in birdwatching usually develop this pastime due to influence of foreign cultures that already practise birding." (WK 2012dd; H. Gómez de Silva 2010)

The Leisure Activity of Fishkeeping

And the leisure activity of "fishkeeping" is "a popular hobby concerned with keeping fish in a home aquarium or garden pond. There is also a fishkeeping industry, as a branch of agriculture." (WK 2012q)

(a) Types of Fishkeeping

Fishkeepers are often called "aquarists," but "the hobby can be broadly divided into three specific disciplines according to the type of water the fish tolerate: freshwater, brackish, and marine (also called saltwater) fishkeeping," as described below.

(a1) Freshwater

In regard to "freshwater fishkeeping"—it is "by far the most popular branch of the hobby, with even small pet stores often selling a variety of freshwater fish, such as goldfish, guppies, and angelfish. While most freshwater aquaria are community tanks containing a variety of compatible species, single-species breeding aquaria are also popular. Livebearing fish such as mollies and guppies are among those most easily raised in captivity, but aquarists also regularly breed many types of cichlid, catfish, characin, and killifish." (WK 2012q)

In addition, "many fishkeepers create freshwater aquascapes where the focus is on aquatic plants as well as fish. These aquaria include 'Dutch Aquaria,' named for European aquarists who designed them. In recent years, one of the most active advocates of the heavily planted aquarium is the Japanese aquarist Takashi Amano. Garden ponds are in some ways similar to freshwater aquaria, but are usually much larger and exposed to ambient weather. In the tropics, tropical fish can be kept in garden ponds, but in the temperate zone species such as goldfish, koi, and orfe work better." (WK 2012q)

(a2) Saltwater

In regard to "saltwater fishkeeping"—"marine aquaria are generally more difficult to maintain and the livestock is significantly more expensive. As a result this branch tends to attract more experienced fishkeepers. Marine aquaria can be exceedingly beautiful, due to the attractive colors and shapes of the corals and the coral reef fish they host. Temperate zone marine fish are not as commonly kept in home aquaria, primarily because they do not thrive at room temperature. Coldwater aquaria must provide cooler temperature via a cool room (such as an unheated basement) or a refrigeration device known as a 'chiller.'" (WK 2012q)

In general, "marine aquarists often attempt to recreate a coral reef in their aquaria using large quantities of living rock, porous calcareous rocks encrusted with coralline algae, sponges, worms, and other small marine organisms. Larger corals as well as shrimps, crabs, echinoderms, and mollusks are added later on, once the aquarium has matured, as well as a variety of small fish. Such aquaria are sometimes called reef tanks." (WK 2012q)

(a3) Brackish Water

In regard to "brackfish water fishkeeping"—"Brackish water aquaria combine elements of the other types, with salinity that must stay between that of freshwater and seawater. Brackish water fish come from habitats with varying salinity, such as mangroves and estuaries, and do not thrive if kept permanently in freshwater. Although brackish water aquaria are not necessarily familiar to inexperienced aquarists, many species prefer brackish water, including some mollies, many gobies, some pufferfish, monos, scats, and virtually all the freshwater soles." (WK 2012q)

(b) History of Fishkeeping

Historically, "fish have been raised as food in pools and ponds for thousands of years. Brightly colored or tame specimens of fish in these pools have sometimes been valued as pets rather than food. Many other cultures kept fish for both functional and decorative purposes." (WK 2012q)

For instance, "ancient Sumerians kept wild-caught fish in ponds, before preparing them for meals. Depictions of the sacred fish of Oxyrhynchus kept in captivity in rectangular temple pools have been found in ancient Egyptian art." (WK 2012q)

By contrast, "Asia has experienced a long history of stocking rice paddies with freshwater fish suitable for eating, including various types of catfish and cyprinid. Selective breeding of carp into today's popular and completely domesticated koi and goldfish began over 2,000 years ago in Japan and China, respectively. The Chinese brought goldfish indoors during the Song Dynasty to enjoy them in large ceramic vessels." (WK 2012q)

In Europe during the medieval period, "carp pools were a standard feature of estates and monasteries, providing an alternative to meat on feast days when meat could not be eaten for religious reasons." (WK 2012q)

And over the centuries, "marine fish have been similarly valued....Wealthy Romans kept lampreys and other fish in salt water pools. Tertullian reports that Asinius Celer paid 8000 sesterces for a particularly fine mullet. Cicero reports that the advocate Quintus Hortensius wept when a favored specimen died. Rather cynically, he referred to these ancient fishkeepers as the Piscinarii, the 'fish-pond owners' or 'fish breeders,' for example when saying that *...the rich (I mean your friends the fish-breeders) did not disguise their jealousy of me*." (WK 2012q; BBC 2007)

Problems with Animal Watching

However, for the critics, animal watching has some major problems. Consider, for illustration, a few examples below (as summarized in *Table 2.7*).

Firstly, one main criticism of animal watching is that there are philosophical disagreements in regard to the use of biodiversity for leisure activities.

For instance, some people argue philosophically that "biodiversity has intrinsic aesthetic and spiritual value to mankind in and of itself," but others disagree, because they instead suggest that "tropical forests and other ecological realms are only worthy of conservation because of the services they provide." (WK 2012i)

Secondly, another main criticism of animal watching is that what is considered as a leisure activity in a culture may not be so at all in another culture, which may choose the use of different animals or natural resources instead.

For instance, for the critics, "the Chinese brought goldfish indoors during the Song Dynasty to enjoy them in large ceramic vessels," but "wealthy Romans kept lampreys and other fish in salt water pools" instead. (WK 2012q)

Thirdly, still another main criticism of animal watching is that it is gendered-relative.

For instance, "a study suggests that males leaned towards competitive birding while females preferred recreational birdwatching." (WK 2012dd; C. Cooper 2010)

And fourthly, an additional main criticism of animal watching is that there is a concern with "animal welfare."

For instance, for the critics, "inexperienced aquarists often keep too many fish in one tank, or add fish too quickly into an immature aquarium, killing many of them. This has given the hobby a bad reputation among some animal welfare groups, such as PETA, who accuse aquarists of treating aquarium fish as cheap toys to be replaced when they die." (WK 2012q' PETA 2007)

Beyond Animal Watching

These criticisms of animal watching, however, do not imply a total rejection of its usefulness but serve to show us the two opposing sides of the debate, such that the possibility (or impossibility) and desirability (or undesirability) of diversity (from the perspective of culture with diversity, animal watching, and leisure activities as a case study here) are not to the extent that the respective defenders would like us to believe.

Furthermore, the analysis of diversity, animal watching, and leisure activities is to reveal the nature of the ontological principles in existential dialectics, and good examples include the formalness-informalness principle, the absoluteness-relativeness principle, the partiality-totality principle, the predictability-unpredictability principle, the explicability-inexplicability principle, the fiction-reality principle, the cognitiveness-noncognitiveness principle, the finiteness-transfiniteness principle, the preciseness-vagueness principle, the simpleness-complicatedness principle, the openness-hiddenness principle, the denseness-emptiness principle, the change-constancy principle, the order-chaos principle, the slowness-quickness principle, the expansion-contraction principle, the theory-praxis principle, the convention-novelty principle, the evolution-transformation principle, the symmetry-asymmetry principle, the softness-hardness principle, the seriousness-playfulness principle, the activeness-inactivess principle, the selfness-otherness principle, the regression-progression principle, the same-difference principle, the stability-reaction principle, the functionality-nonfunctionality principle, and the intentionality-nonintentionality principle.

For instance, in relation to the formalness-informalness principle, if there is formalness (e.g., the formal logical requirement of soundness and completeness in a system of ideas, as in animal watching), there is informalness (e.g., the non-formal existence of unsoundness and incompleteness in animal watching, as shown in the criticisms that there are philosophical disagreements in regard to the use of biodiversity for leisure activities, that what is considered as a leisure activity in a culture may not be so at all in another culture which may choose the use of different animals or natural resources instead, that it is gendered-relative, that there is a concern with "animal welfare," etc.). And the reverse direction also holds true.

In relation to the absoluteness-relativeness principle, if there is absoluteness (e.g., the absolute view of birdwatching "in Britain during the

Victorian Era" mostly for "collection based taxonomy"), there is relativeness (e.g., what is right for thoe birdwatchers "in Britain during the Victorian Era," mostly for "collection based taxonomy," is not necessarily so for "the BOU," which "changed focus to ecology and behaviour...in the 1940s" for "the 'scientification' of the pastime"). And the reverse direction also holds true.

In relation to the partiality-totality principle, if there is partiality (e.g., the partial view of birdwatching "in Britain during the Victorian Era," mostly for "collection based taxonomy"), there is totality (e.g., the more holistic view of birdwatching, such that the whole is not the sum of its parts, that is, it cannot be reduced to the sum of opposing views like those "in Britain during the Victorian Era," those in "the BOU...in the 1940s," etc., since there will emerge new views in the future not yet known today, just as there are already other views in the history of birdwatching nowadays aforecited). And the reverse direction also holds true.

In relation to the predictability-unpredictability principle, if there is predictability (e.g., the predictable tendency of those experienced aquarists in PETA to care about "animal welfare"), there is unpredictability (e.g., the more difficult task to predict exactly the extent to which animal welfare will be successfully maintained on a particular future occasion, since there can be some "inexperienced aquarists" who "often keep too many fish in one tank, or add fish too quickly into an immature aquarium, killing many of them"). And the reverse direction also holds true.

In relation to the explicability-inexplicability principle, if there is explicability (e.g., the explanation, by the advocates of biodiversity for its own sake, in that "biodiversity has intrinsic aesthetic and spiritual value to mankind in and of itself"), there is inexplicability (e.g., the lack of sufficient explanation, by the advocates of biodiversity for its own sake, of why biodiversity necessarily "has intrinsic aesthetic and spiritual value to mankind in and of itself," since, for the critics, others disagree, because they instead suggest that "tropical forests and other ecological realms are only worthy of conservation because of the services they provide"). And the reverse direction also holds true.

In relation to the fiction-reality principle, if there is fiction (e.g., the fictional aspect of the view by the RSPB in the older days, when it falsely "claimed that the 'scientification' of the pastime was 'undesirable,'" as "this stand was to change only in 1936 when the RSPB was taken over by Tom Harrisson and others"), there is reality (e.g., the realistic aspect of the view by the RSPB, when "Harrisson was instrumental in the organization of pioneering surveys of the Great Crested Grebe," instead of dismissing the study of ecology and animal behaviour as was the case in the older days). And the reverse direction also holds true.

In relation to the cognitiveness-noncognitiveness principle, if there is cognitiveness (e.g., a view of things on the basis of reason and evidences, as shown in the study of birds on the basis of "scientification"), there is noncognitiveness (e.g., a view of things on the basis of non-cognitive factors like envy, jealousy, power, nationality, status, race, gender, age, class, greed, lust, status, anger, sadness, joy, fear, wish, etc., as shown in the research finding that, for some birdwatchers, "identification of birds may be a form of gaining status which has been compared with Kula valuables noted in Papua New Guinean cultures"). (WK 2012dd; J. Liep 2001) And the reverse direction also holds true.

In relation to the finiteness-transfiniteness principle, if there is finiteness (e.g., the finite number of the categories of water that fish can live in, namely, "freshwater, brackish, and marine"), there is transfiniteness (e.g., the transfinite number of all the individual fish in history of the Earth that have ever existed, regardless of whether there are in freshwater, brackish, or marine environments). And the reverse direction also holds true.

In relation to the preciseness-vagueness principle, if there is preciseness (e.g., the precise identification of three types of water that fish can live in, namely, "freshwater, brackish, and marine" in the article aforecited), there is vagueness (e.g., the vagueness in the identification of the three types of water that fish can live in, namely, "freshwater, brackish, and marine" in the article aforecited, since it is not clear why there must be only 3, not 4, 5, 6, and so on). And the reverse direction also holds true.

In relation to the simpleness-complicatedness principle, if there is simpleness (e.g., the relatively simple analysis of animal watching by the advocates), there is complicatedness (e.g., the relatively more complicated analysis of animal watching, by challenging the claims and assumptions, as shown in the problems pointed out by the critics as cited). And the reverse direction also holds true.

In relation to the openness-hiddenness principle, if there is openness (e.g., the open exploration of how to protect bird by the RSPB), there is hiddenness (e.g., the hidden bias in bird protection by the RSPB, because it used to claim that "the 'scientification' of the pastime was 'undesirable'" before "this stand was to change only in 1936"). (WK 2012dd; K. Johnson 2004) And the reverse direction also holds true.

In relation to the denseness-emptiness principle, if there is denseness (e.g., the relatively denser concentration of concerns with "collection based taxonomy" by the AOU in the early days of birdwatching), there is emptiness (e.g., the relatively less dense, or more empty, concentration of concerns with "collection based taxonomy" by the BOU, when it "changed focus to ecology and behaviour…in the 1940s"). (WK 2012dd; K. Johnson 2004) And the reverse direction also holds true.

In relation to the change-constancy principle, if there is change (e.g., the ever new ways to understand animal behaviors, as shown in birdwatching and fishkeeping), there is constancy (e.g., the ever constant existence of different problems with the ever new ways to understand animal behaviors, as shown in the problems with birdwatching and fishkeeping pointed out by the critics). And the reverse direction also holds true.

In relation to the order-chaos principle, if there is order (e.g., the more or less orderly view of birdwatching, if looked from the vantage point of "certain locations such as the local patch of forest, wetland and coast," which "may be favoured according to the location and season," like. "seawatching…where observers based at a coastal watch point, such as a headland, watch birds flying over the sea"), there is chaos (e.g., the more or less chaotic view of birdwatching, if looked from the vantage point of often changing weather conditions, such as "in Britain, suitable wind conditions may lead to drift migration, and an influx of birds from the east," and "in North America, birds caught in the tail-end of a hurricane may be blown inland," and so on). (WK 2012dd; S. Moss 1995) And the reverse direction also holds true.

In relation to the slowness-quickness principle, if there is slowness (e.g., the relatively slower speed of fish to be killed by experienced aquarists in PETA who care about "animal welfare"), there is quickness (e.g., the relatively faster speed of fish to be killed by "inexperienced aquarists" who "often keep too many fish in one tank, or add fish too quickly into an immature aquarium, killing many of them"). And the reverse direction also holds true.

In relation to the expansion-contraction principle, if there is expansion (e.g., the relative expansion, nowadays, of different "objects such as the 'AquaBabies Micro Aquaria,' 'Bubble Gear Bubble Bag,' and 'Betta in a Vase,' all of which house live fish in unfiltered and insufficient water"), there is contraction (e.g., the relative decline of "the awarding of goldfish as prizes at funfairs," as shown by "the United Kingdom," which "outlawed live-animal prizes such as goldfish in 2004"). (WK 2012q; BBC 2007) And the reverse direction also holds true.

In relation to the theory-praxis principle, if there is theory (e.g., the theoretical construction of the "scientific" method in epistemology and philosophy of science), there is praxis (e.g., the practical application of the "scientific" method in epistemology and philosophy of science to the field of birdwatching, as shown in the decision by "the BOU," which "changed focus to ecology and behaviour…in the 1940s" so as to study birds in a more "scientfic" way, that is, "the 'scientification' of the pastime"). And the reverse direction also holds true.

In relation to the convention-novelty principle, if there is convention (e.g., the conventional wisdom about birdwatching "in Britain during the Victorian Era," mostly for "collection based taxonomy"), there is novelty (e.g., the alternative novel challenge to the conventional wisdom about about birdwatching "in Britain during the Victorian Era," by the new idea of the BOU," which "changed focus to ecology and behaviour…in the 1940s" so as to study birds in a more "scientfic" way, that is, "the 'scientification' of the pastime"). And the reverse direction also holds true.

In relation to the evolution-transformation principle, if there is evolution (e.g., the natural evolution of humans to deal with animals like fish and birds in the state of nature in a primitive way), there is transformation (e.g., the technical transformation of human ability to study animals by the invention of "birdwatching" and "fishkeeping" in a systematic way, as shown in the decision of the BOU," which "changed focus to ecology and behaviour…in the 1940s" so as to study birds in a more "scientfic" way). And the reverse direction also holds true.

In relation to the symmetry-asymmetry principle, if there is symmetry (e.g., the existence of different names for observing birds), there is asymmetry (e.g., "in North America…the term *birder* is unknown to most lay people"—whereas the term "birdwatching" is more popular). And the reverse direction also holds true.

In relation to the softness-hardness principle, if there is softness (e.g., the care of fish welfare by aquarian activists), there is hardness (e.g., the killing of fish by inexperienced aquarists). And the reverse direction also holds true.

In relation to the seriousness-playfulness principle, if there is seriousness (e.g., the serious business of those who try to classify birdwatching activities), there is playfulness (e.g., the playful part of those who try to classify birdwatching activities, when they play around with different conceptual distinctions over the decades, like "birding," "birdwatching," "twitching," "chasing," etc.). And the reverse direction also holds true.

In relation to the activeness-inactivess principle, if there is activeness (e.g., "the most active times of the year for birding in temperate zones are during the spring or fall migrations when the greatest variety of birds may be seen"), there is inactiveness (e.g., the most inactive times of the year for birding in temperate zones are outside the periods of the spring or fall migrations, when the least variety of birds may be seen). (WK 2012dd) And the reverse direction also holds true.

In relation to the selfness-otherness principle, if there is selfness (e.g., the tendency to be mostly centered on one's self or group, as shown by those who try to win "goldfish as prizes at funfairs"), there is otherness (e.g., the tendency to be mostly centered on others, as shown by those aquarian activists who try to protect fish welfare by campaigning for banning the game of winning "goldfish as prizes at funfairs," since they succeeded to persuade "the United Kingdom," which "outlawed live-animal prizes such as goldfish in 2004"). (WK 2012q) And the reverse direction also holds true.

In relation to the regression-progression principle, if there is regression (e.g., the regression made by animal observing, as shown in the problems with fishkeeping and birdwatching pointed out by the critics), there is progression (e.g., the progress made by animal observing, as shown in its contribution to a better understanding of fish and bird behaviors aforecited). And the reverse direction also holds true.

In relation to the same-difference principle, if there is similarity in outcome (e.g., the contribution to the molding and control of beliefs, values, and behaviors, regardless of whether this be done by way of banning the game of winning "goldfish as prizes at funfairs," or by way of winning "goldfish as prizes at funfairs"), there is difference in outcome (e.g., the contribution to the molding and control of beliefs, values, and behaviors by way of banning the game of winning "goldfish as prizes at funfairs," for a relatively more compassionate lifeworld for animal welfare—but the contribution to the molding and control of beliefs, values, and behaviors by way of winning "goldfish as prizes at funfairs" for a relatively more competitive lifeworld for material gains). And the reverse direction also holds true.

In relation to the stability-reaction principle, if there is stability (e.g., the relatively more stable condition of winning "goldfish as prizes at funfairs" in the older days, as a "tradition…in many parts of the world"), there is reaction (e.g., the tradition of winning "goldfish as prizes at funfairs" in the older days then led to further changes later on, when, nowadays, for example, "the United Kingdom outlawed live-animal prizes such as goldfish in 2004"). And the reverse direction also holds true.

In relation to the functionality-nonfunctionality principle, if there is functionality (e.g., the presence of action for which something is fitted, as shown in "the study of birds and natural history…in Britain during the Victorian Era," which served the function of "collection based taxonomy" at the time), there is nonfunctionality (e.g., the relative lack of action for which something is fitted, as shown in the decision of the BOU," which "changed focus to ecology and behaviour…in the 1940s" so as to study birds in a more "scientfic" way, so that birdwatching solely for "collection based taxonomy" is no longer functional in our time, which has different expectations and needs). And the reverse direction also holds true.

And in relation to the intentionality-nonintentionality principle, if there is intentionality (e.g., the planning or design of something for a certain outcome, as shown by aquarists who keep fish alive so as to enjoy them as a leisure activity), there is nonintentionality (e.g., the relative absence of planning or design of something for a certain outcome, such that the outcome can even be counter-intentional in being the opposite of what was originally planned, as shown by "inexperienced aquarists," who "often keep too many fish in one tank, or add fish too quickly into an immature aquarium, killing many of them," contrary to the original goal of keeping them alive as a leisure activity). And the reverse direction also holds true.

The Undesirability of Diversity

This comprehensive analysis of the possibility (or impossibility) and desirability (or undesirability) of diversity in the context of natural resources, from the perspectives of the mind, nature, society, and culture, is vital in revealing the different ways in which diversity is both possible and desirable, but not to the extent that the spokespersons from each side would like us to believe.

In this sense, diversity thus has its undesirability, just as it has its desirability too, as the other side of the same coin.

But this is only one part of a larger story, since there is the other side of the story, which concerns discontinuity, the opposite of diversity. Since this chapter already deals with diversity, the next chapter will therefore address the issue of discontinuity.

With this in mind, let's now turn to Chapter Three for the study of discontinuity and its dual effects—for which we now turn to chapter three.

Table 2.1. Diversity and Its Ambivalent Characters

- **Diversity and the Mind**
  —Ex: diversity, psychological benefits, and biology
  —Ex: diversity, dietary health, and agriculture

- **Diversity and Nature**
  —Ex: diversity, conservation biology, and environmental protection
  —Ex: diversity, ecology, and adaptive management

- **Diversity and Society**
  —Ex: diversity, cloning, and the role of technology
  —Ex: diversity, natural capital, and ecological economics

- **Diversity and Culture**
  —Ex: diversity, environmental management, and ethics
  —Ex: diversity, animal watching, and leisure activities

Notes: The examples in the categories are solely illustrative (not exhaustive), and the comparison is relative (not absolute), nor are they necessarily mutually exclusive. And some can be easily re-classified elsewhere. As generalities, they allow exceptions.
Source: A summary of Ch.2 of *BNR*

Table 2.2. Diversity, Psychological Benefits, and Biology

- **Types of Biodiversity**
  —Ex: "molecular diversity"
  —Ex: "genetic diversity"
  —Ex: "species diversity"
  —Ex: "ecosystem diversity"

- **Distribution of Biodiversity**
  —Ex: "latitudinal gradients"
  —Ex: "hotspots"

- **Evolution of Biodiversity**
  —Ex: biodiversity has gone through a long history on the Earth, as it is "the result of 3.5 billion years of evolution."

- **Psychological Benefits of Biodiversity**
  —Ex: "some of the health issues influenced by biodiversity include dietary health and nutrition security, infectious disease, medical science and medicinal resources, social and psychological health."

- **Problems with the Study of Psychological Benefits and Biodiversity**
  —First, biodiversity is not stable and subject to different threats.
  —Second, there are major extinctions in the history of biodiveristy.
  —Third, the actual evidences are not sufficient.
  —Fourth, the causal relationship is not so clearcut as the advocates would like us to believe.
  —Fifth, there are many different causes of human health problems.
  —Sixth, there can be alternative resources for health improvement, even when biodiversity occurs.
  —And seventh, different levels of diversity may not correspond.

Notes: The examples in the categories are solely illustrative (not exhaustive), and the comparison is relative (not absolute), nor are they necessarily mutually exclusive. And some can be easily re-classified elsewhere. As generalities, they allow exceptions.
Source: From *Sec. 2.2.1* of *BNR.* See book for citations.

Table 2.3. Diversity, Conservation Biology, and Environmental Protection

- **History of the Protection of Natural Resources**
  —Ex: "natural resource conservation"
  —Ex: "early naturalists"
  —Ex: "19th century conservation "
  —Ex: "20th century conservation"

- **Approaches to to the Protection of Natural Resources**
  —Ex: "voluntary environmental agreements"
  —Ex: "ecosystems approach"
  —Ex: "international environmental agreements"

- **Planning of the Protection of Natural Resources**
  —Ex: "compile data on the biodiversity of the planning region"
  —Ex: "identify conservation goals for the planning region"
  —Ex: "review existing conservation areas"
  —Ex: "select additional conservation areas"
  —Ex: "implement conservation actions"
  —Ex: "maintain the required values of conservation areas"

- **Problems with the Protection of Natural Resources**
  —First, it does not give sufficient attention to the important problem of poor management.
  —Second, it often ignores the factors of power abuse and financial shortage.
  —Third, different approaches have different weaknesses.
  —Fourth, the priority to focus on hotspots for biodiveristy in the protection of natural resources is misleading and dangerous.
  —And fifth, it is too utilitarian in ethical orientation.

Notes: The examples in the categories are solely illustrative (not exhaustive), and the comparison is relative (not absolute), nor are they necessarily mutually exclusive. And some can be easily re-classified elsewhere. As generalities, they allow exceptions.
Source: From *Sec. 2.3.1* of *BNR*. See book for citations.

Table 2.4. Diversity, Ecology, and Adaptive Management

- **Features of Adaptive Management**
  —Ex: "iterative decision-making"
  —Ex: "feedback between monitoring and decisions (learning)"
  —Ex: "system uncertainty through multi-model inference"
  —Ex: "Bayesian [subjective] inference"
  —Ex: "embracing risk and uncertainty"

- **Processes of Adaptive Management**
  —Ex: "appropriate temporal and spatial scales"
  —Ex: "a focus on statistical power and controls"
  —Ex: "computer models" for "an embodied ecological consensus."
  —Ex: "ecological consensus to evaluate strategic alternatives"
  —Ex: "communication of alternatives to political arena"

- **Planning of the Protection of Natural Resources**
  —Ex: "compile data on the biodiversity of the planning region"
  —Ex: "identify conservation goals for the planning region"
  —Ex: "review existing conservation areas"
  —Ex: "select additional conservation areas"
  —Ex: "implement conservation actions"
  —Ex: "maintain the required values of conservation areas"

- **Problems with the Study of Adaptive Management**
  —First, it is risky enough.
  —Second, it suffers from "a common lament" about "significant uncertainty". (WK 2012n; J. Arvai 2006)
  —Third, adaptive management has not yet been widely accepted in the corporate world.
  —And fourth, scientists still disagree about the features of adaptive management.

Notes: The examples in the categories are solely illustrative (not exhaustive), and the comparison is relative (not absolute), nor are they necessarily mutually exclusive. And some can be easily re-classified elsewhere. As generalities, they allow exceptions.
Source: From *Sec. 2.3.2* of *BNR.* See book for citations.

Table 2.5. Diversity, Cloning, and the Role of Technology

- **Cloning, and the Restoration of Life**
  —Ex: with the technology of cloning, it is possible to restore the life of an extinct species for biodiversity.

- **Types of Cloning**
  —Ex: "molecular cloning"
  —Ex: "cell cloning"
  —Ex: "organism cloning"

- **Problems with the Technology of Cloning**
  —Firstly, one main criticism of this study of the restoration of life and the technology of cloning is that it is still technological infeasible in our time.
  —Secondly, another main criticism of this study of the restoration of life and the technology of cloning is that there remain some deep ethical issues to be resolved.
  —Thirdly, still another main criticism of the study of the restoration of life and the technology of cloning is that what counts as extinct species may turn out to be pseudoextinct ones, so the restoration of life through cloning may be misleading.
  —Fourthly, an additional main criticism of the study of the restoration of life and the technology of cloning is that cloning does not restore an exactly identical life.
  —Fifthly, another additional main criticism of the study of the restoration of life and the technology of cloning is that the more difficult question of how to raise the restored species (in the absence of parents) remains unresolved, so it may not be viable in the long run.
  —And sixthly, still another additional main criticism of the study of the restoration of life and the technology of cloning is that cloning distracts the attention and takes away funding from solving the serious issue of animal extinction.

Notes: The examples in the categories are solely illustrative (not exhaustive), and the comparison is relative (not absolute), nor are they necessarily mutually exclusive. And some can be easily re-classified elsewhere. As generalities, they allow exceptions.
Source: From *Sec. 2.4.1* of *BNR*. See book for citations.

Table 2.6. Diversity, Environmental Management, and Ethics

- **Anthropocentrism, and Environmental Management**
  - —Ex: philosophical
  - —Ex: religious
  - —Ex: evolutionary

- **Ecocentrism, and Environmental Management**
  - —Ex: against anthropocentrism
  - —Ex: against technocentrism
  - —Ex: against biocentrism

- **Problems with Environmental Ethics**
  - —Firstly, one main criticism of environmental ethics is that it is divided between competing schools of thought.
  - —Secondly, another main criticism of environmental ethics is that its reasoning is simplistic.
  - —Thirdly, still another main criticism of environmental ethics is that its implementation can be destructive.
  - —Fourthly, an additional main criticism of environmental ethics is that it is self-defeating.
  - —And fifthly, another additional main criticism of environmental ethics is that it focuses too much on the whole without sufficient attention to the individuals.

Notes: The examples in the categories are solely illustrative (not exhaustive), and the comparison is relative (not absolute), nor are they necessarily mutually exclusive. And some can be easily re-classified elsewhere. As generalities, they allow exceptions.
Source: From *Sec. 2.5.1* of *BNR*. See book for citations.

Table 2.7. Diversity, Animal Watching, and Leisure Activities

- **The Leisure Activity of Birdwatching**
  —Ex: birders vs. birdwatchers
  —Ex: history of birdwatching

- **The Leisure Activity of Fishkeeping**
  —Ex: types of fishkeeping
  —Ex: history of fishkeeping

- **Problems with Animal Watching**
  —Firstly, one main criticism of animal watching is that there are philosophical disagreements in regard to the use of biodiversity for leisure activities.
  —Secondly, another main criticism of animal watching is that what is considered as a leisure activity in a culture may not be so at all in another culture, which may choose the use of different animals or natural resources instead.
  —Thirdly, still another main criticism of animal watching is that it is gendered-relative.
  —And fourthly, an additional main criticism of animal watching is that there is a concern with "animal welfare."

Notes: The examples in the categories are solely illustrative (not exhaustive), and the comparison is relative (not absolute), nor are they necessarily mutually exclusive. And some can be easily re-classified elsewhere. As generalities, they allow exceptions.
Source: From *Sec. 2.5.2* of *BNR*. See book for citations.

• PART THREE •

Discontinuity

CHAPTER 3
DISCONTINUITY AND ITS DUAL EFFECTS

> Birth and death are so closely related that one could not destroy either without destroying the other at the same time. It is extinction that makes creation possible."
> —Samuel Butler (TE 2012a)

The Malignity of Discontinuity

As already indicated in *Sec. 1.3*—the study of natural resources involves both diversity and discontinuity.

This dialectic relationship between diversity and discontinuity means that there is no diversity without discontinuity—and vice versa.

One way to explore this dialectic relationship between diversity and discontinuity is by showing how and why the analysis of diversity (in Chapter Two) is inadequate without the inclusion of discontinuity, just as the examination of discontinuity (in Chapter Three) is insufficient without the consideration of diversity.

The issues concerning diversity in the study of natural resources were already addressed in Chapter Two, so what is left is the issues concerning discontinuity in the study of natural resources, which are analyzed in this chapter (Chapter Three).

With this dialectic treatment in mind—a good way to examine discontinuity and its dual effects in the study of natural resources is by way of the evaluation of the extent to which it is in fact both possible (or impossible) and desirable (or undesirable).

This can be done by way of a comprehensive analysis of discontinuity in the context of natural resources from the four perspectives of the mind, nature, society, and culture (in accordance to my sophisticated methodological holism, as explained in *Sec. 1.9*).

In other words, this chapter is organized in four main sections, namely, in relation to (3.2) discontinuity and the mind, (3.3) discontinuity and nature, (3.4) discontinuity and society, and (3.5) discontinuity and culture—to be addressed hereafter, in that order (and summarized in *Table 3.1*).

Discontinuity and the Mind

Discontinuity, when examined from the perspective of the mind, can illuminate its possibility (or impossibility) and desirability (or undesirability), and this can be shown by way of two case studies, namely, (3.2.1) discontinuity, extinction, and overconsumption, and (3.2.2) discontinuity, biodiversity loss, and health risks—to be addressed in what follows, respectively.

Discontinuity, Extinction, and Overconsumption

A good way to evaluate the possibility (or impossibility) and desirability (or undesirability) of discontinuity from the perspective of the mind is to explore discontinuity, extinction, and overconusmption.

An argument here is that there is a relationship between extinction and overconsumption, because human overconsumption can lead to the "overexploitation" of animals and other resources. (WK 2012i)

And the "overexploitation" of animals and other resources can be explained by "the overkill hypothesis" for a relatively recent extinction event, when it "explains why earlier megafaunal extinctions occurred within a relatively short period of time. This can be connected with human migration." (WK 2012i; D. Burney 2005)

In fact, "the overkill hypothesis, a variant of the hunting hypothesis, was proposed 40 years ago by Paul S. Martin, Professor of Geosciences Emeritus at the Desert Laboratory of the University of Arizona. It sparked debate which continues today....The most convincing evidence of his theory is that 80% of the North American large mammal species disappeared within 1000 years of the arrival of humans in the Americas." (WK 2012ee)

History of Extinction

At first glance, extinction serves vital functions (or produces important consequences) for natural evolution. For instance, during the "Cretaceous-Paleogene (or K–Pg) Extinction Event" which "occurred approximately 65.5 million years ago (Ma)," although it is true that "the majority of non-avian dinosaurs became extinct during that time," the important point to remember is that."mammals and birds emerged as dominant land vertebrates in the age of new life." (WK 2012s)

With these dual effects of extinction for natural evolution in mind—in 1982, Jack Sepkoski and David M. Raup wrote "a landmark paper," in which they "identified five mass extinctions," which were already briefly showed in *Sec. 2.2.1* but are more elaborated hereafter (and summarized in *Table 3.2*):

- "The Cretaceous-Paleogene (or K–Pg) Extinction Event"—which is "the most recent" and "occurred approximately 65.5 million years ago (Ma)….About 17% of all families, 50% of all genera and 75% of species became extinct. In the seas it reduced the percentage of sessile animals to about 33%. The majority of non-avian dinosaurs became extinct during that time. The boundary event was severe with a significant amount of variability in the rate of extinction between and among different clades. Mammals and birds emerged as dominant land vertebrates in the age of new life." (WK 2012s; J. Baez 2006; D. Fastovsky 2005; N. Macleod 1997)
- "The Triassic-Jurassic Extinction Event (End Triassic)"—which occurred "205 millioin years ago at the Triassic-Jurassic transition. About 23% of all families and 48% of all genera (20% of marine families and 55% of marine genera) went extinct. Most non-dinosaurian archosaurs, most therapsids, and most of the large amphibians were eliminated, leaving dinosaurs with little terrestrial competition. Non-dinosaurian archosaurs continued to dominate aquatic environments, while non-archosaurian diapsids continued to dominate marine environments. The Temnospondyl lineage of large amphibians also survived until the Cretaceous in Australia (e.g., *Koolasuchus*)." (WK 2012s; J. Baez 2006)
- "The Permian-Triassic Extinction Event (End Permian)"—which occurred "251 million years ago at the Permian-Triassic transition. Earth's largest extinction killed 57% of all families and 83% of all genera (53% of marine families, 84% of marine genera, about 96% of all marine species and an estimated 70% of land species) including insects. The evidence of plants is less clear, but new taxa became dominant after the extinction. The 'Great Dying' had enormous evolutionary significance: on land, it ended the primacy of mammal-like reptiles. The recovery of vertebrates took 30 million years, but the vacant niches created the opportunity for archosaurs to become

ascendant. In the seas, the percentage of animals that were sessile dropped from 67% to 50%. The whole late Permian was a difficult time for at least marine life, even before the 'Great Dying.'" (WK 2012s; C. Labandeira 1993; J. Baez 2006; J. McElwain 2007; S. Sahney 2008)

- "The Late Devonian Extinction"—which occurred "375-360 million years ago near the Devonian-Carboniferous transition. At the end of the Frasnian Age in the later part(s) of the Devonian Period, a prolonged series of extinctions eliminated about 19% of all families, 50% of all genera and 70% of all species. This extinction event lasted perhaps as long as 20 million years, and there is evidence for a series of extinction pulses within this period." (WK 2012s; J. Baez 2006)
- "The Ordovician–Silurian Extinction Event (End Ordovician or O-S)"—which occurred "450-440 million years ago at the Ordovician-Silurian transition. Two events occurred that killed off 27% of all families and 57% of all genera. Together they are ranked by many scientists as the second largest of the five major extinctions in Earth's history in terms of percentage of genera that went extinct." (WK 2012s; J. Baez 2006)

Thus, these extinction events had caused "periodic, massive losses of diversity," while allowing other species to emerge and dominate. (WK 2012s)

History of Overexploitation

So a good question to ask is, What then is the explanation of these extinction events? There are different answers, of course, but one of them has to do "overexploitation."

The phenomenon of "overexploitation" occurs, "when a resource is consumed at an unsustainable rate. This occurs on land in the form of overhunting, excessive logging, poor soil conservation in agriculture and the illegal wildlife trade. Joe Walston, director of the Wildlife Conservation Society's Asian programs, called the latter the 'single largest threat' to biodiversity in Asia. The international trade of endangered species is second in size only to drug trafficking." (WK 2012i; NYT 2010; B. Weirum 2007)

Historically, "concern about overexploitation is relatively recent, though overexploitation itself is not a new phenomenon. It has been observed for millennia. For example, ceremonial cloaks worn by the Hawaiian kings were made from the mamo bird; a single cloak used the feathers of 70,000 birds of this now-extinct species. The dodo, a flightless bird from Mauritius, is another well known example of overexploitation. As with many island species, it was naive about certain predators, allowing humans to approach and kill it with ease." (WK 2012j; J. Fryer 2002)

After all, "hunting for wild mammals and birds has been an important human activity as a means of survival. There is a whole history of overexploitation in the form of overhunting. The overkill hypothesis (quaternary extinction events) explains why the megafaunal extinctions occurred within a relatively short period of time. This can be traced with human migration. The most convincing evidence of this theory is that 80% of the North American large mammal species disappeared within 1000 years of the arrival of humans on the western hemisphere continents. The fastest ever recorded extinction of megafauna occurred in New Zealand, where by 1500 AD, just 200 years after settling the islands, ten species of the giant moa birds were hunted to extinction by the Māori. A second wave of extinctions occurred later with European settlement." (WK 2012j; R. Holdaway 2000)

For illustration, consider three major types of natural resources in what follows, in the context of overexploitation (as summarized in *Table 3.2*).

(a1) Fisheries

The first major type of natural resources in the context of overexploitation has to do with "fisheries."

In regard to fisheries, "overexploitation or overfishing occurs when a fish stock has been fished down 'below the size that, on average, would support the long-term maximum sustainable yield of the fishery'....When a fishery starts harvesting fish from a previously unexploited stock, the biomass of the fish stock will decrease, since harvesting means fish are being removed. For sustainability, the rate at which the fish replenish biomass through reproduction must balance the rate at which the fish are being harvested. If the harvest rate is increased, then the stock biomass will further decrease. At a certain point, the maximum harvest yield that can be sustained will be reached, and further attempts to increase the harvest rate will result in the collapse of the fishery. This point is called the maximum sustainable yield, and in practice, usually occurs when the fishery has been fished down to about 30% of the biomass it had before harvesting started." (WK 2012j) E. Bolden 1999)

Nowadays, "about 25% of world fisheries are...overexploited to the point where their current biomass is less than the level that maximizes their sustainable yield." (WK 2012i; R. Grafton 2007) (a2) Water Resources

The second major type of natural resources in the context of overexploitation has to do with "water resources."

In regard to water resource, "overexploitation occurs if a water resource, such as the Ogallala Aquifer, is mined or extracted at a rate that exceeds the recharge rate, that is, at a rate that exceeds the practical sustained yield. Recharge usually comes from area streams, rivers and lakes. An aquifer which has been overexploited is said to be overdrafted or depleted. Forests enhance the recharge of aquifers in some locales, although generally forests are a major source of aquifer depletion. Depleted aquifers can become polluted with contaminants such as nitrates, or permanently damaged through subsidence or through saline intrusion from the ocean." (WK 2012j)

Unfortunately, "this turns much of the world's underground water and lakes into finite resources with peak usage debates similar to oil. These debates usually centre around agriculture and suburban water usage but generation of electricity from nuclear energy or coal and tar sands mining is also water resource intensive." (WK 2012j; USWN 2006)

(a3) Forest Resources

And the third major type of natural resources in the context of overexploitation has to do with "forest resources."

In regard to forest resources, overexploitation occurs "when they are logged at a rate faster than reforestation takes place. Reforestation competes with other land uses such as food production, livestock grazing, and living space for further economic growth. Historically utilization of forest products, including timber and fuel wood, have played a key role in human societies, comparable to the roles of water and cultivable land." (WK 2012j)

Nowadays, "developed countries continue to utilize timber for building houses, and wood pulp for paper. In developing countries almost three billion people rely on wood for heating and cooking. Short-term economic gains made by conversion of forest to agriculture, or overexploitation of wood products, typically leads to loss of long-term income and long term biological productivity. West Africa, Madagascar, Southeast Asia and many other regions have experienced lower revenue because of overexploitation and the consequent declining timber harvests." (WK 2012j)

Problems with the Study of Extinction and Overconsumption

However, for the critics, the study of extinction and overconsumption has some major problems. Consider, for illustration, a few examples below (as summarized in *Table 3.2*).

Firstly, one main criticism of the study of extinction and overconsumption is that the extinction events cannot be so clearly defined as the advocates would like us to believe.

For example, for the critics, "as more stringent statistical tests have been applied to the accumulating data [by Jack Sepkoski and David M. Raup], the 'Big Five' [extinction events] cannot be so clearly defined." (WK 2012s; J. Alroy 2001)

Secondly, another main criticism of the study of extinction and overconsumption is that there are different threats to species, and overconsumption can be only one of them.

For instance, for the critics, there are other important threats like "habitat destruction," "genetic pollution/erosion," "climate change," "human overpopulation," etc. (WK 2012i)

Thirdly, still another main criticism of the study of extinction and overconsumption is that overexploitation does not necessarily lead to extinction of animals and other resources.

For instance, for the critics, "overexploitation need not necessarily lead to the destruction of the resource, nor is it necessarily unsustainable," although "depleting the numbers or amount of the resource can change its quality. For example, footstool palm is a wild palm tree found in Southeast Asia. Its leaves are used for thatching and food wrapping, and overharvesting has resulted in its leaf size becoming smaller," but not in extinction. (WK 2012j)

Fourthly, an additional main criticism of the study of extinction and overconsumption is that overexploitation can be prevented.

For instance, for the critics, "the tragedy of the commons can be avoided within the context of fisheries if fishing effort and practices are regulated appropriately by fisheries management. One effective approach may be assigning some measure of ownership in the form of individual transferable quotas (ITQs) to fishermen. In 2008, a large scale study of fisheries that used ITQs, and ones that didn't, provided strong evidence that ITQs help prevent collapses and restore fisheries that appear to be in decline." (WK 2012j; ECO 2008; D. MacKenzie 2008)

And fifthly, another additional main criticism of the study of extinction and overconsumption is that extinction has important functions for natural evolution.

For instance, for the critics, during the "Cretaceous-Paleogene (or K–Pg) Extinction Event,...mammals and birds emerged as dominant land vertebrates in the age of new life." (WK 2012s)

Beyond the Study of Extinction and Overconsumption

These criticisms of the study of extinction and overconsumption, however, should not be misunderstood as a total rejection of its validity but serve to show us the two opposing sides of the debate, such that the possibility (or impossibility) and desirability (or undesirability) of discontinuity (from the perspective of the mind with discontinuity, extinction, and overconsumption as a case study here) are not to the extent that the respective defenders would like us to believe.

Besides, the analysis of discontinuity, extinction, and overconsumption can cast a bright light on the working of the ontological principles in existential dialectics, and good examples include the formalness-informalness principle, the absoluteness-relativeness principle, the partiality-totality principle, the predictability-unpredictability principle, the explicability-inexplicability principle, the fiction-reality principle, the cognitiveness-noncognitiveness principle, the finiteness-transfiniteness principle, the preciseness-vagueness principle, the simpleness-complicatedness principle, the openness-hiddenness principle, the denseness-emptiness principle, the change-constancy principle, the order-chaos principle, the slowness-quickness principle, the expansion-contraction principle, the theory-praxis principle, the convention-novelty principle, the evolution-transformation principle, the symmetry-asymmetry principle, the softness-hardness principle, the seriousness-playfulness principle, the activeness-inactivess principle, the selfness-otherness principle, the regression-progression principle, the same-difference principle, the stability-reaction principle, the functionality-nonfunctionality principle, and the intentionality-nonintentionality principle.

For instance, in relation to the formalness-informalness principle, if there is formalness (e.g., the formal logical requirement of soundness and completeness in a system of ideas, as in the study of extinction and overconsumption), there is informalness (e.g., the non-formal existence of unsoundness and incompleteness in the study of extinction and overconsumption, as shown in the criticisms that the extinction events cannot be so clearly defined as the advocates would like us to believe, that there are different threats to species so overconsumption can be only one of them, that overexploitation does not necessarily lead to extinction of animals and other resources, that overexploitation can be prevented, and that extinction has important functions for natural evolution). And the reverse direction also holds true.

In relation to the absoluteness-relativeness principle, if there is absoluteness (e.g., the absolute view by Jack Sepkoski and David M. Raup in regard to "the 'Big Five' [extinction events]"), there is relativeness (e.g., what is true for Jack Sepkoski and David M. Raup in regard to "the 'Big Five' [extinction events]" is not necessarily so for the critics, because, "as more stringent statistical tests have been applied to the accumulating data, the 'Big Five' [extinnction events] cannot be so clearly defined"). And the reverse direction also holds true.

In relation to the partiality-totality principle, if there is partiality (e.g., the partial view of the overkill hypothesis by its advocates), there is totality (e.g., the more holistic view of the overkill hypothesis, such that the whole is not the sum of its parts, that is, it cannot be reduced to the sum of opposing views of the overkill hypothesis by its advocates and the critics, since there will emerge new views in the future not yet known today, as there are already some views nowadays which do not need to exclusively side with any of them, like "some scientists" who instead "support a combination of climate change and overkill"). (WK 2012ee) And the reverse direction also holds true.

In relation to the predictability-unpredictability principle, if there is predictability (e.g., the predictable tendency by the advocates of the overkill hypothesis to claim the occurrence of the overexploitation of natural resources in our time, with its impact on extinction events), there is unpredictability (e.g., the more difficult task to predict exactly the extent to which the overexploitation of natural resources will lead to extinction on a particular future occasion, since, for the critics, "overexploitation need not necessarily lead to the destruction of the resource, nor is it necessarily unsustainable"). (WK 2012j) And the reverse direction also holds true.

In relation to the explicability-inexplicability principle, if there is explicability (e.g., the explanation by the advocates of the overkill hypothesis that overexploitation explains why extinction events occur in history), there is inexplicability (e.g., the lack of sufficient explanation by the advocates of the overkill hypothesis of why overexploitation necessarily explains why extinction events occur in history, since, for the critics, there can be other alternative causes like "habitat destruction," "genetic pollution/erosion," "climate change," "human overpopulation," etc.). And the reverse direction also holds true.

In relation to the fiction-reality principle, if there is fiction (e.g., the fictional aspect of the classification of "the 'Big Five' [extinction events]" by Jack Sepkoski and David M. Raup, insofar as "the 'Big Five' [extinction events]...cannot be so clearly defined" when "more stringent statistical tests have been applied to the accumulating data"), there is reality (e.g., the realistic aspect of the classification of "the 'Big Five' [extinction events]" by Jack Sepkoski and David M. Raup, insofar as extinction events did occur in history). And the reverse direction also holds true.

In relation to the cognitiveness-noncognitiveness principle, if there is cognitiveness (e.g., a view of things on the basis of reason and evidences, as shown in the study of the five extinction events), there is noncognitiveness (e.g., a view of things on the basis of non-cognitive factors like envy, jealousy, power, nationality, race, gender, age, class, greed, lust, status, anger, sadness, joy, fear, wish, etc., as shown in the motivation of some conservation activists who try to use "biodiversity...as a legal standard" to enforce conservation against extinction, without taking into account of "the remaining areas of scientific uncertainty" in regard to extinction and biodiverity loss). (WK 2012i; F. Bosselman 2004) And the reverse direction also holds true.

In relation to the finiteness-transfiniteness principle, if there is finiteness (e.g., the finite number of the categories of lifeforms which suffered from extinction, both partial and complete, during the "Big Five" [extinction events], like insects, worms, fish, amphibians, birds, mammals, etc.), there is transfiniteness (e.g., the transfinite number of all the individual insects, worms, fish, amphibians, birds, mammals, etc. in history of the Earth that had ever suffered from extinction, both partial and complete. during the "Big Five" [extinction events]). And the reverse direction also holds true.

In relation to the preciseness-vagueness principle, if there is preciseness (e.g., the precise identification of three major types of natural resources with the problem of overexploitation, like water, forests, and fish, in the article aforecited), there is vagueness (e.g., the vagueness in the identification of the three major types of natural resources with the problem of overexploitation, like water, forests, and fish, in the article aforecited, since it is not clear why there must be only 3, not 4, 5, 6, 7, 8, and so on). And the reverse direction also holds true.

In relation to the simpleness-complicatedness principle, if there is simpleness (e.g., the relatively simple analysis of the overkill hypothesis by the advocates), there is complicatedness (e.g., the relatively more complicated analysis of the overkill hypothesis, by challenging the claims and assumptions, as shown in the problems pointed out by the critics as cited). And the reverse direction also holds true.

In relation to the openness-hiddenness principle, if there is openness (e.g., the open exploration by Jack Sepkoski and David M. Raup of "the 'Big Five' [extinction events]"), there is hiddenness (e.g., the hidden bias in the classification of "the 'Big Five' [extinction events]"by Jack Sepkoski and David M. Raup, since, for the critics, "as more stringent statistical tests have been applied to the accumulating data, the 'Big Five' [extinnction events] cannot be so clearly defined"). And the reverse direction also holds true.

In relation to the denseness-emptiness principle, if there is denseness (e.g., the relatively denser concentration of extinct species during "the 'Big Five' [extinction events]"), there is emptiness (e.g., the relatively less dense, or more empty, concentration of extinct species outside the historcal eras of "the 'Big Five' [extinction events]"). And the reverse direction also holds true.

In relation to the change-constancy principle, if there is change (e.g., the ever new ways to study discontinuity, as shown in the overkill hypothesis), there is constancy (e.g., the ever constant existence of different problems in the ever new ways to study discontinuity, as shown in the problems with the overkill hypothesis pointed out by the critics). And the reverse direction also holds true.

In relation to the order-chaos principle, if there is order (e.g., the more or less orderly view of natural resources, if looked from the vantage point of other relatively more normal historical periods without all the "extinction events," large or small), there is chaos (e.g., the more or less chaotic view of natural resources, if looked from the vantage point of "earlier megafaunal extinctions" which abruptly "occurred within a relatively short period of time"). (WK 2012i) And the reverse direction also holds true.

In relation to the slowness-quickness principle, if there is slowness (e.g., the relatively slower speed of extinction in those relatively small extinction events, when put within the context of "a relatively smooth continuum of extinction events"), there is quickness (e.g., the relatively quicker speed of extinction in those relatively large extinction events, like "the fastest ever recorded extinction of megafauna" which "occurred in New Zealand, where by 1500 AD, just 200 years after settling the islands, ten species of the giant moa birds were hunted to extinction by the Māori"). (WK 2012j & 2012s)

In relation to the expansion-contraction principle, if there is expansion (e.g., the relative expansion of "mammals and birds" which "emerged as dominant land vertebrates in the age of the Cretaceous-Paleogene"), there is contraction (e.g., the relative decline of "non-avian dinosaurs," which "became extinct" during the age of the Cretaceous-Paleogene). (WK 2012s; J. Baez 2006; D. Fastovsky 2005; N. Macleod 1997) And the reverse direction also holds true.

In relation to the theory-praxis principle, if there is theory (e.g., the theoretical construction of "the overkill hypothesis" in biology), there is praxis (e.g., the practical application of "the overkill hypothesis" in biology to the field of activism, as shown by some conservation activists who try to use "biodiversity…as a legal standard" to enforce conservation against extinction, due to overexploitation). And the reverse direction also holds true.

In relation to the convention-novelty principle, if there is convention (e.g., the conventional wisdom about "the overkill hypothesis" by Paul S. Martin more than 40 years ago), there is novelty (e.g., the alternative novel challenge to the conventional wisdom about "the overkill hypothesis," by the new idea of "the comet hypothesis" which was "first publicly presented at the Spring 2007 joint assembly of the American Geophysical Union in Acapulco, Mexico," in that "the mass extinction was caused by a swarm of comets 12,900 years ago"). (WK 2012ee) And the reverse direction also holds true.

In relation to the evolution-transformation principle, if there is evolution (e.g., the natural evolution of animals to become extinct in the state of nature over time), there is transformation (e.g., the technical transformation of human ability to understand extinction by the invention of "the overkill hypothesis," together with subsequent conservation efforts to save the endangered species). And the reverse direction also holds true.

In relation to the symmetry-asymmetry principle, if there is symmetry (e.g., the existence of different lifeforms like fish and humans), there is asymmetry (e.g., humans can use "one effective approach" like "the individual transferable quotas [ITQs]" which "help prevent collapses and restore fisheries that appear to be in decline"—but fish do not have this ability to save others from extinction). And the reverse direction also holds true.

In relation to the softness-hardness principle, if there is softness (e.g., the saving of fisheries by humans as shown in the use of "one effective approach" known as "the individual transferable quotas [ITQs]"), there is hardness (e.g., the killing of "the dodo, a flightless bird from Mauritius" by humans, which is "another well known example of overexploitation"). And the reverse direction also holds true.

In relation to the seriousness-playfulness principle, if there is seriousness (e.g., the serious business of those who try to explain the occurrence of extinction events in history), there is playfulness (e.g., the playful part of those who try to explain the occurrence of extinction events in history, when they play around with different possible explanations over the decades, namely, "the overkill hypothesis," "the comet hypothesis," "the climate change hypothesis," the "hyperdisease hypothesis," etc.). (WK 2012ee) And the reverse direction also holds true.

In relation to the activeness-inactivess principle, if there is activeness (e.g., the relative activeness of those "non-dinosaurian archosaurs" which did not become sick or were lucky enough to survive in "the Triassic-Jurassic Extinction Event"), there is inactiveness (e.g., the relative inactiveness of those "non-dinosaurian archosaurs" which got sick or could not survive in "the Triassic-Jurassic Extinction Event"). And the reverse direction also holds true.

In relation to the selfness-otherness principle, if there is selfness (e.g., the tendency to be mostly centered on one's self or group, as shown in the killing of "the mamo bird" to the point of overexploitation, for "the ceremonial cloaks worn by the Hawaiian kings"), there is otherness (e.g., the tendency to be mostly centered on others, as shown in the saving of fisheries by those activists who used the "effective approach" known as "the individual transferable quotas [ITQs]"). (WK 2012j) And the reverse direction also holds true.

In relation to the regression-progression principle, if there is regression (e.g., the regression made by the overkill hypothesis, as shown in the problems pointed out by the critics aforecited), there is progression (e.g., the progress made by the overkill hypothesis, as shown in its contribution to a better understanding of the causes of overexploitation aforecited). And the reverse direction also holds true.

In relation to the same-difference principle, if there is similarity in outcome (e.g., the contribution to the molding and control of beliefs, values, and behaviors, regardless of whether this be done by way of overexploitation, or by way of conservation), there is difference in outcome (e.g., the contribution to the molding and control of beliefs, values, and behaviors by way of overexploitation for a relatively more overconsumeristic lifeworld—but the contribution to the molding and control of beliefs, values, and behaviors by way of conservation for a relatively more sustainable lifeworld). And the reverse direction also holds true.

In relation to the stability-reaction principle, if there is stability (e.g., the relatively more stable condition of the overkill hypothesis in the older days, when it was used to explain the extinction event in "the Quaternary period"), there is reaction (e.g., the stable condition of the overkill hypothesis in the older days then led to further changes later on, when nowadays, there are alternative hypotheses, like "the comet hypothesis" "the hyperdisease hypothesis," etc.). And the reverse direction also holds true.

In relation to the functionality-nonfunctionality principle, if there is functionality (e.g., the presence of action for which something is fitted, as shown in the killing of "the mamo bird" in Hawaii in the older days, which served the function of the provision of "the ceremonial cloaks worn by the Hawaiian kings" at the time), there is nonfunctionality (e.g., the relative lack of action for which something is fitted, as shown in the efforts by conservationists in our time to protect endangered species, so that killing a species to the point of overexploitation is no longer functional nowadays, because of the more conscious awareness of the need for conservation for sustainability). And the reverse direction also holds true.

And in relation to the intentionality-nonintentionality principle, if there is intentionality (e.g., the planning or design of something for a certain outcome, as shown in the use of "the hyperdisease hypothesis" as an alternative to "the overkill hypothesis," by claiming that "humans or animals traveling with them [e.g.,dogs]...introduced one or more highly virulent diseases into vulnerable populations of native mammals, eventually causing extinctions"), there is nonintentionality (e.g., the relative absence of planning or design of something for a certain outcome, such that the outcome can even be counter-intentional in being the opposite of what was originally planned, as shown in the failure of "the hyperdisease hypothesis" as an alternative to "the overkill hypothesis," because, for the critics, "it can not account for several major extinction events, notably Australia and North America" when "dogs did not arrive in Australia until approximately 35,000 years after the first humans arrived and approximately 30,000 years after the megafaunal extinction was complete," contary to the origiinal goal of the hyperdisease hypothesis). (WK 2012ee) And the reverse direction also holds true.

Discontinuity, Biodiversity Loss, and Health Risks

Another way to evaluate the possibility (or impossibility) and desirability (or undesirability) of discontinuity from the perspective of the mind is to explore discontinuity, biodiversity loss, and health risks.

An argument here is that there is a relationship between human health and biodiversity.

This relationship between biodiversity loss and health risks is delicate enough, because "many of the anticipated health risks of climate change are associated with changes in biodiversity (e.g. changes in populations and distribution of disease vectors, scarcity of fresh water, impacts on agricultural biodiversity and food resources etc.)." (WK 2012i)

The list of health risks can be long indeed, as it includes, for example, "dietary health and nutrition security, infectious disease, medical science and medicinal resources, social and psychological health. Biodiversity is also known to have an important role in reducing disaster risk, and in post-disaster relief and recovery efforts." (WK 2012i; K. Gaston 2007)

However, just like the dual effects of extinction for natural evolution (in *Sec. 3.2.1*), health risks are important too for natural evolution. For instance, "cycles of infection, disease, morbidity and mortality have played an essential part in natural selection and the evolution of life, and have also driven the evolution of human societies and cultures." (COH 2012)

With these dual effects of health risks for natural evolution in mind—consider, for illustration, three case studies of the relationship between biodiversity loss and health risks below.

The Case of Infectious Disease

The first case study of the relationship between biodiversity loss and health risks has to do with "emerging infectious disease." (COH 2012)

The important point to remember here is that "naturally occurring microbes—including bacteria, viruses, fungi and protozoa—comprise a significant portion of wild biodiversity worldwide. In many ways, these organisms are responsible for supporting and regulating a range of key ecosystem functions, and so provide the foundation for a wide range of ecosystem services. For example, bacteria and fungi are vital to waste decomposition and nutrient cycling, driving primary productivity and affecting climatic patterns on a massive scale. Disease-causing microbes (pathogens) and parasitic invertebrates play an equally important role in ecosystem functioning and productivity, and—though it is perhaps paradoxical and an unpleasant thought—are essential to maintaining the health of ecosystems and populations of wild flora and fauna." (COH 2012)

Other "clear scientific evidence" includes the fact that "outbreaks of many diseases, including SARS, Ebola, hanta viruses, malaria, and the HIV pandemic, have resulted from human impacts on wildlife and ecosystems, including inter alia habitat disturbance, unsustainable trade in bushmeat and other country foods, and the wildlife trade. Urban sprawl, encroachment into wilderness areas, and pollution of oceans and inland waterways have also been linked with the spread of diseases of people and wildlife. Outbreaks of zoonoses (diseases spread from animals to man), including the recent international spread of the H5N1 strain of avian influenza, may also be facilitated by these activities as they affect the integrity of ecosystems and the services they provide. The loss of biodiversity and ecosystem services is increasingly recognized as a risk factor in the transmission of diseases between people, livestock, crops and wildlife." (COH 2012)

Other research studies showed that, "in certain specific areas or situations, disease risk may increase with proximity to wild areas or intact habitats, including for example certain forest areas or wetlands….The occurrence of high biodiversity is often less important as a risk factor than the integrity of the ecosystems involved, and the interactions which take place between people, wildlife and the environment. The influence of climate change on ecosystem processes and disease ecology is a further complicating factor, and it is increasingly clear that each situation must also be assessed at the landscape level, accounting for the role of people and ecosystems in the wider ecological landscape." (COH 2012)

There is a policy implication here, in that "to address these complex issues, whether from the perspective of public health, food safety, wildlife conservation, economics, global security or international development, increased co-operation between disciplines is essential. A failure to address the root causes of disease emergence and spread associated with ecosystem change or impacts on wildlife could have severe global consequences. It is essential that the use of biodiversity indicators and the protection of ecosystems be integrated into international efforts to protect public health and prevent the emergence and spread of infectious diseases. At the same time, biodiversity conservation planning must account for the ecosystem services associated with the regulation of pests and diseases, and for how conservation plans may affect the health or livelihoods of local communities." (COH 2012)

The Case of Medicinal Resources

The second case study of the relationship between biodiversity loss and health risks has to do with "medicinal resources." (COH 2012a)

The reason is that "biodiversity is vital to modern and traditional medicinal practices, particularly in the face of the global problems of increasing antibiotic resistance and emerging infectious diseases. Biodiversity loss severely restricts our potential future resources for the treatment of illness, and threatens the growth of economic activities based on natural products." (COH 2012a)

Especially in "the developing world, many millions of people rely directly upon traditional wisdom and indigenous knowledge for their health and livelihood security. This knowledge is associated with the gathering and cultivation of foods, clothing and building materials, with local cultural traditions, and with systems of traditional medicines, which support entire populations." (COH 2012a)

But "in technologically advanced countries, the vast majority of this indigenous knowledge has been lost to society, and is of little consequence to modern 'Western' lifestyles. However, up to 57% of the top 150 medicinal compounds currently on sale in the U.S. have some basis or origin in exploration from wild species, while 50% of modern prescription medicines are based on compounds originally discovered in plants. Modern drugs derived from wild species include pain killers (e.g. Zinconitide from cone snail toxin), cardiac drugs (e.g. Lanoxin from Digitalis plants), anti-malarials (e.g. Quinine from Cinchona trees), hormone mimetics (e.g. Exenatide from Gila monster venom) and anti-cancer drugs (e.g. Taxol from Taxus trees and Hycamtin from Camptotheca trees). New discoveries continue to be made—recent studies on compounds extracted from sea anemones have shown potential therapeutic applications in treating autoimmune diseases such as diabetes and arthritis. Many other potentially important species are yet to be investigated." (COH 2012a)

In addition, although "bioprospecting—the exploration and exploitation of chemical compounds from natural sources—can represent a valuable tool for both conservation and economic development..., unsustainable patterns of exploitation which impact on biodiversity can jeopardize these resources. The livelihoods and health of indigenous communities in many regions may also be threatened where traditional knowledge systems are exploited without respect for the owners of that knowledge and without due consideration for the principles of equitable access and benefit sharing." (COH 2012a)

There is a policy implication here, in that "there is a need to ensure that bioprospecting respects and preserves traditional knowledge systems, and involves local communities in a manner that values and protects habitats and species, creates local business opportunities and supports the continuance of local cultural traditions. This requires direct engagement and collaboration with the biotech and pharmaceutical industries, national institutes of health and medicine, and research funding agencies worldwide." (COH 2012a)

The Case of Disaster Prevention, Relief, and Recovery

And the third case study of the relationship between biodiversity loss and health risks has to do with "disaster prevention, relief, and recovery." (COH 2012b)

The reasoning here is that "the loss of ecosystem services can increase our vulnerability to the impacts of natural and man-made disasters—flooding, storms, earthquakes, drought, conflict….Those who are poor or sick, or who experience low levels of livelihood security, are likely to be at greatest risk from such events, and are typically more dependent upon ecosystems for their well-being." (COH 2012b)

Some studies showed that "the effects of recent natural disasters—including the tsunami in South East Asia in 2004, the famine in Niger in 2005 and the effects of hurricane Katrina in the Americas in 2005—have been exacerbated by ecosystem change, unsustainable development and biodiversity loss associated with human activities. The impacts of disasters themselves often include disruption of natural habitats and ecosystems and the services they provide. This in turn can have additional consequences for people living in or near the affected areas, and undermine recovery or redevelopment efforts. Man-made disasters—including the effects of war and other conflicts—also have severe implications on natural environments and human well-being." (COH 2012b)

Thus, the danger here is that "a failure to recognise these issues and to account for biodiversity and essential ecosystem services in relief, recovery and redevelopment programmes, may simply negate today's relief efforts if such disasters recur in the future. There is a risk that problems, which may have been created or exacerbated by habitat disturbance, may persist or recur if those programmes fail to recognise the importance of biodiversity to health and livelihood security. Furthermore, in a world where climate change may result in more unpredictable weather patterns, sea level rise and more frequent and more extreme storms, the services provided by biodiversity will be critical for human communities most at risk from such events. Refugees from disasters may be particularly vulnerable to added burdens of disease, food insecurity and malnutrition as a result of their displacement and loss of access to ecosystem benefits." (COH 2012b)

There is a policy implication here, in that "the protection, management, monitoring and restoration of ecosystems should play a central role in disaster prevention, early warning, recovery, and emergency aid programmes. Ecosystems and the benefits they provide, and the impacts of disasters (including refugee crises) often straddle national boundaries, and influence the lives of communities over huge areas.

Collaboration across disciplines and across borders in addressing this issue is therefore essential." (COH 2012b)

Problems with the Study of Biodiversity Loss and Health Risks

However, for the critics, this study of biodiversity and health risks has some major problems. Consider, for illustration, a few examples below.

Firstly, one main criticism of the study of biodiversity loss and health risks is that there are many different causes of human health problems.

For example, for the critics, an individual may suffer from human health problems which are not necessarily related to biodiversity, and good examples include overconsumption, junk diet, personality disorders, etc.

Secondly, another main criticism of the study of biodiversity loss and health risks is that there can be alternative resources for health improvement, even when biodiversity occurs.

For instance, for the critics, although it is true that "biodiversity provides critical support for drug discovery and the availability of medicinal resources" to improve human health—it is also important to point out that "the decline in output from the pharmaceutical sector since the mid-1980s can be attributed to a move away from natural product exploration ('bioprospecting') in favor of genomics and synthetic chemistry" for health improvement instead. (WK 2012i; E. Hawkins 1992)

Thirdly, still another main criticism of the study of biodiversity loss and health risks is that the relationship between biodiversity loss and health risks is more complicated than its advocates would like us to believe.

For instance, for the critics, although "in certain specific areas or situations, disease risk may increase with proximity to wild areas or intact habitats, including for example certain forest areas or wetlands," it is important to remember that "the causes are not always clear and are rarely straightforward." (COH 2012)

Fourthly, an additional main criticism of the study of biodiversity loss and health risks is that there remains a lot of "scientific uncertainty" about the nature of biodiversity loss.

For instance, F. Bosselman (2004) warned that there is still a lot things about biodiveristy loss which remain uncertain, or what he called "the remaining areas of scientific uncertainty" in regard to extinction and biodiverity loss. (WK 2012i) And his warning was targeted at those conservation activists who are too eager to use "biodiversity…as a legal standard" to enforce conservation against biodiversity loss everywhere.

Fifthly, another additional main criticism of the study of biodiversity loss and health risks is that its policy recommendation of "collaboration across disciplines and across borders" is not realistic.

For instance, for the critics, all parties to "the United Nations Framework Convention on Climate Change (UNFCCC)" met at Cancún in 2010 but could not reach any "a binding treaty," but only a voluntary agreement, because different parties have different interests and opposing views. (WK 2012ff) In addition, "organizations such as the libertarian Competitive Enterprise Institute, conservative commentators, and some companies such as ExxonMobil have challenged IPCC climate change scenarios," just as "funded scientists…disagree with the scientific consensus, and provided their own projections of the economic cost of stricter controls." (WK 2012ff; S. Begley 2007; D. Adams 2006; MSNBC 2007; C. Sandell 2007)

Sixthly, still another additional main criticism of the study of biodiversity loss and health risks is that health risks are important too for natural evolution.

For instance, "cycles of infection, disease, morbidity and mortality have played an essential part in natural selection and the evolution of life, and have also driven the evolution of human societies and cultures." (COH 2012)

Beyond the Study of Biodiversity Loss and Health Risks

These criticisms of the study of biodiversity loss and health risks, however, should not be taken as a total rejection of its usefulness but serve to show us the two opposing sides of the debate, such that the possibility (or impossibility) and desirability (or undesirability) of discontinuity (from the perspective of the mind with discontinuity, biodiversity loss and health risks as a case study here) are not to the extent that the respective defenders would like us to believe.

In addition, the analysis of discontinuity, biodiversity loss and health risks is to teach us something important about the ontological principles in existential dialectics, and good examples include the formalness-informalness principle, the absoluteness-relativeness principle, the partiality-totality principle, the predictability-unpredictability principle, the explicability-inexplicability principle, the fiction-reality principle, the cognitiveness-noncognitiveness principle, the finiteness-transfiniteness principle, the preciseness-vagueness principle, the simpleness-complicatedness principle, the openness-hiddenness principle, the denseness-emptiness principle, the change-constancy principle, the order-chaos principle, the slowness-quickness principle, the expansion-contraction principle, the theory-praxis principle, the convention-novelty principle, the evolution-transformation principle, the symmetry-asymmetry principle, the softness-hardness principle, the seriousness-playfulness principle, the activeness-inactivess principle, the selfness-otherness principle, the regression-progression principle, the same-difference principle, the stability-reaction principle, the functionality-nonfunctionality principle, and the intentionality-nonintentionality principle.

For instance, in relation to the formalness-informalness principle, if there is formalness (e.g., the formal logical requirement of soundness and completeness in a system of ideas, as the study of biodiversity loss and health risks), there is informalness (e.g., the non-formal existence of unsoundness and incompleteness in the study of biodiversity loss and health risks, as shown in the criticisms that there are many different causes of human health problems, that there can be alternative resources for health improvement even when biodiversity occurs, that the relationship between biodiversity loss and health risks is more complicated than its advocates would like us to believe, that there remains a lot of "scientific uncertainty" about the nature of biodiversity loss, that its policy recommendation of "collaboration across disciplines and across borders" is not realistic, and that health risks are important too for natural evolution). And the reverse direction also holds true.

In relation to the absoluteness-relativeness principle, if there is absoluteness (e.g., the absolute view of those conservation activists who are too eager to use "biodiversity…as a legal standard" to enforce conservation against biodiversity loss), there is relativeness (e.g., what is right for those conservation activists who are too eager to use "biodiversity…as a legal standard" to enforce conservation against biodiversity loss is not necessarily so for the critics like Bosselman who warned that there is still a lot things about biodiveristy loss which remain uncertain, or what he called "the remaining areas of scientific uncertainty" in regard to extinction and biodiverity loss). And the reverse direction also holds true.

In relation to the partiality-totality principle, if there is partiality (e.g., the partial view of climate change and biodiversity loss by the advocates), there is totality (e.g., the more holistic view of climate change and biodiversity loss, such that the whole is not the sum of its parts, that is, it cannot be reduced to the sum of the opposing views by different parties to the UNFCCC, since there will emerge new views in the future not yet known today, just as there are already "other views" nowadays which do not side exclusively with any of the opposing views, while questioning "aspects of climate-change science"). (WK 2012ff; ROY 2005; S. Weart 2009; SEPP 2008) And the reverse direction also holds true.

In relation to the predictability-unpredictability principle, if there is predictability (e.g., the predictable tendency of the advocates of climate change and biodiversity loss to argue for international efforts against extinction for sustainable development), there is unpredictability (e.g., the more difficult task to predict exactly the extent to which international efforts against extinction for sustainable development will necessarily succeed on a particular future occasion, since, for the critics, "some scientists and non-scientists question aspects of climate-change science" and Bosselman cautioned us about "the remaining areas of scientific uncertainty" in regard to extinction and biodiverity loss). And the reverse direction also holds true.

In relation to the explicability-inexplicability principle, if there is explicability (e.g., the explanation by the advocates of the relationship between biodiversity loss and health risks), there is inexplicability (e.g., the lack of sufficient explanation by the advocates of why the relationship between biodiversity loss and health risks is necessarily valid, since, for the critics, there are many different causes of human health problems, just as there can be alternative resources for health improvement even when biodiversity occurs). And the reverse direction also holds true.

In relation to the fiction-reality principle, if there is fiction (e.g., the fictional aspect of the claim about the relationship between climate change and health risks, insofar as "the Letter from the Royal Society" in 2005 showed some "fictions about climate change"), there is reality (e.g., the realistic aspect of the claim about the relationship between climate change and health risks, insofar as "the Letter from the Royal Society" in 2005 also showed some "facts…about climate change"). (WK 2012ff; ROY 2005) And the reverse direction also holds true.

In relation to the cognitiveness-noncognitiveness principle, if there is cognitiveness (e.g., a view of things on the basis of reason and evidences, as shown in the "facts…about climate change" provided by "the Letter from the Royal Society" in 2005), there is noncognitiveness (e.g., a view of things on the basis of non-cognitive factors like envy, jealousy, power, nationality, race, gender, age, class, greed, lust, status, anger, sadness, joy, fear, wish, etc., as shown by different organizations like "the libertarian Competitive Enterprise Institute, conservative commentators, and some companies such as ExxonMobil" that have different interests or motives and thus "have challenged IPCC climate change scenarios"). (WK 2012ff; S. Begley 2007; D. Adams 2006; MSNBC 2007) And the reverse direction also holds true.

In relation to the finiteness-transfiniteness principle, if there is finiteness (e.g., the finite number of the categories of "naturally occurring microbes" or microorganisms in the world of biodiversity, like "bacteria, viruses, fungi and protozoa"), there is transfiniteness (e.g., the transfinite number of all the individual "bacteria, viruses, fungi and protozoa" in the natural history of the Earth that have ever existed in the world of biodiversity). And the reverse direction also holds true.

In relation to the preciseness-vagueness principle, if there is preciseness (e.g., the precise identification of 3 cases of the relationsihp between biodiversity loss and health risks in the article aforcited, like "infectious disease," "medicinal resources," and "disaster prevention, relief, and recover"), there is vagueness (e.g., the vagueness in the identification of the 3 cases of the relationsihp between biodiversity loss and health risks in the article aforcited, since it is not clear why there must be only 3, not 4, 5, 6, and so on). And the reverse direction also holds true.

In relation to the simpleness-complicatedness principle, if there is simpleness (e.g., the relatively simple analysis of the relationsihp between biodiversity loss and health risks by the advocates), there is complicatedness (e.g., the relatively more complicated analysis of the relationsihp between biodiversity loss and health risks, by challenging the claims and assumptions, as shown in the problems pointed out by the critics as cited). And the reverse direction also holds true.

In relation to the openness-hiddenness principle, if there is openness (e.g., the open exploration of the relationship between biodiversity and human health, in that, "in certain specific areas or situations, disease risk may increase with proximity to wild areas or intact habitats, including for example certain forest areas or wetlands"), there is hiddenness (e.g., the hidden bias in the study of the relationship between biodiversity and human health, because, for the critics, it is important to remember that "the causes are not always clear and are rarely straightforward"). And the reverse direction also holds true.

In relation to the denseness-emptiness principle, if there is denseness (e.g., the relatively denser concentration of traditional medicinal resources in "the developing world," where "many millions of people rely directly upon traditional wisdom and indigenous knowledge for their health and livelihood security"), there is emptiness (e.g., the relatively less dense, or more empty, concentration of traditional medicinal resources in "technologically advanced countries," because "the vast majority of this indigenous knowledge has been lost to society, and is of little consequence to modern 'Western' lifestyles"). And the reverse direction also holds true.

In relation to the change-constancy principle, if there is change (e.g., the ever new ways to study biodiversity, as shown in the study of biodiversity loss and health risks), there is constancy (e.g., the ever constant existence of different problems in the ever new ways to study biodiversity, as shown in the problems with the study of biodiversity loss and health risks pointed out by the critics aforecited). And the reverse direction also holds true.

In relation to the order-chaos principle, if there is order (e.g., the more or less orderly view of biodiversity, if looked from the vantage point of "traditional wisdom and indigenous knowledge for the…health and livelihood security" of a local community," which "is associated with the gathering and cultivation of foods, clothing and building materials, with local cultural traditions, and with systems of traditional medicines, which support entire populations"), there is chaos (e.g., the more or less chaotic view of biodiversity, if looked from the vantage point of "natural and man-made disasters—flooding, storms, earthquakes, drought, conflict"). And the reverse direction also holds true.

In relation to the slowness-quickness principle, if there is slowness (e.g., the relatively slower speed of "the transmission of diseases between people, livestock, crops and wildlife" in the absence of "the loss of biodiversity and ecosystem services"), there is quickness (e.g., the relatively quicker speed of "the transmission of diseases between people, livestock, crops and wildlife" in the presence of "the loss of biodiversity and ecosystem services"). And the reverse direction also holds true.

In relation to the expansion-contraction principle, if there is expansion (e.g., the relative spread of "modern prescription medicines" in "technologically advanced countries" with their "Western" lifestyles in our time), there is contraction (e.g., the relative decline or loss, "in technologically advanced countries" in our time, of "the vast majority of the…indigenous knowledge" for traditional medicines, although "up to 57% of the top 150 medicinal compounds currently on sale in the U.S. have some basis or origin in exploration from wild species"). And the reverse direction also holds true.

In relation to the theory-praxis principle, if there is theory (e.g., the theoretical understanding of the relationship between biodiversity loss and health risks in biology), there is praxis (e.g., the practical application of the understanding of the relationship between biodiversity loss and health risks in biology to the field of socio-political government, like the call by the COHAB Initiative for "internatonal efforts" and "collaboration across disciplines and across borders" to fight against biodiversity loss for the promotion of human health). And the reverse direction also holds true.

In relation to the convention-novelty principle, if there is convention (e.g., the "traditional wisdom and indigenous knowledge for the…health and livelihood security" of a local community in less developed countries), there is novelty (e.g., the alternative novel challenge to the "traditional wisdom and indigenous knowledge for the…health and livelihood security," by the new idea of modern Western prescription medicines "in technologically advanced countries," where "the vast majority of this indigenous knowledge has been lost to society, and is of little consequence to modern 'Western' lifestyles"). And the reverse direction also holds true.

In relation to the evolution-transformation principle, if there is evolution (e.g., the natural evolution of different lifeforms to face the challenge of biodiversity loss in the state of nature), there is transformation (e.g., the technical transformation of human ability to deal with biodiversity loss by the invention of "internatonal efforts" and "collaboration across disciplines and across borders" to fight against biodiversity loss for the promotion of human health, such as the formation of "the United Nations Framework Convention on Climate Change [UNFCCC]"). And the reverse direction also holds true.

In relation to the symmetry-asymmetry principle, if there is symmetry (e.g., the existence of different socioeconomic groups in a society to deal with biodviersity loss and health risks, like the rich and the poor), there is asymmetry (e.g., "those who are poor or sick, or who experience low levels of livelihood security, are likely to be at greatest risk from such events, and are typically more dependent upon ecosystems for their well-being"). And the reverse direction also holds true.

In relation to the softness-hardness principle, if there is softness (e.g., "international efforts" among those in the United Nations Framework Convention on Climate Change [UNFCCC] to fight against biodiversity loss), there is hardness (e.g., "some scientists and non-scientists question aspects of climate-change science"). (WK 2012ff) And the reverse direction also holds true.

In relation to the seriousness-playfulness principle, if there is seriousness (e.g., the serious business of those who try to make medicinal resources), there is playfulness (e.g., the playful part of those who try to make medicinal resources, when they play around with different options over time, like "traditional medicines," "modern prescription medicines," etc.). And the reverse direction also holds true.

In relation to the activeness-inactivess principle, if there is activeness (e.g., the relative activeness of those who are rich and not sick, so they are not "at greatest risk from such events [like natural disasters]" and can engage in acitivities that they prefer), there is inactiveness (e.g., the relative inactiveness of those "who are poor or sick, or who experience low levels of livelihood security" and who "are likely to be at greatest risk from such events, and are typically more dependent upon ecosystems for their well-being," so that they are more likely to be unable to engage much in acitivities that they prefer). And the reverse direction also holds true.

In relation to the selfness-otherness principle, if there is selfness (e.g., the tendency to be mostly centered on one's self or group, as shown by those organizatons like "the libertarian Competitive Enterprise Institute, conservative commentators, and some companies such as ExxonMobil" which "have challenged IPCC climate change scenarios" and pursue their own interests in regard to global warming and biodiversity), there is otherness (e.g., the tendency to be mostly centered on others, as shown by those parties to "the United Nations Framework Convention on Climate Change (UNFCCC)" at Cancún in 2010, which produced a voluntary, though not binding, agreement, for a global campaign against global warming and biodiversity loss). And the reverse direction also holds true.

In relation to the regression-progression principle, if there is regression (e.g., the regression made by the study of biodiversity loss and health risks, as shown in the problems pointed out by the critics aforecited), there is progression (e.g., the progress made by the study of biodiversity loss and health risks, as shown in its contribution to a better understanding of the relationsip between biodiversity loss and health risks aforecited). And the reverse direction also holds true.

In relation to the same-difference principle, if there is similarity in outcome (e.g., the contribution to the molding and control of beliefs, values, and behaviors, regardless of whether this be done by way of the UNFCCC, or by way of individual groups like "the libertarian Competitive Enterprise Institute, conservative commentators, and some companies such as ExxonMobil"), there is difference in outcome (e.g., the contribution to the molding and control of beliefs, values, and behaviors by way of the UNFCCC for a relatively more cooperative lifeworld—but the contribution to the molding and control of beliefs, values, and behaviors by way of individual groups for a relatively more competitive lifeworld). And the reverse direction also holds true.

In relation to the stability-reaction principle, if there is stability (e.g., the relatively more stable condition of "traditional wisdom and indigenous knowledge for the...health and livelihood security" of a local community), there is reaction (e.g., "traditional wisdom and indigenous knowledge for the...health and livelihood security" then led to further changes later on, when, nowadays, especially "in technologically advanced countries, the vast majority of this indigenous knowledge has been lost to society, and is of little consequence to modern 'Western' lifestyles," with their modern prescription medicines). And the reverse direction also holds true.

In relation to the functionality-nonfunctionality principle, if there is functionality (e.g., the presence of action for which something is fitted, as shown in the use of "traditional wisdom and indigenous knowledge" in the older days, which served the function of providing "the…health and livelihood security" of a local community at the time), there is nonfunctionality (e.g., the relative lack of action for which something is fitted, as shown in the use of modern prescription medicines "in technologically advanced countries," where "the vast majority of this indigenous knowledge has been lost to society," so that the "traditional wisdom and indigenous knowledge" is no longer functional in these societies of our time, which has different scientific standards for "modern 'Western' lifestyles"). And the reverse direction also holds true.

And in relation to the intentionality-nonintentionality principle, if there is intentionality (e.g., the planning or design of something for a certain outcome, as shown in the use of "bioprospecting" for "the exploration and exploitation of chemical compounds from natural sources" which is important for human health and economic growth), there is nonintentionality (e.g., the relative absence of planning or design of something for a certain outcome, such that the outcome can even be counter-intentional in being the opposite of what was originally planned, as shown in the overexploitation of bioprospecting," such that "unsustainable patterns of exploitation which impact on biodiversity can jeopardize these resources" and in turn negatively affect economic development—contrary to the original goal of the enhancement of human well being). And the reverse direction also holds true.

Discontinuity and Nature

Discontinuity, when examined from the perspective of nature, can show us its possibility (or impossibility) and desirability (or undesirability), and this can be shown by way of two case studies, namely, (3.3.1) discontinuity, climate change, and water deterioration, and (3.3.2) discontinuity, genetic pollution, and reduction in range—to be addressed in what follows, respectively.

Discontinuity, Climate Change, and Water Deterioration

A good way to evaluate the possibility (or impossibility) and desirability (or undesirability) of discontinuity from the perspective of nature is to explore discontinuity, climate change, and water deterioration.

An argument here is that there is a relationship between climate change and the envrionment in the context of discontinuity, like water deterioration.

The term "climate change" refers to "a significant and lasting change in the statistical distribution of weather patterns over periods ranging from decades to millions of years. It may be a change in average weather conditions, or in the distribution of weather around the average conditions (i.e., more or fewer extreme weather events)." (WK 2012gg)

However, "climate change" should not be confused with two related phenomena. First, "fluctuations over periods shorter than a few decades, such as El Niño, do not represent climate change," and, second, "within scientific journals, global warming refers to surface temperature increases while climate change includes global warming and everything else that increasing greenhouse gas levels will affect." (WK 2012gg; NASA 2011)

In general, "the rate at which energy is received from the sun and the rate at which it is lost to space determine the equilibrium temperature and climate of Earth. This energy is distributed around the globe by winds, ocean currents, and other mechanisms to affect the climates of different regions. Factors that can shape climate are called climate forcings or 'forcing mechanisms,' which "can be either 'internal' or 'external,'" as explained below (and summarized in *Table 3.3*). (WK 2012gg)

Internal Causes of Climate Change

On the one hand, there are "internal" causes of climate change, or "internal forcing mechanisms," which are "natural processes within the climate system itself (e.g., the thermohaline circulation). (WK 2012gg)

These natural processes are closely linked to "the five components of earth's climate system," which "include atmosphere, hydrosphere, cryosphere, lithosphere..., and biosphere," although others also add "pedosphere." (WK 2012h & 2012gg)

(a) Ocean Variability

A good illustration concerns "ocean variability," in that "the ocean is a fundamental part of the climate system, some changes in it occurring at longer timescales than in the atmosphere, massing hundreds of times more and having very high thermal inertia (such as the ocean depths still lagging today in temperature adjustment from the Little Ice Age)." (WK 2012gg; K. Brian 1986)

In this sense, "short-term fluctuations (years to a few decades) such as the El Niño-Southern Oscillation, the Pacific decadal oscillation, the North Atlantic oscillation, and the Arctic oscillation, represent climate variability rather than climate change. On longer time scales, alterations to ocean processes such as thermohaline circulation play a key role in redistributing heat by carrying out a very slow and extremely deep movement of water, and the long-term redistribution of heat in the world's oceans." (WK 2012gg)

External Causes of Climate Change

On the other hand, there are "external" causes of climate change, or "external forcing mechanisms," which can be "either natural (e.g., changes in solar output) or anthropogenic (e.g., increased emissions of greenhouse gases)." (WK 2012gg)

For illustration, consider five of them below (as summarized in *Table 3.3*), namely, (a) orbital variations, (b) solar output, (c) volcanism, (d) plate tectonics, and (e) human influences.

(a) Orbital Variations

The first illustration of the external causes of climate change has to do with "orbital variations."

In regard to this cause, "slight variations in Earth's orbit lead to changes in the seasonal distribution of sunlight reaching the Earth's surface and how it is distributed across the globe. There is very little change to the area-averaged annually averaged sunshine; but there can be strong changes in the geographical and seasonal distribution. The three types of orbital variations are variations in Earth's eccentricity, changes in the tilt angle of Earth's axis of rotation, and precession of Earth's axis. Combined together, these produce Milankovitch cycles which have a large impact on climate and are notable for their correlation to glacial and interglacial periods, their correlation with the advance and retreat of the Sahara, and for their appearance in the stratigraphic record." (WK 2012gg; A. Gale 1989; UM 2011)

In a study by the IPCC, "Milankovitch cycles drove the ice age cycles; CO2 followed temperature change 'with a lag of some hundreds of years'; and that as a feedback amplified temperature change. The depths of the ocean have a lag time in changing temperature (thermal inertia on such scale). Upon seawater temperature change, the solubility of CO2 in the oceans changed, as well as other factors impacting air-sea CO2 exchange." (WK 2012gg; IPCC. 2007)

(b) Solar Output

The second illustration of the external causes of climate change has to do with "solar output."

Solar output is relevant here, because "the sun is the predominant source for energy input to the Earth. Both long- and short-term variations in solar intensity are known to affect global climate." (WK 2012gg)

Historically, "three to four billion years ago the sun emitted only 70% as much power as it does today. If the atmospheric composition had been the same as today, liquid water should not have existed on Earth. However, there is evidence for the presence of water on the early Earth, in the Hadean and Archean eons, leading to what is known as the faint young Sun paradox. Hypothesized solutions to this paradox include a vastly different atmosphere, with much higher concentrations of greenhouse gases than currently exist. Over the following approximately 4 billion years, the energy output of the sun increased and atmospheric composition changed. The Great Oxygenation Event—oxygenation of the atmosphere around 2.4 billion years ago—was the most notable alteration. Over the next five billion years the sun's ultimate death as it becomes a red giant and then a white dwarf will have large effects on climate, with the red giant phase possibly ending any life on Earth that survives until that time." (WK 2012gg; E. Watson 2005; B. Marty 2006; S. Hagemann 1994; C. Sagan 1972 & 1997)

On a smaller time scale, "solar output also varies..., including the 11-year solar cycle and longer-term modulations. Solar intensity variations are considered to have been influential in triggering the Little Ice Age, and some of the warming observed from 1900 to 1950....Research indicates that solar variability has had effects including the Maunder Minimum from 1645 to 1715 A.D., part of the Little Ice Age from 1550 to 1850 A.D. which was marked by relative cooling and greater glacier extent than the centuries before and afterward. Some studies point toward solar radiation increases from cyclical sunspot activity affecting global warming, and climate may be influenced by the sum of all effects (solar variation, anthropogenic radiative forcings, etc.)." (WK 2012gg Willson, Richard 1991 & 2003; E. Bard 2000; NASA 2003; H. Svensmark 2009)

(c) Volcanism

The third illustration of the external causes of climate change has to do with "volcanism."

The reason is that "volcanic eruptions release gases and particulates into the atmosphere. Eruptions large enough to affect climate occur on average several times per century, and cause cooling (by partially blocking the transmission of solar radiation to the Earth's surface) for a period of a few years. The eruption of Mount Pinatubo in 1991, the second largest terrestrial eruption of the 20th century (after the 1912 eruption of Novarupta) affected the climate substantially. Global temperatures decreased by about 0.5 °C (0.9 °F). The eruption of Mount Tambora in 1815 caused the Year Without a Summer. Much larger eruptions, known as large igneous provinces, occur only a few times every hundred million years, but may cause global warming and mass extinctions." (WK 2012gg; N. Adams 2006; M. Diggles 2005; C. Oppenheimer 2003; P. Wignall 2001)

In addition, "volcanoes are…part of the extended carbon cycle. Over very long (geological) time periods, they release carbon dioxide from the Earth's crust and mantle, counteracting the uptake by sedimentary rocks and other geological carbon dioxide sinks. The US Geological Survey estimates are that volcanic emissions are at a much lower level than the effects of current human activities, which generate 100-300 times the amount of carbon dioxide emitted by volcanoes. A review of published studies indicates that annual volcanic emissions of carbon dioxide, including amounts released from mid-ocean ridges, volcanic arcs, and hot spot volcanoes, are only the equivalent of 3 to 5 days of human caused output. The annual amount put out by human activities may be greater than the amount released by supererruptions, the most recent of which was the Toba eruption in Indonesia 74,000 years ago." (WK 2012gg; US 2006; AGU 2011)

(d) Plate Tectonics

The fourth illustration of the external causes of climate change has to do with "plate tectonics."

In a long period time like "over the course of millions of years, the motion of tectonic plates reconfigures global land and ocean areas and generates topography. This can affect both global and local patterns of climate and atmosphere-ocean circulation." (WK 2012gg; C. Forest 1999)

For instance, "the position of the continents determines the geometry of the oceans and therefore influences patterns of ocean circulation. The locations of the seas are important in controlling the transfer of heat and moisture across the globe, and therefore, in determining global climate. A recent example of tectonic control on ocean circulation is the formation of the Isthmus of Panama about 5 million years ago, which shut off direct mixing between the Atlantic and Pacific Oceans. This strongly affected the ocean dynamics of what is now the Gulf Stream and may have led to Northern Hemisphere ice cover. During the Carboniferous period, about 300 to 360 million years ago, plate tectonics may have triggered large-scale storage of carbon and increased glaciation. Geologic evidence points to a 'megamonsoonal' circulation pattern during the time of the supercontinent Pangaea, and climate modeling suggests that the existence of the supercontinent was conducive to the establishment of monsoons." (WK 2012gg; NASA 2003a; G. Haug 2004; P. Bruckschen 1999; J. Parrish 1993)

In addition, "the size of continents is also important. Because of the stabilizing effect of the oceans on temperature, yearly temperature variations are generally lower in coastal areas than they are inland. A larger supercontinent will therefore have more area in which climate is strongly seasonal than will several smaller continents or islands." (WK 2012gg)

(e) Human Influences

And the fifth illustration of the external causes of climate change has to do with "human influences."

In the context of climate change, "anthropogenic factors are human activities which affect the climate. The scientific consensus on climate change is 'that climate is changing and that these changes are in large part caused by human activities,' and it 'is largely irreversible.'" (WK 2012gg; S. Solomon 2009)

Especially, "of most concern in these anthropogenic factors is the increase in CO2 levels due to emissions from fossil fuel combustion, followed by aerosols (particulate matter in the atmosphere) and cement manufacture. Other factors, including land use, ozone depletion, animal agriculture, and deforestation, are also of concern in the roles they play—both separately and in conjunction with other factors—in affecting climate, microclimate, and measures of climate variables." (WK 2012gg; H. Steinfeld 2006)

Climate Change and Water Deterioration

Climate change has major impacts on the environment, and a good illustration concerns its impact on water deterioration in two major ways, namely, (a) temperature and (b) precipitation, as explained below (and summarized in *Table 3.3*).

(a) Climate Change and Temperature

The first way in which climate change has an impact on water deterioration has to do with "temperature."

For instance, "it is predicted that the mean global temperature will rise in the coming years due to a number of forces affecting the climate, the amount of atmospheric CO2 will rise, and both of these will influence water resources; evaporation depends strongly on temperature and moisture availability, which can ultimately affect the amount of water available to replenish groundwater supplies." (WK 2012hh)

In addition, "temperature increase can decrease the length of the snow season in the winter and increase the intensity of snowmelt in warmer seasons, leading to peak runoff of snowmelt earlier in the season, affecting soil moisture, flood and drought risks, and storage capacities depending on the area. Warmer winter temperatures cause a decrease in snowpack, which can result in diminished water resources during the summer. This is especially important at mid-latitudes and in mountain regions that depend on glacial runoff to replenish their river systems and groundwater supplies, making these areas increasingly vulnerable to water shortages over time; an increase in temperature will initially result in a rapid rise in water melting from glaciers in the summer, followed by a retreat in glaciers and a decrease in the melt and consequently the water supply every year as the size of these glaciers get smaller and smaller." (WK 2012hh; C. Prudhomme 2002; CLI 2012)

This water melting from glaciers "from an increase in temperature gives way to a rise in sea level, which can affect the fresh water supply of coastal areas as well; as river mouths and deltas with higher salinity get pushed further inland, an intrusion of saltwater results in an increase of salinity in reservoirs and aquifers. Sea-level rise may also consequently be caused by a depletion of groundwater, as climate change can affect the hydrologic cycle in a number of ways. Uneven distributions of increased temperatures and increased precipitation around the globe results in water surpluses and deficits, but a global decrease in groundwater suggests a rise in sea level, even after meltwater and thermal expansion were accounted for, which can provide a positive feedback to the problems sea-level rise causes to fresh-water supply." (WK 2012hh)

Moreover, "a rise in air temperature results in a rise in water temperature, which is also very significant in water degradation, as the water would become more susceptible to bacterial growth. An increase in water temperature can also affect ecosystems greatly because of a species' sensitivity to temperature, and also by inducing changes in a body of water's self-purification system from decreased amounts of dissolved oxygen in the water due to rises in temperature." (WK 2012hh; CLI 2012; C. Prudhomme 2002; L. Konikow 2011)

(b) Climate Change and Precipitation

And the second way in which climate change has an impact on water deterioration has to do with "precipitation."

In light of this correlation between "a rise in global temperatures" and "an increase in global precipitation," any "increased runoff, floods, increased rates of soil erosion, and mass movement of land" can lead to "a decline in water quality..., while water will carry more nutrients" and "will also carry more contaminants. While most of the attention about climate change is directed towards global warming and greenhouse effect, some of the most severe effects of climate change are likely to be from changes in precipitation, evapotranspiration, runoff, and soil moisture. It is generally expected that, on average, global precipitation will increase, with some areas receiving increases and some decreases." (WK 2012hh; CLI 2012)

Some climate models revealed that "while some regions should expect an increase in precipitation, such as in the tropics and higher latitudes, other areas are expected to see a decrease, such as in the subtropics; this will ultimately cause a latitudinal variation in water distribution. The areas receiving more precipitation are also expected to receive this increase during their winter and actually become drier during their summer, creating even more of a variation of precipitation distribution. Naturally, the distribution of precipitation across the planet is very uneven, causing constant variations in water availability in respective locations." (WK 2012hh; CLI 2012; C. Prudhomme 2002)

Also, "changes in precipitation affect the timing and magnitude of floods and droughts, shift runoff processes, and alter groundwater recharge rates. Vegetation patterns and growth rates will be directly affected by shifts in precipitation amount and distribution, which will in turn affect agriculture as well as natural ecosystems. Decreased precipitation will deprive areas of water, causing water tables to fall and reservoirs and wetlands, rivers, and lakes to empty, and possibly an increase in evaporation and evapotranspiration, depending on the accompanied rise in temperature. Groundwater reserves will be depleted, and the remaining water has a greater chance of being of poor quality from saline or contaminants on the land surface." (WK 2012hh; CLI 2012; C. Prudhomme 2002; K. Frederick 1997)

Problems with the Study of Climate Change and Water Deterioration

However, for the critics, this study of climate change and water deterioration has some major problems. Consider, for illustration, a few examples below (as summarized in *Table 3.3*).

Firstly, one main criticism of the study of climate change and water deterioration is that the causes of climate change are more complicated than the advocates would claim.

For example, for the critics, in regard to "solar output"—"the cyclical nature of the sun's energy output is not yet fully understood." (WK 2012gg) And even the United States National Research Council admitted in 2010 that "much remains to be learned" about climate change. (WK 2012gg)

Secondly, another main criticism of the study of climate change and water deterioration is that there are still areas of scientific "uncertainty."

For instance, for the critics, "interestingly, a 2010 study suggests, 'that the effects of solar variability on temperature throughout the atmosphere may be contrary to current expectations.'" (WK 2012gg; J. Haigh 2010) And even the United States National Research Council recognized in 2010 that "there are still some uncertainties" about "climate change." (WK 2012gg)

Thirdly, still another main criticism of the study of climate change and water deterioration is that water management can minimize the problem of water shortage.

For instance, for the critics, "the issue of the depletion of fresh water can be met by increased efforts in water management....Preventative approaches are necessary to avoid high costs of inefficiency and the need for rehabilitation of water supplies, and innovations to decrease overall demand may be important in planning water sustainability." (WK 2012hh; G. MacDonald 2010; G. Young 2004)

Fourthly, an additional main criticism of the study of climate change and water deterioration is that its objective analysis can be naïve without sufficient attention to political factors.

For instance, for the critics, the issue of environmental degradation, say, in India, involves numerous causes, and part of them has to do with "caste politics," "caste-related violence," etc. (WK 2012ii)

And fifthly, another additional main criticism of the study of climate change and water deterioration is that it is too secular.

For instance, the critic like Dennis Dinger [2010] argued that "the Bible has quite a lot to say" about climate change, because "the Bible says God created the heavens and the earth, the Bible teaches that God maintains the daily workings of the earth,...and finally, it says that God controls the wind and the rain"). (WK 2012gg)

Beyond Climate Change and Water Deterioration

These criticisms of the study of climate change and water deterioration, however, do not ential a total rejection of its usefulness but serve to show us the two opposing sides of the debate, such that the possibility (or impossibility) and desirability (or undesirability) of discontinuity (from the perspective of nature with discontinuity, climate change, and water deterioration as a case study here) are not to the extent that the respective defenders would like us to believe.

Furthermore, the analysis of discontinuity, climate change, and water deterioration can cast a bright light on the nature of the ontological principles in existential dialectics, and good examples include the formalness-informalness principle, the absoluteness-relativeness principle, the partiality-totality principle, the predictability-unpredictability principle, the explicability-inexplicability principle, the fiction-reality principle, the cognitiveness-noncognitiveness principle, the finiteness-transfiniteness principle, the preciseness-vagueness principle, the simpleness-complicatedness principle, the openness-hiddenness principle, the denseness-emptiness principle, the change-constancy principle, the order-chaos principle, the slowness-quickness principle, the expansion-contraction principle, the theory-praxis principle, the convention-novelty principle, the evolution-transformation principle, the symmetry-asymmetry principle, the softness-hardness principle, the seriousness-playfulness principle, the activeness-inactivess principle, the selfness-otherness principle, the regression-progression principle, the same-difference principle, the stability-reaction principle, the functionality-nonfunctionality principle, and the intentionality-nonintentionality principle.

For instance, in relation to the formalness-informalness principle, if there is formalness (e.g., the formal logical requirement of soundness and completeness in a system of ideas, as in the study of climate change and water deterioration), there is informalness (e.g., the non-formal existence of unsoundness in the study of climate change and water deterioration, as shown in the criticisms that the causes of climate change are more complicated than the advocates would claim, that there are still areas of scientific "uncertainty," that water management can minimize the problem of water shortage, that its objective analysis can be naïve without sufficient attention to political factors, and that it is too secular). And the reverse direction also holds true.

In relation to the absoluteness-relativeness principle, if there is absoluteness (e.g., the absolute view of the causes of climate change by the advocates), there is relativeness (e.g., what is sure for of the causes of climate change by the advocates is not necessarily so for the critics, as shown in "a 2010 study [which] suggests, 'that the effects of solar variability on temperature throughout the atmosphere may be contrary to current expectations'"). And the reverse direction also holds true.

In relation to the partiality-totality principle, if there is partiality (e.g., the partial view of those activists who blame human activities on climate change), there is totality (e.g., the more holistic view of cliimate change, such that the whole is not the sum of its parts, that is, it cannot be reduced to the sum of opposing views about different causes of climate change, since there will emerge new views in the future not yet known today, just as there are some other views nowadays which suggest that "much remains to be learned" about climate change, without exclusively siding with any of the opposing views). And the reverse direction also holds true.

In relation to the predictability-unpredictability principle, if there is predictability (e.g., the predictable tendency of the advocates of solar output to claim it as a major cause of climate change), there is unpredictability (e.g., the more difficult task to predict exactly the extent to which solar output will be a major cause of climate change on a particular future occasion, since, for the critics, "the cyclical nature of the sun's energy output is not yet fully understood," and "a 2010 study suggests, 'that the effects of solar variability on temperature throughout the atmosphere may be contrary to current expectations'"). And the reverse direction also holds true.

In relation to the explicability-inexplicability principle, if there is explicability (e.g., the explanation, by the advocates of the human causes of climate change, on the basis of such factors like "emissions from fossil fuel combustion," "aerosols [particulate matter in the atmosphere]," "cement manufacture," "land use," "animal agriculture, and deforestation"), there is inexplicability (e.g., the lack of sufficient explanation by the advocates of the human causes of climate change of why these factors as cited above are enough to understand environmental degradation, since, for example in India, the issue of environmental degradation involves numerous causes, and part of them has to do with "caste politics," "caste-related violence," etc.). And the reverse direction also holds true.

In relation to the fiction-reality principle, if there is fiction (e.g., the fictional aspect of the claim about the "solar output" as a cause of climate change, since, for the critics, "a 2010 study suggests, 'that the effects of solar variability on temperature throughout the atmosphere may be contrary to current expectations,'" and even the United States National Research Council recognized in 2010 that "there are still some uncertainties" about "climate change"), there is reality (e.g., the realistic aspect of the claim about the "solar output" as a cause of climate change, insofar as "solar intensity variations are considered to have been influential in triggering the Little Ice Age, and some of the warming observed from 1900 to 1950"). And the reverse direction also holds true.

In relation to the cognitiveness-noncognitiveness principle, if there is cognitiveness (e.g., a view of things on the basis of reason and evidences, as shown in the scientific report "Advancing the Science of Climate Change" by the United States National Research Council), there is noncognitiveness (e.g., a view of things on the basis of non-cognitive factors like envy, jealousy, power, nationality, race, gender, age, class, greed, lust, status, faith, anger, sadness, joy, fear, wish, etc., as shown in the argument based on "faith" by Dennis Dinger [2010] that "the Bible has quite a lot to say" about global warming, because "the Bible says God created the heavens and the earth, the Bible teaches that God maintains the daily workings of the earth,…and finally, it says that God controls the wind and the rain"). (WK 2012gg) And the reverse direction also holds true.

In relation to the finiteness-transfiniteness principle, if there is finiteness (e.g., the finite number of the categories of particles which have played a major role in global warming, like "carbon dioxide," "aerosol," etc.), there is transfiniteness (e.g., the transfinite number of all the individual particles of "carbon dioxide," "aerosol," etc. in history of the Earth that have played a major role in global warming). And the reverse direction also holds true.

In relation to the preciseness-vagueness principle, if there is preciseness (e.g., the precise identification of 6 causes of climate change in the aritcle aforecited), there is vagueness (e.g., the vagueness in the identification of the 6 causes of climate change in the aritcle aforecited, since it is not clear why there must be only 5, not 6, 7, 8, and so on). And the reverse direction also holds true.

In relation to the simpleness-complicatedness principle, if there is simpleness (e.g., the relatively simple analysis of climate change by environmental activists), there is complicatedness (e.g., the relatively more complicated analysis of climate change, by challenging the claims and assumptions, as shown in the problems pointed out by the critics as cited). And the reverse direction also holds true.

In relation to the openness-hiddenness principle, if there is openness (e.g., the open exploration of the impact of "solar output" on climate change), there is hiddenness (e.g., the hidden aspect of "solar output," insofar as "the cyclical nature of the sun's energy output is not yet fully understood"). And the reverse direction also holds true.

In relation to the denseness-emptiness principle, if there is denseness (e.g., the relatively denser concentration of "CO2 levels" in the atmosphere nowadays, "due to emissions from fossil fuel combustion," as part of the human influences on climate change), there is emptiness (e.g., the relatively less dense, or more empty, concentration of "CO2 levels" in the atmosphere in pre-historical times when humans did not yet produce the massive "emissions from fossil fuel combustion" as is the case in our time). And the reverse direction also holds true.

In relation to the change-constancy principle, if there is change (e.g., the ever new ways to understand climate, as shown in the study of climate change and water deterioration), there is constancy (e.g., the ever constant existence of different problems in the ever new ways to understand climate, as shown in the problems with the study of climate change and water deterioration pointed out by the critics). And the reverse direction also holds true.

In relation to the order-chaos principle, if there is order (e.g., the more or less orderly view of reality, if looked from the vantage point of the 5 causes of climate change in the article aforecited), there is chaos (e.g., the more or less chaotic view of reality, if looked from the vantage point of the critics, like those who shared the view in a report by "a 2010 study" which "suggests, 'that the effects of solar variability on temperature throughout the atmosphere may be contrary to current expectations'" and the view by the United States National Research Council which recognized in 2010 that "there are still some uncertainties" about "climate change"). And the reverse direction also holds true.

In relation to the slowness-quickness principle, if there is slowness (e.g., "whether the initial forcing mechanism is internal or external, the response of the climate system might be…slow [e.g. thermal expansion of warming ocean water]"), there is quickness (e.g., "whether the initial forcing mechanism is internal or external, the response of the climate system might be fast [e.g., a sudden cooling due to airborne volcanic ash reflecting sunlight]"). (WK 2012gg). And the reverse direction also holds true.

In relation to the expansion-contraction principle, if there is expansion (e.g., "climate models [of global warming] show that…some regions should expect an increase in precipitation, such as in the tropics and higher latitudes"), there is contraction (e.g., "climate models [of global warming] show that…other areas are expected to see a decrease, such as in the subtropics"). (WK 2012hh) And the reverse direction also holds true.

In relation to the theory-praxis principle, if there is theory (e.g., the theoretical construction of the impact of climate change on water deterioration in environmental studies), there is praxis (e.g., the practical application of the knowledge of the impact of climate change on water deterioration in environmental studies to the field of politics, as "environmental degradation is one of the Ten Threats officially cautioned by the High Level Threat Panel of the United Nations"). (WK 2012hh) And the reverse direction also holds true.

In relation to the convention-novelty principle, if there is convention (e.g., in orbital variation, there is "thermal inertia," such that "the depths of the ocean have a lag time in changing temperature," like "a lag of some hundreds of years"), there is novelty (e.g., in orbital variation, there are the "Milankovitch cycles" which "drove the ice age cycles," so "CO2 followed temperature change"). (WK 2012gg) And the reverse direction also holds true.

In relation to the evolution-transformation principle, if there is evolution (e.g., the natural evolution of climate change in the state of nature, especially in the context of natural causes, with its impact, say, on water deterioration), there is transformation (e.g., the technical transformation of human ability to deal with the impact of climate change, say, on water deterioration, by the invention of "water management," for the efforts like "preventative approaches…to avoid high costs of inefficiency," "innovations to decrease overall demand," etc.). And the reverse direction also holds true.

In relation to the symmetry-asymmetry principle, if there is symmetry (e.g., the existence of different ways to release carbon dioxide, like volcanic emissions and fossil fuel combustion), there is asymmetry (e.g., "the US Geological Survey estimates are that volcanic emissions are at a much lower level than the effects of current human activities, which generate 100-300 times the amount of carbon dioxide emitted by volcanoes"). (WK 2012gg) And the reverse direction also holds true.

In relation to the softness-hardness principle, if there is softness (e.g., the defense of the scientific evidences of climate change by the United States National Research Council), there is hardness (e.g., the critique of the scientific evidences of climate change by Dennis Dinger [2010], who argued that "the Bible has quite a lot to say" about global warming, because "the Bible says God created the heavens and the earth, the Bible teaches that God maintains the daily workings of the earth,…and finally, it says that God controls the wind and the rain"). (WK 2012gg) And the reverse direction also holds true.

In relation to the seriousness-playfulness principle, if there is seriousness (e.g., the serious business of those who try to understand climate change), there is playfulness (e.g., the playful part of those who try to understand climate change, when they play around with different explanations over the decades, like "ocean variability," "orbital variations," "solar output," "volcanism," "plate tectonics," "human influences," "the Bible"). And the reverse direction also holds true.

In relation to the activeness-inactivess principle, if there is activeness (e.g., the relative activeness of those volcanoes which are still "active"), there is inactiveness (e.g., the relative inactiveness of those volcanoes which are either "dormant" or "extinct"). (WK 2012jj) And the reverse direction also holds true.

In relation to the selfness-otherness principle, if there is selfness (e.g., the tendency to be mostly centered on one's self or group, as shown in "carbon dioxide emissions from fossil fuel combustion," "aerosols [particulate matter in the atmosphere]," and "cement manufacture," by those companies which engaged in these activities for profit-seeking), there is otherness (e.g., the tendency to be mostly centered on others, as shown in the efforts by the High Level Threat Panel of the United Nations to address environmental degradation caused by climate change, for the sustainable development of future humanity). And the reverse direction also holds true.

In relation to the regression-progression principle, if there is regression (e.g., the regression made by the study on climate change, as shown in the problems pointed out by the critics aforecited), there is progression (e.g., the progress made by the study on climate change, as shown in its contribution to the understanding of its causes aforecited). And the reverse direction also holds true.

In relation to the same-difference principle, if there is similarity in outcome (e.g., the contribution to the molding and control of beliefs, values, and behaviors, regardless of whether this be done by way of using the report on climate change by the United States National Research Council, or by way of using the Bible on climate change), there is difference in outcome (e.g., the contribution to the molding and control of beliefs, values, and behaviors by way of using the report on climate change by the United States National Research Council for a relatively more scientific worldview—but the contribution to the molding and control of beliefs, values, and behaviors by way of using the Bible on climate change for a relatively more religious worldview). And the reverse direction also holds true.

In relation to the stability-reaction principle, if there is stability (e.g., the relatively more stable condition of the scientific explanation of climate change in modern times, as shown in the view by the United States National Research Council that there is "a strong, credible body of evidence, based on multiple lines of research, documenting that climate is changing"), there is reaction (e.g., the stable conditon of the scientific explanation of climate change in modern times then led to further changes later on, when nowadays,, there are some people like Dennis Dinger who argue that "the Bible has quite a lot to say" about global warming). (WK 2012gg) And the reverse direction also holds true.

In relation to the functionality-nonfunctionality principle, if there is functionality (e.g., the presence of action for which something is fitted, as shown in the use of the Bible to explain a lot of different things in medieval times, which served the function of understanding the world at the time), there is nonfunctionality (e.g., the relative lack of action for which something is fitted, as shown in the use of science to explain the causes of climate change, such that the appeal to the Bible by Dennis Dinger does not draw a lot of scholarly support nowadays and is no longer much functional in our time, with its more scientific expectations instead). And the reverse direction also holds true.

And in relation to the intentionality-nonintentionality principle, if there is intentionality (e.g., the planning or design of something for a certain outcome, as shown in the study to consider "solar variability" as a cause of climate change), there is nonintentionality (e.g., the relative absence of planning or design of something for a certain outcome, such that the outcome can even be counter-intentional in being the opposite of what was originally planned, as shown in "a 2010 study" which "suggests, 'that the effects of solar variability on temperature throughout the atmosphere may be contrary to current expectations'"). (WK 2012gg) And the reverse direction also holds true.

Discontinuity, Genetic Pollution, and Reduction in Range

Another good way to evaluate the possibility (or impossibility) and desirability (or undesirability) of discontinuity from the perspective of nature is to explore discontinuity, genetic pollution, and reduction in range.

An argument here is that there is a relationship between genetic pollution and reduction in range, in the context of invasive species.

The term "genetic pollution" refers to "uncontrolled gene flow into wild populations," especially "gene flow from domestic, feral, non-native and invasive species into wild indigenous species." (WK 2012kk; N. Ellstrand 2001; P. Boffey 1983; B. Potts 2003; D. Butler 1994)

Conditions of Species Invasion

There are two main types of conditions which lead to invasion of "domestic, feral, non-native and invasive species" into the habitats of "wild indigenous species," namely, (a) "species-based mechanisms" and (b) "ecosystem-based mechanisms," as explained below (and summarized in *Table 3.4*). (WK 2012l)

(a) Species-Based Mechanisms

In "species-based mechanisms"—even though "all species compete to survive,…invasive species appear to have specific traits or specific combinations of traits that allow them to outcompete native species. In some cases the competition is about rates of growth and reproduction. In other cases species interact with each other more directly….Common invasive species traits include" the folowing: (WK 2012l; D. Williams 1998; J. Ewell 1999)

- Ex: "fast growth"
- Ex: "rapid reproduction"
- Ex: "high dispersal ability"
- Ex: "phenotypic plasticity (the ability to alter growth form to suit current conditions)"
- Ex: "tolerance of a wide range of environmental conditions (ecological competence)"
- Ex: "ability to live off of a wide range of food types (generalist)"
- Ex: "association with humans"
- Ex: "prior successful invasions"

In general, "an introduced species must survive at low population densities before it becomes invasive in a new location. At low population densities, it can be difficult for the introduced species to reproduce and maintain itself in a new location, so a species might reach a location multiple times before it becomes established. Repeated patterns of human movement, such as ships sailing to and from ports or cars driving up and down highways, offer repeated opportunities for establishment (also known as a high propagule pressure)." (WK 2012l; D. Tilman 2004; E. Verling 2005)

There are two possible outcomes.

The first possible outcome is that "an introduced species might become invasive if it can outcompete native species for resources, such as nutrients, light, physical space, water or food. If these species evolved under great competition or predation, the new environment may host fewer able competitors, allowing the invader to proliferate quickly. Ecosystems in which all available resources are being used to their fullest capacity by native species can be modeled as zero-sum systems, where any gain for the invader is a loss for the native." (WK 2012l)

And the second possible outcome is that "such unilateral competitive superiority (and extinction of native species with increased populations of the invader) is not the rule. Invasive species often coexist with native species for an extended time, and gradually the superior competitive ability of an invasive species becomes apparent as its population grows larger and denser and it adapts to its new location." (WK 2012l)

(b) Ecosystem-Based Mechanisms

In "ecosystem-based mechanisms"—"the amount of available resources and the extent to which those resources are used by organisms determines the effects of additional species on the ecosystem. In stable ecosystems, equilibrium exists in the use of available resources. These mechanisms describe a situation in which the ecosystem has suffered a disturbance which changes the fundamental nature of the ecosystem." (WK 2012l; J. Byers 2002)

For instance, "when changes such as a forest fire occur, normal succession favors native grasses and forbs. An introduced species that can spread faster than natives can use resources that would have been available to native species, squeezing them out. Nitrogen and phosphorus are often the limiting factors in these situations." (WK 2012l; M. Davis 2000)

Also, "every species occupies a niche in its native ecosystem; some species fill large and varied roles, while others are highly specialized. Some invading species fill niches that are not used by native species, and they also can create new niches." (WK 2012l)

Characteristics of Species Invasion

Besides the two main conditions which lead to invasion, there are some main characteristics of species invasion, namely, (a) native invaders, (b) invaded ecosystems, (c) vectors, and (d) impacts of wildfire, to be addressed hereafter, in that order (and summarized in *Table 3.4*).

(b) Native Invaders

The first characteristic of species invasion is that, "although invasive species have typically been introduced to a habitat, some native species can, under the influence of events, such as long-term rainfall changes or human modifications to the habitat, increase in number and range and become invasive by expanding into new areas and disturbing the balance of species in the new area." (WK 2012l)

After all, "all species experience increases and decreases in numbers, in many cases accompanied by expansion or contraction of range. For example, the Monterey cypress is an endangered endemic, naturally occurring only in two small stands in California. They are being exterminated as exotic invasive species less than 50 miles (80 km) from their natural habitat." (WK 2012l; J. Smith 1988)

(b) Invaded Ecosystems

The second characteristic of species invasion is that "invasion was more likely in ecosystems that were similar to the one in which the potential invader evolved. Island ecosystems may be more prone to invasion because their species faced few strong competitors and predators, or because their distance from colonizing species populations makes them more likely to have 'open' niches." (WK 2012l; J. Williams 1998; J. Stachowicz 2005)

For example, a good illustration is "the decimation of native bird populations on Guam by the invasive brown tree snake. Conversely, invaded ecosystems may lack the natural competitors and predators that check invasives' growth in their native ecosystems, a factor that affected Guam snake populations." (WK 2012l)

In addition, "invaded ecosystems may have experienced disturbance, typically human-induced. Such a disturbance may give invasive species a chance to establish themselves with less competition from natives less able to adapt to a disturbed ecosystem." (WK 2012l; D. Tilman 2004; J. Williams 1998)

(c) Vectors

The third characteristic of species invasion is that "non-native species have many vectors [agents], including biogenic vectors, but most invasions are associated with human activity. Natural range extensions are common in many species, but the rate and magnitude of human-mediated extensions in these species tend to be much larger than natural extensions, and humans typically carry specimens greater distances than natural forces." (WK 2012l; P. Cassey 2005)

For instance, "an early human vector occurred when prehistoric humans introduced the Pacific rat (Rattus exulans) to Polynesia....The pet trade moves animals across borders, where they can escape and become invasive. Organisms stow away on transport vehicles. For example, ballast water taken up at sea and released in port is a major vector for marine invasions. Freshwater zebra mussels, native to the Black, Caspian and

Azov seas, probably reached the Great Lakes via ballast water from a transoceanic vessel." (WK 2012l; E. Matisoo-Smith 1998; UW 2001)

In addition, "species have also been introduced intentionally. For example, to feel more 'at home,' American colonists formed 'Acclimation Societies' that repeatedly imported birds that were native to Europe to North America and other distant lands. In 2008, U.S. postal workers in Pennsylvania noticed noises coming from inside a box from Taiwan; the box contained more than two dozen live beetles. Agricultural Research Service entomologists identified them as rhinoceros beetle, hercules beetle, and king stag beetle. Because these species were not native to the U.S., they could have threatened native ecosystems. To prevent exotic species from becoming a problem in the U.S., special handling and permits are required when living materials are shipped from foreign countries. USDA programs such as Smuggling Interdiction and Trade Compliance (SITC) attempt to prevent exotic species outbreaks in America." (WK 2012l; USDA)

(d) Impacts of Wildfire

And the fourth characteristic of species invasion is that "invasive species often exploit disturbances to an ecosystem (wildfires, roads, foot trails) to colonize an area." (WK 2012l)

For instance, "large wildfires are capable of sterilizing soils, while adding a variety of nutrients. In the resulting free-for-all, formerly entrenched species lose their advantage, leaving more room for invasives. In such circumstances plants that can regenerate from their roots have an advantage. Non-natives with this ability can benefit from a low intensity fire burns that removes surface vegetation, leaving natives that rely on seeds for propagation to find their niches occupied when their seeds finally sprout." (WK 2012l; M. Davis 2000; M. Brooks 2004)

Impacts of Species Invasion

The impacts of species invasion can be many. For illustration, consider two major ones below (as summarized in *Table 3.4*) from the perspective of nature, namely, (a) ecology and (b) biodiversity loss.

(a) Ecology

The first impact of species invasion to be considered here is about "ecology."

For instance, "in the Waterberg region of South Africa, cattle grazing over the past six centuries has allowed invasive scrub and small trees to displace much of the original grassland, resulting in a massive reduction in forage for native bovids and other grazers. Since the 1970s, large scale efforts have been underway to reduce invasive species; partial success has led to re-establishment of many species that had dwindled or left the region. Examples of these species are giraffe, blue wildebeest, impala, kudu and white rhino." (WK 2012l)

In addition, "invasive species can change the functions of ecosystems. For example, invasive plants can alter the fire regimen (cheatgrass, *Bromus tectorum*), nutrient cycling (smooth cordgrass *Spartina alterniflora*), and hydrology (*Tamarix*) in native ecosystems. Invasive species that are closely related to rare native species have the potential to hybridize with the native species. Harmful effects of hybridization have led to a decline and even extinction of native species. For example, hybridization with introduced cordgrass, *Spartina alterniflora*, threatens the existence of California cordgrass (*Spartina foliosa*) in San Francisco Bay." (WK 2012l, R. Mack 2000; C. Hawkes 2005; J. Rhymer 1996; D. Ayres 2004)

(d) Biodiversity Loss

And the second impact of species invasion to be considered here is about "biodversity loss."

For instance, "invasive species may drive local native species to extinction via competitive exclusion, niche displacement, or hybridisation with related native species. Therefore, besides their economic ramifications, alien invasions may result in extensive changes in the structure, composition and global distribution of the biota of sites of introduction, leading ultimately to the homogenisation of the world's fauna and flora and the loss of biodiversity." (WK 2012l, L. Odendaal 2008)

Problems with the Study of Genetic Pollution and Reduction in Range

However, for the critics, this study of genetic pollution and reduction in range (in the context of invasive species) has some major problems. Consider, for illustration, a few examples below (as summarized in *Table 3.4*).

Firstly, one main criticism of the study of genetic pollution and reduction in range is that the use of the term "genetic pollution" is controversial enough, because of its political motivation..

For example, for the critics like Patrick Moore, "the term 'genetic pollution,' otherwise referred to as genetic contamination,…is a propaganda term, not a technical or scientific term Pollution and contamination are both value judgments," because "some conservation biologists and conservationists have used genetic pollution for a number of years as a term to describe gene flow [which they disapprove of]"). (WK 2012kk)

Secondly, another main criticism of the study of genetic pollution and reduction in range is that genetic pollution has unintentional effects not often addressed.

For instance, for the critics, "Jeremy Rifkin in his 1998 book *The Biotech Century* described the problems of "hybridization…between two genetically distinct varieties" and "the subsequent introgression of genes" and warned "what he considered to be problems that might occur due to the unintentional process of (modernly) genetically modified organisms (GMOs) dispersing their genes into the natural environment by breeding with wild plants or animals." (WK 2012kk; A. Otchet 1998)

Thirdly, still another main criticism of the study of genetic pollution and reduction in range is that species invasion is only a cause of genetic pollution.

For instance, "in the field of agriculture, agroforestry and animal husbandry genetic pollution is being used to describe gene flows between GE species and wild relatives. An early use of the term *genetic pollution* in this later sense appears in a wide-ranging review of the potential ecological effects of genetic engineering in *The Ecologist* magazine in July 1989." (WK 2012kk; UN 2002a)

Fourthly, an additional main criticism of the study of genetic pollution and reduction in range is that scientsts disagree about the traits of "invasiveness" in invasive species.

For instance, for the critics, "researchers disagree about the usefulness of traits as invasiveness markers. One study found that of a list of invasive and noninvasive species, 86% of the invasive species could be identified from the traits alone. Another study found invasive species tended only to have a small subset of the presumed traits, and that many such traits were found in noninvasive species, requiring other explanations." (WK 2012l; D. Butler 1994)

Fifthly, another additional main criticism of the study of genetic pollution and reduction in range is that there are inconsistent findings.

For instance, for the critics, "in 1958, Charles S. Elton claimed that ecosystems with higher species diversity were less subject to invasive species because of fewer available niches. Other ecologists later pointed to highly diverse, but heavily invaded ecosystems and argued that ecosystems with high species diversity were more susceptible to invasion." (WK 2012l; C. Elton 2000; T. Stohlgren 1999)

Sixthly, still another additional main criticism of the study of genetic pollution and reduction in range is that invasive species does not necessarily lead to extinction.

For instance, for the critics, "it is difficult to unequivocally attribute extinctions to a species invasion, and the few scientific studies that have done so have been with animal taxa. Concern over the impacts of invasive species on biodiversity must therefore consider the actual evidence (either ecological or economic), in relation to the potential risk." (WK 2012l)

And seventhly, one last additional main criticism of the study of genetic pollution and reduction in range is that genetic pollution is not necessarily undesirable in the way that the activists would like us to believe.

For instance, for the critics, genetic pollution does not have to be undesirable, because there are both winners and losers, in that some species may get squeezed out in the competition, but "others" can "fill niches that are not used by native species, and they also can create new niches." (WK 2012l)

Beyond Genetic Pollution

These criticisms of genetic pollution and reduction in range do not mean, however, a total rejection of its usefulness but serve to show us the two opposing sides of the debate, such that the possibility (or impossibility) and desirability (or undesirability) of discontinuity (from the perspective of nature with discontinuity, genetic pollution and reduction in range as a case study here) are not to the extent that the respective defenders would like us to believe.

Furthermore, the analysis of discontinuity, genetic pollution and reduction in range can teach us a valuable lesson about the ontological principles in existential dialectics, and good examples include the formalness-informalness principle, the absoluteness-relativeness principle, the partiality-totality principle, the predictability-unpredictability principle, the explicability-inexplicability principle, the fiction-reality principle, the cognitiveness-noncognitiveness principle, the finiteness-transfiniteness principle, the preciseness-vagueness principle, the simpleness-complicatedness principle, the openness-hiddenness principle, the denseness-emptiness principle, the change-constancy principle, the order-chaos principle, the slowness-quickness principle, the expansion-contraction principle, the theory-praxis principle, the convention-novelty principle, the evolution-transformation principle, the symmetry-asymmetry principle, the softness-hardness principle, the seriousness-playfulness principle, the activeness-inactivess principle, the selfness-otherness principle, the regression-progression principle, the same-difference principle, the stability-reaction principle, the functionality-nonfunctionality principle, and the intentionality-nonintentionality principle.

For instance, in relation to the formalness-informalness principle, if there is formalness (e.g., the formal logical requirement of soundness, completeness, and consistency in a system of ideas, as shown in the study of genetic pollution and reduction in range), there is informalness (e.g., the non-formal existence of unsoundness, incompleteness, and inconsistency in the study of genetic pollution and reduction in range, as shown in the criticisms that the use of the term "genetic pollution" is controversial, that genetic pollution has unintentional effects, that species invasion is only a cause of genetic pollution, that scientsts disagree about the traits of "invasiveness" in invasive species, that there are inconsistent findings, that invasive species does not necessarily lead to extinction, and that genetic pollution is not necessarily undesirable in the way that the activists would like us to believe). And the reverse direction also holds true.

In relation to the absoluteness-relativeness principle, if there is absoluteness (e.g., the absolute view of those conservation activists that species invastion leads to extinction of native species), there is relativeness (e.g., what is true for those conservation activists who argue that species invastion leads to extinction of native species is not necessarily so for the critics, who instead suggested that "it is difficult to unequivocally attribute extinctions to a species invasion, and the few scientific studies that have done so have been with animal taxa"). And the reverse direction also holds true.

In relation to the partiality-totality principle, if there is partiality (e.g., the partial view of "one study" which "found that, of a list of invasive and noninvasive species, 86% of the invasive species could be identified from the traits alone"), there is totality (e.g., the more holistic view of the traits of invasive species, such that the whole is not the sum of its parts, that is, they cannot be reduced to the sum of the opposing views about the traits of invasive species, since there will emerge new views in the future not yet known today, just as there are already other views nowadays which do not necessarily side exclusively with any of them, as shown in "another study [which] found invasive species tended only to have a small subset of the presumed traits, and that many such traits were found in noninvasive species, requiring other explanations"). And the reverse direction also holds true.

In relation to the predictability-unpredictability principle, if there is predictability (e.g., the predictable tendency of those conservation activists to argue that species invastion leads to extinction of native species), there is unpredictability (e.g., the more difficult task to predict exactly the extent to which species invastion will necessarily lead to extinction of native species on a particular future occasion, since, for the critics, "it is difficult to unequivocally attribute extinctions to a species invasion, and the few scientific studies that have done so have been with animal taxa"). And the reverse direction also holds true.

In relation to the explicability-inexplicability principle, if there is explicability (e.g., the explanation by those in "one study" who "found that, of a list of invasive and noninvasive species, 86% of the invasive species could be identified from the traits alone"), there is inexplicability (e.g., the lack of sufficient explanation by those in "one study" of why its finding about a strong correlation between "a list of invasive and noninvasive species" and the "traits" used, since, for the critics, "another study found invasive species tended only to have a small subset of the presumed traits, and that many such traits were found in noninvasive species, requiring other explanations"). And the reverse direction also holds true.

In relation to the fiction-reality principle, if there is fiction (e.g., the fictional aspect of the argument that invasive species leads to extinction of native species, insofar as the critics pointed out that "it is difficult to unequivocally attribute extinctions to a species invasion, and the few scientific studies that have done so have been with animal taxa"), there is reality (e.g., the realistic aspect of the argument that invasive species leads to extinction of native species, insofar as there are "the few scientific studies…with animal taxa" that backed up the claim). And the reverse direction also holds true.

In relation to the cognitiveness-noncognitiveness principle, if there is cognitiveness (e.g., a view of things on the basis of reason and evidences, as shown in the two scientific studies by "researchers" who "disagree about the usefulness of traits as invasiveness markers"), there is noncognitiveness (e.g., a view of things on the basis of non-cognitive factors like envy, jealousy, power, nationality, race, gender, age, class, greed, lust, status, faith, anger, sadness, joy, fear, wish, etc., as shown in the "activism" of certain groups such as "Greenpeace" and others with their "unequivocal" attribution of "extinctions to a species invasion," such that it is more for propaganda, less for science). (WK 2012kk) And the reverse direction also holds true.

In relation to the finiteness-transfiniteness principle, if there is finiteness (e.g., the finite number of the categories of species involved in genetic pollution, like "domestic, feral, non-native and invasive species"), there is transfiniteness (e.g., the transfinite number of all the individual animals which are "domestic," "feral," "non-native" and "invasive" in history that have ever existed in relation to genetic pollution). And the reverse direction also holds true.

In relation to the preciseness-vagueness principle, if there is preciseness (e.g., the precise identification of 4 main characteristics of species invasion in the article aforecited), there is vagueness (e.g., the vagueness in the identification of the 4 main characteristics of species invasion in the article aforecited, since it is not clear why there must be only 4, not 6, 7, 8. and so on). And the reverse direction also holds true.

In relation to the simpleness-complicatedness principle, if there is simpleness (e.g., the relatively simple analysis of genetic pollution by some conservation activists), there is complicatedness (e.g., the relatively more complicated analysis of genetic pollution, by challenging the claims and assumptions, as shown in the problems pointed out by the critics as cited). And the reverse direction also holds true.

In relation to the openness-hiddenness principle, if there is openness (e.g., the open exploration of the consequences of genetic pollution by some conservation activists like those in Greenpeace), there is hiddenness (e.g., the hidden bias in some conservation activists like those in Greenpeace, since, for the critics like Patrick Moore, "the term 'genetic pollution'…is a propaganda term, not a technical or scientific term," because "some conservation biologists and conservationists have used genetic pollution for a number of years as a term to describe gene flow [which they disapprove of]"). (WK 2012kk) And the reverse direction also holds true.

In relation to the denseness-emptiness principle, if there is denseness (e.g., the relatively denser concentration of an invasive species as "its population grows larger…and it adapts to its new location" with "the superior competitive ability"), there is emptiness (e.g., the relatively less dense, or more empty, concentration of an invasive species at the beginning, when it "must survive at low population densities before it becomes invasive in a new location"). (WK 2012l) And the reverse direction also holds true.

In relation to the change-constancy principle, if there is change (e.g., the ever new ways to study animals, as shown in the study of genetic pollution and invasive species), there is constancy (e.g., the ever constant existence of different problems in the ever new ways to study animals, as shown in the problems with the study of genetic pollution and invasive species pointed out by the critics). And the reverse direction also holds true.

In relation to the order-chaos principle, if there is order (e.g., the more or less orderly view of reality, if looked from the vantage point of some conservation activits who "unequivocally attributes extinctions to a species invasion"), there is chaos (e.g., the more or less chaotic view of reality, if looked from the vantage point of some critics, who argued that "it is difficult to unequivocally attribute extinctions to a species invasion, and the few scientific studies that have done so have been with animal taxa," so "concern over the impacts of invasive species on biodiversity must therefore consider the actual evidence [either ecological or economic], in relation to the potential risk"). And the reverse direction also holds true.

In relation to the slowness-quickness principle, if there is slowness (e.g., the relatively slower speed of some "native species" to survive when invaded by "an introduced species that can spread faster than natives" and therefore "can use resources that would have been available to native species, squeezing them out"), there is quickness (e.g., the relatively quicker speed of "an introduced species that can spread faster than natives" and therefore "can use resources that would have been available to native species, squeezing them out"). (WK 2012l; M. Davis 2000) And the reverse direction also holds true.

In relation to the expansion-contraction principle, if there is expansion (e.g., the relative expansion of those invasive species only "less than 50 miles [80 km] from the…natural habitat" of "the Monterey cypress" in California, when the former succeeded in "exterminating" the latter), there is contraction (e.g., the relative decline of "the Monterey cypress," when "they are being exterminated as exotic invasive species less than 50 miles [80 km] from their natural habitat" by invasive species). (WK 2012l; J. Smith 1988) And the reverse direction also holds true.

In relation to the theory-praxis principle, if there is theory (e.g., the theoretical construction of "genetic pollution" and "invasive species" in ecology and biology), there is praxis (e.g., the practical application of the knowledge of "genetic pollution" and "invasive species" in ecology and biology to the field of conservation politics, when the organizatons like TRAFFIC "promote awareness of the effects of introduced invasive species that may 'hybridize with native species, causing genetic pollution'"). (WK 2012kk) And the reverse direction also holds true.

In relation to the convention-novelty principle, if there is convention (e.g., native species in California used to live, with low density, "on serpentine soils, which have low water-retention, low nutrient levels, a high Magnesium/Calcium ratio, and possible heavy metal toxicity"), there is novelty (e.g., "barbed goatgrass [*Aegilops triuncialis*] was introduced to

California" as an invasive species and, by novel uses of "resources…to live on previously uninhabited soil types," was able to "form dense stands on these soils, crowding out native species that have not adapted well to serpentine soils"). And the reverse direction also holds true.

In relation to the evolution-transformation principle, if there is evolution (e.g., the natural evolution of some native species to become extinct or reduced due to invasive species in the state of nature), there is transformation (e.g., the technical transformation of human ability to deal with biodiversity loss by the invention of nature conservation campaigns, as shown in the efforts by "the Joint Nature Conservation Committee [JNCC] in the UK, which "contributes to maintaining and enriching biological diversity and educating about the effects of the introduction of invasive/non-native species"). (WK 2012kk) And the reverse direction also holds true.

In relation to the symmetry-asymmetry principle, if there is symmetry (e.g., the existence of different plants, like "*Bromus tectorum*, originally from Eurasia" and "native plants" in western North America), there is asymmetry (e.g., "*Bromus tectorum*…is highly fire-adapted" and "not only spreads rapidly after burning, but also increases the frequency and intensity [heat] of fires, by providing large amounts of dry detritus during the fire season in western North America,…so much that native plants cannot survive the frequent fires, allowing *B. tectorum* to further extend and maintain dominance in its introduced range"). (WK 2012l; M. Brooks 2004) And the reverse direction also holds true.

In relation to the softness-hardness principle, if there is softness (e.g., "invasive species often coexist with native species for an extended time"), there is hardness (e.g., "an introduced species…can outcompete native species for resources, such as nutrients, light, physical space, water or food"). (WK 2012l) And the reverse direction also holds true.

In relation to the seriousness-playfulness principle, if there is seriousness (e.g., the serious business of those who try to understand genetic pollution and invasive species), there is playfulness (e.g., the playful part of those who try to understand genetic pollution and invasive species, when they play around with different terms over the decades like "genetic deterioration," "genetic pollution," "genetic swamping," "genetic takeover," "genetic mixing," "genetic aggression," and so on). (WK 2012kk) And the reverse direction also holds true.

In relation to the activeness-inactivess principle, if there is activeness (e.g., the relative activeness of those plants like "*Bromus tectorum*, originally from Eurasia" which "is highly fire-adapted," so that not only they can survive fire but also they "spread…rapidly after burning" and "increase…the frequency and intensity [heat] of fires, by providing large amounts of dry detritus during the fire season in western North America"), there is inactiveness (e.g., the relative inactiveness of those plants like "native plants" in western North America, when they "cannot survive the frequent fires, allowing *B. tectorum* to further extend and maintain dominance in its introduced range"). (WK 2012l; M. Brooks 2004) And the reverse direction also holds true.

In relation to the selfness-otherness principle, if there is selfness (e.g., the tendency to be mostly centered on one's self or group, as shown by "some conservation biologists and conservationists" which "have used genetic pollution for a number of years as a term to describe gene flow [which they disapprove of]," so that "the term 'genetic pollution'…is a propaganda term, not a technical or scientific term," for the pursuit of their own group agenda), there is otherness (e.g., the tendency to be mostly centered on others, as shown by some scientists like J. Rhymer and D. Simberloff [1996] who proposed a more neutral term like "gene mixing," for the enhancement of scientific knowledge, not for the propaganda of some groups). (WK 2012kk) And the reverse direction also holds true.

In relation to the regression-progression principle, if there is regression (e.g., the regression made by the study of genetic pollution and invasive species, as shown in the problems pointed out by the critics), there is progression (e.g., the progress made by the study of genetic pollution and invasive species, as shown in the contribution to a better understanding about the relationship between invasive species and native species in a given habitat aforecited). And the reverse direction also holds true.

In relation to the same-difference principle, if there is similarity in outcome (e.g., the contribution to the molding and control of beliefs, values, and behaviors, regardless of whether this be done by way of the use of a propaganda term like "genetic pollution," or by way of the use of a technical term like "genetic mixing"), there is difference in outcome (e.g., the contribution to the molding and control of beliefs, values, and behaviors by way of the use of a propaganda term like "genetic pollution" for a relatively more politicized worldview—but the contribution to the molding and control of beliefs, values, and behaviors by way of the use of a technical term like "genetic mixing" for a relatively more scientific worldview). And the reverse direction also holds true.

In relation to the stability-reaction principle, if there is stability (e.g., the relatively more stable condition of "part of an ecosystem" which is not yet "disturbed" by human activities like land use, agricutlure, and so on, or by different species), there is reaction (e.g., the undistrubed part of an ecosystems can then lead to further changes later on, when "an ecosystem is disturbed as when land is cleared for agriculture," so that "the boundary between remaining undisturbed habitat and the newly cleared land itself forms a distinct habitat, creating new winners and losers and possibly hosting species that would not thrive outside the boundary habitat"). (WK 2012l) And the reverse direction also holds true.

In relation to the functionality-nonfunctionality principle, if there is functionality (e.g., the presence of action for which something is fitted, as shown in the use of the term "invasive species" in the older days, which served the function of providing "a nomenclature term and categorization phrase used for flora and fauna, and for specific restoration-preservation processes in native habitats"), there is nonfunctionality (e.g., the relative lack of action for which something or someone is fitted, as shown in "the new nomenclature system based on biogeography rather than on taxa" to understand invasive species by R. Colautti and Hugh MacIsaac [2004], in that "the model evaluated individual populations rather than entire species" and "classified each population based on its success in that environment," so that the old usage of "invasive species" is no longer functional much in our time, because nowadays "the phrase *invasive species* is often criticized as an imprecise term for the scientific field of ecology"). (WK 2012l) And the reverse direction also holds true.

And in relation to the intentionality-nonintentionality principle, if there is intentionality (e.g., the planning or design of something for a certain outcome, as shown in "hybridization," when it is intended to "crossbreed…two genetically distinct varieties" in genetic engineering), there is nonintentionality (e.g., the relative absence of planning or design of something for a certain outcome, such that the outcome can even be counter-intentional in being the opposite of what was originally planned, as shown by the work of Jeremy Rifkin which showed that "intentional crossbreeding between two genetically distinct varieties" can end up producing "problems that might occur due to the unintentional process of [modernly] genetically modified organisms [GMOs] dispersing their genes into the natural environment by breeding with wild plants or animals," with extinction or reduction of species, contrary to the original goal of creating new ones). (WK 2012kk) And the reverse direction also holds true.

Discontinuity and Society

Discontinuity, when examined from the perspective of society, can reveal to us its possibility (or impossibility) and desirability (or undesirability), and this can be shown by way of two case studies, namely, (3.4.1) discontinuity, habitat fragmentation, and deforestation, and (3.4.2) discontinuity, oil depletion, and resource curse—to be addressed hereafter.

Discontinuity, Habitat Fragmentation, and Deforestation

A good way to evaluate the possibility (or impossibility) and desirability (or undesirability) of discontinuity from the perspective of society is to explore discontinuity, habitat fragmentation, and deforestation.

An argument here is that there is a relationship between habitat fragmentation and human activities like deforestation.

The term "habitat fragmentation" refers to "the emergence of discontinuities (fragmentation) in an organism's preferred environment (habitat), causing population fragmentation." (WK 2012f)

Features of Habitat Fragmentation

There are some main features of habitat fragmentation, and hereafter are some of them, for illustration (as summarized in *Table 3.5*): (WK 2012f)

- Ex: "reduction in the total area of the habitat"
- Ex: "decrease of the interior: edge ratio"
- Ex: "isolation of one habitat fragment from other areas of habitat"
- Ex: "breaking up of one patch of habitat into several smaller patches"
- Ex: "decrease in the average size of each patch of habitat"

Of course, there can be other features, so the ones here are illustrative, not exhaustive.

Causes of Deforestation

There are different causes of habitat fragmentation, which "can be caused by geological processes that slowly alter the layout of the physical environment (suspected of being one of the major causes of speciation), or by human activity such as land conversion, which can alter the environment much faster and causes extinctions of many species." (WK 2012f; S. Sahney 2010a)

One good case study of habitat fragmentation here concerns "deforestation," which is a human encroachment by "the removal of a forest or stand of trees where the land is thereafter converted to a nonforest use" like the "conversion of forestland to farms, ranches, or urban use." (WK 2012ll)

According to "the United Nations Framework Convention on Climate Change (UNFCCC) secretariat, the overwhelming direct cause of deforestation is agriculture. Subsistence farming is responsible for 48% of deforestation; commercial agriculture is responsible for 32% of deforestation; logging is responsible for 14% of deforestation and fuel wood removals make up 5% of deforestation." (WK 2012ll; UN 2007)

In addition, the historical causes of deforestation can be divided into three important eras, namely, (a) pre-historical, (b) pre-industrial and (c) industrial, as described below (and summarized in *Table 3.5*). (WK 2012ll)

(a) Pre-Historical

The first historical era in regard to the causes of deforestation is "pre-historical."

For instance, "carboniferous Rainforest Collapse was an event that occurred 300 million years ago. Climate change devastated tropical rainforests causing the extinction of many plant and animal species. The change was abrupt, specifically, at this time climate became cooler and drier, conditions that are not favorable to the growth of rainforests and much of the biodiversity within them. Rainforests were fragmented forming shrinking 'islands' further and further apart. This sudden collapse affected several large groups, effects on amphibians were particularly devastating, while reptiles fared better, being ecologically adapted to the drier conditions that followed." (WK 2012ll; S. Sahney 2010a)

Human activities also had a role to play. For instance, "small scale deforestation was practiced by some societies for tens of thousands of years before the beginnings of civilization. The first evidence of deforestation appears in the Mesolithic period. It was probably used to convert closed forests into more open ecosystems favorable to game animals. With the advent of agriculture, larger areas began to be deforested, and fire became the prime tool to clear land for crops." (WK 2012ll; T. Flannery 1994; T. Brown 1997)

Later, "the Neolithic period saw extensive deforestation for farming land. Stone axes were being made from about 3000 BC not just from flint, but from a wide variety of hard rocks from across Britain and North America as well." (WK 2012ll; ENC 2012)

(b) Pre-Industrial

The second historical era in regard to the causes of deforestation is "pre-industrial."

For instance, "throughout most of history, humans were hunter gatherers who hunted within forests. In most areas, such as the Amazon, the tropics, Central America, and the Caribbean, only after shortages of wood and other forest products occur are policies implemented to ensure forest resources are used in a sustainable manner." (WK 2012ll)

Later, when agriculture was more used, with "most of the population remaining active in (or indirectly dependent on) the agricultural sector, the main pressure in most areas remained land clearing for crop and cattle farming. Enough wild green was usually left standing (and partially used, for example, to collect firewood, timber and fruits, or to graze pigs) for wildlife to remain viable. The elite's (nobility and higher clergy) protection of their own hunting privileges and game often protected significant woodlands." (WK 2012ll)

Then, "from 1100 to 1500 AD, significant deforestation took place in Western Europe as a result of the expanding human population. The large-scale building of wooden sailing ships by European (coastal) naval owners since the 15th century for exploration, colonisation, slave trade—and other trade on the high seas consumed many forest resources. Piracy also contributed to the over harvesting of forests, as in Spain. This led to a weakening of the domestic economy after Columbus' discovery of America, as the economy became dependent on colonial activities (plundering, mining, cattle, plantations, trade, etc.)." (WK 2012ll)

In general, "a typical progress trap was that cities were often built in a forested area, which would provide wood for some industry (for example, construction, shipbuilding, pottery). When deforestation occurs without proper replanting, however; local wood supplies become difficult to obtain near enough to remain competitive, leading to the city's abandonment, as happened repeatedly in Ancient Asia Minor. Because of fuel needs, mining and metallurgy often led to deforestation and city abandonment." (WK 2012ll)

(c) Industrial

And the third historical era in regard to the causes of deforestation is "industrial."

For instance, "the massive use of charcoal on an industrial scale in Early Modern Europe was a new type of consumption of western forests; even in Stuart England, the relatively primitive production of charcoal has already reached an impressive level. Stuart England was so widely deforested that it depended on the Baltic trade for ship timbers, and looked to the untapped forests of New England to supply the need. Each of Nelson's Royal Navy war ships at Trafalgar (1805) required 6,000 mature oaks for its construction. In France, Colbert planted oak forests to supply the French navy in the future. When the oak plantations matured in the mid-19th century, the masts were no longer required because shipping had changed." (WK 2012ll)

By 1852, "global deforestation sharply accelerated….It has been estimated that about half of the Earth's mature tropical forests—between 7.5 million and 8 million km2 (2.9 million to 3 million sq mi) of the original 15 million to 16 million km2 (5.8 million to 6.2 million sq mi) that until 1947 covered the planet—have now been destroyed." (WK 2012ll; E. Wilson 2002; GUARD 2008; R. Nielsen 2006; NAT 2006)

Since 1900, "up to 90% of West Africa's coastal rainforests have disappeared….In South Asia, about 88% of the rainforests have been lost. Much of what remains of the world's rainforests is in the Amazon basin, where the Amazon Rainforest covers approximately 4 million square kilometres. The regions with the highest tropical deforestation rate between 2000 and 2005 were Central America—which lost 1.3% of its forests each year—and tropical Asia. In Central America, two-thirds of lowland tropical forests have been turned into pasture since 1950 and 40% of all the rainforests have been lost in the last 40 years. Brazil has lost 90–95% of its Mata Atlântica forest. Paraguay was losing its natural semi humid forests in the country's western regions at a rate of 15.000 hectares at a randomly studied 2 month period in 2010." (WK 2012ll; NAT 2012; J. Revington 2012)

Around the world, "Mexico, India, the Philippines, Indonesia, Thailand, Burma, Malaysia, Bangladesh, China, Sri Lanka, Laos, Nigeria, the Democratic Republic of the Congo, Liberia, Guinea, Ghana and the Côte d'Ivoire, have lost large areas of their rainforest….The World Wildlife Fund's ecoregion project catalogues habitat types throughout the world, including habitat loss such as deforestation, showing for example that even in the rich forests of parts of Canada such as the Mid-Continental Canadian forests of the prairie provinces half of the forest cover has been lost or altered." (WK 2012ll; MON 2012)

In fact, "some scientists have predicted that unless significant measures (such as seeking out and protecting old growth forests that have not been disturbed) are taken on a worldwide basis, by 2030 there will only be 10% remaining, with another 10% in a degraded condition. 80% will have been lost, and with them hundreds of thousands of irreplaceable species." (WK 2012ll; R. Nielsen 2006)

Stages of Deforestation

Besides the causes, deforestation has some stages, as suggested by the "forest transition (FT) theory." (P. Meyfroidt 2011)

At the early stages of deforestation, "a country is characterized by high forest cover and low deforestation rates (HFLD countries)." (WK 2012ll; T. Rudel 2005)

Then, "deforestation rates accelerate (HFHD, high forest cover—high deforestation rate), and forest cover is reduced (LFHD. low forest cover—high deforestation rate)." (WK 2012ll)

Then, "the deforestation rate slows (LFLD, low forest cover—low deforestation rate)." (WK 2012ll)

Finally, "forest cover stabilizes and eventually starts recovering." (WK 2012ll)

In fact, "studies of forest transitions have been conducted for several nations as well as sub-national regions" around the world. (WK 2012mm; P. Meyfroidt 2011)

Impact of Deforestation

The impacts of deforestation can be multiple, and since this section is about society, let's illustrative the economic one hereafter.

For instance, "the forest products industry is a large part of the economy in both developed and developing countries. Short-term economic gains made by conversion of forest to agriculture, or over-exploitation of wood products, typically leads to loss of long-term income and long-term biological productivity. West Africa, Madagascar, Southeast Asia and many other regions have experienced lower revenue because of declining timber harvests. Illegal logging causes billions of dollars of losses to national economies annually." (WK 2012ll; R. Butler 2007)

In fact, "the new procedures to get amounts of wood are causing more harm to the economy and overpower the amount of money spent by people employed in logging. According to a study, 'in most areas studied, the various ventures that prompted deforestation rarely generated more than US$5 for every ton of carbon they released and frequently returned far less than US$1.'" (WK 2012ll)

In one estimate at the Convention on Biological Diversity (CBD) meeting in Bonn, "damage to forests and other aspects of nature could halve living standards for the world's poor and reduce global GDP by about 7% by 2050." (WK 2012ll; R. Black 2008)

Problems with the Study of Habitat Fragmentation and Deforestation

However, for the critics, this study of habitat fragmentation and deforestation has some major problems. Consider, for illustration, a few examples below (as summarized in *Table 3.5*).

Firstly, one main criticism of the study of habitat fragmentation and deforestation is that there is the contentious North-South divide to be resolved.

For example, "from the perspective of the developing world, the benefits of forest as carbon sinks or biodiversity reserves go primarily to richer developed nations and there is insufficient compensation for these services. Developing countries feel that some countries in the developed world, such as the United States of America, cut down their forests centuries ago and benefited greatly from this deforestation, that it is hypocritical to deny developing countries the same opportunities, and that the poor shouldn't have to bear the cost of preservation when the rich created the problem." (WK 2012ll)

Secondly, another main criticism of the study of habitat fragmentation and deforestation is that scientists disagree about the causes of deforestaton.

For instance, for the critics, "experts do not agree on whether industrial logging is an important contributor to global deforestation. Some argue that poor people are more likely to clear forest because they have no alternatives, others that the poor lack the ability to pay for the materials and labour needed to clear forest. One study found that population increases due to high fertility rates were a primary driver of tropical deforestation in only 8% of cases." (WK 2012ll; A. Angelsen 1999; W. Laurance 1999; H. Geist 2002)

Thirdly, still another main criticism of the study of habitat fragmentation and deforestation is that scientists also disagree about the rate of deforestation around the world.

For instance, "a 2002 analysis of satellite imagery suggested that the rate of deforestation in the humid tropics (approximately 5.8 million hectares per year) was roughly 23% lower than the most commonly quoted rates. Conversely, a newer analysis of satellite images reveals that deforestation of the Amazon rainforest is twice as fast as scientists previously estimated." (WK 2012ll; F. Achard 2002; A. Jha 2005; P. Spotts 2005)

Fourthly, an additional main criticism of the study of habitat fragmentation and deforestation is that scientists disagree about the definition of forest types, with implications for measurement.

For instance, "critics of the FAO data point out that they do not distinguish between forest types, and that they are based largely on reporting from forestry departments of individual countries, which do not take into account unofficial activities like illegal logging." (WK 2012ll)

Fifthly, another additional main criticism of the study of habitat fragmentation and deforestation is that the proposed stages of deforestation are questionable.

For instance, for the critics, the forest transition theory or "FT is not a 'law of nature,' and the pattern is influenced by national context (for example, human population density, stage of development, structure of the economy), global economic forces, and government policies. A country may reach very low levels of forest cover before it stabilizes, or it might, through good policies, be able to 'bridge' the forest transition." (WK 2012ll)

And sixthly, still another additional main criticism of the study of habitat fragmentation and deforestation is that deforestation can be controlled by different mechanisms.

For instance, for the critics, there have been various mechanisms used to control deforestation around the world, and good examples include "reducing emissions," "cyclic agriculture," "monitoring deforestation," "forest management," "reforestation," "forest plantations," and so on. (WK 2012ll)

Beyond Habitat Fragmentation

These criticisms of the study of habitat fragmentation and deforestation, however, do not imply a total rejection of its usefulness but serve to show us the two opposing sides of the debate, such that the possibility (or impossibility) and desirability (or undesirability) of discontinuity (from the perspective of society with discontinuity, habitat fragmentation, and deforestation as a case study here) are not to the extent that the respective defenders would like us to believe.

In fact, the analysis of discontinuity, habitat fragmentation, and deforestation reveals some insights about the ontological principles in existential dialectics, and good examples include the formalness-informalness principle, the absoluteness-relativeness principle, the partiality-totality principle, the predictability-unpredictability principle, the explicability-inexplicability principle, the fiction-reality principle, the cognitiveness-noncognitiveness principle, the finiteness-transfiniteness principle, the preciseness-vagueness principle, the simpleness-complicatedness principle, the openness-hiddenness principle, the denseness-emptiness principle, the change-constancy principle, the order-chaos principle, the slowness-quickness principle, the expansion-contraction principle, the theory-praxis principle, the convention-novelty principle, the evolution-transformation principle, the symmetry-asymmetry principle, the softness-hardness principle, the seriousness-playfulness principle, the activeness-inactivess principle, the selfness-otherness principle, the regression-progression principle, the same-difference principle, the stability-reaction principle, the functionality-nonfunctionality principle, and the intentionality-nonintentionality principle.

For instance, in relation to the formalness-informalness principle, if there is formalness (e.g., the formal logical requirement of soundness, consistency, and completeness in a system of ideas, as in the study of habitat fragmentation and deforestation), there is informalness (e.g., the non-formal existence of unsoundness, inconsistency, and incompleteness in the study of habitat fragmentation and deforestation, as shown in the criticisms that there is the contentious North-South divide to be resolved, that scientists disagree about the causes of deforestaton, that scientists also disagree about the rate of deforestation around the world, that scientists disagree about the definition of forest types with implications for measurement, that the proposed stages of deforestation are questionable, and that deforestation can be controlled by different mechanisms). And the reverse direction also holds true.

In relation to the absoluteness-relativeness principle, if there is absoluteness (e.g., the absolute view of "some countries in the developed world, such as the United States of America" which "cut down their forests centuries ago and benefited greatly from this deforestation" but now "deny developing countries the same opportunities"), there is relativeness (e.g., what is right for "some countries in the developed world, such as the United States of America" to "deny developing countries the same opportunities" about deforestation is not necessarily so for "developing countries [which] feel…that it is hypocritical to deny developing countries the same opportunities" and "that the poor shouldn't have to bear the cost of preservation when the rich created the problem"). (WK 2012ll) And the reverse direction also holds true.

In relation to the partiality-totality principle, if there is partiality (e.g., the partial view of those in "a 2002 analysis of satellite imagery," which "suggested that the rate of deforestation in the humid tropics [approximately 5.8 million hectares per year] was roughly 23% lower than the most commonly quoted rates"), there is totality (e.g., the more holistic view of the rate of deforestation in the world, such that the whole is not the sum of its parts, that is, it cannot be reduced to the sum of the opposing views of the "2002 analysis of satellite imagery" and its critics, since there will emerge new views in the future not yet known today, just as there are already alternative views nowadays, like "a newer analysis of satellite images" which "reveals that deforestation of the Amazon rainforest is twice as fast as scientists previously estimated"). And the reverse direction also holds true.

In relation to the predictability-unpredictability principle, if there is predictability (e.g., the predictable tendency of those in the "2002 analysis of satellite imagery" to suggest that "the rate of deforestation in the humid tropics [approximately 5.8 million hectares per year] was roughly 23% lower than the most commonly quoted rates"), there is unpredictability (e.g., the more difficult task to predict exactly the extent to which "the rate of deforestation in the humid tropics," if relied on the "2002 analysis of satellite imagery," will be so on a particular future occasion, since "a newer analysis of satellite images reveals that deforestation of the Amazon rainforest is twice as fast as scientists previously estimated"). And the reverse direction also holds true.

In relation to the explicability-inexplicability principle, if there is explicability (e.g., the explanation by those who "argue that poor people are more likely to clear forest because they have no alternatives"), there is inexplicability (e.g., the lack of sufficient explanation, by those who blame the poor for deforestation, of why this is necessarily true, since, for the critics, "the poor lack the ability to pay for the materials and labour needed to clear forest"). And the reverse direction also holds true.

In relation to the fiction-reality principle, if there is fiction (e.g., the fictional aspect of "the FAO data" about deforestation, insofar as "they do not distinguish between forest types, and that they are based largely on reporting from forestry departments of individual countries, which do not take into account unofficial activities like illegal logging"), there is reality (e.g., the realistic aspect of "the FAO data," insofar as they pointed out some instances of deforestation). And the reverse direction also holds true.

In relation to the cognitiveness-noncognitiveness principle, if there is cognitiveness (e.g., a view of things on the basis of reason and evidences, as shown in the historical study which showed that "some countries in the developed world, such as the United States of America cut down their forests centuries ago and benefited greatly from this deforestation"), there is noncognitiveness (e.g., a view of things on the basis of non-cognitive factors like envy, jealousy, power, nationality, race, gender, age, class, greed, lust, status, faith, anger, sadness, joy, fear, wish, etc., as shown in the hypocrisy of "some countries in the developed world, such as the United States of America" which "cut down their forests centuries ago and benefited greatly from this deforestation" but now "deny developing countries the same opportunities," on the basis of the power politics of the North-South divide). And the reverse direction also holds true.

In relation to the finiteness-transfiniteness principle, if there is finiteness (e.g., the finite number of the categories of entities affected by deforestation, like soils, water, trees, species, etc.), there is transfiniteness (e.g., the transfinite number of all the individual pieces of minerals in soils, drops of water, leaves on trees, animals in species, etc. in history that have ever existed and been affected by deforestation). And the reverse direction also holds true.

In relation to the preciseness-vagueness principle, if there is preciseness (e.g., the precise identification of 3 historical eras in regard to the causes of deforestation in the article aforecited), there is vagueness (e.g., the vagueness in the identification of the 3 historical eras in regard to the causes of deforestation in the article aforecited, since it is not clear why there must be only 3, not 4, 5, 6, and so on). And the reverse direction also holds true.

In relation to the simpleness-complicatedness principle, if there is simpleness (e.g., the relatively simple analysis of deforestiation by the conservation activists), there is complicatedness (e.g., the relatively more complicated analysis of deforestiation, by challenging the claims and assumptions, as shown in the problems pointed out by the critics as cited). And the reverse direction also holds true.

In relation to the openness-hiddenness principle, if there is openness (e.g., the open exploration of deforestation around the world by "the FAO data"), there is hiddenness (e.g., the hidden bias in "the FAO data," because "critics of the FAO data point out that they do not distinguish between forest types, and that they are based largely on reporting from forestry departments of individual countries, which do not take into account unofficial activities like illegal logging"). And the reverse direction also holds true.

In relation to the denseness-emptiness principle, if there is denseness (e.g., the relatively denser concentration of forests around the world in the pre-historical era), there is emptiness (e.g., the relatively less dense, or more empty, concentration of forests around the world in our time). And the reverse direction also holds true.

In relation to the change-constancy principle, if there is change (e.g., the ever new ways to understand discontinuity, as shown in the study of habitat fragmentation in the context deforestation), there is constancy (e.g., the ever constant existence of different problems in the ever new ways to understand discontinuity, as shown in the problems with the study of habitat fragmentation in the context of deforestation pointed out by the critics). And the reverse direction also holds true.

In relation to the order-chaos principle, if there is order (e.g., the more or less orderly view of deforestation, if looked only from the vantage point like the "2002 analysis of satellite imagery" which suggested that "the rate of deforestation in the humid tropics [approximately 5.8 million hectares per year] was roughly 23% lower than the most commonly quoted rates"), there is chaos (e.g., the more or less chaotic view of deforestation, if looked from the multiple opposing and inconsistent vantage points, like the "2002 analysis of satellite imagery" which suggested that "the rate of deforestation in the humid tropics [approximately 5.8 million hectares per year] was roughly 23% lower than the most commonly quoted rates"—and "a newer [but opposing] analysis of satellite images" which reveals that deforestation of the Amazon rainforest is twice as fast as scientists previously estimated"). And the reverse direction also holds true.

In relation to the slowness-quickness principle, if there is slowness (e.g., the relatively slower speed of deforestation in the pre-historical era, because only "small scale deforestation was practiced by some societies for tens of thousands of years before the beginnings of civilization"), there is quickness (e.g., the relatively quicker speed of deforestation in the industrial era, because "by 1852, 'global deforestation sharply accelerated….It has been estimated that about half of the Earth's mature tropical forests…have now been destroyed'"). And the reverse direction also holds true.

In relation to the expansion-contraction principle, if there is expansion (e.g., the relative expansion of urbanization, agriculture, mining, etc. in the industrial era), there is contraction (e.g., the relative decline of forests in the industrial era). And the reverse direction also holds true.

In relation to the theory-praxis principle, if there is theory (e.g., the theoretical construction of deforestation, as shown by the "forest transition theory"), there is praxis (e.g., the practical application of the "forest transition theory" to the field of policy, as shown in the policy implication that "a global forest transition would require major policy and technological innovations as well as shifts in demands for fiber, fuel, and food"). (WK 2012mm; P. Meyfroidt 2011) And the reverse direction also holds true.

In relation to the convention-novelty principle, if there is convention (e.g., the conventional practice of uncontrolled deforestation which "resulted in damage to habitat, biodiversity loss and aridity" in the "pre-industrial" and "industrial" eras), there is novelty (e.g., the alternative novel challenge to the conventional practice of uncontrolled deforestation, by the new idea of forest management with such diverse mechanisms as "reducing emissions," "cyclic agriculture," "monitoring deforestation," "reforestation," "forest plantations," and so on). (WK 2012ll) And the reverse direction also holds true.

In relation to the evolution-transformation principle, if there is evolution (e.g., the natural evolution of deforestation, by natural causes, in the state of nature, like climate change which led to "the Carboniferous Rainforest Collapse"), there is transformation (e.g., the technical transformation of human ability to deal with deforestation by the invention of forest management, with such diverse mechanisms as "cyclic agriculture," "monitoring deforestation," "reforestation," and so on). And the reverse direction also holds true.

In relation to the symmetry-asymmetry principle, if there is symmetry (e.g., the existence of different countries to deal with deforestation around the world), there is asymmetry (e.g., "developing countries feel that some countries in the developed world, such as the United States of America, cut down their forests centuries ago and benefited greatly from this deforestation, that it is hypocritical to deny developing countries the same opportunities, and that the poor shouldn't have to bear the cost of preservation when the rich created the problem"). And the reverse direction also holds true.

In relation to the softness-hardness principle, if there is softness (e.g., the defense of the forest transition theory by the advocates), there is hardness (e.g., the critique of the forest transition theory by the critics aforecited). And the reverse direction also holds true.

In relation to the seriousness-playfulness principle, if there is seriousness (e.g., the serious business of those who try to deal with the problem of deforestation), there is playfulness (e.g., the playful part of those who try to deal with the problem of deforestation, when they play around with different mechanisms over the decades, like "reducing emissions," "cyclic agriculture," "monitoring deforestation," "forest management," "reforestation," "forest plantations," and so on). And the reverse direction also holds true.

In relation to the activeness-inactivess principle, if there is activeness (e.g., the relative activeness of biodiversity in in the Amazon basin, since "much of what remains of the world's rainforests is in the Amazon basin, where the Amazon Rainforest covers approximately 4 million square kilometres"), there is inactiveness (e.g., the relative inactiveness of biodiversity "in South Asia," where "about 88% of the rainforests have been lost"). (WK 2012ll; BBC 2003a)

In relation to the selfness-otherness principle, if there is selfness (e.g., the tendency to be mostly centered on one's self or group, as shown in those "countries in the developed world, such as the United States of America," which "cut down their forests centuries ago and benefited greatly from this deforestation" but now "deny developing countries the same opportunities," on the basis of the power politics of the North-South divide), there is otherness (e.g., the tendency to be mostly centered on others, as shown in the efforts of some "international organizations including the United Nations and the World Bank," which "have begun to develop programs aimed at curbing deforestation," with different "monetary or other incentives to encourage developing countries to limit and/or roll back deforestation," for a better human future). (WK 2012ll) And the reverse direction also holds true.

In relation to the regression-progression principle, if there is regression (e.g., the regression made by the study on deforestation, as shown in the problems pointed out by the critics aforecited), there is progression (e.g., the progress made by the study on deforestation, as shown in its contribution to a better understanding of the causes and impacts aforecited). And the reverse direction also holds true.

In relation to the same-difference principle, if there is similarity in outcome (e.g., the contribution to the molding and control of beliefs, values, and behaviors, regardless of whether this be done by way of the North-South divide, or by way of the curbing mechanisms of forest management), there is difference in outcome (e.g., the contribution to the molding and control of beliefs, values, and behaviors by way of the North-South divide for a relatively more politicized lifeworld—but the contribution to the molding and control of beliefs, values, and behaviors by way of the curbing mechanisms of forest management for a relatively more technocratic lifeworld). And the reverse direction also holds true.

In relation to the stability-reaction principle, if there is stability (e.g., the relatively more stable condition of deforestation in the pre-historical era, since there was only "small scale deforestation…practiced by some societies for tens of thousands of years before the beginnings of civilization"), there is reaction (e.g., "small scale deforestation" in the pre-historical era led to further changes later on, when in the industrial era, "global deforestation sharply accelerated around 1852"). And the reverse direction also holds true.

In relation to the functionality-nonfunctionality principle, if there is functionality (e.g., the presence of action for which something is fitted, as shown in the "small scale deforestation" in the pre-historical era, which served the function of providing the needs of "some societies for tens of thousands of years before the beginnings of civilization" in that era), there is nonfunctionality (e.g., the relative lack of action for which something is fitted, as shown in the massive deforestation in the industrial era, like "the massive use of charcoal on an industrial scale in Early Modern Europe," such that the "small scale deforestation" in the pre-historical era is no longer function in our itme, because of different needs in the industrial era). And the reverse direction also holds true.

And in relation to the intentionality-nonintentionality principle, if there is intentionality (e.g., the planning or design of something for a certain outcome, as shown in "a 2002 analysis of satellite imagery," which tried to find out the true rate of massive deforestation in our time), there is nonintentionality (e.g., the relative absence of planning or design of something for a certain outcome, such that the outcome can even be counter-intentional in being the opposite of what was originally planned, as shown in the "2002 analysis of satellite imagery," which "suggested that the rate of deforestation in the humid tropics [approximately 5.8 million hectares per year] was roughly 23% lower than the most commonly quoted rates," contrary to the original expectation of a higher rate of deforestation in our time). And the reverse direction also holds true.

Discontinuity, Oil Depletion, and Resource Curse

Another good way to evaluate the possibility (or impossibility) and desirability (or undesirability) of discontinuity from the perspective of society is to explore discontinuity, oil depletion and resource curse .

An argument here is that there is a relationship between oil depletion and resource curse.

The term "oil depletion" refers to "the second half of the production curve of an oil well, oil field, or the average of total world oil production." (WK 2012nn)

Predictions of Production Decline

Oil depletion is due to the fact that "the world's oil supply is fixed because petroleum is naturally formed far too slowly to be replaced at the rate at which it is being extracted. Over many millions of years, plankton, bacteria, and other plant and animal matter become buried in sediments on the ocean floor. When conditions are right—a lack of oxygen for decomposition, and sufficient depth and temperature of burial—these organic remains are converted into petroleum compounds, while the sediment accompanying them is converted into sandstone, siltstone, and other porous sedimentary rock. When capped by impermeable rocks such as shale, salt, or igneous intrusions, they form the petroleum reservoirs which are exploited today." (WK 2012nn)

So, there have been different models over time to predict oil depletion, and three major models are presented below (and summarized in *Table 3.6*), namely, (a) oil well production decline, (b) oil field production decline, and (c) world oil production.

(a) Oil Well Production Decline

The first prediction model to be presented here has to do with "oil well prodution decline."

According to this model, "oil well production curves typically end in an exponential decline. At natural rates, oil well production curves appear similar to a bell curve, a phenomenon known as the Hubbert curve. The typical decline is a rapid drop in production, and eventually a leveling off to a point at which they no longer produce profitable amounts. Such wells are referred to as marginal or stripper wells." (WK 2012nn)

The Hubbert curve is based on "the Hubbert peak theory" which "makes predictions of production rates based on prior discovery rates and anticipated production rates. Hubbert curves predict that the production curves of non-renewing resources approximate a bell curve. Thus, when the peak of production is passed, production rates enter an exponential decline." (WK 2012nn; M. Hubbert 1956)

However, "the shape of production curve of an oil well can be affected by a number of factors," as shown below: (WK 2012nn).

- Ex: "well may be restricted by choice by lack of market demand or government regulation. This flattens the peak of the curve, but will not change the well's total production significantly.
- Ex: "hydraulic fracturing (*fracking*) or acidizing may be used to cause a sharp spike in production, and may increase the recoverable reserves of a given well."
- Ex: "the field may undergo a secondary or tertiary recovery project" (to be addressed in the next section).

Of course, there can be other factors, so the ones above are illustrative.

(b) Oil Field Production Decline

The second prediction model to be presented here has to do with "oil field prodution decline."

After all, "each individual oil well is a portion of a larger fixed area oil field. As with individual wells, discovery and production amounts of oil fields generally average to a similar bell shaped production curve.

Eventually, when the field is completely drilled out, a field's production goes into a sharp decline as the average production of its wells enter decline. As this decline levels off, production can continue at relatively low rates. A number of oil fields in the U.S. have been producing for over 100 years." (WK 2012nn)

In general, "oil field production curves can be modified by a number of factors," as shown below: (WK 2012nn)

- Ex: "production may be restricted by market conditions or government regulation."
- Ex: "a secondary recovery project, such as water or gas injection, can repressurize the field and improve the production rate temporarily. However, it will not change the total production amount over the life of the field. Eventually the field will go into a steeper than normal decline."
- Ex: "the field may undergo an enhanced oil recovery project, such as drilling of wells for injection of solvents, carbon dioxide, or steam. This can be very expensive but allows more oil to be coaxed out of the rock, increasing the ultimate production of the field."

Again, there can be other factors, so the ones above are illustrative.

(c) World Oil Production

And the third prediction model to be presented here has to do with "world oil production."

Just like oil well production decline (and also oil field production decline)—"world oil production has followed a typical Hubbert curve, rising over the past century with only a few dips. The 1970 production peak in the U.S. caused many people to begin to question when the world production peak would occur. The peak of world production is known as Peak oil. By the mid-2000s, all of the world's major oil producing countries except Saudi Arabia were producing at maximum capacity (many having peaked in production), and some experts such as Matthew Simmons were questioning whether even Saudi Arabia had any reserve capacity left." (WK 2012nn)

So, "industry observers have pointed to the similarities between the global production curve in mid-2000s and that of the United States in the 1970s," with the following predictions: (WK 2012nn)

- Ex: "the oil price increases since 2003 were preceded by a decade of production cutbacks in OPEC countries in an attempt to keep prices high despite an oil glut. This is similar to production cutbacks in

Texas and other states to maintain prices despite an oil glut in the decade prior to the 1973 oil crisis."

- Ex: "world oil prices reached record inflation adjusted highs beginning in 2008, but new oil did not appear on the market, as the theory of supply and demand would predict. This is reminiscent of price increases in the United States in the 1970s when U.S. oil production started to decline despite record high prices and record drilling by oil companies."
- Ex: "there are serious doubts about whether OPEC countries really have the oil reserves they claim. This is similar to the illusionary oil reserves that U.S. oil companies claimed to have in the decade prior to the 1973 and 1979 oil crisis. In the 1970s, those companies were unable to produce as much oil as they had predicted, and production went down instead of up."

But others are more optimistic; for instance, "a Harvard report estimates 110 million barrels per day in 2020, with significant contributions from shale oil and other unconventional resources." (WK 2012nn; L. Maugeri 2012)

Implications of Production Decline

In any event, oil depletion can have major impacts on the world. Consider, for illustration, three major scenarios below (as summarized in *Table 3.6*), namely, (a) catastrophe, (b) recession, and (c) rising food prices.

(a) Catastrophe

The first scenario of a major impact of oil depletion on the world to be illustrated here has to do with "catastrophe."

The underlying problem here is that "economic growth and prosperity since the industrial revolution have, in large part, been due to increased efficiencies in the use of better and higher concentrations of energy in fossil fuels. The use of fossil fuels allows humans to participate in takedown, which is the consumption of energy at a greater rate than it is being replaced." (WK 2012nn)

Thus, "some envisage a Malthusian catastrophe occurring as oil becomes increasingly inefficient to produce; others have learned from the examples demonstrated in mature basins and applied those operational procedures to these basins to preserve their operational tempo. Since the 1940s, agriculture has dramatically increased its productivity, due largely to the use of chemical pesticides, fertilizers, and increased mechanisation.

This process has been called the Green Revolution. The increase in food production has allowed world population to grow dramatically over the last 50 years. Pesticides rely upon oil as a critical ingredient, and fertilizers require natural gas. Farm machinery also requires oil." (WK 2012nn)

It is therefore not surprising for some people to argue that, since "in today's world every joule one eats requires 5–15 joules to produce and deliver,...decreasing supply of oil will cause modern industrial agriculture to collapse, leading to a drastic decline in food production, food shortages and possibly even mass starvation." (WK 2012nn)

Another major possible effect has to do with "modern transportation and housing infrastructure. A large proportion of the developed world's population live in suburbs, a type of low-density settlement designed with the automobile in mind. Commentators such as James Howard Kunstler argue that because of its reliance on the automobile, the suburb is an unsustainable living arrangement; the implications of peak oil would leave many suburb-dwellers unable to afford fuel for their cars, and force them to move to higher density, more walkable areas. In effect, surburbia would comprise the 'slums of the future.' A movement to deal with this problem early, called 'New Urbanism,' seeks to develop the suburbs into higher density neighborhoods and use high density, mixed-use forms for new building projects." (WK 2012nn)

(b) Recession

The second scenario of a major impact of oil depletion on the world to be illustrated here has to do with "recession."

This second scenario is more modest, not so extreme as the first one, since it assumes "a slower rate of depletion and a smooth transition to alternative energy sources," which "could cause substantial economic hardship such as a recession or depression due to higher energy prices." (WK 2012nn)

If history has taught us anything, "there is a close correlation in the timing of oil price spikes and economic downturns. Inflation has also been linked to oil price spikes. However, economists disagree on the strength and causes of this association. Conversely, the recessions of the early 1970s and early 1980s were associated with a relatively brief period of somewhat dwindling energy availability; the possible future increase in oil prices might be much higher and last longer." (WK 2012nn)

(c) Rising Food Prices

And the third scenario of a major impact of oil depletion on the world to be illustrated here has to do with "rising food prices."

Any rise of oil prices because of oil depletion can eaisly "cause rising food prices in three ways. First, increased equipment fuel costs drive higher prices. Second, transportation costs increase retail prices. Third, higher oil prices are causing farmers to switch from producing food crops to producing biofuel crops." (WK 2012nn; T. Brown 1997; E. Swanepoel 2007)

This then means that "the law of supply and demand predicts that if fewer farmers are producing food the price of food will rise." (WK 2012nn; ECO 2007)

Worsening Effects of Resource Curse

However, the impacts of oil depletion can become worsen, if the effects of resource curse (from the vantage point of society) are taken into account; in other words, resource-cursed countries are generally not prepared to deal with the negative impacts of oil depletion.

The word "resource curse" (also known as "paradox of plenty") refers to "the paradox that countries and regions with an abundance of natural resources, specifically point-source non-renewable resources like minerals and fuels, tend to have less economic growth and worse development outcomes than countries with fewer natural resources." (WK 2012oo)

There are different reasons for this resource curse; consider, for illustration, the following 10 main ones (which can worsen all the more the impacts of oil depletion in the long run), as summarized in *Table 3.6*.

(a) Tug of War between People and Government

The first reason for resource curse (which can worsen the effects of oil depletion in the long run) has to do with "tug of war between people and government."

This happens, because "the ambitions of the people and the government conflict, due to the large amount of resources and money a country's government amass for their own luxuries rather than for the people. Thus natural resources serve as a curse for the people, who then have a lower relative standard of living." (WK 2012oo)

In fact, "it has...been argued that one can correlate rises and falls in the price of petroleum with rises and falls in the implementation of human rights in major oil-producing countries. Human rights throughout resource-cursed countries are dismal or completely lacking. Most normally resource-cursed countries are ruled by either authoritarian or other types of highly repressive regimes. These regimes are kept in power by a select elite such as high-ranking politicians and military leaders. As long as the existing government keeps these few happy they can rule without fear of consequence. This system is set up so that the common folk those most in need of the protection are left to fend for themselves. In places like the Democratic Republic of the Congo human life has almost no value and slave labor is common....The resource curse of the Congo takes the form of diamonds. While no true government exists, each faction exploits the people they rule to harvest diamonds and other rare minerals." (WK 2012oo; P. O'Neil 2004)

(b) Conflict

The second reason for resource curse (which can worsen the effects of oil depletion in the long run) has to do with "conflict."

This is because "different groups and factions fight for their share. Sometimes these emerge openly as separatist conflicts in regions where the resources are produced (such as in Angola's oil-rich Cabinda province) but often the conflicts occur in more hidden forms, such as fights between different government ministries or departments for access to budgetary allocations. This tends to erode governments' abilities to function effectively." (WK 2012oo)

In general, "there are several main types of relationships between natural resources and armed conflicts. First, resource curse effects can undermine the quality of governance and economic performances, thereby increasing the vulnerability of countries to conflicts (the 'resource curse' argument). Second, conflicts can occur over the control and exploitation of resources and the allocation of their revenues (the 'resource war' argument). Third, access to resource revenues by belligerents can prolong conflicts (the 'conflict resource' argument). According to one academic study, a country that is otherwise typical but has primary commodity exports around 25% of GDP has a 33% risk of conflict, but when exports are 5% of GDP the chance of conflict drops to 6%." (WK 2012oo; C. Norman 2009; P. Le Billon 2006; I. Bannon 2003; P. Collier 2003)

(c) International Pressure

The third reason for resource curse (which can worsen the effects of oil depletion in the long run) has to do with "international pressure."

This is because "there is often international pressure on resource-rich countries of the Third World to avoid reinvesting resource revenues in social investment purposes, or even in developmental efforts towards economic diversification. It is alleged that this pressure comes from powerful states such as the United States and other leading western countries as well as pro-liberalization institutions such as the World Trade Organization and the International Monetary Fund." (WK 2012oo)

(d) Taxation

The fourth reason for resource curse (which can worsen the effects of oil depletion in the long run) has to do with "taxation."

This is because "in many economies that are not resource-dependent, governments tax citizens, who demand efficient and responsive government in return. This bargain establishes a political relationship between rulers and subjects. In countries whose economies are dominated by natural resources, however, rulers don't need to tax their citizens because they have a guaranteed source of income from natural resources." (WK 2012oo; D. Bräutigam 2008)

One consequence is that, "because the country's citizens aren't being taxed, they have less incentive to be watchful with how government spends its money. In addition, those benefiting from mineral resource wealth may perceive an effective and watchful civil service and civil society as a threat to the benefits that they enjoy, and they may take steps to thwart them. As a result, citizens are often poorly served by their rulers, and if the citizens complain, money from the natural resources enables governments to pay for armed forces to keep the citizens in check. Countries whose economies are dominated by resource extraction industries tend to be more repressive, corrupt and badly managed." (WK 2012oo; M. Moore 2007; T. Palley 2003)

(e) Dutch Disease

The fifth reason for resource curse (which can worsen the effects of oil depletion in the long run) has to do with the "Dutch disease."

The term "Dutch disease" refers to "an economic phenomenon in which the revenues from natural resource exports damage a nation's productive economic sectors by causing an increase of the real exchange rate and wage increase. This makes tradable sectors, notably agriculture and manufacturing, less competitive in world markets." (WK 2012oo)

So, "the increasing national revenue will often result in higher government spending (health, welfare, military) that increases the real exchange rate and raises wages. The decrease in the sectors exposed to international competition and consequently even greater dependence on natural resource revenue leaves the economy vulnerable to price changes in the natural resource. Also, since productivity generally increases faster in the manufacturing sector, the economy will lose out on some of those productivity gains." (WK 2012oo)

The word "Dutch" here is because the "Dutch Disease first became apparent after the Dutch discovered a massive natural gas field in Groningen in 1959. The Netherlands sought to tap this resource in an attempt to export the gas for profit. However when the gas began to flow out of the country so too did its ability to compete against other countries' exports. With the Netherlands' focus primarily on the new gas exports, the Dutch currency grew at a very quick rate which harmed the country's ability to export other products. With the growing gas market and the shrinking export economy, the Netherlands began to experience a recession. This process has been witnessed in multiple countries around the world including but not limited to Venezuela (oil), Angola (diamonds, oil), the Democratic Republic of the Congo (diamonds), and various other nations. All of these resources are considered 'resource-cursed.'" (WK 2012oo; P. O'Neil 2004)

(f) Revenue Volatility

The sixth reason for resource curse (which can worsen the effects of oil depletion in the long run) has to do with the "revenue volatility."

This is because "prices for some natural resources are subject to wide fluctuation: for example crude oil prices rose from around $3 per barrel to $12/bbl in 1974 following the 1973 oil crisis and fell from $27/bbl to below $10/bbl during the 1986 glut. In the decade from 1998 to 2008, it rose from $10/bbl to $145/bbl, before falling by more than half to $60/bbl over a few months. When government revenues are dominated by inflows from natural resources (for example, 99.3% of Angola's exports came from just oil and diamonds in 2005), this volatility can play havoc with government planning and debt service. Abrupt changes in economic realities that result from this often provoke widespread breaking of contracts or curtailment of social programs, eroding the rule of law and popular support." (WK 2012oo; IMF 2007)

(g) Excessive Borrowing

The seventh reason for resource curse (which can worsen the effects of oil depletion in the long run) has to do with the "excessive borrowing."

This is because, "since governments expect more income in the future, they start accumulating debt, even though they are receiving natural resource revenues as well. This is encouraged, since, if the real exchange rate increases, through capital inflows or the Dutch disease, this makes the interest payments on the debt cheaper. In addition, the country's natural resources act as collateral leading to more credit." (WK 2012oo)

But the drawback here is that, "if the natural resources' prices begin to fall, and if the real exchange rate falls, a government would have less money with which to pay a more expensive debt. For example, many oil-rich countries like Nigeria and Venezuela saw rapid expansions of their debt burdens during the 1970s oil boom; however, when oil prices fell in the 1980s, bankers stopped lending to them and many of them fell into arrears, triggering penalty interest charges that made their debts grow even more" (WK 2012oo)

(h) Corruption

The eighth reason for resource curse (which can worsen the effects of oil depletion in the long run) has to do with the "corruption."

This is because, "in resource-rich countries, it is often easier to maintain authority through allocating resources to favoured constituents than through growth-oriented economic policies and a level, well-regulated playing field. Huge flows of money from natural resources fuel this political corruption. The government has less need to build up the institutional infrastructure to regulate and tax a productive economy outside the resource sector, so the economy may remain undeveloped. The presence of offshore tax havens provide widespread opportunities for corrupt politicians to hide their wealth." (WK 2012oo)

In addition, "many extractive operations are illegal and encouraged by corrupt multi-national corporations in collusion with national governments. Objections made by indigenous inhabitants are usually ignored. Ed Ayres of Worldwatch Institute reported on these collusive operations in many parts of the world. The United States Senate Foreign Relations Committee report entitled 'Petroleum and Poverty Paradox' states that 'too often, oil money that should go to a nation's poor ends up in the pockets of the rich, or it may be squandered on grand palaces and massive showcase projects instead of being invested productively.'" (WK 2012oo; WW 2011)

(i) Lack of Diversification and Enclave Effects

The ninth reason for resource curse (which can worsen the effects of oil depletion in the long run) has to do with the "lack of diversification and enclave effects."

This is because "economic diversification may be neglected by authorities or delayed in the light of the temporary high profitability of the limited natural resources. The attempts at diversification that do occur are often grand public works projects which may be misguided or mismanaged. However, even if the authorities try to diversify the economy, this is made difficult because the resource extraction is vastly more lucrative and out-competes other industry. Successful natural-resource-exporting countries often become more dependent on extractive industries over time. The abundance of revenue from natural resources discourages long-term investment in infrastructure which would support a more diverse economy, increasing the negative impact of sudden resource-price drops. While the resource sectors tend to provide large financial revenues, they often provide few jobs, and tend to operate as enclaves with few forward and backward connections to the rest of the economy." (WK 2012oo)

(j) Human Resources

And the tenth reason for resource curse (which can worsen the effects of oil depletion in the long run) has to do with the "natural resources" themselves.

This is because, "in many poor countries, natural resource industries tend to pay far higher salaries than what would be available elsewhere in the economy. This tends to attract the best talent from both private and government sectors, damaging these sectors by depriving them of their best skilled personnel. Another possible effect of the resource curse is the crowding out of human capital; countries that rely on natural resource exports may tend to neglect education because they see no immediate need for it." (WK 2012oo)

By contrast, "resource-poor economies like Singapore, Taiwan or South Korea,…spent enormous efforts on education, and this contributed in part to their economic success." (WK 2012oo)

Problems with the Study of Oil Depletion and Resource Curse

However, for the critics, this study of oil depletion and resource curse has some major problems. Consider, for illustration, a few examples below (as summarized in *Table 3.6*).

Firstly, one main criticism of the study of oil depletion and resource curse is that scientists disagree about the prediction of oil depletion.

For example, "the American Petroleum Institute estimated in 1999 the world's oil supply would be depleted between 2062 and 2094….A study published in the journal *Energy Policy* by researchers from Oxford University, however, predicted demand would surpass supply by 2015 (unless constrained by strong recession pressures caused by reduced supply or government intervention)." (WK 2012nn; N. Owen 2010)

Secondly, another main criticism of the study of oil depletion and resource curse is that some prediction models are more optimistic (or conversely, more pessimistic) than others.

For instance, for the critics, there are different scenarios of the implications of oil depletion, ranging from the extreme to the moderate, like the three examples of "catastrophe," "recession," and "increasing food prices" as discussed earlier.

Thirdly, still another main criticism of the study of oil depletion and resource curse is that some predictions are based on wishful thinking.

For instance, "there are serious doubts about whether OPEC countries really have the oil reserves they claim. This is similar to the illusionary oil reserves that U.S. oil companies claimed to have in the decade prior to the 1973 and 1979 oil crisis. In the 1970s, those companies were unable to produce as much oil as they had predicted, and production went down instead of up." (WK 2012nn)

Fourthly, an additional main criticism of the study of oil depletion and resource curse is that there are alternatives to oil dependency.

For instance, for the critics, "most or all of the uses of fossil fuels in agriculture can be replaced with alternatives. For example, by far the biggest fossil fuel input to agriculture is the use of natural gas as a hydrogen source for the Haber-Bosch fertilizer-creation process. Natural gas is used simply because it is the cheapest currently available source of hydrogen; were that to change, other sources, such as electrolysis powered by solar energy, could be used to provide the hydrogen for creating fertilizer without relying on fossil fuels. (WK 2012nn)

And fifthly, another additional main criticism of the study of oil depletion and resource curse is that researchers disagree on the causal relationship between natural resources and economic development in the long run.

For instance, some argued that "resource-poor economies like Singapore, Taiwan or South Korea,…spent enormous efforts on education, and this contributed in part to their economic success"—but "other researchers, however, dispute this conclusion; they argue that natural resources generate easily taxable rents that more often than not result in increased spending on education." (WK 2012oo; J. Stijns 2006) Some others are even more assertive, because, for them, "that a high dependency on resource exports correlates with bad policies and effects is not caused by the large degree of resource exportation. The causation goes in the opposite direction: conflicts and bad policies created the heavy dependence on exports of natural resources." (WK 2012oo)

Beyond Resource Curse

These criticisms of the study on oil depletion and resource curse, however, do not claim a total rejection of its usefulness but serve to show us the two opposing sides of the debate, such that the possibility (or impossibility) and desirability (or undesirability) of discontinuity (from the perspective of society with discontinuity, oil depletion, and resource curse as a case study here) are not to the extent that the respective defenders would like us to believe.

Furthermore, the analysis of discontinuity, oil depletion, and resource curse reveals the inner workings of the ontological principles in existential dialectics, and good examples include the formalness-informalness principle, the absoluteness-relativeness principle, the partiality-totality principle, the predictability-unpredictability principle, the explicability-inexplicability principle, the fiction-reality principle, the cognitiveness-noncognitiveness principle, the finiteness-transfiniteness principle, the preciseness-vagueness principle, the simpleness-complicatedness principle, the openness-hiddenness principle, the denseness-emptiness principle, the change-constancy principle, the order-chaos principle, the slowness-quickness principle, the expansion-contraction principle, the theory-praxis principle, the convention-novelty principle, the evolution-transformation principle, the symmetry-asymmetry principle, the softness-hardness principle, the seriousness-playfulness principle, the activeness-inactivess principle, the selfness-otherness principle, the regression-progression principle, the same-difference principle, the stability-reaction principle, the functionality-nonfunctionality principle, and the intentionality-nonintentionality principle.

For instance, in relation to the formalness-informalness principle, if there is formalness (e.g., the formal logical requirement of soundness, completeness, and consistency in a system of ideas, as shown in the study on oil depletion and resource curse), there is informalness (e.g., the non-formal existence of unsoundness, incompleteness, and inconsistency in the study on oil depletion and resource curse, as shown in the criticisms that scientists disagree about the prediction of oil depletion, that some prediction models are more optimistic [or conversely, more pessimistic] than others, that some predictions are based on wishful thinking, that there are alternatives to oil dependency, and that researchers disagree on the causal relationship between natural resources and economic development in the long run). And the reverse direction also holds true.

In relation to the absoluteness-relativeness principle, if there is absoluteness (e.g., the absolute view of those rulers in "resource-cursed countries…ruled by either authoritarian or other types of highly repressive regimes," in that "these regimes are kept in power by a select elite such as high-ranking politicians and military leaders," so that "the common folk, those most in need of the protection, are left to fend for themselves," as "in places like the Democratic Republic of the Congo"), there is relativeness (e.g., what is beneficial for these authoritarian rulers "in places like the Democratic Republic of the Congo" is not necessarily so for "the common folk, those most in need of the protection," who "are left to fend for themselves," and "if the citizens complain, money from the natural resources enables governments to pay for armed forces to keep the citizens in check"). (WK 2012oo) And the reverse direction also holds true.

In relation to the partiality-totality principle, if there is partiality (e.g., the partial view of those rulers in "resource-cursed countries…ruled by either authoritarian or other types of highly repressive regimes"), there is totality (e.g., the more holistic view of governance in resource-cursed countries, such that the whole is not the sum of its parts, that is, it cannot be reduced to the sum of opposing views of those authoritarian rulers in "resource-cursed countries" and "the common folk," since there will emerge new views in the future not yet known today, just as there are already some researchers who do not side exclusively with any of them and hold alternative views evaluating the pros and cons of the debate aforecited). And the reverse direction also holds true.

In relation to the predictability-unpredictability principle, if there is predictability (e.g., the predictable tendency of those in the American Petroleum Institute, who "estimated in 1999 the world's oil supply would be depleted between 2062 and 2094"), there is unpredictability (e.g., the more difficult task to predict exactly the extent to which the estimate by the American Petroleum Institute will be accurate, since there was another "study published in the journal *Energy Policy* by researchers from Oxford University," who, "however, predicted demand would surpass supply by 2015" instead). And the reverse direction also holds true.

In relation to the explicability-inexplicability principle, if there is explicability (e.g., the explanation by those who argued that, unlike resource-rich countries, "resource-poor economies like Singapore, Taiwan or South Korea,…spent enormous efforts on education, and this contributed in part to their economic success"), there is inexplicability (e.g., the lack of sufficient explanation by those about "resource-poor economies like Singapore, Taiwan or South Korea" of why this correlation between resource-poor economies and economic success is necessarily valid, because "other researchers, however, dispute this conclusion," as "they argue that natural resources generate easily taxable rents that more often than not result in increased spending on education" in resource-rich economies). And the reverse direction also holds true.

In relation to the fiction-reality principle, if there is fiction (e.g., the fictional aspect of the estimate by the 1999 study published in the journal *Energy Policy*, insofar as its prediction that the world's oil demand "would surpass supply by 2015" is too pessimistic), there is reality (e.g., the realistic aspect of the estimate by the 1999 study published in the journal

Energy Policy, insofar as oil depletion is a long-term problem which is to be taken seriously). And the reverse direction also holds true.

In relation to the cognitiveness-noncognitiveness principle, if there is cognitiveness (e.g., a view of things on the basis of reason and evidences, as shown in the data from OPEC in 2011, which showed that the major oil resreves were in Venezuela, Saudi Arabia, Canada, and so on), there is noncognitiveness (e.g., a view of things on the basis of non-cognitive factors like envy, jealousy, power, nationality, race, gender, age, class, greed, lust, status, faith, anger, sadness, joy, fear, wish, etc., as shown in the wishful thinking embedded in some estimates, as the critics pointed out that "there are serious doubts about whether OPEC countries really have the oil reserves they claim," and "this is similar to the illusionary oil reserves that U.S. oil companies claimed to have in the decade prior to the 1973 and 1979 oil crisis," but "in the 1970s, those companies were unable to produce as much oil as they had predicted, and production went down instead of up"). (WK 2012nn & 2012pp) And the reverse direction also holds true

In relation to the finiteness-transfiniteness principle, if there is finiteness (e.g., the finite number of the categories of entities which were converted into petroleum compounds over many millions of years, like "plankton," "bacteria," and "other plant and animal matter"), there is transfiniteness (e.g., the transfinite number of all the individual organic remains from "plankton," "bacteria," and "other plant and animal matter" in history that have ever existed and been converted into petroleum compounds). And the reverse direction also holds true.

In relation to the preciseness-vagueness principle, if there is preciseness (e.g., the precise identification of 3 scenarios for the prediction of the impacts of oil depletion in the article aforecited, namely, "catastrophe," "recession," and "increasing food prices"), there is vagueness (e.g., the vagueness in the identification of the 3 scenarios for the prediction of the impacts of oil depletion in the article aforecited, since it is not clear why there must be only 3, not 4, 5, 6, and so on, since, for the critics, there can be another alternative like the replacement with renewable energy). And the reverse direction also holds true.

In relation to the simpleness-complicatedness principle, if there is simpleness (e.g., the relatively simple analysis of oil depletion by the advocates), there is complicatedness (e.g., the relatively more complicated analysis of oil depletion, by challenging the claims and assumptions, as shown in the problems pointed out by the critics as cited). And the reverse direction also holds true.

In relation to the openness-hiddenness principle, if there is openness (e.g., "different groups and factions fight for their share" of oil revenues openly, in that "these emerge openly as separatist conflicts in regions where the resources are produced…such as in Angola's oil-rich Cabinda province"), there is hiddenness (e.g., "different groups and factions fight for their share" of oil revenues hiddenly, in that "often the conflicts occur in more hidden forms, such as fights between different government ministries or departments for access to budgetary allocations"). (WK 2012oo) And the reverse direction also holds true.

In relation to the denseness-emptiness principle, if there is denseness (e.g., the relatively denser concentration of oil use in the sector of transportation, as "over 90% of transportation in the United States relies on oil"), there is emptiness (e.g., the relatively less dense, or more empty, concentration of oil use in the sector of "electric cars"). And the reverse direction also holds true.

In relation to the change-constancy principle, if there is change (e.g., the ever new ways to study natural resources, as shown in the study on oil depletion and resource curse), there is constancy (e.g., the ever constant existence of different problems in the ever new ways to study natural resources, as shown in the problems with the study on oil depletion and resource curse pointed out by the critics). And the reverse direction also holds true.

In relation to the order-chaos principle, if there is order (e.g., the more or less orderly view of reality, if looked from the vantage point of the U.S. nowadays, which remains one of the most stable countries on earth, when contrasted with, say, "the Democratic Republic of the Congo," which "is perhaps one of the volatile places on the planet, suffering from disease, civil war, and the resource curse"), there is chaos (e.g., the more or less chaotic view of reality, if looked from the vantage point of "the Democratic Republic of the Congo," which "is perhaps one of the volatile places on the planet, suffering from disease, civil war, and the resource curse"). And the reverse direction also holds true.

In relation to the slowness-quickness principle, if there is slowness (e.g., the relatively slower speed of the world's oil supply to deplete, if, according to an estimate by the American Petroleum Institute in 1999, "the world's oil supply would be depleted between 2062 and 2094, assuming total world oil reserves at between 1.4 and 2 trillion barrels [220 and 320 km3] and consumption at 80 million barrels per day [13,000,000 m3/d]"), there is quickness (e.g., the relatively quicker speed of the world's oil supply to deplete, if, according to an estimate in 2004, "total world reserves were estimated to be 1.25 trillion barrels [199 km3] and daily consumption was about 85 million barrels [13,500,000 m3]," so as to "shift…the estimated oil depletion year to 2057"). (WK 2012nn) And the reverse direction also holds true.

In relation to the expansion-contraction principle, if there is expansion (e.g., "U.S. oil production reached its peak in 1970"), there is contraction (e.g., "in 1950, the United States produced over half the world's oil, but by 2005 that proportion had dropped to about 8%"). (WK 2012nn) And the reverse direction also holds true.

In relation to the theory-praxis principle, if there is theory (e.g., the theoretical construction of "the Hubbert curves" based on "the Hubbert peak theory"), there is praxis (e.g., the practical application of "the Hubbert curves" by researchers to the field of various other resources, such as "natural gas," coal," "fissionable materials," "Helium," "transition metals," and "water"). (WK 2012qq) And the reverse direction also holds true.

In relation to the convention-novelty principle, if there is convention (e.g., the conventional wisdom about oil in the older days, when it was assumed that "countries and regions with an abundance of natural resources" can lead to higher "economic growth"), there is novelty (e.g., the alternative novel challenge to the conventional wisdom about about oil in the older days, by the new idea of "resource curse" nowadays, in that, "countries and regions with an abundance of natural resources, specifically point-source non-renewable resources like minerals and fuels, tend to have less economic growth and worse development outcomes than countries with fewer natural resources"). And the reverse direction also holds true.

In relation to the evolution-transformation principle, if there is evolution (e.g., the natural evolution of petroleum compounds to form from "plankton," "bacteria," and "other plant and animal matter" over millions of years in the state of nature), there is transformation (e.g., the technical transformation of human ability to deal with oil depletion by the invention of alternative solutions, like the use of "solar power and "wind power"). (WK 2012nn) And the reverse direction also holds true.

In relation to the symmetry-asymmetry principle, if there is symmetry (e.g., the existence of different countries like "resource-rich countries of the Third World" and "powerful states such as the United States and other leading western countries"), there is asymmetry (e.g., "there is often international pressure on resource-rich countries of the Third World to avoid reinvesting resource revenues in social investment purposes, or even in developmental efforts towards economic diversification," and "this pressure comes from powerful states such as the United States and other leading western countries as well as pro-liberalization institutions such as the World Trade Organization and the International Monetary Fund"). And the reverse direction also holds true.

In relation to the softness-hardness principle, if there is softness (e.g., the defense of the resource-curse thesis, as shown by those who argued that, unlike resource-rich countries, "resource-poor economies like Singapore, Taiwan or South Korea,…spent enormous efforts on education, and this contributed in part to their economic success"), there is hardness (e.g., the critique of the resource-curse thesis, as shown by those who argued that "natural resources generate easily taxable rents that more often than not result in increased spending on education" in resource-rich countries). And the reverse direction also holds true.

In relation to the seriousness-playfulness principle, if there is seriousness (e.g., the serious business of those who try to predict the impacts of oil depletion in the future), there is playfulness (e.g., the playful part of those who try to predict the impacts of oil depletion in the future, when they play around with different scenarios over time, like "catastrophe," "recession," "increasing food prices," and "replacement with renewable energy"). And the reverse direction also holds true.

In relation to the activeness-inactivess principle, if there is activeness (e.g., the relative activeness of those in the U.S. nowadays, which remains one of the most stable countries on earth and attracts a lot of immigrants seeking for opportunities, when contrasted with those in "the Democratic Republic of the Congo," who "suffer" or die "from disease, civil war, and the resource curse), there is inactiveness (e.g., the relative inactiveness of those common folks like those in "the Democratic Republic of the Congo," who "suffer" or die "from disease, civil war, and the resource curse"). And the reverse direction also holds true.

In relation to the selfness-otherness principle, if there is selfness (e.g., the tendency to be mostly centered on one's self or group, as shown by those "authoritarian…high-ranking politicians and military leaders" in resource-cursed countries, who "rule without fear of consequence" for their own greed), there is otherness (e.g., the tendency to be mostly centered on others, as shown by "the Center for Global Development," which "advocates direct distribution of revenues," such that "a government would transfer some or all of the revenue from natural resource extraction to citizens in a universal, transparent, and regular payment,…thus forcing…the government to be transparent and accountable in its management of natural resource revenues," for the future of better global development). And the reverse direction also holds true.

In relation to the regression-progression principle, if there is regression (e.g., the regression made by the availiability of natural resources, as shown in the problems of "resource curse" pointed out by the critics aforecited), there is progression (e.g., the progress made by the availiability of natural resources, as shown in the contribution to "economic growth and...development" aforecited). And the reverse direction also holds true.

In relation to the same-difference principle, if there is similarity in outcome (e.g., the contribution to the molding and control of beliefs, values, and behaviors, regardless of whether this be done by way of the "high-ranking politicians and military leaders" in resource-cursed countries who "rule without fear of consequence," or by way of "the Center for Global Development" for "direct distribution of revenues"), there is difference in outcome (e.g., the contribution to the molding and control of beliefs, values, and behaviors by way of the "high-ranking politicians and military leaders" in resource-cursed countries for a relatively more authoritarian lifeworld—but the contribution to the molding and control of beliefs, values, and behaviors by way of "the Center for Global Development" for a relatively more egalitarian lifeworld). And the reverse direction also holds true.

In relation to the stability-reaction principle, if there is stability (e.g., the relatively more stable condition of oil dependency in the older days, when, for example, "over 90% of transportation in the United States relies on oil"), there is reaction (e.g., oil dependency in the older days then led to further changes later on, when, nowadays, there is an increasing call for the replacement with renewable resources, like "solar power and "wind power"). And the reverse direction also holds true.

In relation to the functionality-nonfunctionality principle, if there is functionality (e.g., the presence of action for which something is fitted, as shown in the use of oil in modern times, which served the function of "economic growth and prosperity since the industrial revolution"), there is nonfunctionality (e.g., the relative lack of action for which something is fitted, as shown in an increasing call for the replacement with renewable resources like "solar power," "wind power," and so on, such that the excessive oil dependency is no longer functional in our time, because of different needs and challenges). And the reverse direction also holds true.

And in relation to the intentionality-nonintentionality principle, if there is intentionality (e.g., the planning or design of something for a certain outcome, as shown in the attempt by the Dutch, when they "discovered a massive natural gas field in Groningen in 1959" and tried to "tap this resource...to export the gas for profit"), there is nonintentionality (e.g., the relative absence of planning or design of something for a certain outcome, such that the outcome can even be counter-intentional in being the opposite of what was originally planned, as shown in the "Dutch Disease," when the Dutch discovered, to their surprise, that, "when the gas began to flow out of the country so too did its ability to compete against other countries' exports," contrary to their original expectations). And the reverse direction also holds true.

Discontinuity and Culture

Discontinuity, when examined from the perspective of culture, can illuminate its possibility (or impossibility) and desirability (or undesirability), and this can be shown by way of two case studies, namely, (3.5.1) discontinuity, misused traditional knowledge, and biopiracy, and (3.5.2) discontinuity, extinction, and the history of scientific ideology—to be addressed hereafter, in that order.

Discontinuity, Misused Traditional Knowledge, and Biopiracy

An excellent case study here to understand discontinuity from the perspective of culture has to do with discontinuity, misused traditional knowledge, and biopiracy

An argument here is that there is a relationship between misused traditional knowledge and biopiracy.

The term "biopiracy" refers to "a situation where indigenous knowledge of nature, originating with indigenous people, is used by others for profit, without permission from and with little or no compensation or recognition to the indigenous people themselves. For example when bioprospectors draw on indigenous knowledge of medicinal plants which is later patented by medical companies without recognizing the fact that the knowledge is not new, or invented by the patenter, and depriving the indigenous community to the rights to commercial exploitation of the technology that they themselves had developed.'" (WK 2012r)

For this reason, in 1992, "tte Convention on Biological Diversity (CBD), signed at the United Nations Conference on Environment and Development (UNCED) in 1992, was the first international environmental convention to develop measures for the use and protection of traditional knowledge, related to the conservation and sustainable use of biodiversity." (WK 2012rr)

In other words, "traditional knowledge" is treated as part of biodiversity which is endangered and requires protection.

Famous Cases of Biopiracy

There are some well-known cases of biopiracy in recent history, and hereafter are some good examples (as summarized in *Table 3.7*): (WK 2012r)

(a) The Maya ICBG Controversy

This incident took place "in 1999-2000, when the International Cooperative Biodiversity Group led by Ethnobiologist Dr. Brent Berlin was accused of being engaged in unethical forms of bioprospecting by several NGOs and indigenous organizations. The ICBG aimed to document the biodiversity of Chiapas, Mexico and the ethnobotanical knowledge of the indigenous Maya people—in order to ascertain whether there were possibilities of developing medical products based on any of the plants used by the indigenous groups." (WK 2012r; C. Hayden 2003; D. Feinholz-Klip 2009)

The significance of this case is that "the Maya ICBG case was among the first to draw attention to the problems of distinguishing between benign forms of bioprospecting and unethical biopiracy...." (WK 2012r; J. Lavery 2007)

(b) The Rosy Periwinkle

This incident took place in the 1950s, when "the rosy periwinkle, while native to Madagascar [in Africa], had been widely introduced into other tropical countries around the world well before the discovery of vincristine. This meant that researchers could obtain local knowledge from one country and plant samples from another. The locally known medical properties of the plant were not the same as the medical properties discovered and commercially used by Eli Lilly. The use of the plant as a cure for diabetes was the original stimulus for research, but cures for cancer were the most important results. Different countries are reported as having acquired different beliefs about the medical properties of the plant. The Hodgkin's lymphoma chemotherapeutic drug vinblastine is also derivable from the rosy periwinkle." (WK 2012r; SCEMS 2012; C. Karasov 2001)

(c) The Neem Tree

This incident took place in 1995, when "the U.S. Department of Agriculture and a pharmaceutical research firm received a patent on a technique to extract an antifungal agent from the neem tree (*Azadirachta indica*), which grows throughout India and Nepal; Indian villagers have long understood the tree's medicinal value. Although the patent had been granted on an extraction technique, the Indian press described it as a patent on the neem tree itself; the result was widespread public outcry, echoed throughout the developing world. Legal action by the Indian government followed, with the patent eventually being overturned in 2005." (WK 2012r; V. Shiva 2012; PIB 2010)

Outrageously, "the pharmaceutical company involved in the neem case argued that as traditional Indian knowledge of the properties of the neem tree had never been published in an academic journal, such knowledge did not amount to 'prior art' (prior art is the term used when previously existing knowledge bars a patent)," but "public knowledge and public disclosure (including oral or written descriptions) are considered prior art in most countries." (WK 2012r)

(d) The Enola Bean

This incident took place in 1990s, when "the Enola bean (not a traditional medicine) is a variety of Mexican yellow bean, so called after the wife of the man who patented it in 1999. The allegedly distinguishing feature of the variety is seeds of a specific shade of yellow. The patent-holder subsequently sued a large number of importers of Mexican yellow beans with the following result: '...export sales immediately dropped over 90% among importers that had been selling these beans for years, causing economic damage to more than 22,000 farmers in northern Mexico who depended on sales of this bean.' A lawsuit was filed on behalf of the farmers, and on April 14, 2005 the US-PTO ruled in favor of the farmers. An appeal was heard on 16 January 2008, and the patent was revoked in May 2008. An appeal to the court against the revocation was unsuccessful" in 2009. (WK 2012r; L. Pallottinia 2004)

(e) Basmati Rice

This incident took place in 2000, when "the US corporation RiceTec (a subsidiary of RiceTec AG of Liechtenstein) attempted to patent certain hybrids of basmati rice and semidwarf long-grain rice (U.S. Patent No. 5,663,484). The Indian government intervened and several claims of the patent were invalidated. Meanwhile, the European Commission has agreed to protect basmati rice under its regulations pertaining to geographical indications." (WK 2012r)

(f) Hoodia

This incident took place in 1996, as "*Hoodia*, a succulent plant, originates from the Kalahari Desert of South Africa. For generations it has been known to the traditionally living San people as an appetite suppressant. In 1996 South Africa's Council for Scientific and Industrial Research began working with medical companies to develop dietary supplements based on hoodia. Originally the San people were not planned to receive any benefits from the commercialization of their traditional knowledge, but in 2003 the South African San Council made an agreement with CSIR in which they would receive from 6 to 8% of the revenue from the sale of Hoodia products." (WK 2012r; V. Maharaj 2008)

(g) Further Cases

There are other cases too, as shown in the following examples: (WK 2012r)

- Ex: "a large selection of African biopiracy cases are discussed at GhanaWeb's *Public Agenda* (March 31, 2006)."
- Ex: "the case of the Maya people's pozol drink is discussed in *The New England Journal of International and Comparative Law* and at Global Exchange."
- Ex: "the case of the Maya and other people's use of Mimosa tenuifolia, including many other such cases in general are discussed at GRAIN."
- Ex: "the case of the Andean maca radish is discussed in the American University's Trade Environment Database."
- Ex: "*the United Kingdom Select Committee on Environmental Audit 1999; Appendices to the Minutes of Evidence, Appendix 7: Trade Related Intellectual Property Rights (TRIPs) and Farmers' Rights* interestingly lists and describes the cases of turmeric (India), karela (India), quinoa (Bolivia), brazzein berries (Gabon), and others."

Of course, these examples are illustrative, not exhaustive.

Characteristics of Traditional Knowledge

With these famous case of biopiracy in mind—traditional knowledge has some distinctive characteristics, when contrasted with other forms of knowledge.

As a start, "traditional knowledge has been defined as 'a cumulative body of knowledge, know-how, practices and representations maintained and developed by peoples with extended histories of interaction with the natural environment. These sophisticated sets of understandings, interpretations and meanings are part and parcel of a cultural complex that encompasses language, naming and classification systems, resource use practices, ritual, spirituality and worldview.'" (WK 2012rr; ICS 2002)

(a) Communal

The first distinctive characteristic of traditional knowledge is "communal."

In general, "traditional knowledge typically distinguishes one community from another. For some communities, traditional knowledge takes on a personal and spiritual meaning. Traditional knowledge can also reflect a community's interests. Some communities depend on their traditional knowledge for survival. This is particularly true of traditional environmental knowledge, which refers to a 'particular form of place-based knowledge of the diversity and interactions among plant and animal species, landforms, watercourses, and other qualities of the biophysical environment in a given place.' An example of a society with a wealth of TEK is the South American Kayapo people, who have developed an extensive classification system of ecological zones of the Amazonian tropical savannah (i.e., campo / cerrado) to better manage the land." (WK 2012rr; D. Peña 2005; D. Posey 2008)

(b) Accumulative

The second distinctive characteristic of traditional knowledge is "accumulative."

Traditional knowledge is different from scientific knowledge, because "some social scientists conceptualize knowledge within a naturalistic framework, and emphasize the gradation of recent knowledge into knowledge acquired over many generations. These accounts use terms like 'adaptively acquired knowledge,' 'socially constructed knowledge,' and other terms that emphasize the social aspects of knowledge. Local knowledge and traditional knowledge may be thought of as distinguished by the length of time they have existed—decades to centuries versus millennia." (WK 2012rr; V. Wasongo 2011)

(c) Spiritual

And the third distinctive characteristic of traditional knowledge is "spiritual."

For its proponents, "traditional knowledge…is often embedded in a cosmology, and the distinction between 'intangible' knowledge and physical things is often blurred. Indigenous peoples often say that 'our knowledge is holistic, and cannot be separated from our lands and resources.' Traditional knowledge in these cosmologies is inextricably bound to ancestors, and ancestral lands. Knowledge may not be acquired by naturalistic trial and error, but through direct revelation through conversations with 'the creator,' spirits, or ancestors. As Chamberlin (2003) writes of a Gitksan elder from British Columbia confronted by a government land claim: 'If this is your land,' he asked, 'where are your stories?.'" (WK 2012rr)

In addition, "indigenous and local communities often do not have strong traditions of ownership over knowledge that resemble the modern forms of private ownership. Many have clear traditions of custodianship over knowledge, and customary law may guide who may use different kinds of knowledge at particular times and places, and obligations that accompany the use of knowledge. From their perspective, misappropriation and misuse of knowledge may be offensive to traditions, and may have spiritual and physical repercussions in their cosmological systems. Subsequently, indigenous and local communities argue that others' use of their traditional knowledge warrants respect and sensitivity." (WK 2012rr)

Mechanisms to Protect Traditional Knowledge

Because "traditional knowledge" is treated as part of biodiversity which is endangered and requires protection, there have been different mechanisms set up over time for the protection.

Consider, for illustration, some main examples below (as summarized in *Table 3.7*).

(a) Indigenous Intellectual Property

The first mechanism to protect traditional knowledge is "indigenous intellectual property."

The "indigenous intellectual property" refers to "an umbrella legal term used in national and international forums to identify indigenous peoples' special rights to claim (from within their own laws) all that their indigenous groups know now, have known, or will know. It is a concept that has developed out of a predominantly western legal tradition, and has most recently been promoted by the World Intellectual Property Organisation, as part of a more general United Nations push to see the diverse wealth of this world's indigenous, intangible cultural heritage better valued and better protected against probable, ongoing misappropriation and misuse." (WK 2012rr; UN 2007a)

In addition, "in the lead up to and during the United Nations International Year for the World's Indigenous Peoples (1993), then during the following United Nations Decade of the World's Indigenous Peoples (1995–2004), a number of conferences of both indigenous and non-indigenous specialists were held in different parts of the world, resulting in a number of declarations and statements identifying, explaining, refining, and defining 'indigenous intellectual property.'" (WK 2012rr; I. Watson 1992)

(b) Agreement on Trade-Related Aspects of Intellectual Property Rights (TRIPs)

The second mechanism to protect traditional knowledge is "TRIPs."

In particular, "Article 27. 3(b) of the Agreement on Trade-Related Aspects of Intellectual Property Rights (TRIPs) sets out certain conditions under which certain biological materials or intellectual innovations may be excluded from patenting. The Article also contains a requirement that Article 27 be reviewed. In the TRIPs-related Doha Declaration of 2001, Paragraph 19 expanded the review to a review of Article 27 and the rest of the TRIPs agreement to include the relationship between the TRIPS Agreement and the 1992 Convention on Biological Diversity (CBD) and the protection of traditional knowledge and folklore." (WK 2012rr)

(c) The Convention on Biological Diversity (CBD)

And the third mechanism to protect traditional knowledge is "CBD."

In other words, "the Convention on Biological Diversity (CBD), signed at the United Nations Conference on Environment and Development (UNCED) in 1992, was the first international environmental convention to develop measures for the use and protection of traditional knowledge, related to the conservation and sustainable use of biodiversity. By 2006, 188 had ratified the Convention and agreed to be bound by its provisions, the largest number of nations to accede to any existing treaty (the United States is one of the few countries that has signed, but not ratified, the CBD)." (WK 2012rr)

Later, "at the Convention on Biological Diversity meeting, in Buenos Aires, in 1996, emphasis was put on local knowledge. Key players, such as local communities and indigenous peoples, should be recognized by States, and have their sovereignty recognised over the biodiversity of their territories, so that they can continue protecting it." (WK 2012rr; B. Roussel 2003)

But then, "the parties to the Convention set a 2010 target to negotiate an international legally binding regime on access and benefit sharing (ABS) at the Eighth meeting (COP8), 20–31 March 2006 in Curitiba, Brazil. This target was met in October 2010 in Nagoya, Japan, by conclusion of the Nagoya Protocol to the CBD. The agreement is now open for ratification, and will come into force when 50 signatories have ratified it. So far five have done so (Mexico is the largest to date) but many will follow. The Protocol treats of inter-governmental obligations related to genetic resources, and includes measures related to the rights of indigenous and local communities to control access to and derive benefits from the use of genetic resources and associated traditional knowledge." (WK 2012rr)

Problems with the Study of Biopiracy and Traditional Knowledge

However, for the critics, the study of biopiracy and traditional knowledge has some major problems. Consider, for illustration, a few examples below (as summarized in *Table 3.7*).

Firstly, one main criticism of the study of biopiracy and traditional knowledge is that it can become quickly political in tone (without examining the actual evidences more carefully).

For instance, for the critics, "one common misunderstanding is that pharmaceutical companies patent the plants they collect. While obtaining a patent on a naturally occurring organism as previously known or used is not possible, patents may be taken out on specific chemicals isolated or developed from plants. Often these patents are obtained with a stated and researched use of those chemicals....As a result, even if the indigenous medical knowledge is taken..., that knowledge does not by itself make the active chemical compound 'obvious,' which is the standard applied under patent law." (WK 2012r)

Secondly, another main criticism of the study of biopiracy and traditional knowledge is that there is still no "international legally binding regime" on the protection of traditional knowledge against biopiracy.

This is true, even though "the parties to the Convention set a 2010 target to negotiate an international legally binding regime on access and benefit sharing (ABS) at the Eighth meeting (COP8), 20–31 March 2006 in Curitiba, Brazil....The agreement is now open for ratification, and will come into force when 50 signatories have ratified it," but "so far," only "five have done so (Mexico is the largest to date)." (WK 2012r)

Thirdly, still another main criticism of the study of biopiracy and traditional knowledge is that the existing mechanisms against biopiracy are not effective.

For instance, for the critics, a good case in point is the "Convention on Biological Diversity (CBD)," which "came into force in 1993. It secured rights to control access to genetic resources for the countries in which those resources are located. One objective of the CBD is to enable lesser-developed countries to better benefit from their resources and traditional knowledge. Under the rules of the CBD, bioprospectors are required to obtain informed consent to access such resources, and must share any benefits with the biodiversity-rich country." (WK 2012r) But the problem here is that, for the opponents of CBD, "the CBD has failed to establish appropriate regulations to prevent biopiracy." (WK 2012r)

Fourthly, an additional main criticism of the study of biopiracy and traditional knowledge is that the defense of traditional knowledge is not critical to its own version of destructiveness against biodiversity.

For instance, the critics of "'traditional knowledge'...maintain that such demands for 'respect' [by its advocates] are really an attempt to prevent unsubstantiated beliefs from being subjected to the same scrutiny as other knowledge claims. This has particular significance for environmental management because the spiritual component of 'traditional knowledge' can be used to justify any activity, including the unsustainable harvesting of resources." (WK 2012rr)

And fifthly, another additional main criticism of the study of biopiracy and traditional knowledge is that traditional knowledge is not so naturalistic as its advocates suggested, because of inherent power struggles.

For instance, for the critics, "a large number of scholarly studies in the naturalistic tradition demonstrate that traditional knowledge is not a natural category, and may reflect power struggles and relationships for land, resources and social control rather than adherence to a claimed ancestry or heritage." (WK 2012rr)

Beyond Biopiracy and Traditional Knowledge

These criticisms of biopiracy and traditional knowledge should not be misunderstood, however, as a total rejection of its usefulness but serve to show us the two opposing sides of the debate, such that the possibility (or impossibility) and desirability (or undesirability) of discontinuity (from the perspective of culture with discontinuity, misused traditional knowledge, and biopiracy as a case study here) are not to the extent that the respective defenders would like us to believe.

Besides, the analysis of discontinuity, misused traditional knowledge, and biopiracy can teach us something important about the ontological principles in existential dialectics, and good examples include the formalness-informalness principle, the absoluteness-relativeness principle, the partiality-totality principle, the predictability-unpredictability principle, the explicability-inexplicability principle, the fiction-reality principle, the cognitiveness-noncognitiveness principle, the finiteness-transfiniteness principle, the preciseness-vagueness principle, the simpleness-complicatedness principle, the openness-hiddenness principle, the denseness-emptiness principle, the change-constancy principle, the order-chaos principle, the slowness-quickness principle, the expansion-contraction principle, the theory-praxis principle, the convention-novelty principle, the evolution-transformation principle, the symmetry-asymmetry principle, the softness-hardness principle, the seriousness-playfulness principle, the activeness-inactivess principle, the selfness-otherness principle, the regression-progression principle, the same-difference principle, the stability-reaction principle, the functionality-nonfunctionality principle, and the intentionality-nonintentionality principle.

For instance, in relation to the formalness-informalness principle, if there is formalness (e.g., the formal logical requirement of soundness and completeness in a system of ideas, as in the study of biopiracy and traditional knowledge), there is informalness (e.g., the non-formal existence of unsoundness and incompleteness in the study of biopiracy and traditional knowledge, as shown in the criticisms that it can become quickly political in tone without examining the actual evidences more carefully, that there is still no "international legally binding regime" on the protection of traditional knowledge against biopiracy, that the existing mechanisms against biopiracy are not effective, that the defense of traditional knowledge is not critical to its own version of destructiveness against biodiversity, and that traditional knowledge is not so naturalistic as its advocates suggested, because of inherent power struggles). And the reverse direction also holds true.

In relation to the absoluteness-relativeness principle, if there is absoluteness (e.g., the absolute view of those biopirates, in regard to the use, for profit, of "indigenous knowledge…originating with indigenous people…without permission from and with little or no compensation or recognition to the indigenous people themselves"), there is relativeness (e.g., what is profitable for those biopirates in regard to the use, for profit, of "indigenous knowledge" is not necessarily so for "the indigenous people themselves," who did not grant any "permission and received "little or no compensation" from the biopirates). And the reverse direction also holds true.

In relation to the partiality-totality principle, if there is partiality (e.g., the partial view of knowledge from the scientific vantage point, insofar as it is to be "acquired by naturalistic trial and error" and emphasizes "the gradation of recent knowledge"), there is totality (e.g., the more holistic view of knowledge, such that the whole is not the sum of its parts, that is, it cannot be reduced to the sum of the opposing views of scientific knowledge in the developed world and traditional knowledge in the developing world, since there will emerge new views in the future not yet known today, just as there are already those nowadays who do not side exclusively with any of them and are critical to traditional knowledge too, as cited earlier). And the reverse direction also holds true.

In relation to the predictability-unpredictability principle, if there is predictability (e.g., the predictable tendency of those for the "Convention on Biological Diversity [CBD]," to try to "to enable lesser-developed countries to better benefit from their resources and traditional knowledge"), there is unpredictability (e.g., the more difficult task to predict exactly the extent to which CBD will be successful in any attempt "to enable lesser-developed countries to better benefit from their resources and traditional knowledge" on a particular future occasion, because, for the critics, "the CBD has failed to establish appropriate regulations to prevent biopiracy"). And the reverse direction also holds true.

In relation to the explicability-inexplicability principle, if there is explicability (e.g., the explanation by the advocates of traditional knowledge about its naturalistic basis, that is, "a 'particular form of place-based knowledge of the diversity and interactions among plant and animal species, landforms, watercourses, and other qualities of the biophysical environment in a given place'"), there is inexplicability (e.g., the lack of sufficient explanation by the advocates of traditional knowledge of why it is necessarily true that traditional knowledge is naturalistic, because, for the critics, "traditional knowledge is not a natural category, and may reflect power struggles and relationships for land, resources and social control rather than adherence to a claimed ancestry or heritage"). And the reverse direction also holds true.

In relation to the fiction-reality principle, if there is fiction (e.g., the fictional aspect of biopiracy, insofar as there is "one common misunderstanding," in that "pharmaceutical companies patent the plants they collect," but the reality is that "patents may be taken out on specific chemicals isolated or developed from plants" and "often these patents are obtained with a stated and researched use of those chemicals," so "even if the indigenous medical knowledge is taken…, that knowledge does not by itself make the active chemical compound 'obvious,' which is the standard applied under patent law"), there is reality (e.g., the realistic aspect of biopiracy, insofar as there are other cases when biopirates used, for profit, "indigenous knowledge…originating with indigenous people…without permission from and with little or no compensation or recognition to the indigenous people themselves"). And the reverse direction also holds true.

In relation to the cognitiveness-noncognitiveness principle, if there is cognitiveness (e.g., a view of things on the basis of reason and evidences, as shown in the study of traditional knowledge on the basis of its "naturalistic" representation, that is, "a 'particular form of place-based knowledge of the diversity and interactions among plant and animal species, landforms, watercourses, and other qualities of the biophysical environment in a given place'"), there is noncognitiveness (e.g., a view of things on the basis of non-cognitive factors like envy, jealousy, power, nationality, race, gender, age, class, greed, lust, status, faith, anger, sadness, joy, fear, wish, etc., as shown in the criticism that "traditional knowledge is not a natural category, and may reflect power struggles and relationships for land, resources and social control rather than adherence to a claimed ancestry or heritage," that is, on the basis of power). And the reverse direction also holds true.

In relation to the finiteness-transfiniteness principle, if there is finiteness (e.g., the finite number of the categories of entities addressed by traditional knowledge, namely, "plants," "animals," etc.), there is transfiniteness (e.g., the transfinite number of all the individual "plants," "animals," etc. in history that have ever existed and are included under traditional knowledge). And the reverse direction also holds true.

In relation to the preciseness-vagueness principle, if there is preciseness (e.g., the precise identification of 3 characteristics of traditional knowledge in the article aforecited), there is vagueness (e.g., the vagueness in the identification of the 3 characteristics of traditional knowledge in the article aforecited, since it is not clear why there must be only 3, not 4, 5, 6, and so on). And the reverse direction also holds true.

In relation to the simpleness-complicatedness principle, if there is simpleness (e.g., the relatively simple analysis of traditional knowledge by the advocates), there is complicatedness (e.g., the relatively more complicated analysis of traditional knowledge, by challenging the claims and assumptions, as shown in the problems pointed out by the critics as cited). And the reverse direction also holds true.

In relation to the openness-hiddenness principle, if there is openness (e.g., the open exploration, within traditional knowledge, of "the diversity and interactions among plant and animal species, landforms, watercourses, and other qualities of the biophysical environment in a given place"), there is hiddenness (e.g., the hidden bias in traditional knowledge, since, for the critics, "traditional knowledge is not a natural category, and may reflect power struggles and relationships for land, resources and social control rather than adherence to a claimed ancestry or heritage"). And the reverse direction also holds true.

In relation to the denseness-emptiness principle, if there is denseness (e.g., the relatively denser concentration of concerns with the protection of traditional knowledge among those in CBD), there is emptiness (e.g., the relatively less dense, or more empty, concentration of concerns with the protection of traditional knowledge among biopirates). And the reverse direction also holds true.

In relation to the change-constancy principle, if there is change (e.g., the ever new ways to knowledge, as shown in the study of traditional knowledge), there is constancy (e.g., the ever constant existence of different problems in the ever new ways to knowledge, as shown in the problems with traditional knowledge pointed out by the critics). And the reverse direction also holds true.

In relation to the order-chaos principle, if there is order (e.g., the more or less orderly view of traditional knowledge, if looked from the vantage point of its three characteristics, namely, "communal," "accumulative," and "spiritual"), there is chaos (e.g., the more or less chaotic view of traditional knowledge, if looked from the vantage points that "traditional knowledge is not a natural category" and "reflect[s] power struggles and relationships for land, resources and social control rather than adherence to a claimed ancestry or heritage," that "the spiritual component of 'traditional knowledge' can be used to justify any activity, including the unsustainable harvesting of resources," and so on). And the reverse direction also holds true.

In relation to the slowness-quickness principle, if there is slowness (e.g., the relatively slower speed of knowledge to be formed, if looked from the vantage point of traditional knowledge being accumulated over "millennia"), there is quickness (e.g., the relatively quicker speed of knowledge to be formed, if looked from the vantage point of scientific knowledge with its emphasis on "the gradation of recent knowledge"). And the reverse direction also holds true.

In relation to the expansion-contraction principle, if there is expansion (e.g., the relative expansion of the mechanisms to protect traditional knowledge nowadays, as shown in the three types aforecited), there is contraction (e.g., the relative decline of blatant biopiracy nowadays, because of different legal mechanisms for the protection of traditional knowledge). And the reverse direction also holds true.

In relation to the theory-praxis principle, if there is theory (e.g., the theoretical construction of "intellectual property" in economics), there is praxis (e.g., the practical application of the understanding of "intellectual property" in economics to the field of public policy, as shown by "the World Intellectual Property Organisation" which has pushed "to see the diverse wealth of this world's indigenous, intangible cultural heritage better valued and better protected against probable, ongoing misappropriation and misuse"). (WK 2012rr) And the reverse direction also holds true.

In relation to the convention-novelty principle, if there is convention (e.g., the conventional wisdom about traditional knowledge on a naturalistic basis), there is novelty (e.g., the alternative novel challenge to the conventional wisdom about traditional knowledge on a naturalistic basis, by the new idea that "traditional knowledge is not a natural category, and may reflect power struggles and relationships for land, resources and social control rather than adherence to a claimed ancestry or heritage"). And the reverse direction also holds true.

In relation to the evolution-transformation principle, if there is evolution (e.g., the natural evolution of humans to form some understanding of the world in the state of nature in a primitive way), there is transformation (e.g., the technical transformation of human ability to understand the world by the invention of "traditional knowledge" and "scientifc knowledge"). And the reverse direction also holds true.

In relation to the symmetry-asymmetry principle, if there is symmetry (e.g., the existence of different countries, like "developing" and "developed" countries), there is asymmetry (e.g., "these practices [of biopiracy] contribute to inequality between developing countries rich in biodiversity, and developed countries hosting companies that engage in 'biopiracy'"). (WK 2012r) And the reverse direction also holds true.

In relation to the softness-hardness principle, if there is softness (e.g., the occurrence of cooperation among parties to CBD), there is hardness (e.g., the occurrence of biopiracy against traditional knowledge). And the reverse direction also holds true.

In relation to the seriousness-playfulness principle, if there is seriousness (e.g., the serious business of those who try to protect traditional knowledge against biopiracy), there is playfulness (e.g., the playful part of those who try to protect traditional knowledge against biopiracy, when they play around with different mechanisms over time, like "TRIPs," "CBD," "indegeneous intellectual property," etc.). And the reverse direction also holds true.

In relation to the activeness-inactivess principle, if there is activeness (e.g., the relative activeness of those biopirates who are able to use, for profit, "indigenous knowledge…originating with indigenous people…without permission from and with little or no compensation or recognition to the indigenous people themselves"), there is inactiveness (e.g., the relative inactiveness of other biopirates who do not succeed to exploit traditional knowledge without permission or compensation, because of the mechanisms to protect traditional knowledge nowadays). And the reverse direction also holds true.

In relation to the selfness-otherness principle, if there is selfness (e.g., the tendency to be mostly centered on one's self or group, as shown by the exploitation of traditional knowledge by biopirates), there is otherness (e.g., the tendency to be mostly centered on others, as shown by "the World Intellectual Property Organisation," which had pushed "to see the diverse wealth of this world's indigenous, intangible cultural heritage better valued and better protected against probable, ongoing misappropriation and misuse," so as to protect traditional knowledge as part of biodiversity which is endangered and requires protection, for the betterment of future humanity). And the reverse direction also holds true.

In relation to the regression-progression principle, if there is regression (e.g., the regression made by the mechanisms against biopiracy, as shown in the problems pointed out by the critics), there is progression (e.g., the progress made by the mechanisms against biopiracy, as shown in the contribution to the protection of traditional knowledge aforecited). And the reverse direction also holds true.

In relation to the same-difference principle, if there is similarity in outcome (e.g., the contribution to the molding and control of beliefs, values, and behaviors, regardless of whether this be done by way of traditional knowledge, or by way of scientific knowledge), there is difference in outcome (e.g., the contribution to the molding and control of beliefs, values, and behaviors by way of traditional knowledge for a relatively more spiritual worldview—but the contribution to the molding and control of beliefs, values, and behaviors by way of scientific knowledge for a relatively more secular worldview). And the reverse direction also holds true.

In relation to the stability-reaction principle, if there is stability (e.g., the relatively more stable condition of traditional knowledge in the older days, when it showed "the diversity and interactions among plant and animal species, landforms, watercourses, and other qualities of the biophysical environment in a given place'"), there is reaction (e.g., traditional knowledge then led to further changes later on, when nowadays, there is scientific knowledge which is "acquired by naturalistic trial and error," but not "through direct revelation through conversations with 'the creator,' spirits, or ancestors" as in traditional knowledge). And the reverse direction also holds true.

In relation to the functionality-nonfunctionality principle, if there is functionality (e.g., the presence of action for which something is fitted, as shown in the belief in traditional knowledge in developing countries, which serves the function of providing a body of knowledge to understand "the diversity and interactions among plant and animal species, landforms, watercourses, and other qualities of the biophysical environment in a given place"), there is nonfunctionality (e.g., the relative lack of action for which something is fitted, as shown in the appeal to scientific knowledge in developed countries nowadays, such that traditional knowledge is no longer functional much in our time, because of different needs and expectations for a naturalistic basis of understanding, although it tries to make good use of traditional knowledge for some patents as shown in the mechanisms for its protection as part of endangered biodiversity). And the reverse direction also holds true.

And in relation to the intentionality-nonintentionality principle, if there is intentionality (e.g., the planning or design of something for a certain outcome, as shown in the formation of traditional knowledge for a naturalistic understanding of the world, especially in regard to "the diversity and interactions among plant and animal species, landforms, watercourses, and other qualities of the biophysical environment in a given place"), there is nonintentionality (e.g., the relative absence of planning or design of something for a certain outcome, such that the outcome can even be counter-intentional in being the opposite of what was originally planned, as shown in the criticism that "the spiritual component of 'traditional knowledge' can be used to justify any activity, including the unsustainable harvesting of resources," contrary to its original naturalistic claims). And the reverse direction also holds true.

Discontinuity, Extinction, and the History of Scientific Ideology

Another way to evaluate the possibility (or impossibility) and desirability (or undesirability) of discontunity from the perspective of culture concerns discontinuity, extinction, and the history of scientific ideology.

An argument here is that there is a relationship between the understanding about extinction and the history of scientific ideology.

As a reminder from *Sec. 1.3*, the term "extinction," refers to, "especially in biology and ecology,...the end of an organism or of a group of organisms (taxon), normally a species." (WK 2012e)

But the idea of "extinction" can be understood in different ways in its history, and one reason has to do with ideology in science, and the term "ideology" here refers to "a set of ideas that constitute one's goals, expectations, and actions" or "a comprehensive vision, a way of looking at things (compare worldview)." (WK 2012ss)

Consider, for illustration, the following four major ways in which "extinction" can be understood in its history.

Extinction, and the Great Chain of Being

The first major way in which the idea of "extinction" can be understood concerns "the great chain of being."

At the beginning of its history, "when it was first described in the 1750's, the idea of extinction was threatening to those who held a belief in the Great Chain of Being, a theological position that did not allow for 'missing links.'" (WK 2012e)

The idea of "the great chain of being (Latin: *scala naturae*, literally 'ladder or stair-way of nature'), is a concept derived from Plato and Aristotle, and developed fully in Neoplatonism. It details a strict, religious hierarchical structure of all matter and life, believed to have been decreed by God. The chain starts from God and progresses downward to angels, demons (fallen/renegade angels), stars, moon, kings, princes, nobles, men, wild animals, domesticated animals, trees, other plants, precious stones, precious metals, and other minerals." (WK 2012tt; A. Lovejoy 1964)

This idea of the "stair-way of nature" is traced to "Plato's division of the world into the Forms, which are full beings, and sensible things, which are imitations of the Forms and are both being and not being. Aristotle's teleology recognized a perfect being, and he also arranges all animals by a single natural scale according to the degree of perfection of their souls. The idea of the great chain of being was fully developed in Neoplatonism and in the Middle Ages." (Wk 2012tt; BDWP 2004)

In the chain, "earth (rock) is at the bottom of the chain: this element possesses only the attribute of existence. Each link succeeding upward contains the positive attributes of the previous link and adds (at least) one other. Rocks, as above, possess only existence; the next link up, plants, possess life and existence. Animals add not only motion, but appetite as well." (WK 2012tt)

In addition, "each link in the chain might be divided further into its component parts"; for instance, "at the top of the animals are wild beasts (such as lions), which were seen as superior as they defied training and domestication. Below them are domestic animals, further sub-divided so that useful animals (such as dogs and horses) are higher than docile creatures, such as sheep. Birds are also sub-divided, with eagles above pigeons, for example. Fish come below birds and are sub-divided between actual fish and other sea creatures. Below them come insects, with useful insects such as spiders and bees and attractive creatures such as ladybirds and dragonflies at the top, and unpleasant insects such as flies and beetles at the bottom. At the very bottom of the animal sector are snakes, which are relegated to this position as punishment for the serpent's actions in the

Garden of Eden"; and "below animals comes the division for plants, which is further sub-divided." (WK 2012tt)

So, "the central concept of the chain of being is that everything imaginable fits into it somewhere, giving order and meaning to the universe." (WK 2012tt)

But the idea of "extinction" contradicts this inherent continuity in the concept of the great chain of being. It is no wonder that "the possibility of extinction was not widely accepted before the 1800s. The devoted naturalist Carl Linnaeus, could 'hardly entertain' the idea that humans could cause the extinction of a species. When parts of the world had not been thoroughly examined and charted, scientists could not rule out that animals found only in the fossil record were not simply 'hiding' in unexplored regions of the Earth." (WK 2012e; L. Koerner 1999)

Extinction, and Creation Theory

The second major way in which the idea of "extinction" can be understood concerns "creation theory."

Both Christians and Jews hold a different view about "extinction," in the context of creation theory, especially the version known as "young Earth creationism (YEC)," which is "the religious belief that the Universe, Earth, and all life on Earth were created by direct acts of the Abrahamic God during a relatively short period, sometime between 5,700 and 10,000 years ago. Its primary adherents are those Christians and Jews who believe that God created the Earth in six 24-hour days, using a literal interpretation of the Genesis creation narrative as a basis." (WK 2012uu; E. Scott 2004; B. Freundel 2004; P. James-Griffiths 2012)

Thus, the idea of extinction is reframed within the work of creation by God, and "Christians up until the 19th century held dominant scientific beliefs that were founded on the biblical narratives of Creation and the universal deluge. Other ancient deluge myths have been discovered since then, explaining why the flood story was 'stated in scientific methods with surprising frequency among the Greeks,' an example being Plutarch's account of the Ogygian flood." (WK 2012vv; C. King 1877)

In this way, "Earth's history was viewed as the result of an accumulation of catastrophic events over a relatively short time period, before the depth of geological time was appreciated. In this way they were able to explain the observations of early geologists within the framework of a short Earth history." (WK 2012vv)

Extinction events are thus reframed as part of "catastrophic events over a relatively short time period" in Earth's history as designed by God, in the context of creation theory.

Extinction, and Catastrophism

The third major way in which the idea of "extinction" can be understood concerns "catastrophism" in the scientific direction.

Although catastrophism can be understood within the context of "creation theory" in its earlier history (as described above), another main version of catastrophism is scientific, especially in the work of "the French anatomist and paleontologist Georges Cuvier. His motivation was to explain the patterns of extinction and faunal succession that he and others were observing in the fossil record." (WK 2012vv)

What is interesting about Cuvier's work is that, "while he did speculate that the catastrophe responsible for the most recent extinctions in Eurasia might have been the result of the inundation of low-lying areas by the sea, he did not make any reference to Noah's flood. Nor did he ever make any reference to divine creation as the mechanism by which repopulation occurred following the extinction event. In fact Cuvier, influenced by the ideas of the Enlightenment and the intellectual climate of the French revolution, avoided religious or metaphysical speculation in his scientific writings." (WK 2012vv; C. McGowan 2001; M. Rudwick 1972)

Moreover, "Cuvier…believed that the stratigraphic record indicated that there had been several of these revolutions, which he viewed as recurring natural events, amid long intervals of stability during the history of life on earth. This led him to believe the Earth was several million years old." (WK 2012vv; M. Rudwick 1972)

In this way, extinction events are reframed within the context of "revolutions…amid long intervals of stability during the history of life on earth."

Extinction, and Uniformitarianism

And the fourth major way in which the idea of "extinction" can be understood concerns "uniformitarianism."

Extinction events are reframed within uniformitarianism as part of evolution over a long period of time. For instance, "James Hutton, sometimes known as the father of geology, in the late 18th century" came up with the idea of "deep time," so as to explain "the formation of sedimentary rock" and to provide "an understanding of the immense stretch of geological time." (WK 2012vv)

Later, "the geologist Charles Lyell built upon Hutton's ideas during the first half of 19th century and amassed observations in support of the uniformitarian idea that the Earth's features had been shaped by same geological processes that could be observed in the present acting gradually over an immense period of time. Lyell presented his ideas in the influential three volume work, *Principles of Geology*, published in the 1830s, which challenged theories about geological cataclysms proposed by proponents of catastrophism like Cuvier and Buckland." (WK 2012vv; M. Rudwick 1972)

And "from around 1850 to 1980, most geologists endorsed uniformitarianism ('The present is the key to the past') and gradualism (geologic change occurs slowly over long periods of time) and rejected the idea that cataclysmic events such as earthquakes, volcanic eruptions, or floods of vastly greater power than those observed at the present time, played any significant role in the formation of the Earth's surface. Instead they believed that the earth had been shaped by the long term action of forces such as volcanism, earthquakes, erosion, and sedimentation, that could still be observed in action today." (WK 2012vv)

The rejection of catastrophism by these uniformitarians "was fostered by their impression that the catastrophists of the early nineteenth century believed that God was directly involved in determining the history of Earth. Some of the theories about Catastrophism in the nineteenth and early twentieth centuries were connected with religion and catastrophic origins were sometimes considered miraculous rather than natural events." (WK 2012vv; M. Rudwick 1972)

In this way, extinction events are incorporated into uniformitarianism along the line of "gradualism," as part of evolution over "the immense stretch of geological time." (WK 2012vv)

Problems with the Study of Extinction

However, for the critics, this study of extinction has some major problems. Consider, for illustration, a few examples below.

Firstly, one main criticism of the study of extinction is that each view is relative to its own assumptions and claims.

For example, extinction events are incompatible with the idea of the great chain of being—but can be explained as part of "catastrophic events over a relatively short time period" in Earth's history as designed by God in the context of creation theory.

Secondly, another main criticism of the study of extinction is that some of the views can be criticized as false, albeit in different ways.

For instance, for the critics, creation theory is false, because "the scientific consensus, supported by a 2006 statement by 68 national and international science academies, is that it is evidence-based fact derived from observations and experiments in multiple scientific disciplines that the universe has existed for around 13 billion years, that the Earth was formed about 4.5 billion years ago with life first appearing at least 2.5 billion years ago. Although many young Earth creationists (YECs) are active in the development of creation science, an endeavor that holds that the events associated with supernatural creation can be evidenced and modeled through an interpretation of the scientific method, the consensus among scientists is that creation science is unscientific in both conception and methodology." (WK 2012uu; IAP 2006; W. Overton 1982)

Thirdly, still another main criticism of the study of extinction is that the truth about extinction can be a mixture of some of the views.

For instance, "today most geologists combine catastrophist and uniformitarianist standpoints, taking the view that Earth's history is a slow, gradual story punctuated by occasional natural catastrophic events that have affected Earth and its inhabitants." (WK 2012vv; COL 2007)

And fourthly, an additional main criticism of the study of extinction is that extinction can be prevented.

For instance, for the critics, "ongoing technological advances have encouraged the hypothesis that by using DNA from the remains of an extinct species, through the process of cloning, the species may be 'brought back to life'" (as already discussed in *Sec. 2.4.1*). (WK 2012e)

Beyond Scientific Ideology

These criticisms of the study of extinction, however, do not mean a total rejection of its usefulness but serve to show us the two opposing sides of the debate, such that the possibility (or impossibility) and desirability (or undesirability) of discontinuity (from the perspective of culture with discontinuity, extinction, and the history of scientific ideology as a case study here) are not to the extent that the respective defenders would like us to believe.

Furthermore, the analysis of discontinuity, extinction, and the history of scientific ideology is to reveal the insights of the ontological principles in existential dialectics, and good examples include the formalness-informalness principle, the absoluteness-relativeness principle, the partiality-totality principle, the predictability-unpredictability principle, the explicability-inexplicability principle, the fiction-reality principle, the cognitiveness-noncognitiveness principle, the finiteness-transfiniteness principle, the preciseness-vagueness principle, the simpleness-complicatedness principle, the openness-hiddenness principle, the denseness-emptiness principle, the change-constancy principle, the order-chaos principle, the slowness-quickness principle, the expansion-contraction principle, the theory-praxis principle, the convention-novelty principle, the evolution-transformation principle, the symmetry-asymmetry principle, the softness-hardness principle, the seriousness-playfulness principle, the activeness-inactivess principle, the selfness-otherness principle, the regression-progression principle, the same-difference principle, the stability-reaction principle, the functionality-nonfunctionality principle, and the intentionality-nonintentionality principle.

For instance, in relation to the formalness-informalness principle, if there is formalness (e.g., the formal logical requirement of soundness and completeness in a system of ideas, as in the study of extinction and the history of scientific ideology), there is informalness (e.g., the non-formal existence of unsoundness and incompleteness in the study of extinction and the history of scientific ideology, as shown in the criticisms that each view is relative to its own assumptions and claims, that some of the views can be criticized as false [albeit in different ways], that the truth about extinction can be a mixture of some of the views, and that extinction can be prevented). And the reverse direction also holds true.

In relation to the absoluteness-relativeness principle, if there is absoluteness (e.g., the absolute view of the advocates of the great chain of being, in that it is "a theological position that did not allow for 'missing links'"), there is relativeness (e.g., what is right for the advocates of the great chain of being in regard to its "theological position that did not allow for 'missing links'" is not necessarily so for those in the scientific version of catastrophism that does allow the occurrence of extinction events). And the reverse direction also holds true.

In relation to the partiality-totality principle, if there is partiality (e.g., the partial view of the advocates of the great chain of being, in regard to their rejection of extinction), there is totality (e.g., the more holistic view of extinction, such that the whole is not the sum of its parts, that is, it cannot be reduced to the sum of opposing views like the four versions aforediscussed, since there will emerge new views in the future not yet known today, just as there are already some nowadays who do not side exclusively with any of them). And the reverse direction also holds true.

In relation to the predictability-unpredictability principle, if there is predictability (e.g., the predictable tendency of those in uniformitarianism to regard extinction events as part of evolution over a long period of time), there is unpredictability (e.g., the more difficult task to predict exactly the extent to which certain catastrophes will occur in a particular future historical period, since, for those in the scientific version of catastrophism, catastrophes are essentially "sudden"). (WK 2012vv) And the reverse direction also holds true.

In relation to the explicability-inexplicability principle, if there is explicability (e.g., the explanation by those in creation theory that "the Universe, Earth, and all life on Earth were created by direct acts of the Abrahamic God during a relatively short period, sometime between 5,700 and 10,000 years ago"), there is inexplicability (e.g., the lack of sufficient explanation by those in creation theory of why "the Universe, Earth, and all life on Earth" were necessarily "created by direct acts of the Abrahamic God during a relatively short period," because, for the critics, "the consensus among scientists is that creation science is unscientific in both conception and methodology"). And the reverse direction also holds true.

In relation to the fiction-reality principle, if there is fiction (e.g., the fictional aspect of uniformitarianism, insofar as "Earth's history is...punctuated by occasional natural catastrophic events," contrary to the uniformitarian claim), there is reality (e.g., the realistic aspect of uniformitarianism, insofar as there is some truth that "Earth's history is a slow, gradual story"). And the reverse direction also holds true.

In relation to the cognitiveness-noncognitiveness principle, if there is cognitiveness (e.g., a view of things on the basis of reason and evidences, as shown in the rational understanding by "most theologians" who are willing to "reject a literal interpretation of the creation narrative in Genesis" and "propose there are statements in the creation week itself which render the historical interpretation of Genesis incompatible with scientific evidence"), there is noncognitiveness (e.g., a view of things on the basis of non-cognitive factors like envy, jealousy, power, nationality, race, gender, age, class, greed, lust, status, faith, anger, sadness, joy, fear, wish, etc., as shown in the reliance on "faith" by "most young Earth creationists today" who still believe in "the apparent age concept," which "is still used in young Earth creationist literature"). (WK 2012uu; K. Butt 2001) And the reverse direction also holds true.

In relation to the finiteness-transfiniteness principle, if there is finiteness (e.g., the finite number of the categories of things in the great chain of being, like God, angels, humans, animals, plants, and minerals), there is transfiniteness (e.g., the transfinite number of all the individual humans, animals, plants, and minerals in history that have ever existed). And the reverse direction also holds true.

In relation to the preciseness-vagueness principle, if there is preciseness (e.g., the precise identification of 4 views about extinction in the article aforecited), there is vagueness (e.g., the vagueness in the identification of the 4 views about extinction in the article aforecited, since it is not clear why there must be only 4, not 5, 6, 7, and so on). And the reverse direction also holds true.

In relation to the simpleness-complicatedness principle, if there is simpleness (e.g., the relatively simple analysis of creation theory by the advocates), there is complicatedness (e.g., the relatively more complicated analysis of creation theory, by challenging the claims and assumptions, as shown in the problems pointed out by the critics as cited). And the reverse direction also holds true.

In relation to the openness-hiddenness principle, if there is openness (e.g., the open exploration of the "hierarchical structure of all matter and life" in the world by the idea of the great chain of being), there is hiddenness (e.g., the hidden bias in the idea of the great chain of being, because of its rigid religious assumption which does not "allow for 'missing links'"). And the reverse direction also holds true.

In relation to the denseness-emptiness principle, if there is denseness (e.g., the relatively denser concentration of concerns with God in creation theory), there is emptiness (e.g., the relatively less dense, or more empty, concentration of concerns with God in uniformitarianism). And the reverse direction also holds true.

In relation to the change-constancy principle, if there is change (e.g., the ever new ways to study matter and life in the world, as shown in the study of extinction and the history of scientific ideology), there is constancy (e.g., the ever constant existence of different problems in the ever new ways to study matter and life in the world, as shown in the problems with the study of extinction and the history of scientific ideology pointed out by the critics aforecited). And the reverse direction also holds true.

In relation to the order-chaos principle, if there is order (e.g., the more or less orderly view of reality, if looked from the vantage point of uniformitarianism, with its focus on "processes that could be observed in the present acting gradually over an immense period of time"), there is chaos (e.g., the more or less chaotic view of reality, if looked from the vantage point of catastrophism, with its focus on "sudden, short-lived, violent events"). (WK 2012vv) And the reverse direction also holds true.

In relation to the slowness-quickness principle, if there is slowness (e.g., the relatively slower speed of events if they follow the processes as understood in uniformitariansim, that is, "acting gradually over an immense period of time"), there is quickness (e.g., the relatively quicker speed of events, if they follow the processes as understood in catastrophism, which are "sudden, short-lived, violent events"). And the reverse direction also holds true.

In relation to the expansion-contraction principle, if there is expansion (e.g., the relative spread of the ideas of catastrophism and uniformitarianism in modern times), there is contraction (e.g., the relative decline of the idea of the great chain of being in modern times). And the reverse direction also holds true.

In relation to the theory-praxis principle, if there is theory (e.g., the theoretical construction of the great chain of being by Plato and Aristotle in philosophy), there is praxis (e.g., the practical application of the idea of the great chain of being in philosophy to the field of geology, when "the possibility of extinction was not widely accepted before the 1800s," because "the idea of extinction was threatening to those who held a belief in the Great Chain of Being, a theological position that did not allow for 'missing links'"). And the reverse direction also holds true.

In relation to the convention-novelty principle, if there is convention (e.g., the conventional wisdom of rejecting "the idea of extinction" before 1750's, when many "held a belief in the Great Chain of Being, a theological position that did not allow for 'missing links'"), there is novelty (e.g., the alternative novel challenge to the conventional wisdom of rejecting "the idea of extinction" before 1750's, by the new idea of uniformitarianism in the 19th century, which accepts extinction events as part of evolution over a long period of time). And the reverse direction also holds true.

In relation to the evolution-transformation principle, if there is evolution (e.g., the natural evolution of humans to emerge in the history of the Earth), there is transformation (e.g., the technical transformation of human ability to understand the history of the Earth by the invention of different theories like catastrophism and uniformitarianism). And the reverse direction also holds true.

In relation to the symmetry-asymmetry principle, if there is symmetry (e.g., the existence of different lifeforms in the world like humans, animals, plants, and minerals), there is asymmetry (e.g., the "hierarchical structure of all matter and life" for those with the belief in the great chain of being, such that humans are ranked higher than animals, plants, and minerals in the chain). And the reverse direction also holds true.

In relation to the softness-hardness principle, if there is softness (e.g., the inclusion of extinction events by those in catastrophism), there is hardness (e.g., the exclusion of extinction events by those with the belief in the idea of the great chain of being). And the reverse direction also holds true.

In relation to the seriousness-playfulness principle, if there is seriousness (e.g., the serious business of those who try to understand extinction events), there is playfulness (e.g., the playful part of those who try to understand extinction events, when they play around with different stories to tell over time, like "the great chain of being," "creation theory," "catastrohism," and "uniformitarianis"). And the reverse direction also holds true.

In relation to the activeness-inactivess principle, if there is activeness (e.g., the relative activeness of those animals which survived catastrophic events in history and were able to engage in different life activities), there is inactiveness (e.g., the relative inactiveness of those animals which did not survive catastrophic events in history and became so injured, sick, or hurt that they were unable to engage in different life activities).

In relation to the selfness-otherness principle, if there is selfness (e.g., the tendency to be mostly centered on one's self or group, as shown by some humans in "poaching," such that "species were deliberately driven to extinction"), there is otherness (e.g., the tendency to be mostly centered on others, as shown in the design of creation by God, according to those with the belief in creation theory). (WK 2012e) And the reverse direction also holds true.

In relation to the regression-progression principle, if there is regression (e.g., the regression made by the study on extinction, as shown in the problems pointed out by the critics), there is progression (e.g., the progress made by the study on extinction, as shown in the contribution to a better understanding of extinction events on Earth aforecited). And the reverse direction also holds true.

In relation to the same-difference principle, if there is similarity in outcome (e.g., the contribution to the molding and control of beliefs, values, and behaviors, regardless of whether this be done by way of creation theory, or by way of uniformitarianism), there is difference in outcome (e.g., the contribution to the molding and control of beliefs, values, and behaviors by way of creation theory for a relatively more religious worldview—but the contribution to the molding and control of beliefs, values, and behaviors by way of uniformitarianism for a relatively more secular worldview). And the reverse direction also holds true.

In relation to the stability-reaction principle, if there is stability (e.g., the relatively more stable condition of the idea of the great chain of being in the older days, when "the possibility of extinction was not widely accepted before the 1800s"), there is reaction (e.g., the idea of the great chain of being in the older days then led to further changes later on, when "today most geologists combine catastrophist and uniformitarianist standpoints, taking the view that Earth's history is a slow, gradual story punctuated by occasional natural catastrophic events that have affected Earth and its inhabitants"). And the reverse direction also holds true.

In relation to the functionality-nonfunctionality principle, if there is functionality (e.g., the presence of action for which something is fitted, as shown in the belief in the idea of the great chain of being in the older days, which served the function of explaining the "hierarchical structure of all matter and life" on the basis of the Platonic Forms at the time), there is nonfunctionality (e.g., the relative lack of action for which something is fitted, as shown in the belief in the combination of catastrophism and uniformitarianism nowadays, such that the idea of the great chain of being is no longer functional in our time, because of different scientific standards and expectations). And the reverse direction also holds true.

And in relation to the intentionality-nonintentionality principle, if there is intentionality (e.g., the planning or design of something for a certain outcome, as shown in the intention of Georges Cuvier to use catastrophism "to explain the patterns of extinction and faunal succession that he and others were observing in the fossil record," but without "any reference to Noah's flood" or to "divine creation"), there is nonintentionality (e.g., the relative absence of planning or design of something for a certain outcome, such that the outcome can even be counter-intentional in being the opposite of what was originally planned, as shown in the work of Robert Jameson, who "translated the introduction Cuvier wrote for a collection of his papers on fossil quadrupeds that discussed his ideas on catastrophic extinction into English" but "added extensive editorial notes to the translation that explicitly linked the latest of Cuvier's revolutions with the biblical flood," contrary to the original non-religious focus of Cuvier). And the reverse direction also holds true.

The Benignity of Discontinuity

This comprehensive analysis of the possibility (or impossibility) and desirability (or undesirability) of discontinuity in the context of natural resources, from the perspectives of the mind, nature, society, and culture, has the scholarly virtue to show the different ways in which discontinuity is both possible and desirable, but not to the extent that the spokespersons from each side would like us to believe.

Discontinuity thus has its malignity, just as it has its benignity too, as the other side of the same coin.

But this is only one part of a larger story, since there is the other side of the story, which concerns diversity, the opposite of discontinuity (as already discussed in Chapter Two).

These dual analyses should not be underestimated, because those on the side of diversity often downgrade discontinuity merely as the opposite without really appreciating it from the vantage point of discontinuity too—and vice versa.

Now that we reach the end of this long intellectual journey, what then can be concluded about the future of natural resources in the context of diversity and discontinuity?

This is the question which is to be answered in the next chapter, namely, the future of natural resources.

Table 3.1. Discontinuity and Its Dual Effects

- **Discontinuity and the Mind**
  —Ex: discontinuity, extinction, and overconsumption
  —Ex: discontinuity, biodiversity loss, and health risks

- **Discontinuity and Nature**
  —Ex: discontinuity, climate change, and water deterioration
  —Ex: discontinuity, genetic pollution, and reduction in range

- **Discontinuity and Society**
  —Ex: discontinuity, habitat fragmentation, and deforestation
  —Ex: discontinuity, oil depletion, and resource curse

- **Discontinuity and Culture**
  —Ex: discontinuity, misused traditional knowledge, and biopiracy
  —Ex: discontinuity, extinction, and the history of scientific ideology

Notes: The examples in the categories are solely illustrative (not exhaustive), and the comparison is relative (not absolute), nor are they necessarily mutually exclusive. And some can be easily re-classified elsewhere. As generalities, they allow exceptions.
Source: A summary of Ch.3 of *BNR*

Table 3.2. Discontinuity, Extinction, and Overconsumption

- **History of Extinction**
  —Ex: "the Cretaceous-Paleogene extinction event"
  —Ex: "the Triassic-Jurassic extinction event"
  —Ex: "the Permian-Triassic extinction event"
  —Ex: "the late Devonian extinction"
  —Ex: "the Ordovician–Silurian extinction event"

- **History of Overexploitation**
  —Ex: "fisheries"
  —Ex: "water resources"
  —Ex: "forest resources"

- **Problems with the Study of Extinction and Overconsumption**
  —Firstly, one main criticism of the study of extinction and overconsumption is that the extinction events cannot be so clearly defined as the advocates would like us to believe.
  —Secondly, another main criticism of the study of extinction and overconsumption is that there are different threats to species, and overconsumption can be only one of them.
  —Thirdly, still another main criticism of the study of extinction and overconsumption is that overexploitation does not necessarily lead to extinction of animals and other resources.
  —Fourthly, an additional main criticism of the study of extinction and overconsumption is that overexploitation can be prevented.
  —And fifthly, another additional main criticism of the study of extinction and overconsumption is that extinction has important functions for natural evolution.

Notes: The examples in the categories are solely illustrative (not exhaustive), and the comparison is relative (not absolute), nor are they necessarily mutually exclusive. And some can be easily re-classified elsewhere. As generalities, they allow exceptions.
Source: From *Sec. 2.2.1* of *BNR*. See book for citations.

Table 3.3. Discontinuity, Climate Change, and Water Deterioration

- **Internal Causes of Climate Change**
  —Ex: "ocean variability"

- **External Causes of Climate Change**
  —Ex: "orbital variations"
  —Ex: "solar output"
  —Ex: "volcanism"
  —Ex: "plate tectonics"
  —Ex: "human influences"

- **Climate Change and Water Deterioration**
  —Ex: "climate change and temperature"
  —Ex: "climate change and precipitation"

- **Problems with the Study of Climate Change**
  —Firstly, one main criticism of the study of climate change and water deterioration is that the causes of climate change are more complicated than the advocates would claim.
  —Secondly, another main criticism of the study of climate change and water deterioration is that there are still areas of scientific "uncertainty."
  —Thirdly, still another main criticism of the study of climate change and water deterioration is that water management can minimize the problem of water shortage.
  —Fourthly, an additional main criticism of the study of climate change and water deterioration is that its objective analysis can be naïve without sufficient attention to political factors.
  —And fifthly, another additional main criticism of the study of climate change and water deterioration is that it is too secular.

Notes: The examples in the categories are solely illustrative (not exhaustive), and the comparison is relative (not absolute), nor are they necessarily mutually exclusive. And some can be easily re-classified elsewhere. As generalities, they allow exceptions.
Source: From *Sec. 3.3.1* of *BNR*. See book for citations.

Table 3.4. Discontinuity, Genetic Pollution, and Reduction in Range

- **Conditions of Species Invasion**
  —Ex: "species-based mechanisms"
  —Ex: "ecosystem-based mechanisms"

- **Characteristics of Species Invasion**
  —Ex: "native invaders"
  —Ex: "invaded ecosystems"
  —Ex: "vectors"
  —Ex: "impacts of wildfire"

- **Impacts of Species Invasion**
  —Ex: "ecology"
  —Ex: "biodiversity loss"

- **Problems with the Study of Genetic Pollution and Species Invasion**
  —First, the use of the term "genetic pollution" is
  —Second, genetic pollution has unintentional effects not often addressed.
  —Third, species invasion is only a cause of genetic pollution.
  —Fourth, scientsts disagree about the traits of "invasiveness" in invasive species.
  —Fifth, there are inconsistent findings.
  —Sixth, invasive species does not necessarily lead to extinction.
  —And seventh, genetic pollution is not necessarily undesirable in the way that the activists would like us to believe.

Notes: The examples in the categories are solely illustrative (not exhaustive), and the comparison is relative (not absolute), nor are they necessarily mutually exclusive. And some can be easily re-classified elsewhere. As generalities, they allow exceptions.
Source: From *Sec. 3.3.2* of *BNR*. See book for citations.

Table 3.5. Discontinuity, Habitat Fragmentation, and Deforestation

- **Features of Habitat Fragmentation**
  —Ex: "reduction in the total area of the habitat"
  —Ex: "decrease of the interior: edge ratio"
  —Ex: "isolation of one habitat fragment from other areas of habitat"
  —Ex: "breaking up of one patch of habitat into several smaller"
  —Ex: "decrease in the average size of each patch of habitat"

- **Causes of Deforestation**
  —Ex: "pre-historical"
  —Ex: "pre industrial"
  —Ex: "industrial"

- **Problems with the Study of Deforestation**
  —Firstly, one main criticism of the study of habitat fragmentation and deforestation is that there is the contentious North-South divide to be resolved.
  —Secondly, another main criticism of the study of habitat fragmentation and deforestation is that scientists disagree about the causes of deforestaton.
  —Thirdly, still another main criticism of the study of habitat fragmentation and deforestation is that scientists also disagree about the rate of deforestation around the world.
  —Fourthly, an additional main criticism of the study of habitat fragmentation and deforestation is that scientists disagree about the definition of forest types, with implications for measurement.
  —Fifthly, another additional main criticism of the study of habitat fragmentation and deforestation is that the proposed stages of deforestation are questionable.
  —And sixthly, still another additional main criticism of the study of habitat fragmentation and deforestation is that deforestation can be controlled by different mechanisms.

Notes: The examples in the categories are solely illustrative (not exhaustive), and the comparison is relative (not absolute), nor are they necessarily mutually exclusive. And some can be easily re-classified elsewhere. As generalities, they allow exceptions.
Source: From *Sec. 3.4.1* of *BNR*. See book for citations.

Table 3.6. Discontinuity, Oil Depletion, and Resource Curse

- **Predictions of Production Decline**
  —Ex: "oil well production decline"
  —Ex: "oil field production decline"
  —Ex: "world oil production"

- **Implications of Production Decline**
  —Ex: "catastrophe"
  —Ex: "recession"
  —Ex: "rising food prices"

- **Worsening Effects of Resource Curse**
  —Ex: "tug of war between people and government"
  —Ex: "conflict"
  —Ex: "international pressure"
  —Ex: "taxation"
  —Ex: "Dutch disease"
  —Ex: "revenue volatility"
  —Ex: "excessive borrowing"
  —Ex: "corruption"
  —Ex: "lack of diversification and enclave effects"

- **Problems with the Study of Oil Depletion and Resource Curse**
  —First, scientists disagree about the prediction of oil depletion.
  —Second, some prediction models are more optimistic (or conversely, more pessimistic) than others.
  —Third, some predictions are based on wishful thinking.
  —Fourth, there are alternatives to oil dependency.
  —And fifth, researchers disagree on the causal relationship between natural resources and economic development in the long run.

Notes: The examples in the categories are solely illustrative (not exhaustive), and the comparison is relative (not absolute), nor are they necessarily mutually exclusive. And some can be easily re-classified elsewhere. As generalities, they allow exceptions.
Source: From *Sec. 3.4.2* of *BNR*. See book for citations.

Table 3.7. Discontinuity, Misused Traditional Knowledge, and Biopiracy

- **Famous Cases of Biopiracy**
  —Ex: "the Maya ICBG controversy"
  —Ex: "the rosy periwinkle"
  —Ex: "the neem tree"
  —Ex: "the Enola bean"
  —Ex: "the basmati rice"
  —Ex: "hoodia"

- **Characteristics of Traditional Knowledge**
  —Ex: communal
  —Ex: accumulative
  —Ex: spiritual

- **Mechanisms to Protect Traditional Knowledge**
  —Ex: "indigenous intellectual property"
  —Ex: "Agreement on Trade-Related Aspects of Intellectual Property Rights"
  —Ex: "the Convention on Biological Diversity"

- **Problems with the Study of Biopiracy and Traditional Knowledge**
  —First, it can become quickly political in tone (without examining the actual evidences more carefully).
  —Second, there is still no "international legally binding regime" on the protection of traditional knowledge against biopiracy.
  —Third, the existing mechanisms against biopiracy are not effective.
  —Fourth, the defense of traditional knowledge is not critical to its own version of destructiveness against biodiversity.
  —And fifth, traditional knowledge is not so naturalistic as its advocates suggested, because of inherent power struggles.

Notes: The examples in the categories are solely illustrative (not exhaustive), and the comparison is relative (not absolute), nor are they necessarily mutually exclusive. And some can be easily re-classified elsewhere. As generalities, they allow exceptions.
Source: From *Sec. 3.5.1* of *BNR*. See book for citations.

• Part Four •

Conclusion

CHAPTER 4
CONCLUSION—THE FUTURE OF NATURAL RESOURCES

I think natural resources will help us come up with a bit of a gain at the opening, but the question is how long is it going to stay.
__Fred Ketchen (TE 2012)

Beyond Diversity and Discontinuity

The analyses of diversity (in Chapter Two) and discontinuity (in Chapter Three) show the extent of their possibility (or impossibility) and desirability (or undesirability) in the study of natural resources.

These two analyses are essential for the evaluation of the opposing views on natural resources (and other views as already introduced in *Sec. 1.5* and the rest of the book), such that natural resources (in relation to both diversity and discontinuity) are neither possible (or impossible) nor desirable (or undesirable) to the extent that the respective ideologues (on different sides) would like us to believe.

As already indicated in *Sec.1.1*, this challenge to the opposing views on natural resources does not imply that natural resources are unimportant, or that those interdisciplinary fields (related to natural resources) like conservation biology, environmental management, ecological economics, political ecology, environmental ethics, deep ecology, adaptive management, genetic engineering, Malthusianism, and so on are not worth studying. Of course, neither of these extreme views is reasonable.

Rather, this book offers an alternative (better) way to understand the future of natural resources, especially in the dialectic context of diversity and discontinuity—while learning from different approaches in the literature but without favoring any one of them (nor integrating them, since they are not necessarily compatible with each other).

More specifically, this book offers a new theory (that is, *the resilient theory of natural resources*) to go beyond the existing approaches in a novel way.

As a recap from Chapter One, my *resilient theory of natural resources* has four major distinct features.

Firstly, my theory makes good use of all theoretical approaches to the study of natural resources in relation to diversity and discontinuity, be they about the mental argument, the natural argument, the societal argument, or the cultural argument—especially from the most comprehensive combined perspectives of the mind, nature, society, and culture (as already dealt with in Chapter Two and Chapter Three).

Secondly, just like many other theories of mine in my previous books, my theory here does not heavily favor any specific theory over others in the literature, nor trying to integrate them (as they are not necessarily compatible with each other).

Thirdly, my theory treats the issue of natural resources in the distinctive dialectic context of no diversity without discontinuity, and vice versa.

And fourthly, it contains thirty major theses, namely, (a) the first thesis: the formalness-informalness principle, (b) the second thesis: the absoluteness-relativeness principle, (c) the third thesis: the partiality-totality principle, (d) the fourth thesis: the predictability-unpredictability principle, (e) the fifth thesis: the explicability-inexplicability principle, (f) the sixth thesis: the fiction-reality principle, (g) the seventh thesis: the cognitiveness-noncognitiveness principle, (h) the eighth thesis: the finiteness-transfiniteness principle, (i) the ninth thesis: the preciseness-vagueness principle, (j) the tenth thesis: the simpleness-complicatedness principle, (k) the eleventh thesis: the openness-hiddenness principle, (l) the twelfth thesis: the denseness-emptiness principle, (m) the thirteenth thesis: the change-constancy principle, (n) the fourteenth thesis: the order-chaos principle, (o) the fifteenth thesis: the slowness-quickness principle, (p) the sixteenth thesis: the expansion-contraction principle, (q) the seventeenth thesis: the theory-praxis principle, (r) the eighteenth thesis: the convention-novelty principle, (s) the nineteenth thesis: the evolution-transformation principle, (t) the twentieth thesis: the symmetry-asymmetry principle, (u) the twenty-first thesis: the softness-hardness principle, (v) the twenty-second thesis: the seriousness-playfulness principle, (w) the twenty-third thesis: the activeness-inactiveness principle, (x) the twenty-fourth thesis: the selfness-otherness principle, (y) the twenty-fifth thesis: the regression-progression principle, (z) the twenty-sixth thesis: the sameness-difference principle, (aa) the twenty-seventh thesis: the stability-reaction principle, (bb) the twenty-eighth thesis: the functionality-nonfunctionality principle, (cc) the twenty-ninth thesis: the intentionality-nonintentionality principle, and (dd) the thirtieth thesis: the post-human resilience—to be elaborated in this chapter.

Because of the last thesis on "the post-human resilience," the word "resilient" in the title of my theory are suggestive, because it refers to the adaptive uses of different resources in face of changed circumstances over time—which then require a multi-faceted exploration of the dialectic relations between diversity and discontinuity, as well as an understanding of the different views regarding these dialectic relations in the study of natural resources (together with other debates as already discussed in the rest of the book) and in the end it calls for a need to transcend them by future advanced humans and later post-humans in the context of the great transformations of natural resources in the future world (as will be discussed in the rest of this chapter).

In addition, this book on natural resources is a sequel to my previous books on diversity and discontinuity. The difference, however, is that this book focuses more on diversity and discontinuity in the natural world, whereas those previous books of mine (like *FHC*, *FCD*, *FPHC*, *FPHU*, *FC*, *FPHMIG*, and *BDPD*) which addressed, among many other things, diversity and discontinuity focus more on the human world instead.

With this clarification in mind—the seminal project here, if successful, will fundamentally change the way that we think about diveristy and discontinuity in the study of natural resources together with other debates as already discussed in the rest of the book), from the combined perspectives of the mind, nature, society, and culture, with enormous implications for the human future and what I originally called its "post-human" fate.

In Relation to Method

The first seven theses are in relation to method in existential dialectics, as shown below.

1st Thesis: The Formalness-Informalness Principle

The first thesis refers to the formalness-informalness principle (on the formal requirements of logical systems) in the category of "method" in existential dialectics, which was first proposed in *FPHML* (and later, *FPHH*, *FPHGAM*, *FPHSEM*, and other books of mine), in that the formal requirements of a logical system (e.g., consistency, soundness, and completeness) have both usefulness and non-usefulness, to the extent that, if there are formal systems requiring them, there are alternative ones which do not.

The formalness-informalness principle reveals something interesting in existential dialectics, in that it goes beyond both classical and non-classical logics (while learning something from both).

Thus, it does *not* exclude classical logics but simply goes beyond both classical and non-classical logics, while learning something from each. There is no formalness without informalness—and vice versa.

For illustration, existential dialectics can make use of both classical logics under certain conditions (e.g., especially, though not exclusively, when they are clear-cut, etc.) and non-classical logics under alternative conditions (especially, though not exclusively, when they are "unknown," "irrelevant," "ambiguous," "possible," with "different degrees of truth," empirically inconsistent in a desirable way, etc.). (WK 2008u)

For instance, on the one hand, there is formalness (e.g., the formal logical requirement of soundness, consistency, and completeness in a system of ideas, as in the study of psychological benefits and biodiversity, in *Sec. 2.2.1*; the formal logical requirement of soundness, consistency, and completeness in a system of ideas, as shown in the study on dietary health and agrobiodiversity, in *Sec. 2.2.2*; the formal logical requirement of soundness and completeness in a system of ideas, as in the study of conservation biology and environmental protection in *Sec. 2.3.1*; the formal logical requirement of soundness and completeness in a system of ideas, as in the study of adaptive management, in *Sec. 2.3.2*; the formal logical requirement of soundness and completeness in a system of ideas, as in the study on the restoration of life and the technology of cloning, in *Sec. 2.4.1*; the formal logical requirement of soundness, completeness, and consistency in a system of ideas, as in ecological economics, in *Sec. 2.4.2*; the formal logical requirement of soundness, consistency, and completeness in a system of ideas, as in environmental ethics, in *Sec. 2.5.1*; the formal logical requirement of soundness and completeness in a system of ideas, as in animal watching, in *Sec. 2.5.2*; the formal logical requirement of soundness and completeness in a system of ideas, as in the study of extinction and overconsumption, in *Sec. 3.2.1*; the formal logical requirement of soundness and completeness in a system of ideas, as the study of biodiversity loss and health risks, in *Sec. 3.2.2*; the formal logical requirement of soundness and completeness in a system of ideas, as in the study of climate change and water deterioration, in *Sec. 3.3.1*; the formal logical requirement of soundness, completeness, and consistency in a system of ideas, as shown in the study of genetic pollution and reduction in range, in *Sec. 3.3.2*; the formal logical requirement of soundness, consistency, and completeness in a system of ideas, as in the study of habitat fragmentation and deforestation, in *Sec. 3.4.1*; the formal logical requirement of soundness, completeness, and consistency in a system of ideas, as shown in the study on oil depletion and resource curse, in *Sec. 3.4.2*; the formal logical requirement of soundness and completeness in a system of ideas, as in the study of biopiracy and traditional knowledge, in *Sec. 3.5.1*; and the formal logical requirement of soundness and completeness in a system of ideas, as in the study of extinction and the history of scientific ideology, in *Sec. 3.5.2*).

On the other hand, there is also informalness (e.g., the non-formal existence of unsoundness, inconsistency, and incompleteness in the study of psychological benefits and biodiversity, as shown in the criticisms that biodiversity is not stable and subject to different threats, that there are major extinctions in the history of biodiveristy, that the actual evidences are not sufficient, that the causal relationship is not so clearcut as the advocates would like us to believe, that there are many different causes of human health problems, that there can be alternative resources for health improvement even when biodiversity occurs, and that different levels of diversity may not correspond, in *Sec. 2.2.1*; the non-formal existence of unsoundness, inconsistency, and incompleteness in the study on dietary health and agrobiodiversity, as shown in the criticisms that agrodiversity is not assured for human dependency in regard to dietary health, that the relationship between agrobiodiversity and the Irish Potato Famine is much exaggerated, that there are alternative causes of human dietary health, that the nature of agrodiversity is still not well known, and that agrobiodiveristy can contribute to dietary problems [via crop diseases], in *Sec. 2.2.2*; the non-formal existence of unsoundness and incompleteness in the study of conservation biology and environmental protection, as shown in the criticisms that it does not give sufficient attention to the important problem of poor management, that it often ignores the factors of power abuse and financial shortage, that different approaches have different weaknesses, that the priority to focus on hotspots for biodiveristy in the protection of natural resources is misleading and dangerous, and that it is too utilitarian in ethical orientation, in *Sec. 2.3.1*; the non-formal existence of unsoundness and incompleteness in the study of adaptive management, as shown in the criticisms that it is risky enough, that it suffers from "a common lament" about "significant uncertainty," that adaptive management has not yet been widely accepted in the corporate world, that scientists still disagree about the nature of adaptive management, etc., in *Sec. 2.3.2*; the non-formal existence of unsoundness and incompleteness in the study on the restoration of life and the technology of cloning, as shown in the criticisms that it is still technological infeasible in our time, that there remain some deep ethical issues to be resolved, that what counts as extinct species may turn out to be pseudoextinct ones so the restoration of life through cloning may be misleading, that cloning does not restore an exactly identical life, that the more difficult question of how to raise the restored species remains unresolved so it may not be viable in the long run, that cloning distracts the attention and takes away funding from solving the serious issue of animal extinction, etc., in *Sec. 2.4.1*; the non-formal existence of unsoundness, incompleteness, and inconsistency in ecological economics, as shown in the criticisms that its estimates are highly speculative and thus questionable, that it is essentially ideological just as neoclassical economics has its own share of ideological bias too, that there are alternative schools of thought so it does not have the last word, and that it evolves over time to become more mainstream and betrays itself, contrary to its original goal," in *Sec. 2.4.2*; the non-formal existence of unsoundness, inconsistency, and incompleteness in environmental ethics, as shown in the criticisms that it is divided between competing schools of thought, that its reasoning is simplistic, that its implementation can be destructive, that it is self-defeating, that it focuses too much on the whole without sufficient attention to the individuals, etc., in *Sec. 2.5.1*; the non-formal existence of unsoundness and incompleteness in animal watching, as shown in the criticisms that there are philosophical disagreements in regard to the use of biodiversity for leisure activities, that what is considered as a leisure activity in a culture may not be so at all in another culture which may choose the use of different animals or natural resources instead, that it is gendered-relative, that there is a concern with "animal welfare," etc., in *Sec. 2.5.2*; the non-formal existence of unsoundness and incompleteness in the study of extinction and overconsumption, as shown in the criticisms that the extinction events cannot be so clearly defined as the advocates would like us to believe, that there are different threats to species so overconsumption can be only one of them, that overexploitation does not necessarily lead to extinction of animals and other resources, that overexploitation can be prevented, and that extinction has important functions for natural evolution, in *Sec. 3.2.1*; the non-formal existence of unsoundness and incompleteness in the study of biodiversity loss and health risks, as shown in the criticisms that there are many different causes of human health problems, that there can be alternative resources for health improvement even when biodiversity occurs, that the relationship between biodiversity loss and health risks is more complicated than its advocates would like us to believe, that there remains a lot of "scientific uncertainty" about the nature of biodiversity loss, that its policy recommendation of "collaboration across disciplines and across borders" is not realistic, and that health risks are important too for natural evolution, in *Sec. 3.2.2*; the non-formal existence of unsoundness in the study of climate change and water deterioration, as shown in the criticisms that the causes of climate change are more complicated than the advocates would claim, that there are still areas of scientific "uncertainty," that water management can minimize the problem of water shortage, that its objective analysis can be naïve without sufficient attention to political factors, and that it is too secular, in *Sec. 3.3.1*; the non-formal existence of unsoundness, incompleteness, and inconsistency in the study of genetic pollution and reduction in range, as shown in the criticisms that the use of the term "genetic pollution" is controversial, that genetic pollution has unintentional effects, that species invasion is only a cause of genetic pollution, that scientsts disagree about the traits of "invasiveness" in invasive species, that there are inconsistent findings, that invasive species does not necessarily lead to extinction, and that genetic pollution is not necessarily undesirable in the way that the activists would like us to believe, in *Sec. 3.3.2*; the non-formal existence of unsoundness, inconsistency, and incompleteness in the study of habitat fragmentation and deforestation, as shown in the criticisms that there is the contentious North-South divide to be resolved, that scientists disagree about the causes of deforestaton, that scientists also disagree about the rate of deforestation around the world, that scientists disagree about the definition of forest types with implications for measurement, that the proposed stages of deforestation are questionable, and that deforestation can be controlled by different mechanisms, in *Sec. 3.4.1*; the non-formal existence of unsoundness, incompleteness, and inconsistency in the study on oil depletion and resource curse, as shown in the criticisms that scientists disagree about the prediction of oil depletion, that some prediction models are more optimistic [or conversely, more pessimistic] than others, that some predictions are based on wishful thinking, that there are alternatives to oil dependency, and that researchers disagree on the causal relationship between natural resources and economic development in the long run, in *Sec. 3.4.2*; the non-formal existence of unsoundness and incompleteness in the study of biopiracy and traditional knowledge, as shown in the criticisms that it can become quickly political in tone without examining the actual evidences more carefully, that there is still no "international legally binding regime" on the protection of traditional knowledge against biopiracy, that the existing mechanisms against biopiracy are not effective, that the defense of traditional knowledge is not critical to its own version of destructiveness against biodiversity, and that traditional knowledge is not so naturalistic as its advocates suggested, because of inherent power struggles, in *Sec. 3.5.1*; and the non-formal existence of unsoundness and incompleteness in the study of extinction and the history of scientific ideology, as shown in the criticisms that each view is relative to its own assumptions and claims, that some of the views can be criticized as false [albeit in different ways], that the truth about extinction can be a mixture of some of the views, and that extinction can be prevented, in *Sec. 3.5.2*).

And the reverse direction also holds true.

However, it should be stressed (as this is something that I regularly did in my previous books, *almost verbatim*) that there are different shades of gray (or different degrees of truth) in the two opposites, and the classification is not necessarily mutually exclusive either. In fact, even when some combinations of the two occur, they only end up sharing the same dialectic relationship, but in a different degree.

Also, there are some other relationships (as a kind of family resemblance) which have something in common with the principle, but they are not exactly the same but only more or less comparable, which varies from case to case.

Good examples of family resemblance in relation to the principle are logicalness-nonlogicalness, rationality-nonrationality, etc.

2nd Thesis: The Absoluteness-Relativeness Principle

The second thesis refers to the absoluteness-relativeness principle (on the multiplicity of things) in the category of "method" in existential dialectics, which was first proposed in *FPHK*, in that there is the multiplicity of things in reality, be they about entities, qualities (or properties), and relationships. If there is something absolute, there is likewise something relative. And there is no absoluteness without relativeness—and vice versa.

Both absoluteness and relativeness here are also relevant to different modalities often cited in the literature on ontology, such as possibility (e.g., something "can" happen) and its opposite (e.g., impossibility), probability (e.g., something "will" happen) and its opposite (e.g., improbability), and necessity (e.g., something "should" happen) and its opposite (e.g., contingency).

For instance, on the one hand, there is absoluteness (e.g., the absolute view of those advocates of the "latitudinal gradients" in biodiversity, in that "there is an increase in biodiversity from the poles to the tropics," in

Sec. 2.2.1; the absolute view of "a meta-analysis of studies comparing biodiversity, which "noted that, when compared to organic cropping systems, conventional systems had significantly lower species richness and abundance," in *Sec. 2.2.2*; the absolute view of K. Chan, who argued that "conservationists must advocate for biodiversity and can do so" for "the greatest good for the greatest number of people for the longest time," in *Sec. 2.3.1*; the absolute view of C. Holling and C. Walters in regard to "passive and adaptive management practice," in *Sec. 2.3.2*; the absolute view of the advocates about cloning for biodivesity, in *Sec. 2.4.1*; the absolute view of ecological economics, in regard to its own ideology, in *Sec. 2.4.2*; the absolute view of ecocentrism in regard to its claim about "the intrinsic value of nature," in *Sec. 2.5.1*; the absolute view of birdwatching "in Britain during the Victorian Era" mostly for "collection based taxonomy," in *Sec. 2.5.2*; the absolute view by Jack Sepkoski and David M. Raup in regard to "the 'Big Five' [extinction events]," in *Sec. 3.2.1*; the absolute view of those conservation activists who are too eager to use "biodiversity…as a legal standard" to enforce conservation against biodiversity loss, in *Sec. 3.2.2*; the absolute view of the causes of climate change by the advocates, in *Sec. 3.3.1*; the absolute view of those conservation activists that species invastion leads to extinction of native species, in *Sec. 3.3.2*; the absolute view of "some countries in the developed world, such as the United States of America" which "cut down their forests centuries ago and benefited greatly from this deforestation" but now "deny developing countries the same opportunities," in *Sec. 3.4.1*; the absolute view of those rulers in "resource-cursed countries…ruled by either authoritarian or other types of highly repressive regimes," in that "these regimes are kept in power by a select elite such as high-ranking politicians and military leaders," so that "the common folk, those most in need of the protection, are left to fend for themselves," as "in places like the Democratic Republic of the Congo," in *Sec. 3.4.2*; the absolute view of those biopirates, in regard to the use, for profit, of "indigenous knowledge…originating with indigenous people…without permission from and with little or no compensation or recognition to the indigenous people themselves, " in *Sec. 3.5.1*; and the absolute view of the advocates of the great chain of being, in that it is "a theological position that did not allow for 'missing links,'" in *Sec. 3.5.2*).

On the other hand, there is relativeness (e.g., what is true for those advocates of the "latitudinal gradients" in biodiversity is not necessarily so for the critics, who "claim that this characteristic is unverified in aquatic ecosystems, especially in marine ecosystems," that "the latitudinal distribution of parasites does not follow this rule," and that "other instances of great diversity in higher latitudes have also been recorded," in *Sec. 2.2.1*; what is right for "a meta-analysis of studies comparing biodiversity" is not necessarily so for for the critics, since different studies offer different insights, as shown in other studies ["16% of studies"] which "did find a greater level of species richness in conventional systems" instead, in *Sec. 2.2.2*; what is right for K. Chan in regard to his utilitarian ethics for conservation is not necessarily so for the critics, who argue instead that "nature has an intrinsic value that is independent of anthropocentric usefulness or utilitarianism," in *Sec. 2.3.1*; what is right for C. Holling and C. Walters in regard to "passive and adaptive management practice" is not necessarily so for the critics like R. Hilbourne, who proposed instead "three learning models..., around which adaptive management approaches could be developed," and "these are reactive, passive and active, in *Sec. 2.3.2*; what is right for the advocates of cloning for biodivesity is not necessarily so for the critics, since there are deep "bioethical and philosophical objections" to "the cloning of extinct creatures," which have yet to be satisfactorily answered, as indicated earlier, in *Sec. 2.4.1*; what is right for ecological economists in regard to their own ideology is not necessarily so for neoclassical economists, who maintain their own version instead, that is, with their focus on "the Economic Man" as "a utility maximizing consumer" and on technological optimism, in *Sec. 2.4.2*; what is right for ecocentrism is not necessarily so for anthropocentrism, since they constitute two opposing schools of environmental ethics, in *Sec. 2.5.1*; what is right for thoe birdwatchers "in Britain during the Victorian Era," mostly for "collection based taxonomy," is not necessarily so for "the BOU," which "changed focus to ecology and behaviour...in the 1940s" for "the 'scientification' of the pastime," in *Sec. 2.5.2*; what is true for Jack Sepkoski and David M. Raup in regard to "the 'Big Five' [extinction events]" is not necessarily so for the critics, because, "as more stringent statistical tests have been applied to the accumulating data, the 'Big Five' [extinnction events] cannot be so clearly defined," in *Sec. 3.2.1*; what is right for those conservation activists who are too eager to use "biodiversity...as a legal standard" to enforce conservation against biodiversity loss is not necessarily so for the critics like Bosselman who warned that there is still a lot things about biodiveristy loss which remain uncertain, or what he called "the remaining areas of scientific uncertainty" in regard to extinction and biodiverity loss, in *Sec. 3.2.2*; what is sure for of the causes of climate change by the advocates is not necessarily so for the critics, as shown in "a 2010 study [which] suggests, 'that the effects of solar variability on temperature throughout the atmosphere may be contrary to current expectations,'" in *Sec. 3.3,1*; what is true for those conservation activists who argue that species invastion leads to extinction of native species is not necessarily so for the critics, who instead suggested that "it is difficult to unequivocally attribute extinctions to a species invasion, and the few scientific studies that have done so have been with animal taxa," in *Sec. 3.3.2*; what is right for "some countries in the developed world, such as the United States of America" to "deny developing countries the same opportunities" about deforestation is not necessarily so for "developing countries [which] feel...that it is hypocritical to deny developing countries the same opportunities" and "that the poor shouldn't have to bear the cost of preservation when the rich created the problem," in *Sec. 3.4.1*; what is beneficial for these authoritarian rulers "in places like the Democratic Republic of the Congo" is not necessarily so for "the common folk, those most in need of the protection," who "are left to fend for themselves," and "if the citizens complain, money from the natural resources enables governments to pay for armed forces to keep the citizens in check," in *Sec. 3.4.2*; what is profitable for those biopirates in regard to the use, for profit, of "indigenous knowledge" is not necessarily so for "the indigenous people themselves," who did not grant any "permission and received "little or no compensation" from the biopirates, in *Sec. 3.5.1*; and what is right for the advocates of the great chain of being in regard to its "theological position that did not allow for 'missing links'" is not necessarily so for those in the scientific version of catastrophism that does allow the occurrence of extinction events, in *Sec. 3.5.2*).

And the reverse direction also holds true.

However, it should be stressed (as this is something that I regularly did in my previous books, *almost verbatim*) that there are different shades of gray (or different degrees of truth) in the two opposites, and the classification is not necessarily mutually exclusive either. In fact, even when some combinations of the two occur, they only end up sharing the same dialectic relationship, but in a different degree.

Also, there are some other relationships (as a kind of family resemblance) which have something in common with the principle, but they are not exactly the same but only more or less comparable, which varies from case to case.

Good examples of family resemblance in relation to the principle are internalness-externalness, immanence-transcendence, independence-dependence, isolatedness-interconnectiveness, internality-externality, autonomy-interdependency, universalness-relativeness, and so forth.

3rd Thesis: The Partiality-Totality Principle

The third thesis refers to the partiality-totality principle (on the relationships between whole and parts) in the category of "method" in existential dialectics, which was first proposed in *FC* and later *FPHGEOL*, based on earlier works like *FHC* and *FPHC*), in that any analysis of a phenomenon in the world requires both the examination of the individual parts and the whole of them (with the whole as being not the sum of its parts), without privileging the whole as more important than the parts, or vice-versa (that is, the parts as more important than the whole).

In fact, this principle targets two kinds of misleading methodologies as used over the ages, that is, again in the absence of better words, what I called (in *FHC*, *FPHC* and *FC*) "reductionism" and its opposite, "reverse-reductionism," which can take four major forms in relation to concept, theory, methodology, and ontology (as summarized in *Sec. 1.8*).

Thus, reductionism and reverse-reductionism can be (i) conceptual, (ii) theoretical, (iii) methodological, and (iv) ontological. And the whole is not the sum of its parts, and there is no partiality without totality—and vice versa.

Works on emergence (as in *FC*), gestalt psychology (as in *FPHGEOL*), and the hermeneutic circle (as in *FPHDA*), just to name a few of them, have contributed to the understanding of whole and parts.

For instance, on the one hand, there is partiality (e.g., the partial view of the advocates of biodiversity on the basis of fossil records, in *Sec. 2.2.1*; the partial view of "a meta-analysis of studies comparing biodiversity," which "noted that, when compared to organic cropping systems, conventional systems had significantly lower species richness and abundance," in *Sec. 2.2.2*; the partial view of the volunary approach to the protection of natural resources, because there are alternative ones too, as cited earlier, in *Sec. 2.3.1*; the partial view of "the Yap people of Micronesia" in regard to their "adaptive management techniques to sustain high population densities in the face of resource scarcity" at the local level, in *Sec. 2.3.2*; the partial view of cloning, if looked from the vantage point of cloning an individual animal in question, like the cloning of the "a cow named Bessie" in 2001, in *Sec. 2.4.1*; the partial view of "neoclassical economics," with its narrow focus on "the Economic Man" as "a utility maximizing consumer," in *Sec. 2.4.2*; the partial view of anthropocentrism in regard to man-nature interactions, in *Sec. 2.5.1*; the partial view of birdwatching "in Britain during the Victorian Era," mostly for "collection based taxonomy," in *Sec. 2.5.2*; the partial view of the overkill hypothesis by its advocates, in *Sec. 3.2.1*; the partial view of climate change and biodiversity loss by the advocates, in *Sec. 3.2.2*; the partial view of those activists who blame human activities on climate change, in *Sec. 3.3.1*; the partial view of "one study" which "found that, of a list of invasive and noninvasive species, 86% of the invasive species could be identified from the traits alone," in *Sec. 3.3.2*; the partial view of those in "a 2002 analysis of satellite imagery," which "suggested that the rate of deforestation in the humid tropics [approximately 5.8 million hectares per year] was roughly 23% lower than the most commonly quoted rates," in *Sec. 3.4.1*; the partial view of those rulers in "resource-cursed countries…ruled by either authoritarian or other types of highly repressive regimes," in *Sec. 3.4.2*; the partial view of knowledge from the scientific vantage point, insofar as it is to be "acquired by naturalistic trial and error" and emphasizes "the gradation of recent knowledge," in *Sec. 3.5.1*; and the partial view of the advocates of the great chain of being, in regard to their rejection of extinction, in *Sec. 3.5.2*).

On the other hand, there is also totality (e.g., the more holistic view of biodiversity, such that the whole is not the sum of its parts, that is, it cannot be reduced to the sum of opposing views of the advocates and their critics, since there will emerge new views in the future not yet known today, just as there are already some skeptical views nowadays which suggest that "there is uncertainty as to how strongly the fossil record is biased by the greater availability and preservation of recent geologic sections," whereas other views are more in-between, in *Sec. 2.2.1*; the more holistic view of agrobiodiversity, such that the whole is not the sum of its parts, that is, it cannot be reduced to the sum of opposing views about differrent cropping systems, since there will emerge new views in the future not yet known today, just as there are already some alternative views nowadays which warn us that "other cropping systems like permaculture have undergone little study to determine relative levels of biodiversity compared to other cropping systems," so the overall nature of agrobiodiversity remains uncertain, in *Sec. 2.2.2*; the more holistic view of the protection of natural resources, such that the whole is not the sum of its parts, that is, it cannot be reduced to the sum of the opposing views in the three different approaches as discussed earlier, since there will emerge new views in the future not yet known today, just as there are others which do not identify with any of them, in *Sec. 2.3.1*; the more holistic view of adaptive management techniques, such that the whole is not the sum of its parts, that is, they cannot be reduced to the sum of the opposing views of the Yap people, modern scientists like C. Holling and C. Walters, etc., since there will emerge new views in the future not yet known today, just as there are already nowadays alternative views like the ones by

Hilbourne and the critics, in *Sec. 2.3.2*; the more holistic view of cloning, such that the whole is not the sum of its parts, that is, it cannot be reduced to the sum of different cases of individual cloning, since, as the critics pointed out, "for such a program to succeed, a sufficient number of individuals would have to be cloned, from the DNA of different individuals [in the case of sexually reproducing organisms] to create a viable population," which also means that there will emerge new problems and viewpoints in the future not yet known today, in *Sec. 2.4.1*; the more holistic view of economic thought, such that the whole is not the sum of its parts, that is, it cannot be reduced to the sum of opposing views like those of neoclassical economics and ecological economics, since there will emerge new approaches in the future not yet known today, just as there are already critics nowadays who question them all, as shown earlier, in *Sec. 2.4.2*; the more holistic view of man-nature interactions, such that the whole is not the sum of its parts, that is, it cannot be reduced to the sum of opposing views like ecocentrism, anthropocentrism, technocentrism, etc., since there will emerge new views in the future not yet known today, just as there are critics nowadays who do not side exclusively with any of them, in *Sec. 2.5.1*; the more holistic view of birdwatching, such that the whole is not the sum of its parts, that is, it cannot be reduced to the sum of opposing views like those "in Britain during the Victorian Era," those in "the BOU…in the 1940s," etc., since there will emerge new views in the future not yet known today, just as there are already other views in the history of birdwatching nowadays aforecited, in *Sec. 2.5.2*; the more holistic view of the overkill hypothesis, such that the whole is not the sum of its parts, that is, it cannot be reduced to the sum of opposing views of the overkill hypothesis by its advocates and the critics, since there will emerge new views in the future not yet known today, as there are already some views nowadays which do not need to exclusively side with any of them, like "some scientists" who instead "support a combination of climate change and overkill," in *Sec. 3.2.1*; the more holistic view of climate change and biodiversity loss, such that the whole is not the sum of its parts, that is, it cannot be reduced to the sum of the opposing views by different parties to the UNFCCC, since there will emerge new views in the future not yet known today, just as there are already "other views" nowadays which do not side exclusively with any of the opposing views, while questioning "aspects of climate-change science," in *Sec. 3.2.2*; the more holistic view of cliimate change, such that the whole is not the sum of its parts, that is, it cannot be reduced to the sum of opposing views about different causes of climate change, since there will emerge new views in the future not yet known today, just as there are some other views nowadays which suggest that "much remains to be learned" about climate change, without exclusively siding with any of the opposing views, in *Sec. 3.3.1*; the more holistic view of the traits of invasive species, such that the whole is not the sum of its parts, that is, they cannot be reduced to the sum of the opposing views about the traits of invasive species, since there will emerge new views in the future not yet known today, just as there are already other views nowadays which do not necessarily side exclusively with any of them, as shown in "another study [which] found invasive species tended only to have a small subset of the presumed traits, and that many such traits were found in noninvasive species, requiring other explanations," in *Sec. 3.3.2*; the more holistic view of the rate of deforestation in the world, such that the whole is not the sum of its parts, that is, it cannot be reduced to the sum of the opposing views of the "2002 analysis of satellite imagery" and its critics, since there will emerge new views in the future not yet known today, just as there are already alternative views nowadays, like "a newer analysis of satellite images" which "reveals that deforestation of the Amazon rainforest is twice as fast as scientists previously estimated," in *Sec. 3.4.1*; the more holistic view of governance in resource-cursed countries, such that the whole is not the sum of its parts, that is, it cannot be reduced to the sum of opposing views of those authoritarian rulers in "resource-cursed countries" and "the common folk," since there will emerge new views in the future not yet known today, just as there are already some researchers who do not side exclusively with any of them and hold alternative views evaluating the pros and cons of the debate aforecited, in *Sec. 3.4.2*; the more holistic view of knowledge, such that the whole is not the sum of its parts, that is, it cannot be reduced to the sum of the opposing views of scientific knowledge in the developed world and traditional knowledge in the developing world, since there will emerge new views in the future not yet known today, just as there are already those nowadays who do not side exclusively with any of them and are critical to traditional knowledge too, as cited earlier, in *Sec. 3.5.1*; and the more holistic view of extinction, such that the whole is not the sum of its parts, that is, it cannot be reduced to the sum of opposing views like the four versions aforediscussed, since there will emerge new views in the future not yet known today, just as there are already some nowadays who do not side exclusively with any of them, in *Sec. 3.5.2*).

And the reverse direction also holds true.

However, it should be stressed (as this is something that I regularly did in my previous books, *almost verbatim*) that there are different shades of gray (or different degrees of truth) in the two opposites, and the

classification is not necessarily mutually exclusive either. In fact, even when some combinations of the two occur, they only end up sharing the same dialectic relationship, but in a different degree.

Also, there are some other relationships (as a kind of family resemblance) which have something in common with the principle, but they are not exactly the same but only more or less comparable, which varies from case to case.

Good examples of family resemblance in relation to the principle are individualisticness-holisticness, partness-wholeness, analysis-synthesis, and the like.

4th Thesis: The Predictability-Unpredictability Principle

The fourth thesis is called the predictability-unpredictability principle (on the occurrence of events) in the category of "method" in existential dialectics (which was first proposed in *FC* and, later, other books of mine), in that both predictability and unpredictability have a major role to play in the occurrence of things, so that neither determinism nor indeterminism wins the centuries-old fight. There is no predictability without unpredictability—and vice versa.

There are events which are predictable, just as there are those which are not. Or what is regarded as unpredictable at one point in time may turn out to be predictable later, and, conversely, what is deemed as predictable may turn out to not be so predictable. Even in predictability, outcomes are subject to uncertainty, the degree of which varies from case to case.

Works on ecological uncertainty (as in *BNR*), complexity theory (as in *FC*), quantum mechanics (as in *FPHST* and *FPHG*), frequency probability, subjective (or Bayesian) probability, and "special causes" (as in *FPHGAM*), just to name a few of them, have contributed to the understanding of uncertainity.

For instance, on the one hand, there is predictability (e.g., the predictable tendency of many scientists to focus on biodiversity at the macro level, since "contemporary biodiversity physics is 'firmly fixated on the visible [macroscopic] world,'" in *Sec. 2.2.1*; the predictable tendency of the advocates of agrobiodiversity to stress the positive impact of biodiversity on dietary health, in *Sec. 2.2.2*; the predictable tendency of those for the voluntary approach to target developing countries with these agreements so as "to remedy significant levels of non-compliance with mandatory regulation," in *Sec. 2.3.1*; the predictable tendency of those practitioners of adaptive management systems to call "for near-term action" with "a structured, iterative process of robust decision making in the face of uncertainty," in *Sec. 2.3.2*; the predictable tendency of some scientists to continue experimenting with cloning in our time, in *Sec. 2.4.1*; the predictable tendency of ecological economists to focus on investment in natural capital, in *Sec. 2.4.2*; the predictable tendency of anthropocentrists to "support…an understanding of nature as existing solely for the benefit of man and as a commodity to be used for the good of humanity and improved human quality of life," in *Sec. 2.5.1*; the predictable tendency of those experienced aquarists in PETA to care about "animal welfare," in *Sec. 2.5.2*; the predictable tendency by the advocates of the overkill hypothesis to claim the occurrence of the overexploitation of natural resources in our time, with its impact on extinction events, in *Sec. 3.2.1*; the predictable tendency of the advocates of climate change and biodiversity loss to argue for international efforts against extinction for sustainable development, in *Sec. 3.2.2*; the predictable tendency of the advocates of solar output to claim it as a major cause of climate change, in *Sec. 3.3.1*; the predictable tendency of those conservation activists to argue that species invastion leads to extinction of native species, in *Sec. 3.3.2*; the predictable tendency of those in the "2002 analysis of satellite imagery" to suggest that "the rate of deforestation in the humid tropics [approximately 5.8 million hectares per year] was roughly 23% lower than the most commonly quoted rates," in *Sec. 3.4.1*; the predictable tendency of those in the American Petroleum Institute, who "estimated in 1999 the world's oil supply would be depleted between 2062 and 2094," in *Sec. 3.4.2*; the predictable tendency of those for the "Convention on Biological Diversity [CBD]," to try to "to enable lesser-developed countries to better benefit from their resources and traditional knowledge," in *Sec. 3.5.1*; and the predictable tendency of those in uniformitarianism to regard extinction events as part of evolution over a long period of time, in *Sec. 3.5.2*).

On the other hand, there is unpredictability (e.g., the more difficult task to predict exactly how biodiversity works in the nonvisible, microscopic world, since "less than 1% of all species that have been described have been studied beyond simply noting their existence," while "the vast majority of Earth's species are microbial," in *Sec. 2.2.1*; the more difficult task to predict exactly the extent to which agrobiodiversity is contritutive to better dietary health on a particular future occasion, since scientists also discovered that agrobiodiveristy can contribute to dietary problems [via crop diseases] like the case of "coffee rust," in *Sec. 2.2.2*; the more difficult task to predict exactly the extent to which with the voluntary approach will be successful on a particular future occasion, "due to the difficulties inherent in evaluating effectiveness," such that "their use is often questioned and, indeed, the environment may well be adversely affected as a result," in *Sec. 2.3.1*; the more difficult task to predict exactly the extent to which the practitioners of adaptive management systems will be successful on a particular future occasion, because of the three kinds of "ecological uncertainty," namely, "structural uncertainty," "parameter uncertainty," and "stochastic uncertainty," in *Sec. 2.3.2*; the more difficult task to predict exactly the extent to which cloning will be successful on a particular future occasion, since, for the critics, there are already quite a number of failed attempts in the past, as David Ehrenfeld once warned that "cloning in animal conservation is an experimental technology that, at its state in 2006, could not be expected to work except by pure chance," in *Sec. 2.4.1*; the more difficult task to predict exactly the extent to which ecological economists will achieve its own objectives on a particular future occasion, since, for the critics, it can evolve over time to become more mainstream and betrays itself, contrary to its original goal, in *Sec. 2.4.2*; the more difficult task to predict exactly the extent to which anthropocentrism will succeed in a particular future era, since, for the critics, it would lead "to rampant selfishness and behaviour viewed as sinful," among other problems aforediscussed, in *Sec. 2.5.1*; the more difficult task to predict exactly the extent to which animal welfare will be successfully maintained on a particular future occasion, since there can be some "inexperienced aquarists" who "often keep too many fish in one tank, or add fish too quickly into an immature aquarium, killing many of them," in *Sec. 2.5.2*; the more difficult task to predict exactly the extent to which the overexploitation of natural resources will lead to extinction on a particular future occasion, since, for the critics, "overexploitation need not necessarily lead to the destruction of the resource, nor is it necessarily unsustainable," in *Sec. 3.2.1*; the more difficult task to predict exactly the extent to which international efforts against extinction for sustainable development will necessarily succeed on a particular future occasion, since, for the critics, "some scientists and non-scientists question aspects of climate-change science" and Bosselman cautioned us about "the remaining areas of scientific uncertainty" in regard to extinction and biodiverity loss, in *Sec. 3.2.2*; the more difficult task to predict exactly the extent to which solar output will be a major cause of climate change on a particular future occasion, since, for the critics, "the cyclical nature of the sun's energy output is not yet fully understood," and "a 2010 study suggests, 'that the effects of solar variability on temperature throughout the atmosphere may be contrary to current expectations,'" in *Sec. 3.3.1*; the more difficult task to predict exactly the extent to which species invastion will necessarily lead to extinction of native species on a particular future occasion, since, for the critics, "it is difficult to unequivocally attribute extinctions to a species invasion, and the few scientific studies that have done so have been with animal taxa," in *Sec. 3.3.2*; the more difficult task to predict exactly the extent to which "the rate of deforestation in the humid tropics," if relied on the "2002 analysis of satellite imagery," will be so on a particular future occasion, since "a newer analysis of satellite images reveals that deforestation of the Amazon rainforest is twice as fast as scientists previously estimated," in *Sec. 3.4.1*; the more difficult task to predict exactly the extent to which the estimate by the American Petroleum Institute will be accurate, since there was another "study published in the journal *Energy Policy* by researchers from Oxford University," who, "however, predicted demand would surpass supply by 2015" instead, in *Sec. 3.4.2*; the more difficult task to predict exactly the extent to which CBD will be successful in any attempt "to enable lesser-developed countries to better benefit from their resources and traditional knowledge" on a particular future occasion, because, for the critics, "the CBD has failed to establish appropriate regulations to prevent biopiracy" in *Sec. 3.5.1*; and the more difficult task to predict exactly the extent to which certain catastrophes will occur in a particular future historical period, since, for those in the scientific version of catastrophism, catastrophes are essentially "sudden," in *Sec. 3.5.2*).

And the reverse direction also holds true.

Again, it should be stressed, however, that there are different shades of gray (or different degrees of truth) in the two opposites, and the classification is not necessarily mutually exclusive either. In fact, even when some combinations of the two occur, they only end up sharing the same dialectic relationship, but in a different degree.

Also, there are some other relationships (as a kind of family resemblance) which have something in common with the principle, but they are not exactly the same but only more or less comparable, which varies from case to case.

Good examples of family resemblance in relation to the principle are sureness-arbitrariness, deterministicness-randomness, likeliness-riskiness, sureness-doubtfulness, deterministicness-indeterministicness, certainty-uncertainty, and so forth.

5th Thesis: The Explicability-Inexplicability Principle

The fifth thesis is called the explicability-inexplicability principle (on the underlying mechanisms of things) in the category of "method" in existential dialectics (which was proposed in *FPHU* and other books of mine), in that both explicability and inexplicability are part of the understanding of things. There is no explicability without inexplicability—and vice versa.

This principle tells us the dual nature of the research dilemma, in that, if reality can be explained in some ways, it also has its other ways which are not quite explainable, at a given point in time.

For instance, on the one hand, there is explicability (e.g., the explanation by the advocates of the "latitudinal gradients" in biodiversity on the basis of "the greater mean temperature at the equator compared to that of the poles," in *Sec. 2.2.1*; the explanation by those who advocated the causal relationship between agrobiodiversity and the Irish Potato Famine on the basis of the potato disease or blight, in *Sec. 2.2.2*; the explanation by K. Chan of the need for conservation on the basis of "the greatest good for the greatest number of people for the longest time," in *Sec. 2.3.1*; the explanation by the advocates of adaptive management that it is desirable on the basis of being able to carry out "near-term action" with "a structured, iterative process of robust decision making in the face of uncertainty," in *Sec. 2.3.2*; the explanation by the advocates of cloning that it is desirable on the basis of biodiversity, in *Sec. 2.4.1*; the explanation by the advocates of ecological economists of their ideology on the basis of stressing natural capital for sustainable development, in *Sec. 2.4.2*; the explanation by the advocates of environmental ethics of the importance of focusing on "the whole" of man-nature interactions, in *Sec. 2.5.1*; the explanation, by the advocates of biodiversity for its own sake, in that "biodiversity has intrinsic aesthetic and spiritual value to mankind in and of itself," in *Sec. 2.5.2*; the explanation by the advocates of the overkill hypothesis that overexploitation explains why extinction events occur in history, in *Sec. 3.2.1*; the explanation by the advocates of the relationship between biodiversity loss and health risks, in *Sec. 3.2.2*; the explanation, by the advocates of the human causes of climate change, on the basis of such factors like "emissions from fossil fuel combustion," "aerosols [particulate matter in the atmosphere]," "cement manufacture," "land use," "animal agriculture, and deforestation," in *Sec. 3.3.1*; the explanation by those in "one study" who "found that, of a list of invasive and noninvasive species, 86% of the invasive species could be identified from the traits alone," in *Sec. 3.3.2*; the explanation by those who "argue that poor people are more likely to clear forest because they have no alternatives," in *Sec. 3.4.1*; the explanation by those who argued that, unlike resource-rich countries, "resource-poor economies like Singapore, Taiwan or South Korea,…spent enormous efforts on education, and this contributed in part to their economic success," in *Sec. 3.4.2*; the explanation by the advocates of traditional knowledge about its naturalistic basis, that is, "a 'particular form of place-based knowledge of the diversity and interactions among plant and animal species, landforms, watercourses, and other qualities of the biophysical environment in a given place,'" in *Sec. 3.5.1*; and the explanation by those in creation theory that "the Universe, Earth, and all life on Earth were created by direct acts of the Abrahamic God during a relatively short period, sometime between 5,700 and 10,000 years ago," in *Sec. 3.5.2*).

On the other hand, there is inexplicability (e.g., the lack of sufficient explanation by the advocates of the "latitudinal gradients" in biodiversity, since the critics had pointed out some studies which instead showed that "this characteristic [of latitudinal gradients] is unverified in aquatic ecosystems, especially in marine ecosystems," that "the latitudinal distribution of parasites does not follow this rule," and "other instances of great diversity in higher latitudes have also been recorded," in *Sec. 2.2.1*; the lack of sufficient explanation, by those who advocated the causal relationship between agrobiodiversity and the Irish Potato Famine, of why it is necessarily true, since, for the critics, there are other main causes of the Irsh Potato Famine, and important factors include "laws that restricted the rights of Irish Catholics," the "middleman system," "tenants, subdivisions, and bankruptcy," "potato dependency," and so on, in *Sec. 2.2.2*; the lack of sufficient explanation by K. Chan of why utilitarian ethics is necessarily valid for conservation, since, for the critics, "nature has an intrinsic value that is independent of anthropocentric usefulness or utilitarianism," in *Sec. 2.3.1*; the lack of sufficient explanation by the advocates of adaptive management of why it is necessarily desirable, since, for the critics, adaptive management has not yet been widely accepted in the corporate world, because "it is so difficult to "model…an economically viable case for the switch," and, in addition, "deriving actively adaptive policies is technically very difficult, which prevents it being more commonly applied," in *Sec. 2.3.2*; the lack of sufficient explanation by the advocates of cloning of why it is necessarily desirable, because, for the critics, there remain some deep ethical and philosophical issues to be resolved, in *Sec. 2.4.1*; the lack of sufficient explanation by the advocates of ecological economists of why their own ideology is necessarily desirable, since, for the critics, there are alternative schools of thought, so it does not have the last word, in *Sec. 2.4.2*; the lack of sufficient explanation by the advocates of environmental ethics of why focusing on "the whole" of man-nature interactions is necessarily desirable, since, for the critics like Marti Kheel, "self-realization and identification with all nature places too much emphasis on the whole, at the expense of the independent being," in *Sec. 2.5.1*; the lack of sufficient explanation, by the advocates of biodiversity for its own sake, of why biodiversity necessarily "has intrinsic aesthetic and spiritual value to mankind in and of itself," since, for the critics, others disagree, because they instead suggest that "tropical forests and other ecological realms are only worthy of conservation because of the services they provide," in *Sec. 2.5.2*; the lack of sufficient explanation by the advocates of the overkill hypothesis of why overexploitation necessarily explains why exttinction events occur in history, since, for the critics, there can be other alternative causes like "habitat destruction," "genetic pollution/erosion," "climate change," "human overpopulation," etc., in *Sec. 3.2.1*; the lack of sufficient explanation by the advocates of why the relationship between biodiversity loss and health risks is necessarily valid, since, for the critics, there are many different causes of human health problems, just as there can be alternative resources for health improvement even when biodiversity occurs, in *Sec. 3.2.2*; the lack of sufficient explanation by the advocates of the human causes of climate change of why these factors as cited above are enough to understand environmental degradation, since, for example in India, the issue of environmental degradation involves numerous causes, and part of them has to do with "caste politics," "caste-related violence," etc., in *Sec. 3.3.1*; the lack of sufficient explanation by those in "one study" of why its finding about a strong correlation between "a list of invasive and noninvasive species" and the "traits" used, since, for the critics, "another study found invasive species tended only to have a small subset of the presumed traits, and that many such traits were found in noninvasive species, requiring other explanations," in *Sec. 3.3.2*; the lack of sufficient explanation, by those who blame the poor for deforestation, of why this is necessarily true, since, for the critics, "the poor lack the ability to pay for the materials and labour needed to clear forest," in *Sec. 3.4.1*; the lack of sufficient explanation by those about "resource-poor economies like Singapore, Taiwan or South Korea" of why this correlation between resource-poor economies and economic success is necessarily valid, because "other researchers, however, dispute this conclusion," as "they argue that natural resources generate easily taxable rents that more often than not result in increased spending on education" in resource-rich economies, in *Sec. 3.4.2*; the lack of sufficient explanation by the advocates of traditional knowledge of why it is necessarily true that traditional knowledge is naturalistic, because, for the critics, "traditional knowledge is not a natural category, and may reflect power struggles and relationships for land, resources and social control rather than adherence to a claimed ancestry or heritage," in *Sec. 3.5.1*; and the lack of sufficient explanation by those in creation theory of why "the Universe, Earth, and all life on Earth" were necessarily "created by direct acts of the Abrahamic God during a relatively short period," because, for the critics, "the consensus among scientists is that creation science is unscientific in both conception and methodology," in *Sec. 3.5.2*).

And the reverse direction also holds true.

Once more, it should be stressed, however, that there are different shades of gray (or different degrees of truth) in the two opposites, and the classification is not necessarily mutually exclusive either. In fact, even when some combinations of the two occur, they only end up sharing the same dialectic relationship, but in a different degree.

Also, there are some other relationships (as a kind of family resemblance) which have something in common with the principle, but they are not exactly the same but only more or less comparable, which varies from case to case.

Good examples of family resemblance in relation to the principle are underlyingness-regularness, causation-regularity, causation-correlation, explanation-description, explanation-narration, explanation-interpretation, and so forth.

6th Thesis: The Fiction-Reality Principle

The sixth thesis is called the fiction-reality principle (on the different faces of reality) in the category of "method" in existential dialectics (as first proposed in *FPHLIT*), in that both fiction and reality co-exist. There is no reality without fiction—and vice versa.

This principle tells us the different categories of reality, in that, if there is reality, there is also some fiction in it—just as, if there is fiction, there is also some reality in it.

For instance, on the one hand, there is fiction (e.g., the fictional aspect of the "latitudinal gradients" in biodiversity, insofar as "the latitudinal distribution of parasites does not follow this rule," and "other instances of great diversity in higher latitudes have also been recorded," in *Sec. 2.2.1*; the fictional aspect of the argument for agrobiodiversity as contritutive to better dietary health, insofar as scientists also discovered that agrobiodiveristy can contribute to dietary problems [via crop diseases] like the case of "coffee rust," in *Sec. 2.2.2*; the fictional aspect of the ecosystems approach, insofar as it is "idealistic" in calling for "a broad range of stakeholders across all relevant governmental departments, as well as representatives of industry, environmental groups and community," in *Sec. 2.3.1*; the fictional aspect of adaptive management, insofar as, if any of its processes do not work out as planned, the consequences are not pretty at all, as the failures "can seriously compromise effective adaptive management decision making," in *Sec. 2.3.2*; the fictional aspect of cloning, insofar as there were already various failed attempts in the past, especially in regard to the artificial cloning of organisms, including human cloning, in *Sec. 2.4.1*; the fictional aspect of the "attempted valuation of the global ecosystem in 1997" by Robert Costanza, insofar as it "concluded on $33 trillion with a range from $16 trillion to $54 trillion," which the critics rejected as speculative in *Ecological Economics*, Volume 25, Issue 1," in *Sec. 2.4.2*; the fictional aspect of ecocentrism in deep ecology, insofar as its reasoning is not "deep" enough, because, for the critic like Murray Bookchin [1987], "deep ecology fails to link environmental crises with authoritarianism and hierarchy" and therefore fails to see that "environmental problems are firmly rooted in the manner of human social interaction," so "an ecologically sustainable society could still be socially exploitative," in *Sec. 2.5.1*; the fictional aspect of the view by the RSPB in the older days, when it falsely "claimed that the 'scientification' of the pastime was 'undesirable,'" as "this stand was to change only in 1936 when the RSPB was taken over by Tom Harrisson and others," in *Sec. 2.5.2*; the fictional aspect of the classification of "the 'Big Five' [extinction events]" by Jack Sepkoski and David M. Raup, insofar as "the 'Big Five' [extinction events]…cannot be so clearly defined" when "more stringent statistical tests have been applied to the accumulating data," in *Sec. 3.2.1*; the fictional aspect of the claim about the relationship between climate change and health risks, insofar as "the Letter from the Royal Society" in 2005 showed some "fictions about climate change," in *Sec. 3.2.2*; the fictional aspect of the claim about the "solar output" as a cause of climate change, since, for the critics, "a 2010 study suggests, 'that the effects of solar variability on temperature throughout the atmosphere may be contrary to current expectations,'" and even the United States National Research Council recognized in 2010 that "there are still some uncertainties" about "climate change," in *Sec. 3.3.1*; the fictional aspect of the argument that invasive species leads to extinction of native species, insofar as the critics pointed out that "it is difficult to unequivocally attribute extinctions to a species invasion, and the few scientific studies that have done so have been with animal taxa," in *Sec. 3.3.2*; the fictional aspect of "the FAO data" about deforestation, insofar as "they do not distinguish between forest types, and that they are based largely on reporting from forestry departments of individual countries, which do not take into account unofficial activities like illegal logging," in *Sec. 3.4.1*; the fictional aspect of the estimate by the 1999 study published in the journal *Energy Policy*, insofar as its prediction that the world's oil demand "would surpass supply by 2015" is too pessimistic, in *Sec. 3.4.2*; the fictional aspect of biopiracy, insofar as there is "one common misunderstanding," in that "pharmaceutical companies patent the plants they collect," but the reality is that "patents may be taken out on specific chemicals isolated or developed from plants" and "often these patents are obtained with a stated and researched use of those chemicals," so "even if the indigenous medical knowledge is taken..., that knowledge does not by itself make the active chemical compound 'obvious,' which is the standard applied under patent law," in *Sec. 3.5.1*; and the fictional aspect of uniformitarianism, insofar as "Earth's history is...punctuated by occasional natural catastrophic events," contrary to the uniformitarian claim, in *Sec. 3.5.2*).

On the other hand, there is reality (e.g., the realistic aspect of the "latitudinal gradients" in biodiversity, insofar as many terrestrial ecosystems follow the rule of latitudinal gradients, in *Sec. 2.2.1*; the realistic aspect of the argument for agrobiodiversity as contritutive to better dietary health, insofar as, in some estimates, "about 80 percent of humans' food supply comes from just 20 kinds of plants" and "humans use at least 40,000 species," in *Sec. 2.2.2*; the realistic aspect of the ecosystems approach, insofar as it stresses the need to address "the complex interrelationships of an entire ecosystem in decision making rather than simply responding to specific issues and challenges," in *Sec. 2.3.1*; the realistic aspect of adaptive management, insofar as it is willing to deal with "decision making" in regard to ecological systems, while acknowledging the existence of high "uncertainty," in *Sec. 2.3.2*; the realistic aspect of cloning, insofar as there are successful cases of cloning, especially in regard to horticultural cloning, in *Sec. 2.4.1*; the realistic aspect of the "attempted valuation of the global ecosystem in 1997" by Robert Costanza, insofar as "the critics acknowledged the positive potential for economic valuation of the global ecosystem," in *Sec. 2.4.2*; the realistic aspect of ecocentrism in deep ecology, insofar as it points out the need to consider "the intrinsic value of nature while maintaining an understanding that 'human beings must use and even exploit nature to survive and live," in *Sec. 2.5.1*; the realistic aspect of the view by the RSPB, when "Harrisson was instrumental in the organization of pioneering surveys of the Great Crested Grebe," instead of dismissing the study of ecology and animal behaviour as was the case in the older days, in *Sec. 2.5.2*; the realistic aspect of the classification of "the 'Big Five' [extinction events]" by Jack Sepkoski and David M. Raup, insofar as extinction events did occur in history, in *Sec. 3.2.1*; the realistic aspect of the claim about the relationship between climate change and health risks, insofar as "the Letter from the Royal Society" in 2005 also showed some "facts…about climate change," in *Sec. 3.2.2*; the realistic aspect of the claim about the "solar output" as a cause of climate change, insofar as "solar intensity variations are considered to have been influential in triggering the Little Ice Age, and some of the warming observed from 1900 to 1950," in *Sec. 3.3.1*; the realistic aspect of the argument that invasive species leads to extinction of native species, insofar as there are "the few scientific studies…with animal taxa" that backed up the claim, in *Sec. 3.3.2*; the realistic aspect of "the FAO data," insofar as they pointed out some instances of deforestation, in *Sec. 3.4.1*; the realistic aspect of the estimate by the 1999 study published in the journal *Energy Policy*, insofar as oil depletion is a long-term problem which is to be taken seriously, in *Sec. 3.4.2*; the realistic aspect of biopiracy, insofar as there are other cases when biopirates used, for profit, "indigenous knowledge…originating with indigenous people…without permission from and with little or no compensation or recognition to the indigenous people themselves," in *Sec. 3.5.1*; and the realistic aspect of uniformitarianism, insofar as there is some truth that "Earth's history is a slow, gradual story"in *Sec. 3.5.2*).

And the reverse direction also holds true.

Once more, it should be stressed, however, that there are different shades of gray (or different degrees of truth) in the two opposites, and the classification is not necessarily mutually exclusive either. In fact, even when some combinations of the two occur, they only end up sharing the same dialectic relationship, but in a different degree.

Also, there are some other relationships (as a kind of family resemblance) which have something in common with the principle, but they are not exactly the same but only more or less comparable, which varies from case to case.

Good examples of family resemblance in relation to the principle are fiction-nonfiction, fiction-fact, imagination-fact, imagination-nonfiction, subjectivity-objectivity, and so forth.

7th Thesis: The Cognitiveness-Noncognitiveness Principle

The seventh thesis is called the cognitiveness-noncognitiveness principle (on the understanding of meaning) in the category of "method" in existential dialectics (as first proposed in *BNR*), in that, if a view of things is based on reason and evidences, it is also affected by non-cognitive factors (such that it is neither true nor false in the latter case)—

and vice versa. In other words, there is no cognitiveness without noncognitiveness—and vice versa.

Noncognitive factors can come in all shapes and sizes, and good examples include envy, jealousy, power, nationality, race, gender, age, class, greed, lust, status, faith, anger, sadness, joy, fear, wish, etc.

Communication, even in its ideal type, is seldom based on reason and evidences only, as falsely assumed during the Age of the Enlightenment and its continued supporters nowadays. What is understood about something is often twisted on the basis of something else too.

For instance, on the one hand, there is cognitiveness (e.g., a view of things on the basis of reason and evidences, as shown in the empirical study by Fuller and his colleagues on "the…associations [between psychological benefits and biodversity]," in *Sec. 2.2.1*; a view of things on the basis of reason and evidences, as shown in the study of the Irish Potato Famine, which found that "the proximate cause of famine was a potato disease commonly known as potato blight," in *Sec. 2.2.2*; a view of things on the basis of reason and evidences, as shown in the study of environmental protection of natural resources in Tanzania in the article aforecited, in *Sec. 2.3.1*; a view of things on the basis of reason and evidences, as shown in the use of adaptive management for dealing with ecological isues in the face of uncertainty, in *Sec. 2.3.2*; a view of things on the basis of reason and evidences, as shown in some well known cases of cloning like the ones about "carp" in China in 1963, "Dolly the sheep" in Scotland in 1996, "water buffalo" in the Philippines in 2007, and so on, in *Sec. 2.4.1*; a view of things on the basis of reason and evidences, as shown in the use of evidences in ecological economics, like the consideration of "the Earth's carrying capacity," in that, for example, "food production may become a problem, as erosion, an impending water crisis, and soil salinity [from irrigation] reduce the productivity of agriculture," in *Sec. 2.4.2*; a view of things on the basis of reason and evidences, as shown in the case of anthropocentrism, when its advocates "point out that maintenance of a healthy, sustainable environment is necessary for human well-being," in *Sec. 2.5.1*; a view of things on the basis of reason and evidences, as shown in the study of birds on the basis of "scientification," in *Sec. 2.5.2*; a view of things on the basis of reason and evidences, as shown in the study of the five extinction events, in *Sec. 3.2.1*; a view of things on the basis of reason and evidences, as shown in the "facts…about climate change" provided by "the Letter from the Royal Society" in 2005, in *Sec. 3.2.2*; a view of things on the basis of reason and evidences, as shown in the scientific report "Advancing the Science of Climate Change" by the United States National Research Council, in *Sec.*

3.3.1; a view of things on the basis of reason and evidences, as shown in the two scientific studies by "researchers" who "disagree about the usefulness of traits as invasiveness markers," in *Sec. 3.3.2*; a view of things on the basis of reason and evidences, as shown in the historical study which showed that "some countries in the developed world, such as the United States of America cut down their forests centuries ago and benefited greatly from this deforestation," in *Sec. 3.4.1*; a view of things on the basis of reason and evidences, as shown in the data from OPEC in 2011, which showed that the major oil resreves were in Venezuela, Saudi Arabia, Canada, and so on, in *Sec. 3.4.2*; a view of things on the basis of reason and evidences, as shown in the study of traditional knowledge on the basis of its "naturalistic" representation, that is, "a 'particular form of place-based knowledge of the diversity and interactions among plant and animal species, landforms, watercourses, and other qualities of the biophysical environment in a given place,'" in *Sec. 3.5.1*; and a view of things on the basis of reason and evidences, as shown in the rational understanding by "most theologians" who are willing to "reject a literal interpretation of the creation narrative in Genesis" and "propose there are statements in the creation week itself which render the historical interpretation of Genesis incompatible with scientific evidence" in *Sec. 3.5.2*).

On the other hand, there is noncognitiveness (e.g., a view of things on the basis of non-cognitive factors like envy, jealousy, power, nationality, race, gender, age, class, greed, lust, anger, sadness, joy, fear, wish, etc., as shown in the study by R. Fuller and his colleagues, when they found that, "it is unlikely that people were randomly distributed among parks, perhaps preferring those that enhance their psychological well-being," so "selection by individuals of environments that enhance personal well-being makes interpretation of the causal structure of these relationships difficult," because of this factor of personal preference, in *Sec. 2.2.1*; a view of things on the basis of non-cognitive factors like envy, jealousy, power, nationality, race, gender, age, class, greed, lust, anger, sadness, joy, fear, wish, etc., as shown in the study of the Irish Potato Famine, when "the impact and human cost in Ireland…was exacerbated by a host of political, social and economic factors which remain the subject of historical debate" among different groups today, in *Sec. 2.2.2*; a view of things on the basis of non-cognitive factors like envy, jealousy, power, nationality, race, gender, age, greed, lust, class, anger, sadness, joy, fear, wish, etc., as shown in the study of environmental protection of natural resources in Tanzania, when researchers discovered that "African governments face several challenges in implementing environmental protection mechanisms," and "in Tanzania for example these include…corruption" because of greed in *Sec. 2.3.1*; a view of things on the basis of non-cognitive factors like envy, jealousy, power, nationality, race, gender, age, class, greed, lust, anger, sadness, joy, fear, wish, etc., as shown in the use of adaptive management in environmental management by some companies, which accept the switch, not because of the soundness of the techniques, but because it can promote its business reputation as environmentally friendly, that is, as a "strategic advantage" for its profit-seeking purpose, in *Sec. 2.3.2*; a view of things on the basis of non-cognitive factors like envy, jealousy, power, nationality, race, gender, age, class, greed, lust, anger, sadness, joy, fear, wish, etc., as shown in the fraudulent 2005 research on "the world's first cloned dog" by South Korea's top human cloning scientist Hwang Woo-suk, which was later found to be "fabricated" instead, in *Sec. 2.4.1*; a view of things on the basis of non-cognitive factors like envy, jealousy, power, nationality, race, gender, age, class, greed, lust, anger, sadness, joy, fear, wish, etc., as shown in the fear embedded in many ecological economists who "are inclined to be technological pessimists," in *Sec. 2.4.2*; a view of things on the basis of non-cognitive factors like envy, jealousy, power, nationality, race, gender, age, class, greed, lust, anger, sadness, joy, fear, wish, etc., as shown in the case of anthropocentrism, when "some Christian proponents of anthropocentrism base their belief on the Bible, such as the verse 1:26 in the Book of Genesis," that is, on the basis of religious faith, not of empirical proof, in *Sec. 2.5.1*; a view of things on the basis of non-cognitive factors like envy, jealousy, power, nationality, status, race, gender, age, class, greed, lust, anger, sadness, joy, fear, wish, etc., as shown in the research finding that, for some birdwatchers, "identification of birds may be a form of gaining status which has been compared with Kula valuables noted in Papua New Guinean cultures," in *Sec. 2.5.2*; a view of things on the basis of non-cognitive factors like envy, jealousy, power, nationality, race, gender, age, class, greed, lust, status, anger, sadness, joy, fear, wish, etc., as shown in the motivation of some conservation activists who try to use "biodiversity…as a legal standard" to enforce conservation against extinction, without taking into account of "the remaining areas of scientific uncertainty" in regard to extinction and biodiverity loss, in *Sec. 3.2.1*; a view of things on the basis of non-cognitive factors like envy, jealousy, power, nationality, race, gender, age, class, greed, lust, status, anger, sadness, joy, fear, wish, etc., as shown by different organizations like "the libertarian Competitive Enterprise Institute, conservative commentators, and some companies such as ExxonMobil" that have different interests or motives and thus "have challenged IPCC climate change scenarios," in *Sec. 3.2.2*; a view of things on the basis of non-cognitive factors like envy, jealousy, power, nationality, race, gender, age, class, greed, lust, status, faith, anger, sadness, joy, fear, wish, etc., as shown in the argument based on "faith" by Dennis Dinger [2010] that "the Bible has quite a lot to say" about global warming, because "the Bible says God created the heavens and the earth, the Bible teaches that God maintains the daily workings of the earth,...and finally, it says that God controls the wind and the rain," in *Sec. 3.3.1*; a view of things on the basis of non-cognitive factors like envy, jealousy, power, nationality, race, gender, age, class, greed, lust, status, faith, anger, sadness, joy, fear, wish, etc., as shown in the "activism" of certain groups such as "Greenpeace" and others with their "unequivocal" attribution of "extinctions to a species invasion," such that it is more for propaganda, less for science, in *Sec. 3.3.2*; a view of things on the basis of non-cognitive factors like envy, jealousy, power, nationality, race, gender, age, class, greed, lust, status, faith, anger, sadness, joy, fear, wish, etc., as shown in the hypocrisy of "some countries in the developed world, such as the United States of America" which "cut down their forests centuries ago and benefited greatly from this deforestation" but now "deny developing countries the same opportunities," on the basis of the power politics of the North-South divide, in *Sec. 3.4.1*; a view of things on the basis of non-cognitive factors like envy, jealousy, power, nationality, race, gender, age, class, greed, lust, status, faith, anger, sadness, joy, fear, wish, etc., as shown in the wishful thinking embedded in some estimates, as the critics pointed out that "there are serious doubts about whether OPEC countries really have the oil reserves they claim," and "this is similar to the illusionary oil reserves that U.S. oil companies claimed to have in the decade prior to the 1973 and 1979 oil crisis," but "in the 1970s, those companies were unable to produce as much oil as they had predicted, and production went down instead of up," in *Sec. 3.4.2*; a view of things on the basis of non-cognitive factors like envy, jealousy, power, nationality, race, gender, age, class, greed, lust, status, faith, anger, sadness, joy, fear, wish, etc., as shown in the criticism that "traditional knowledge is not a natural category, and may reflect power struggles and relationships for land, resources and social control rather than adherence to a claimed ancestry or heritage," that is, on the basis of power, in *Sec. 3.5.1*; and a view of things on the basis of non-cognitive factors like envy, jealousy, power, nationality, race, gender, age, class, greed, lust, status, faith, anger, sadness, joy, fear, wish, etc., as shown in the reliance on "faith" by "most young Earth creationists today" who still believe in "the apparent age concept," which "is still used in young Earth creationist literature," in *Sec. 3.5.2*).

And the reverse direction also holds true.

Once more, it should be stressed, however, that there are different shades of gray (or different degrees of truth) in the two opposites, and the classification is not necessarily mutually exclusive either. In fact, even when some combinations of the two occur, they only end up sharing the same dialectic relationship, but in a different degree.

Also, there are some other relationships (as a kind of family resemblance) which have something in common with the principle, but they are not exactly the same but only more or less comparable, which varies from case to case.

Good examples of family resemblance in relation to the principle are proposition-attitude, cognition-emotion, cognition-prescription, theory-extratheory, and so forth.

In Relation to Structure

The next five theses are in relation to structure in existential dialectics, as shown below.

8th Thesis: The Finiteness-Transfiniteness Principle

The eighth thesis is called the finiteness-transfiniteness principle (on the nature of numbers) in the category of "structure" in existential dialectics (as first proposed in *FPHG*), in that, if there are finite things, there are likewise transfinite ones—and vice versa.

To avoid confusion, my usage of the word "transfinite" here differs radically from the one used by Cantor (and other mathematicians) for "relative" infinity—and is more limited, in light of the problems confronting any attempt to understand the idea of infinity, be it by intuition, imagination, and conception (as detailedly analyzed in *Sec. 2.2.3* of *FPHG*).

Instead, by "transfinity," I allow numbers which can be many times larger—or smaller, for that matter—than the finite things that we encounter in daily life, but they do not have to be related to the idea of infinity at all (which may not exist).

Of course, there may be some *borderline* cases, in which it is not clear whether the number in question is transfinite (in my usage) or simply a mathematical convenience. A good example of a borderline case is the Planck unit of length for "the smallest space possibly measured in nature," which is “less than billionths of trillionths of trillionths of an inch” (or something like 1.6 × 10^{-35} meters). (P. Baofu 2006a; N. McAleer 1987: 219; WKV 2008; D. Corbett 2008a)

That said—my usage of “transfinity” can also resolve (or better, dissolve) an age-old problem in philosophy known as “Zeno's paradoxes” (as already explained in *Sec. 2.2.3* of *FPHG*).

In any event, in relation to the finiteness-transfiniteness principle, on the other hand, there is finiteness (e.g., the finite number of the levels of biodiversity, namely, “molecular,” “genetic,” “species,” and “ecosystem,” in *Sec. 2.2.1*; the finite number of the kinds of support by agrobiodiversity, like the 3 kinds in the article aforecited, in *Sec. 2.2.2*; the finite number of the categories of concerns with biodiversity in conservation efforts, like “genes,” “species,” and “ecosystems” in the article aforecited, in *Sec. 2.3.1*; the finite number of the features and processes of adpative management, as shown in the 5 features and 5 processes aforecited, in *Sec. 2.3.2*; the finite number of the categories of cloning, like “molecular cloning,” “cell cloning,” and “organism cloning,” in *Sec. 2.4.1*; the finite number of the categories of capital in economics, like “natural capital,” “human-made capital,” etc., in *Sec. 2.4.2*; the finite number of the categories of lifeforms in the man-nature interactions as understood in ecocentrism, such as microorganisms, plants, animals, humans, etc., in *Sec. 2.5.1*; the finite number of the categories of water that fish can live in, namely, “freshwater, brackish, and marine,” in *Sec. 2.5.2*; the finite number of the categories of lifeforms which suffered from extinction, both partial and complete, during the “Big Five” [extinction events], like insects, worms, fish, amphibians, birds, mammals, etc., in *Sec. 3.2.1*; the finite number of the categories of “naturally occurring microbes” or microorganisms in the world of biodiversity, like “bacteria, viruses, fungi and protozoa,” in *Sec. 3.2.2*; the finite number of the categories of particles which have played a major role in global warming, like “carbon dioxide,” “aerosol,” etc., in *Sec. 3.3.1*; the finite number of the categories of species involved in genetic pollution, like “domestic, feral, non-native and invasive species,” in *Sec. 3.3.2*; the finite number of the categories of entities affected by deforestation, like soils, water, trees, species, etc., in *Sec. 3.4.1*; the finite number of the categories of entities which were converted into petroleum compounds over many millions of years, like “plankton,” “bacteria,” and “other plant and animal matter,” in *Sec. 3.4.2*; the finite number of the categories of entities addressed by traditional knowledge, namely, “plants,” “animals,” etc., in *Sec. 3.5.1*; and the finite number of the categories of things in the great chain of being, like God, angels, humans, animals, plants, and minerals, in *Sec. 3.5.2*).

On the other hand, there is transfinitty (e.g., the transfinite number of all the individual organisms, species, and ecosystems that have ever existed at the four levels of biodiversity in the history of the Earth, in *Sec. 2.2.1*; the transfinite number of all the individual microorganisms, species, and ecosystems that have ever existed in the history of agrobiodiversity on Earth, in *Sec. 2.2.2*; the transfinite number of all the individual "genes," "species," and "ecosystems" that have ever existed in the history of biodiversity on Earth, in *Sec. 2.3.1*; the transfinite number of all the individual genes, microorganisms, species, habitats, ecosystems, etc. that have ever existed in the history of ecology on Earth and which can be studied for adaptive management, in *Sec. 2.3.2*; the transfinite number of all the individual molecules, cells, and organisms that have ever been cloned in history, both naturally and artifically, in *Sec. 2.4.1*; the transfinite number of all the individual units of natural capital, human-made capital, etc. in history that have ever existed on Earth, in *Sec. 2.4.2*; the transfinite number of all the individual microorganisms, plants, animals, humans, etc. in the history of man-nature interaction on Earth that have ever existed, in *Sec. 2.5.1*; the transfinite number of all the individual fish in history of the Earth that have ever existed, regardless of whether there are in freshwater, brackish, or marine environments, in *Sec. 2.5.2*; the transfinite number of all the individual insects, worms, fish, amphibians, birds, mammals, etc. in history of the Earth that had ever suffered from extinction, both partial and complete during the "Big Five" [extinction events], in *Sec. 3.2.1*; the transfinite number of all the individual "bacteria, viruses, fungi and protozoa" in the natural history of the Earth that have ever existed in the world of biodiversity, in *Sec. 3.2.2*; the transfinite number of all the individual particles of "carbon dioxide," "aerosol," etc. in history of the Earth that have played a major role in global warming, in *Sec. 3.3.1*; the transfinite number of all the individual animals which are "domestic," "feral," "non-native" and "invasive" in history that have ever existed in relation to genetic pollution, in *Sec. 3.3.2*; the transfinite number of all the individual pieces of minerals in soils, drops of water, leaves on trees, animals in species, etc. in history that have ever existed and been affected by deforestation, in *Sec. 3.4.1*; the transfinite number of all the individual organic remains from "plankton," "bacteria," and "other plant and animal matter" in history that have ever existed and been converted into petroleum compounds, in *Sec. 3.4.2*; the transfinite number of all the individual "plants," "animals," etc. in history that have ever existed and are included under traditional knowledge, in *Sec. 3.5.1*; and the transfinite number of all the individual humans, animals, plants, and minerals in history that have ever existed, in *Sec. 3.5.2*).

And the reverse direction also holds true.

Once more, it should be stressed, however, that there are different shades of gray (or different degrees of truth) in the two opposites, and the classification is not necessarily mutually exclusive either. In fact, even when some combinations of the two occur, they only end up sharing the same dialectic relationship, but in a different degree.

Also, there are some other relationships (as a kind of family resemblance) which have something in common with the principle, but they are not exactly the same but only more or less comparable, which varies from case to case.

Good examples of family resemblance in relation to the principle are boundedness-quasiunboundedness, smallness-largeness, microness-macroness, boundedness-unboundedness, and the like.

9th Thesis: The Preciseness-Vagueness Principle

The ninth thesis is called the preciseness-vagueness principle (on the refinement of things) in the category of "structure" in existential dialectics (which was first out worked out in *FIA*), in that both preciseness and vagueness are important, not that one is better than the other, but that both are used, in different degrees of preference, in accordance to the contextual application from the perspectives of nature, the mind, culture, and society. There is no preciseness without vagueness—and vice versa.

For instance, on the one hand, there is preciseness (e.g., the precise identification of four levels of biodiversity in the article aforecited, in *Sec. 2.2.1*; the precise identification of three major scopes of agrobiodiversity in the article aforecited, in *Sec. 2.2.2*; the precise identification of six stages in the systematic planning" of the protection of natural resources in the article aforecited, in *Sec. 2.3.1*; the precise identification of 5 features of adaptive management in the article aforecited, in *Sec. 2.3.2*; the precise identification of 4 steps in molecular cloning in the article aforecited, like "fragmentation," "ligation," "transfection" and "screening/selection," in *Sec. 2.4.1*; the precise identification of 3 opposing schools of economic thought in the context of natural resources in the article aforecited, namely, "neoclassical economics," "environmental economics," and "ecological economics," in *Sec. 2.4.2*; the precise identification of 3 roots of anthropocentrism in the article aforecited, in *Sec. 2.5.1*; the precise identification of three types of water that fish can live in, namely, "freshwater, brackish, and marine" in the article aforecited, in *Sec. 2.5.2*;

the precise identification of three major types of natural resources with the problem of overexploitation, like water, forests, and fish, in the article aforecited, in *Sec. 3.2.1*; the precise identification of 3 cases of the relationsihp between biodiversity loss and health risks in the article aforcited, like "infectious disease," "medicinal resources," and "disaster prevention, relief, and recover," in *Sec. 3.2.2*; the precise identification of 6 causes of climate change in the aritcle aforecited, in *Sec. 3.3.1*; the precise identification of 4 main characteristics of species invasion in the article aforecited, in *Sec. 3.3.2*; the precise identification of 3 historical eras in regard to the causes of deforestation in the article aforecited, in *Sec. 3.4.1*; the precise identification of 3 scenarios for the prediction of the impacts of oil depletion in the article aforecited, namely, "catastrophe," "recession," and "increasing food prices," in *Sec. 3.4.2*; the precise identification of 3 characteristics of traditional knowledge in the article aforecited, in *Sec. 3.5.1*; and the precise identification of 4 views about extinction in the article aforecited, in *Sec. 3.5.2*).

On the other hand, there is vagueness (e.g., the vagueness in the identification of the four levels of biodiversity in the article aforecited, since it is not clear why there must be only 4, not 5, 6, 7, and so on, in *Sec. 2.2.1*; the vagueness in the identification of the three major scopes of agrobiodiversity in the article aforecited, since it is not clear why there must be only 3, not 4, 5, 6, and so on, in *Sec. 2.2.2*; the vagueness in the identification of the six stages in the systematic planning" of the protection of natural resources in the article aforecited, since it is not clear why there must be only 6, not 7, 8, 9, and so on, in *Sec. 2.3.1*; the vagueness in the identification of the 5 features of adpative management in the article aforecited, since it is not clear why there must be only 5, not 6, 7, 8, and so on, in *Sec. 2.3.2*; the vagueness in the identification of the 4 steps in molecular cloning in the article aforecited, since it is not clear why there must be only 4, not 5, 6, and so on, since, for the critics, "a number of alternative routes can be selected," in *Sec. 2.4.1*; the vagueness in the identification of the 3 opposing schools of economic thought in the context of natural resources in the article aforecited, namely, "neoclassical economics," "environmental economics," and "ecological economics, since it is not clear why there must be only 3, not 4, 5, 6, and so on in *Sec. 2.4.2*; the vagueness in the identification of the 3 roots of anthropocentrism in the article aforecited, since it is not clear why there must be only 3, not 4, 5, 6, and so on, in *Sec. 2.4.2*; the vagueness in the identification of the three types of water that fish can live in, namely, "freshwater, brackish, and marine" in the article aforecited, since it is not clear why there must be only 3, not 4, 5, 6, and so on, in *Sec. 2.5.2*; the vagueness in the identification of the three major types of natural resources with the problem of overexploitation, like water, forests, and fish, in the article aforecited, since it is not clear why there must be only 3, not 4, 5, 6, 7, 8, and so on, in *Sec. 3.2.1*; the vagueness in the identification of the 3 cases of the relationsihp between biodiversity loss and health risks in the article aforcited, since it is not clear why there must be only 3, not 4, 5, 6, and so on, in *Sec. 3.2.2*; the vagueness in the identification of the 6 causes of climate change in the aritcle aforecited, since it is not clear why there must be only 5, not 6, 7, 8, and so on, in *Sec. 3.3.1*; the vagueness in the identification of the 4 main characteristics of species invasion in the article aforecited, since it is not clear why there must be only 4, not 6, 7, 8. and so on, in *Sec. 3.3.2*; the vagueness in the identification of the 3 historical eras in regard to the causes of deforestation in the article aforecited, since it is not clear why there must be only 3, not 4, 5, 6, and so on, in *Sec. 3.4.1*; the vagueness in the identification of the 3 scenarios for the prediction of the impacts of oil depletion in the article aforecited, since it is not clear why there must be only 3, not 4, 5, 6, and so on, since, for the critics, there can be another alternative like the replacement with renewable energy, in *Sec. 3.4.2*; the vagueness in the identification of the 3 characteristics of traditional knowledge in the article aforecited, since it is not clear why there must be only 3, not 4, 5, 6, and so on, in *Sec. 3.4.2*; and the vagueness in the identification of the 4 views about extinction in the article aforecited, since it is not clear why there must be only 4, not 5, 6, 7, and so on, in *Sec. 3.5.2*).

And the reverse direction also holds true.

Once more, it should be stressed, however, that there are different shades of gray (or different degrees of truth) in the two opposites, and the classification is not necessarily mutually exclusive either. In fact, even when some combinations of the two occur, they only end up sharing the same dialectic relationship, but in a different degree.

Also, there are some other relationships (as a kind of family resemblance) which have something in common with the principle, but they are not exactly the same but only more or less comparable, which varies from case to case.

Good examples of family resemblance in relation to the principle are clarity-ambiguity, directness-indirectness, quantitativeness-qualitativeness, specificity-obscurity, thickness-thinness, describability-nondescribability, concretness-abstractness, translatability-untranslatability, specificity-generality, clearcutness-transitionalness, and the like.

10th Thesis: The Simpleness-Complicatedness Principle

The tenth thesis refers to the simpleness-complicatedness principle (on the interconnection among things) in the category of "structure" in existential dialectics (which was first out worked out in *FIA*), in that both simpleness and complicatedness are vital, without favoring one over the other, and each is utilized, depending on the basis of the perspectives of nature, the mind, culture, and society. There is no simpleness without complicatedness—and vice versa.

For instance, on the one hand, there is simpleness (e.g., the relatively simple analysis of biodiversity, when "less than 1% of all species that have been described have been studied beyond simply noting their existence," and "contemporary biodiversity physics is 'firmly fixated on the visible [macroscopic] world,'" but "the vast majority of Earth's species are microbial" and non-visible, in *Sec. 2.2.1*; the relatively simple analysis of the Irish Potato Famine on the basis of the potato disease or blight only, in *Sec. 2.2.2*; the relatively simple analysis of utilitarian ethics for conservation efforts by K. Chan, in *Sec. 2.3.1*; the relatively simple analysis of adaptive management by its advocates, in *Sec. 2.3.2*; the relatively simple analysis of cloning by the advocates, in *Sec. 2.4.1*; the relatively simple analysis of ecological economics by its advocates, in *Sec. 2.4.2*; the relatively simple analysis of each school of environmental ethics by its advocates, in *Sec. 2.5.1*; the relatively simple analysis of animal watching by the advocates, in *Sec. 2.5.2*; the relatively simple analysis of the overkill hypothesis by the advocates, in *Sec. 3.2.1*; the relatively simple analysis of the relationsihp between biodiversity loss and health risks by the advocates, in *Sec. 3.2.2*; the relatively simple analysis of climate change by environmental activists, in *Sec. 3.3.1*; the relatively simple analysis of genetic pollution by some conservation activists, in *Sec. 3.3.2*; the relatively simple analysis of deforestiation by the conservation activists, in *Sec. 3.4.1*; the relatively simple analysis of oil depletion by the advocates, in *Sec. 3.4.2*; the relatively simple analysis of traditional knowledge by the advocates, in *Sec. 3.5.1*; and the relatively simple analysis of creation theory by the advocates, in *Sec. 3.5.2*).

On the other hand, there is complicatedness (e.g., the relatively more complicated analysis of biodiversity, if one wants to study in depth not only "the visible [macroscopic] world'" but also the non-visible "microbial" world, in *Sec. 2.2.1*; the relatively more complicated analysis of the Irish Potato Famine, by challenging the claims and assumptions, as shown in the more complicated analysis of other important contributing factors like "laws that restricted the rights of Irish Catholics," the "middleman system," "tenants, subdivisions, and bankruptcy," "potato dependency," and so on, in *Sec. 2.2.2*; the relatively more complicated analysis of utilitarian ethics for conservation efforts, by challenging the claims and assumptions, as shown in the problems pointed out by the critics as cited, in *Sec. 2.3.1*; the relatively more complicated analysis of adaptive management, by challenging the claims and assumptions, as shown in the problems pointed out by the critics as cited, in *Sec. 2.3.2*; the relatively more complicated analysis of cloning, by challenging the claims and assumptions, as shown in the problems pointed out by the critics as cited, in *Sec. 2.4.1*; the relatively more complicated analysis of ecological economics, by challenging the claims and assumptions, as shown in the problems pointed out by the critics as cited, in *Sec. 2.4.2*; the relatively more complicated analysis of each school of environmental ethics, by challenging the claims and assumptions, as shown in the problems pointed out by the critics as cited, in *Sec. 2.5.1*; the relatively more complicated analysis of animal watching, by challenging the claims and assumptions, as shown in the problems pointed out by the critics as cited, in *Sec. 2.5.2*; the relatively more complicated analysis of the overkill hypothesis, by challenging the claims and assumptions, as shown in the problems pointed out by the critics as cited, in *Sec. 3.2.1*; the relatively more complicated analysis of the relationsihp between biodiversity loss and health risks, by challenging the claims and assumptions, as shown in the problems pointed out by the critics as cited, in *Sec. 3.2.2*; the relatively more complicated analysis of climate change, by challenging the claims and assumptions, as shown in the problems pointed out by the critics as cited, in *Sec. 3.3.1*; the relatively more complicated analysis of genetic pollution, by challenging the claims and assumptions, as shown in the problems pointed out by the critics as cited, in *Sec. 3.3.2*; the relatively more complicated analysis of deforestiation, by challenging the claims and assumptions, as shown in the problems pointed out by the critics as cited, in *Sec. 3.4.1*; the relatively more complicated analysis of oil depletion, by challenging the claims and assumptions, as shown in the problems pointed out by the critics as cited, in *Sec. 3.4.2*; the relatively more complicated analysis of traditional knowledge, by challenging the claims and assumptions, as shown in the problems pointed out by the critics as cited, in *Sec. 3.5.1*; and the relatively more complicated analysis of creation theory, by challenging the claims and assumptions, as shown in the problems pointed out by the critics as cited, in *Sec. 3.5.2*).

And the reverse direction also holds true.

Again, it should be stressed, however, that there are different shades of gray (or different degrees of truth) in the two opposites, and the

classification is not necessarily mutually exclusive either. In fact, even when some combinations of the two occur, they only end up sharing the same dialectic relationship, but in a different degree.

Also, there are some other relationships (as a kind of family resemblance) which have something in common with the principle, but they are not exactly the same but only more or less comparable, which varies from case to case.

Good examples of family resemblance in relation to the principle are inflexibility-flexibility, standardization-specialization, imperfectness-perfectness, superficiality-depth, shallowness-deepness, economicalness-elaboratedness, plainness-circumspection, onesidedness-multisidedness, succinctness-duplicatedness, and the like.

11th Thesis: The Openness-Hiddenness Principle

The eleventh thesis refers to the openness-hiddenenss principle (on the detection of things) in the category of "structure" in existential dialectics (which was already worked out in my previous books, especially in *FPHU*), in that reality has its hidden face, just as it is open to outside view in some other ways. There is no openness without hiddenness—and vice versa.

For instance, on the one hand, there is openness (e.g., the open exploration by some scientists of biodiversity in nature, in *Sec. 2.2.1*; the open exploration of different cropping systems in agrobiodiversity, as shown by "a meta-analysis of studies comparing biodiversity," which "noted that, when compared to organic cropping systems, conventional systems had significantly lower species richness and abundance," in *Sec. 2.2.2*; the open exploration of the priority to emphasize hotspots for the understanding of biodiversity, in *Sec. 2.3.1*; the open exploration of different kinds of adaptive management, as shown by the work of C. Holling and C. Walters on "passive and adaptive management practice," in *Sec. 2.3.2*; the open exploration of the artificial cloning of organisms, in *Sec. 2.4.1*; the open exploration of the "limited carrying capacity" of the Earth, so that "its resources may run out," as studied by those in ecological economics, in *Sec. 2.4.2*; the open exploration of the need for a "maintenance of a healthy, sustainable environment…for human well-being" in anthropocentrism, in *Sec. 2.5.1*; the open exploration of how to protect bird by the RSPB, in *Sec. 2.5.2*; the open exploration by Jack Sepkoski and David M. Raup of "the 'Big Five' [extinction events]," in *Sec. 3.2.1*; the open exploration of the relationship between biodiversity and human health, in that, "in certain specific areas or situations, disease risk may increase with proximity to wild areas or intact habitats, including for example certain forest areas or wetlands," in *Sec. 3.2.2*; the open exploration of the impact of "solar output" on climate change, in *Sec. 3.3.1*; the open exploration of the consequences of genetic pollution by some conservation activists like those in Greenpeace, in *Sec. 3.3.2*; the open exploration of deforestation around the world by "the FAO data," in *Sec. 3.4.1*; "different groups and factions fight for their share" of oil revenues openly, in that "these emerge openly as separatist conflicts in regions where the resources are produced…such as in Angola's oil-rich Cabinda province," in *Sec. 3.4.2*; the open exploration, within traditional knowledge, of "the diversity and interactions among plant and animal species, landforms, watercourses, and other qualities of the biophysical environment in a given place," in *Sec. 3.5.1*; and the open exploration of the "hierarchical structure of all matter and life" in the world by the idea of the great chain of being, in *Sec. 3.5.2*).

On the other hand, there is hiddenness (e.g., the hidden bias by some scientists in the exploration of biodiversity in nature, because "selection bias amongst researchers may contribute to biased empirical research for modern estimates of biodiversity," as "Rev. Gilbert White [1909] succinctly observed of his Selborne, Hampshire [that] 'all nature is so full, that that district produces the most variety which is the most examined,'" in *Sec. 2.2.1*; the hidden bias in the "meta-analysis of studies comparing biodiversity," since different studies offer different insights, as shown in other studies ["16% of studies"] which "did find a greater level of species richness in conventional systems" instead, in *Sec. 2.2.2*; the hidden bias in the priority to emphasize hotspots for the understanding of biodiversity, because the critics point out that "emphasizing hotspots downplays the importance of the social and ecological connections to vast areas of the Earth's ecosystems where biomass, not biodiversity, reigns supreme," so "a hotspot priority approach would not invest so heavily in places such as steppes, the Serengeti, the Arctic, or taiga," but "these areas contribute a great abundance of population [not species] level biodiversity and ecosystem services, including cultural value and planetary nutrient cycling," in *Sec. 2.3.1*; the hidden bias in C. Holling and C. Walters on "passive and adaptive management practice," since the critics like R. Hilbourne proposed instead "three learning models…, around which adaptive management approaches could be developed," and "these are reactive, passive and active," in *Sec. 2.3.2*; the hidden bias in the artificial cloning of organisms, since "researchers have found several abnormalities in cloned organisms," because "the cloned organism may be born normal and resemble its non-cloned counterpart, but majority of the time will express changes in its genome later on in life," in *Sec. 2.4.1*; the hidden bias in ecological economics, since, "whereas mainstream economists tend to be technological optimists, ecological economists are inclined to be technological pessimists," in *Sec. 2.4.2*; the hidden bias in anthropocentrism, since, for the critics, "anthropocentrism" is "a root cause of the ecological crisis, human overpopulation, and the extinctions of many non-human species," in *Sec. 2.5.1*; the hidden bias in bird protection by the RSPB, because it used to claim that "the 'scientification' of the pastime was 'undesirable'" before "this stand was to change only in 1936," in *Sec. 2.5.2*; the hidden bias in the classification of "the 'Big Five' [extinction events]"by Jack Sepkoski and David M. Raup, since, for the critics, "as more stringent statistical tests have been applied to the accumulating data, the 'Big Five' [extinnction events] cannot be so clearly defined," in *Sec. 3.2.1*; the hidden bias in the study of the relationship between biodiversity and human health, because, for the critics, it is important to remember that "the causes are not always clear and are rarely straightforward," in *Sec. 3.2.2*; the hidden aspect of "solar output," insofar as "the cyclical nature of the sun's energy output is not yet fully understood," in *Sec. 3.3.1*; the hidden bias in some conservation activists like those in Greenpeace, since, for the critics like Patrick Moore, "the term 'genetic pollution'...is a propaganda term, not a technical or scientific term," because "some conservation biologists and conservationists have used genetic pollution for a number of years as a term to describe gene flow [which they disapprove of]," in *Sec. 3.3.2*; the hidden bias in "the FAO data," because "critics of the FAO data point out that they do not distinguish between forest types, and that they are based largely on reporting from forestry departments of individual countries, which do not take into account unofficial activities like illegal logging," in *Sec. 3.4.1*; "different groups and factions fight for their share" of oil revenues hiddenly, in that "often the conflicts occur in more hidden forms, such as fights between different government ministries or departments for access to budgetary allocations," in *Sec. 3.4.2*; the hidden bias in traditional knowledge, since, for the critics, "traditional knowledge is not a natural category, and may reflect power struggles and relationships for land, resources and social control rather than adherence to a claimed ancestry or heritage," in *Sec. 3.5.1*; and the hidden bias in the idea of the great chain of being, because of its rigid religious assumption which does not "allow for 'missing links,'" in *Sec. 3.5.2*).

And the reverse direction also holds true.

Again, it should be stressed, however, that there are different shades of gray (or different degrees of truth) in the two opposites, and the

classification is not necessarily mutually exclusive either. In fact, even when some combinations of the two occur, they only end up sharing the same dialectic relationship, but in a different degree.

Also, there are some other relationships (as a kind of family resemblance) which have something in common with the principle, but they are not exactly the same but only more or less comparable, which varies from case to case.

Good examples of family resemblance in relation to the principle are overtness-covertness, participation-obstruction, publicness-privateness, openness-closedness, surface-depth, transparency-secrecy, openness-biasedness, , exposition-illusion, and so on.

12th Thesis: The Denseness-Emptiness Principle

The twelfth thesis concerns the denseness-emptiness principle (on the distribution of entities in space) in the category of "structure" in existential dialectics, which is first proposed in *FPHUP*, in that both density and void are needed, in relation to the mind, nature, culture, and society, albeit in different ways. There is no denseness without emptiness—and vice versa.

Lest any misunderstanding occurs, the term "void" is used here only as an approximation of emptiness (depending on the degree of the lack of density), since, in physics, it is well known that "empty" space is not really empty all the way, because it can be full of energy (e.g., random quantum fluctuations at the sub-atomic level, and, for that matter, dark energy in the universe) and matter (e.g., different versions of sub-atomic particles, and, for that matter, dark matter in the universe). (F. Wilczek 2008)

For instance, on the one hand, there is denseness (e.g., the relatively denser concentration of species in the tropical regions, in *Sec. 2.2.1*; the relatively denser concentration of potatoes in Irish diet during the early 19th century, in *Sec. 2.2.2*; the relatively denser concentration of concerns with "endemism" in a "hotspot priority approach," in *Sec. 2.3.1*; the relatively denser concentration of concerns with decision making about ecological issues in the face of uncertainty among the practitioners of adaptive management, in *Sec. 2.3.2*; the relatively denser concentration of successful cases of cloning, as shown in the finding that the "European cultivars of grapes represent clones that have been propagated for over two millennia," in *Sec. 2.4.1*; the relatively denser concentration of concerns with natural capital in ecological economics, in *Sec. 2.4.2*; the relatively denser concentration of concerns with "the conservation of the environment solely for the environment's sake" in ecocentrism, in *Sec.*

2.5.1; the relatively denser concentration of concerns with "collection based taxonomy" by the AOU in the early days of birdwatching, in *Sec. 2.5.2*; the relatively denser concentration of extinct species during "the 'Big Five' [extinction events]," in *Sec. 3.2.1*; the relatively denser concentration of traditional medicinal resources in "the developing world," where "many millions of people rely directly upon traditional wisdom and indigenous knowledge for their health and livelihood security," in *Sec. 3.2.2*; the relatively denser concentration of "CO2 levels" in the atmosphere nowadays, "due to emissions from fossil fuel combustion," as part of the human influences on climate change, in *Sec. 3.3.1*; the relatively denser concentration of an invasive species as "its population grows larger…and it adapts to its new location" with "the superior competitive ability," in *Sec. 3.3.2*; the relatively denser concentration of forests around the world in the pre-historical era, in *Sec. 3.4.1*; the relatively denser concentration of oil use in the sector of transportation, as "over 90% of transportation in the United States relies on oil," in *Sec. 3.4.2*; the relatively denser concentration of concerns with the protection of traditional knowledge among those in CBD, in *Sec. 3.5.1*; and the relatively denser concentration of concerns with God in creation theory, in *Sec. 3.5.2*).

On the other hand, there is emptiness (e.g., the relatively less dense, or more empty, concentration of species in the polar regions, in *Sec. 2.2.1*; the relatively less dense, or more empty, concentration of potatoes in Chinese diet during the early 19th century, in *Sec. 2.2.2*; the relatively less dense, or more empty, concentration of concerns with "endemism" in a "coldspot priority approach," in *Sec. 2.3.1*; the relatively less dense, or more empty, concentration of concerns with decision making about ecological issues in the face of uncertainty among those in the corporate world, because business people, unlike the practitioners of adaptive management, are more interested in models which are "economically viable" and "technically" feasible instead, for their profit-seeking purpose, in *Sec. 2.3.2*; the relatively less dense, or more empty, concentration of successful cases of human cloning, as of 2012, in *Sec. 2.4.1*; the relatively less dense, or more empty, concentration of concerns with natural capital in neoclassical economics, in *Sec. 2.4.2*; the relatively less dense, or more empty, concentration of concerns with "the conservation of the environment solely for the environment's sake" in anthropocentrism, in *Sec. 2.5.1*; the relatively less dense, or more empty, concentration of concerns with "collection based taxonomy" by the BOU, when it "changed focus to ecology and behaviour…in the 1940s," in *Sec. 2.5.2*; the relatively less dense, or more empty, concentration of extinct species outside the historcal eras of “the 'Big Five' [extinction events],” in *Sec. 3.2.1*; the relatively less dense, or more empty, concentration of traditional medicinal resources in “technologically advanced countries,” because “the vast majority of this indigenous knowledge has been lost to society, and is of little consequence to modern 'Western' lifestyles,” in *Sec. 3.2.2*; the relatively less dense, or more empty, concentration of “CO2 levels” in the atmosphere in pre-historical times when humans did not yet produce the massive “emissions from fossil fuel combustion” as is the case in our time, in *Sec. 3.3.1*; the relatively less dense, or more empty, concentration of an invasive species at the beginning, when it “must survive at low population densities before it becomes invasive in a new location,” in *Sec. 3.3.2*; the relatively less dense, or more empty, concentration of forests around the world in our time, in *Sec. 3.4.1*; the relatively less dense, or more empty, concentration of oil use in the sector of “electric cars,” in *Sec. 3.4.2*; the relatively less dense, or more empty, concentration of concerns with the protection of traditional knowledge among biopirates, in *Sec. 3.5.1*; and the relatively less dense, or more empty, concentration of concerns with God in uniformitarianism in *Sec. 3.5.2*).

And the reverse direction also holds true.

Again, it should be stressed, however, that there are different shades of gray (or different degrees of truth) in the two opposites, and the classification is not necessarily mutually exclusive either. In fact, even when some combinations of the two occur, they only end up sharing the same dialectic relationship, but in a different degree.

Also, there are some other relationships (as a kind of family resemblance) which have something in common with the principle, but they are not exactly the same but only more or less comparable, which varies from case to case.

Good examples of family resemblance in relation to the principle are fullness-voidness, nearness-farness, concentration-dispersion, denseness-nothingness, something-nothing, heaviness-lightness, composition-emptiness, solidness-void, presence-absence, focus-unfocus, emphasis-disemphasis, and the like.

In Relation to Process

The next four theses are in relation to process in existential dialectics, as shown below.

13th Thesis: The Change-Constancy Principle

The thirteenth thesis concerns the change-constancy principle (on the alteration of things) in the category of "process" in existential dialectics, which is first named in *BCPC* (and were worked out in my books like *FHC*, *FCD*, and *FPHC*), in that change occurs over time, although constancy is also allowed.

Old players fade away, and new ones emerges, with ever new causes and ever new forms—and asymmetry undergoes changes over time too, so does symmetry. There is no change without constancy—and vice versa.

For instance, on the one hand, there is change (e.g., the ever new ways to study diversity, as shown in the study of psychological benefits and biodiversity, in *Sec. 2.2.1*; the ever new ways to understand biodiversity, as shown in the study of dietary health and agrobiodiversity, in *Sec. 2.2.2*; the ever new ways to understand ecosystems, as shown in the study of conservation biology and environmental protection, in *Sec. 2.3.1*; the ever new ways to study ecology, as shown in the work on adaptive management, in *Sec. 2.3.2*; the ever new ways to deal with biodiversity, as shown in the case of cloning with the use of technology, in *Sec. 2.4.1*; the ever new ways to study economics, as shown in the rise of ecological economics, in *Sec. 2.4.2*; the ever new ways to understand ethics, as shown in environmental ethics, in *Sec. 2.5.1*; the ever new ways to understand animal behaviors, as shown in birdwatching and fishkeeping, in *Sec. 2.5.2*; the ever new ways to study discontinuity, as shown in the overkill hypothesis, in *Sec. 3.2.1*; the ever new ways to study biodiversity, as shown in the study of biodiversity loss and health risks, in *Sec. 3.2.2*; the ever new ways to understand climate, as shown in the study of climate change and water deterioration, in *Sec. 3.3.1*; the ever new ways to study animals, as shown in the study of genetic pollution and invasive species, in *Sec. 3.3.2*; the ever new ways to understand discontinuity, as shown in the study of habitat fragmentation in the context deforestation, in *Sec. 3.4.1*; the ever new ways to study natural resources, as shown in the study on oil depletion and resource curse, in *Sec. 3.4.2*; the ever new ways to knowledge, as shown in the study of traditional knowledge, in *Sec. 3.5.1*; and the ever new ways to study matter and life in the world, as shown in the study of extinction and the history of scientific ideology, in *Sec. 3.5.2*).

On the other hand, there is constancy (e.g., the ever constant existence of different problems in the ever new ways to study diversity, as shown in the problems with the study of psychological benefits and biodiversity pointed out by the critics, in *Sec. 2.2.1*; the ever constant existence of different problems in the ever new ways to understand biodiversity, as shown in the problems with the study of dietary health and agrobiodiversity pointed out by the critics, in *Sec. 2.2.2*; the ever constant existence of different problems in the ever new ways to understand ecosystems, as shown in the problems with the study of conservation biology and environmental protection pointed out by the critics, in *Sec. 2.3.1*; the ever constant existence of different problems in the ever new ways to study ecology, as shown in the problems with the work on adaptive management pointed out by the critics, in *Sec. 2.3.2*; the ever constant existence of different problems in the ever new ways to deal with biodiversity, as shown in the problems with the technology of cloning pointed out by the critics aforecited, in *Sec. 2.4.1*; the ever constant existence of different problems in the ever new ways to study economics, as shown in the problems with ecological economics pointed out by the critics, in *Sec. 2.4.2*; the ever constant existence of different problems with the ever new ways to understand ethics, as shown in the problems with environmental ethics pointed out by the critics as cited earlier, in *Sec. 2.5.1*; the ever constant existence of different problems with the ever new ways to understand animal behaviors, as shown in the problems with birdwatching and fishkeeping pointed out by the critics, in *Sec. 2.5.2*; the ever constant existence of different problems in the ever new ways to study discontinuity, as shown in the problems with the overkill hypothesis pointed out by the critics, in *Sec. 3.2.1*; the ever constant existence of different problems in the ever new ways to study biodiversity, as shown in the problems with the study of biodiversity loss and health risks pointed out by the critics aforecited, in *Sec. 3.2.2*; the ever constant existence of different problems in the ever new ways to understand climate, as shown in the problems with the study of climate change and water deterioration pointed out by the critics, in *Sec. 3.3.1*; the ever constant existence of different problems in the ever new ways to study animals, as shown in the problems with the study of genetic pollution and invasive species pointed out by the critics, in *Sec. 3.3.2*; the ever constant existence of different problems in the ever new ways to understand discontinuity, as shown in the problems with the study of habitat fragmentation in the context of deforestation pointed out by the critics, in *Sec. 3.4.1*; the ever constant existence of different problems in the ever new ways to study natural resources, as shown in the problems with the study on oil depletion and resource curse pointed out by the critics, in *Sec. 3.4.2*; the ever constant existence of different problems in the ever new ways to knowledge, as shown in the problems with traditional knowledge pointed out by the critics, in *Sec. 3.5.1*; and the ever constant existence of different problems in the ever new ways to study matter and life in the world, as shown in the problems with the study of extinction and the history of scientific ideology pointed out by the critics aforecited, in *Sec. 3.5.2*).

And the reverse direction also holds true.

Again, it should be stressed, however, that there are different shades of gray (or different degrees of truth) in the two opposites, and the classification is not necessarily mutually exclusive either. In fact, even when some combinations of the two occur, they only end up sharing the same dialectic relationship, but in a different degree.

Also, there are some other relationships (as a kind of family resemblance) which have something in common with the principle, but they are not exactly the same but only more or less comparable, which varies from case to case.

Good examples of family resemblance in relation to the principle are dynamicness-staticness, instability-stability, and the like.

14th Thesis: The Order-Chaos Principle

The fourteenth thesis is called the the order-chaos principle (on the pattern of things) in the category of "process" in existential dialectics (as first worked out in *FC*), in that both order and chaos are vital in the process of change in the world.

The preference for order is biased, since it does not give sufficient attention to the vital role of chaos in the transformation of the world (without somehow reducing it for the understanding of order). There is no order without chaos—and vice versa.

The scientific search for order in the world is often a hidden bias in its ontological obsession with order, since chaos is often treated as the "bad" guy, with order as the "good" guy (for the end goal of science).

Neither order nor chaos is the final end of the world, and one is not to be treated as the means for the other in the transformation of things. Both are fundamental in their recurrent dialectical interactions with each other over time, without reducing one for the other.

For instance, on the one hand, there is order (e.g., the more or less orderly appearance of reality, if looked from the vantage point of the period before "the Cambrian explosion," when "most organisms were simple, composed of individual cells occasionally organized into colonies," in *Sec. 2.2.1*; the more or less orderly appearance of the Irish dietary situation in the early 19th century, when it was dependent on potatoes, in *Sec. 2.2.2*; the more or less orderly appearance of conservation efforts, if looked from the sole vantage point of ecological issues like biodiversity, in *Sec. 2.3.1*; the more or less orderly view of reality, if looked from the perspective of those in the corporate world who are more interested in models which are "economically viable" and "technically" feasible, in *Sec. 2.3.2*; the more or less orderly view of cloning, if looked from the vantage point of its proponents who promise successful cloning for biodiversity, in *Sec. 2.4.1*; the more or less orderly view of economics, if looked from the sole vantage point of ecological economics, with its focus on natural capital, in *Sec. 2.4.2*; the more or less orderly view of nature, if looked from the Christian vantage point of "stewardship" on the basis of the Book of Genesis, in that "mankind should take care of the earth and its various forms of life," in *Sec. 2.5.1*; the more or less orderly view of birdwatching, if looked from the vantage point of "certain locations such as the local patch of forest, wetland and coast," which "may be favoured according to the location and season," like. "seawatching…where observers based at a coastal watch point, such as a headland, watch birds flying over the sea," in *Sec. 2.5.2*; the more or less orderly view of natural resources, if looked from the vantage point of other relatively more normal historical periods without all the "extinction events," large or small, in *Sec. 3.2.1*; the more or less orderly view of biodiversity, if looked from the vantage point of "traditional wisdom and indigenous knowledge for the…health and livelihood security" of a local community," which "is associated with the gathering and cultivation of foods, clothing and building materials, with local cultural traditions, and with systems of traditional medicines, which support entire populations," in *Sec. 3.2.2*; the more or less orderly view of reality, if looked from the vantage point of the 5 causes of climate change in the article aforecited, in *Sec. 3.3.1*; the more or less orderly view of reality, if looked from the vantage point of some conservation activits who "unequivocally attributes extinctions to a species invasion," in *Sec. 3.3.2*; the more or less orderly view of deforestation, if looked only from the vantage point like the "2002 analysis of satellite imagery" which suggested that "the rate of deforestation in the humid tropics [approximately 5.8 million hectares per year] was roughly 23% lower than the most commonly quoted rates," in *Sec. 3.4.1*; the more or less orderly view of reality, if looked from the vantage point of the U.S. nowadays, which remains one of the most stable countries on earth, when contrasted with, say, "the Democratic Republic of the Congo," which "is perhaps one of the volatile places on the planet, suffering from disease, civil war, and the resource curse," in *Sec. 3.4.2*; the more or less orderly view of traditional knowledge, if looked from the vantage point of its three characteristics, namely, "communal," "accumulative," and "spiritual," in *Sec. 3.5.1*; and the more or less orderly view of reality, if looked from the vantage point of uniformitarianism, with its focus on "processes that could be observed in the present acting gradually over an immense period of time," in *Sec. 3.5.2*).

On the other hand, there is chaos (e.g., the more or less chaotic appearance of reality, if looked from the vantage point of the period known as "the Cambrian explosion," especially in regard to "the long-running puzzlement about the appearance of the Cambrian fauna, seemingly abruptly and from nowhere," in *Sec. 2.2.1*; the more or less chaotic appearance of the Irsh dietary situation during the Irish Potato Famine, when "blight ravaged potato crops throughout Europe during the 1840s" and "the impact and human cost in Ireland…was exacerbated by a host of political, social and economic factors which remain the subject of historical debate," in *Sec. 2.2.2*; the more or less chaotic appearance of conservation efforts, if looked from the multifaceted vantage point of myriad linkages, including "encroachment and poor management" as in the case studies of Brazil and Mexico, "lack of financial resources" as well as "poor governance and corruption" as in the case studies of African countries, in *Sec. 2.3.1*; the more or less chaotic view of reality, if looked from the practitioners of adaptive management who are more interested in models which deal with several dimensions of ecological uncertainty," like "structural uncertainty," "parameter uncertainty," and "stochastic uncertainty," in *Sec. 2.3.2*; the more or less chaotic view of cloning, if looked from the vantage point of its critics like David Ehrenfeld, who "concluded that cloning in animal conservation is an experimental technology that, at its state in 2006, could not be expected to work except by pure chance," in *Sec. 2.4.1*; the more or less chaotic view of economics, if looked from all the conflicting, competing approaches, such as "neoclassical economics," "ecological economics," "resource and environmental economics," "some of the controversy between American and European schools of thought," and so on, in *Sec. 2.4.2*; the more or less chaotic view of nature, if looked from the vantage point of laissez-faire capitalists, who prefer to compete with each other for resources in the spirit of social Darwinism, where chance or luck also has a role to play, in *Sec. 2.5.1*; the more or less chaotic view of birdwatching, if looked from the vantage point of often changing weather conditions, such as "in Britain, suitable wind conditions may lead to drift migration, and an influx of birds from the east," and "in North America, birds caught in the tail-end of a hurricane may be blown inland," and so on, in *Sec. 2.5.2*; the more or less chaotic view of natural resources, if looked from the vantage point of "earlier megafaunal extinctions" which abruptly "occurred within a relatively short period of time," in *Sec. 3.2.1*; the more or less chaotic view of biodiversity, if looked from the vantage point of "natural and man-made disasters—flooding, storms, earthquakes, drought, conflict," in *Sec. 3.2.2*; the more or less chaotic view of reality, if looked from the vantage point of the critics, like those who shared the view in a report by "a 2010 study" which "suggests, 'that the effects of solar variability on temperature throughout the atmosphere may be contrary to current expectations'" and the view by the United States National Research Council which recognized in 2010 that "there are still some uncertainties" about "climate change," in *Sec. 3.3.1*; the more or less chaotic view of reality, if looked from the vantage point of some critics, who argued that "it is difficult to unequivocally attribute extinctions to a species invasion, and the few scientific studies that have done so have been with animal taxa," so "concern over the impacts of invasive species on biodiversity must therefore consider the actual evidence [either ecological or economic], in relation to the potential risk," in *Sec. 3.3.2*; the more or less chaotic view of deforestation, if looked from the multiple opposing and inconsistent vantage points, like the "2002 analysis of satellite imagery" which suggested that "the rate of deforestation in the humid tropics [approximately 5.8 million hectares per year] was roughly 23% lower than the most commonly quoted rates"—and "a newer [but opposing] analysis of satellite images" which "reveals that deforestation of the Amazon rainforest is twice as fast as scientists previously estimated," in *Sec. 3.4.1*; the more or less chaotic view of reality, if looked from the vantage point of "the Democratic Republic of the Congo," which "is perhaps one of the volatile places on the planet, suffering from disease, civil war, and the resource curse," in *Sec. 3.4.2*; the more or less chaotic view of traditional knowledge, if looked from the vantage points that "traditional knowledge is not a natural category" and "reflect[s] power struggles and relationships for land, resources and social control rather than adherence to a claimed ancestry or heritage," that "the spiritual component of 'traditional knowledge' can be used to justify any activity, including the unsustainable harvesting of resources," and so on, in *Sec. 3.5.1*; and the more or less chaotic view of reality, if looked from the vantage point of catastrophism, with its focus on "sudden, short-lived, violent events," in *Sec. 3.5.2*).

And the reverse direction also holds true.

Once more, it should be stressed, however, that there are different shades of gray (or different degrees of truth) in the two opposites, and the classification is not necessarily mutually exclusive either. In fact, even when some combinations of the two occur, they only end up sharing the same dialectic relationship, but in a different degree.

Also, there are some other relationships (as a kind of family resemblance) which have something in common with the principle, but they are not exactly the same but only more or less comparable, which varies from case to case.

Good examples of family resemblance in relation to the principle are lawfulness-disorder, order-disorder, pattern-chaos, linearity-nonlinearity, unity-disunity, linearity-irreversalness, etc.

15th Thesis: The Slowness-Quickness Principle

The fifteenth thesis concerns the slowness-quickness principle (on the speed of change) in the category of "process" in existential dialectics (which was first worked out in *FIA*), in that both slowness and quickness co-exist, with their own internal tension, to the extent that each fights for its own relevance with the other, in accordance to the perspectives of nature, the mind, culture, and society, without one being the victor and the other being the vanquished in the long haul. There is no slowness without quickness—and vice versa.

For instance, on the one hand, there is slowness (e.g., the relatively slower speed of the emergence of new life forms before the period known as "the Cambrian explosion," in *Sec. 2.2.1*; the relatively slower speed of the fall of the Irish population before the Irish Potato Famine, when the demographic situation was more normal, in *Sec. 2.2.2*; the relatively slower speed of a decision making process in conservation if the approach is to require "the complex interrelationships of an entire ecosystem in decision making…that involves a broad range of stakeholders across all relevant governmental departments, as well as representatives of industry, environmental groups and community," in *Sec. 2.3.1*; the relatively slower speed of the techniques of adaptive management to be accepted by "the majority of global companies operating in the markets," in *Sec. 2.3.2*; the relatively slower speed of human cloning to be achieved at all in reality nowadays, although it was successful in science fiction works like Aldous Huxley's *Brave New World* in 1932, in *Sec. 2.4.1*; the relatively slower speed of natural resources to be consumed, if operated under the guidance of ecological economics, with its focus on sustainaiblity, in *Sec. 2.4.2*; the relatively slower speed of the environment to be affected or controlled by technology under the guidance of ecocentrism, in *Sec. 2.5.1*; the relatively slower speed of fish to be killed by experienced aquarists in PETA who care about "animal welfare," in *Sec. 2.5.2*; the relatively slower speed of extinction in those relatively small extinction events, when put within the context of "a relatively smooth continuum of extinction events," in *Sec. 3.2.1*; the relatively slower speed of "the transmission of diseases between people, livestock, crops and wildlife" in the absence of "the loss of biodiversity and ecosystem services," in *Sec. 3.2.2*; "whether the initial forcing mechanism is internal or external, the response of the climate system might be...slow [e.g. thermal expansion of warming ocean water]," in *Sec. 3.3.1*; the relatively slower speed of some "native species" to survive when invaded by "an introduced species that can spread faster than natives" and therefore "can use resources that would have been available to native species, squeezing them out," in *Sec. 3.3.2*; the relatively slower speed of deforestation in the pre-historical era, because only "small scale deforestation was practiced by some societies for tens of thousands of years before the beginnings of civilization," in *Sec. 3.4.1*; the relatively slower speed of the world's oil supply to deplete, if, according to an estimate by the American Petroleum Institute in 1999, "the world's oil supply would be depleted between 2062 and 2094, assuming total world oil reserves at between 1.4 and 2 trillion barrels [220 and 320 km3] and consumption at 80 million barrels per day [13,000,000 m3/d]," in *Sec. 3.4.2*; the relatively slower speed of knowledge to be formed, if looked from the vantage point of traditional knowledge being accumulated over "millennia," in *Sec. 3.5.1*; and the relatively slower speed of events if they follow the processes as understood in uniformitariansim, that is, "acting gradually over an immense period of time," in *Sec. 3.5.2*).

On the other hand, there is quickness (e.g., the relatively quicker speed of the emergence of new life forms, as shown in "the relatively rapid appearance...of most major animal phyla" in the period known as "the Cambrian explosion," in *Sec. 2.2.1*; the relatively quicker speed of the fall of the Irish population during the Irish Potato Famine and its aftermath, when "approximately 1 million people died and a million more emigrated from Ireland, causing the island's population to fall by between 20% and 25%," in *Sec. 2.2.2*; the relatively quicker speed of a decision making process in conservation if the approach is to "simply respond...to specific issues and challenges" locally at a time, in *Sec. 2.3.1*; the relatively quicker speed of the techniques of adaptive management to be accepted by the Nature Conservancy, the World Resources Institute, etc., in *Sec. 2.3.2*; the relatively quicker speed of horticultural cloing to be achieved, as shown in the case that the "European cultivars of grapes represent clones that have been propagated for over two millennia," in *Sec. 2.4.1*; the relatively quicker speed of natural resources to be consumed, if operated under the guidance of neoclassical economics, with its focus on technology, in that "every technology can be improved upon or replaced by innovation, and that there is a substitute for any and all scarce materials" in the natural world, in *Sec. 2.4.2*; the relatively quicker speed of the environment to be affected or controlled by technology under the guidance of technocentrism, in *Sec. 2.5.1*; the relatively faster speed of fish to be killed by "inexperienced aquarists" who "often keep too many fish in one tank, or add fish too quickly into an immature aquarium, killing many of them," in *Sec. 2.5.2*; the relatively quicker speed of extinction in those relatively large extinction events, like "the fastest ever recorded extinction of megafauna" which "occurred in New Zealand, where by 1500 AD, just 200 years after settling the islands, ten species of the giant moa birds were hunted to extinction by the Māori," in *Sec. 3.2.1*; the relatively quicker speed of "the transmission of diseases between people, livestock, crops and wildlife" in the presence of "the loss of biodiversity and ecosystem services," in *Sec. 3.2.2*; "whether the initial forcing mechanism is internal or external, the response of the climate system might be fast [e.g., a sudden cooling due to airborne volcanic ash reflecting sunlight]," in *Sec. 3.3.1*; the relatively quicker speed of "an introduced species that can spread faster than natives" and therefore "can use resources that would have been available to native species, squeezing them out," in *Sec. 3.3.2*; the relatively quicker speed of deforestation in the industrial era, because "by 1852, 'global deforestation sharply accelerated....It has been estimated that about half of the Earth's mature tropical forests...have now been destroyed,'" in *Sec. 3.4.1*; the relatively quicker speed of the world's oil supply to deplete, if, according to an estimate in 2004, "total world reserves were estimated to be 1.25 trillion barrels [199 km3] and daily consumption was about 85 million barrels [13,500,000 m3]," so as to "shift...the estimated oil depletion year to 2057," in *Sec. 3.4.2*; the relatively quicker speed of knowledge to be formed, if looked from the vantage point of scientific knowledge with its emphasis on "the gradation of recent knowledge," in *Sec. 3.5.1*; and the relatively quicker speed of events, if they follow the processes as understood in catastrophism, which are "sudden, short-lived, violent events," in *Sec. 3.5.2*).

And the reverse direction also holds true.

Again, it should be stressed, however, that there are different shades of gray (or different degrees of truth) in the two opposites, and the classification is not necessarily mutually exclusive either. In fact, even when some combinations of the two occur, they only end up sharing the same dialectic relationship, but in a different degree.

Also, there are some other relationships (as a kind of family resemblance) which have something in common with the principle, but they are not exactly the same but only more or less comparable, which varies from case to case.

Good examples of family resemblance in relation to the principle are inconvenience-convenience, passiveness-activenessness, gradualness-abruptness, deceleration-acceleration, difficulty-ease, nonresilience-resilience, and the like.

16th Thesis: The Expansion-Contraction Principle

The sixteenth thesis is called the expansion-contraction principle in the category of "process" in existential dialectics, in that entities in the world can both expand in some ways and contract in other ones, as part of their nature. There is no expansion without contraction—and vice versa.

This principle, although not so explicitly called, was already used in my previous works on different topics (e.g., the theory of floating consciousness in *FCD* and *FPHC*, the union of the unions in *BWT*, and the cyclical progression of system fragmentation and integration in *FCD*).

For instance, on the one hand, there is expansion (e.g., the relative expansion of life forms, as shown in "the Phanerozoic eon [the last 540 million years]," which "marked a rapid growth in biodiversity via the Cambrian explosion," in *Sec. 2.2.1*; the relative expansion of the American population in the 19th century, when it gained from the arrival of Irish immigrants and others, in *Sec. 2.2.2*; the relative expansion of the commitment to "the principles of conservation of biological diversity" in the 1990s, as shown in "the Convention on Biological Diversity" in 1992 accepted by "most of the countries of the world," in *Sec. 2.3.1*; the relative expansion of "the demand for lateral decision making that supports affective participation" in environmental management compatible with adaptive management nowadays, in *Sec. 2.3.2*; the increase of different cases of cloning for biodiversity nowadays, in *Sec. 2.4.1*; the relative expansion of the approach to economics with the inclusion of the ecosystems for analysis nowadays, in *Sec. 2.4.2*; the relative expanson, in our time, of the influence of anthropocentrism, since it is used by some defenders like Wesley J. Smith, who argued that it "gives rise to human duties to each other, the natural world, and to treat animals humanely," in *Sec. 2.5.1*; the relative expansion, nowadays, of different "objects such as the 'AquaBabies Micro Aquaria,' 'Bubble Gear Bubble Bag,' and 'Betta in a Vase,' all of which house live fish in unfiltered and insufficient water," in *Sec. 2.5.2*; the relative expansion of "mammals and birds" which "emerged as dominant land vertebrates in the age of the Cretaceous-Paleogene," in *Sec. 3.2.1*; the relative spread of "modern prescription medicines" in "technologically advanced countries" with their "Western" lifestyles in our time, in *Sec. 3.2.2*; "climate models [of global warming] show that…some regions should expect an increase in precipitation, such as in the tropics and higher latitudes," in *Sec. 3.3.1*; the relative expansion of those invasive species only "less than 50 miles [80 km] from the…natural habitat" of "the Monterey cypress" in California, when the former succeeded in "exterminating" the latter, in *Sec. 3.3.2*; the relative expansion of urbanization, agriculture, mining, etc. in the industrial era, in *Sec. 3.4.1*; "U.S. oil production reached its peak in 1970," in *Sec. 3.4.2*; the relative expansion of the mechanisms to protect traditional knowledge nowadays, as shown in the three types aforecited, in *Sec. 3.5.1*; and the relative spread of the ideas of catastrophism and uniformitarianism in modern times, in *Sec. 3.5.2*).

On the other hand, there is contraction (e.g., the relative decrease of life forms, as shown in the period after the Cambrian explosion, when "the next 100 million years included repeated, massive biodiversity losses classified as mass extinction events," in *Sec. 2.2.1*; the relative decline of the Irish population, when it fell "by between 20% and 25%" during the Irish Potato Famine and its aftermath, in *Sec. 2.2.2*; the relative decline of the opponents of "the principles of conservation of biological diversity" in the 1990s, in *Sec. 2.3.1*; the relative decline of the "traditional…hierarchal decision making" in environmental management compatible with adaptive management nowadays, in *Sec. 2.3.2*; the decrease of some species nowadays, because they become extinct or endangered, in *Sec. 2.4.1*; the relative decline of the approach to economics with the neglect of the ecosystems for analysis nowadays, in *Sec. 2.4.2*; the relative decline, in our time, of the elitist idea that "groups of superior men" should "be able to justify their enslavement, exploitation, or even genocide of inferior human groups on factual and moral grounds akin to those we now rely on to justify our treatment of the animals we harness as beasts of burden, that we butcher for food and clothing, or that we destroy as disease-bearing pests or as dangerous predators?," in *Sec. 2.5.1*; the relative decline of "the awarding of goldfish as prizes at funfairs," as shown by "the United Kingdom," which "outlawed live-animal prizes such as goldfish in 2004," in *Sec. 2.5.2*; the relative decline of "non-avian dinosaurs," which "became extinct" during the age of the Cretaceous-Paleogene, in *Sec. 3.2.1*; the relative decline or loss, "in technologically advanced countries" in our time, of "the vast majority of the…indigenous knowledge" for traditional medicines, although "up to 57% of the top 150 medicinal compounds currently on sale in the U.S. have some basis or origin in exploration from wild species," in *Sec. 3.2.2*; "climate models [of global warming] show that…other areas are expected to see a decrease, such as in the subtropics," in *Sec. 3.3.1*; the relative decline of "the Monterey cypress," when "they are being exterminated as exotic invasive species less than 50 miles [80 km] from their natural habitat" by invasive species, in *Sec. 3.3.2*; the relative decline of forests in the industrial era, in *Sec. 3.4.1*; "in 1950, the United States produced over half the world's oil, but by 2005 that proportion had dropped to about 8%," in *Sec. 3.4.2*; the relative decline of blatant biopiracy nowadays, because of different legal mechanisms for the protection of traditional knowledge, in *Sec. 3.5.1*; and the relative decline of the idea of the great chain of being in modern times, in *Sec. 3.5.2*).

And the reverse direction also holds true.

Again, it should be stressed, however, that there are different shades of gray (or different degrees of truth) in the two opposites, and the classification is not necessarily mutually exclusive either. In fact, even when some combinations of the two occur, they only end up sharing the same dialectic relationship, but in a different degree.

Also, there are some other relationships (as a kind of family resemblance) which have something in common with the principle, but they are not exactly the same but only more or less comparable, which varies from case to case.

Good examples of family resemblance in relation to the principle are conquest-autarky, rise-fall, spread-shrink, extendingness-shorteningness, widening-narrowing, construction-deconstruction, expansion-destruction, expansion-decline, construction-demolition, creation-destruction, and so forth.

In Relation to Agency

The next eight theses are in relation to agency in existential dialectics, as shown below.

17th Thesis: The Theory-Praxis Principle

The seventeenth thesis is the theory-praxis principle (on the duality of knowledge) in the category of "agency" in existential dialectics (which was first worked out in *FPHE*), in that, if there is theoretical construction, there is likewise its practical application, both technical and normative. There is no theory without praxis—and vice versa.

For instance, on the one hand, there is theory (e.g., the theoretical construction of the relationship between psychological benefits and biodiversity, in *Sec. 2.2.1*; the theoretical understanding of the history of the Irish Potato Famine, like its causes and impacts, as shown in the study aforecited, in *Sec. 2.2.2*; the theoretical construction of the "ecosystems approach," which requires, at least on paper, "the complex interrelationships of an entire ecosystem in decision making…that involves a broad range of stakeholders across all relevant governmental departments, as well as representatives of industry, environmental groups and community," in *Sec. 2.3.1*; the theoretical construction of the features and processes of adaptive management, in *Sec. 2.3.2*; the theoretical understanding of the process of cloning in biology, in *Sec. 2.4.1*; the theoretical construction of "natural capital" in ecological economics, in *Sec. 2.4.2*; the theoretical construction of anthropocentrism in environmental ethics, in *Sec. 2.5.1*; the theoretical construction of the "scientific" method in epistemology and philosophy of science, in *Sec. 2.5.2*; the theoretical construction of "the overkill hypothesis" in biology in *Sec. 3.2.1*; the theoretical understanding of the relationship between biodiversity loss and health risks in biology, in *Sec. 3.2.2*; the theoretical construction of the impact of climate change on water deterioration in environmental studies, in *Sec. 3.3.1*; the theoretical construction of "genetic pollution" and "invasive species" in ecology and biology, in *Sec. 3.3.2*; the theoretical construction of deforestation, as shown by the "forest transition theory," in *Sec. 3.4.1*; the theoretical construction of "the Hubbert curves" based on "the Hubbert peak theory," in *Sec. 3.4.2*; the theoretical construction of "intellectual property" in economics in *Sec. 3.5.1*; and the theoretical construction of the great chain of being by Plato and Aristotle in philosophy in *Sec. 3.4.2*).

On the other hand, there is praxis (e.g., the practical application of the understanding of the relationship between psychological benefits and biodiversity to the field of pharmaceutical industry, as shown in the statistics that "a significant proportion of drugs are derived, directly or indirectly, from biological sources: at least 50% of the pharmaceutical compounds on the US market are derived from plants, animals, and micro-organisms," in *Sec. 2.2.1*; the practical application of the understanding of the history of the Irish Potato Famine like its causes and impacts to the field of politics, when nowadays, there are different memorials throughout Ireland to memorialise the tragedy, as shown by the memorial at "Custom House Quays, Dublin," with "the thin sculptural figures, by artist Rowan Gillespie, who stand as if walking towards the emigration ships on the Dublin Quayside," in *Sec. 2.2.2*; the practical application of the "ecosystems approach to the real world of international relations, such that the approach turns out to be "idealistic" in practice, as shown by the problems confronted by many countries in implementing conservation efforts aforecited, in *Sec. 2.3.1*; the practical application of the features and processes of adaptive management to the real world, like its application by such organizations as DFG, CMR, LACPR, and so on, as cited earlier, in *Sec. 2.3.2*; the practical application of the understanding of the process of cloning in biology to the field of food science, as shown in 2006, when "the U.S. Food and Drug Administration [FDA] approved the consumption of meat and other products from cloned animals," in *Sec. 2.4.1*; the practical application of "natural capital" in ecological economics to the field of politics, as shown in the policy of "the government of Ecuador" to "preserve the rainforest" of "the Yasuni region," together with the political support of "Bill Clinton, Paul Martin and other former world leaders," in *Sec. 2.4.2*; the practical application of anthropocentrism in environmental ethics to the field of politics, as shown in the use of anthropocentrism as "the grounding for some naturalistic concepts of human rights," as "defenders of anthropocentrism argue that it is the necessary fundamental premise to defend universal human rights," in *Sec. 2.5.1*; the practical application of the "scientific" method in epistemology and philosophy of science to the field of birdwatching, as shown in the decision by "the BOU," which "changed focus to ecology and behaviour…in the 1940s" so as to study birds in a more "scientfic" way, that is, "the 'scientification' of the pastime," in *Sec. 2.5.2*; the practical application of "the overkill hypothesis" in biology to the field of activism, as shown by some conservation activists who try to use "biodiversity…as a legal standard" to enforce conservation against extinction, due to overexploitation, in *Sec. 3.2.1*; the practical application of the understanding of the relationship between biodiversity loss and health risks in biology to the field of socio-political government, like the call by the COHAB Initiative for "internatonal efforts" and "collaboration across disciplines and across borders" to fight against biodiversity loss for the promotion of human health, in *Sec. 3.2.2*; the practical application of the knowledge of the impact of climate change on water deterioration in environmental studies to the field of politics, as "environmental degradation is one of the Ten Threats officially cautioned by the High Level Threat Panel of the United Nations," in *Sec. 3.3.1*; the practical application of the knowledge of "genetic pollution" and "invasive species" in ecology and biology to the field of conservation politics, when the organizatons like TRAFFIC "promote awareness of the effects of introduced invasive species that may 'hybridize with native species, causing genetic pollution,'" in *Sec. 3.3.2*; the practical application of the "forest transition theory" to the field of policy, as shown in the policy implication that "a global forest transition would require major policy and technological innovations as well as shifts in demands for fiber, fuel, and food," in *Sec. 3.4.1*; the practical application of "the Hubbert curves" by researchers to the field of various other resources, such as "natural gas," coal," "fissionable materials," "Helium," "transition metals," and "water," in *Sec. 3.4.2*; the practical application of the understanding of "intellectual property" in economics to the field of public policy, as shown by "the World Intellectual Property Organisation" which has pushed "to see the diverse wealth of this world's indigenous, intangible cultural heritage better valued and better protected against probable, ongoing misappropriation and misuse," in *Sec. 3.5.1*; and the practical application of the idea of the great chain of being in philosophy to the field of geology, when "the possibility of extinction was not widely accepted before the 1800s," because "the idea of extinction was threatening to those who held a belief in the Great Chain of Being, a theological position that did not allow for 'missing links,'" in *Sec. 3.4.2*).

And the reverse direction also holds true.

Again, it should be stressed, however, that there are different shades of gray (or different degrees of truth) in the two opposites, and the classification is not necessarily mutually exclusive either. In fact, even when some combinations of the two occur, they only end up sharing the same dialectic relationship, but in a different degree.

Also, there are some other relationships (as a kind of family resemblance) which have something in common with the principle, but they are not exactly the same but only more or less comparable, which varies from case to case.

Good examples of family resemblance in relation to the principle are discovery-application, knowledge-action, invention-innovation, theory-intentionality, discourse-action, knowledge-interest, knowledge-motivation, universality-particularity, potentiality-actuality, and so forth.

18th Thesis: The Convention-Novelty Principle

The eighteenth thesis is the convention-novelty principle (on the nature of creative thinking) in the category of "agency" in existential dialectics (which was first worked out in *FPHCT*), in that, if there is conventional wisdom, there is likewise novel challenge, to the extent that both convergent and divergent thinking are part of life. There is no convention without novelty—and vice versa.

As srummarized from *FPHCT* (almost verbatim here), there are (a) "creative techniques" and (b) "creative traits," which, when satisfied—in relation to the larger context of the mind, nature, society, and culture—can be used to enhance creative works.

In addition, creative thinking has its own possibilities and limits (in relation to invention), just as it has its own promises and pitfalls (in relation to innovation)— as already analyzed in *FPHCT*).

In the end, creative thinking has its own desirability and dark sides (as also already analyzed in *FPHCT*).

With these clarifications in mind—there are good empirical examples for the convention-novelty princple.

For instance, on the one hand, there is convention (e.g., the conventional wisdom about the "latitudinal gradients" in biodiversity, in *Sec. 2.2.1*; the conventional wisdom about the positive impact of agrobiodiversity on dietary health, in *Sec. 2.2.2*; the conventional wisdom about the "hotspot priority approach, in *Sec. 2.3.1*; the conventional wisdom about the two types of adaptive management by C. Holling and C. Walters, namely, "passive and adaptive management practice," in *Sec. 2.3.2*; the conventional wisdom about the process of "somatic cell nuclear transfer" [SCNT], in *Sec. 2.4.1*; the conventional wisdom about "natural resources,...by considering [capital] as interchangeable with human-made capital—labor and technology," in *Sec. 2.4.2*; the conventional wisdom of anthropocentrism, since it is "an ethic reflected in the major interpretations of Western religions and the dominant economic paradigms of the industrialised world," in *Sec. 2.5.1*; the conventional wisdom about birdwatching "in Britain during the Victorian Era," mostly for "collection based taxonomy," in *Sec. 2.5.2*; the conventional wisdom about "the overkill hypothesis" by Paul S. Martin more than 40 years ago, in *Sec. 3.2.1*; the "traditional wisdom and indigenous knowledge for the...health and livelihood security" of a local community in less developed countries, in *Sec. 3.2.2*; in orbital variation, there is "thermal inertia," such that "the depths of the ocean have a lag time in changing temperature," like "a lag of some hundreds of years," in *Sec. 3.3.1*; native species in California used to live, with low density, "on serpentine soils, which have low water-retention, low nutrient levels, a high Magnesium/Calcium ratio, and possible heavy metal toxicity," in *Sec. 3.3.2*; the conventional practice of uncontrolled deforestation which "resulted in damage to habitat, biodiversity loss and aridity" in the "pre-industrial" and "industrial" eras, in *Sec. 3.4.1*; the conventional wisdom about oil in the older days, when it was assumed that "countries and regions with an abundance of natural resources" can lead to higher "economic growth, in *Sec. 3.4.2*; the conventional wisdom about traditional knowledge on a naturalistic basis, in *Sec. 3.5.1*; and the conventional wisdom of rejecting "the idea of extinction" before 1750's, when many "held a belief in the Great Chain of

Being, a theological position that did not allow for 'missing links,'" in *Sec. 3.5.2*).

On the other hand, there is novelty (e.g., the alternative novel challenge to the conventional wisdom about the "latitudinal gradients" in biodiversity, by the new idea that "this characteristic [of latitudinal gradients] is unverified in aquatic ecosystems, especially in marine ecosystems," that "the latitudinal distribution of parasites does not follow this rule," and "other instances of great diversity in higher latitudes have also been recorded," in *Sec. 2.2.1*; the alternative novel challenge to the conventional wisdom about the positive impact of agrobiodiversity on dietary health, by the new idea of "coffee rust" which "attacked coffee plantations in Sri Lanka, Brazil, and Central America in 1970," when scientists discovered that "the diseases are themselves a form of biodiversity," in *Sec. 2.2.2*; the alternative novel challenge to the conventional wisdom about the "hotspot" priority approach, by the new idea of the "coldspot" priority approach, in *Sec. 2.3.1*; the alternative novel challenge to the conventional wisdom about the two types of adaptive management, by the new idea of R. Hilbourne, who proposed instead "three learning models," and "these are reactive, passive and active," in *Sec. 2.3.2*; the alternative novel challenge to the conventional wisdom about the process of "somatic cell nuclear transfer" [SCNT], by the new idea of "artificial embryo splitting or embryo twinning," which "may also be used as a method of cloning," in *Sec. 2.4.1*; the alternative novel challenge to the conventional wisdom about "natural resources" on the basis of "human-made capital—labor and technology," by the new idea of "ecological economics" which focuses on natural capital instead, in *Sec. 2.4.2*; the alternative novel challenge to the conventional wisdom of anthropocentrism, by the new idea of ecocentrism, which is "a radical challenge to long-standing and deeply rooted anthropocentric attitudes in Western culture, science, and politics," in *Sec. 2.5.1*; the alternative novel challenge to the conventional wisdom about about birdwatching "in Britain during the Victorian Era," by the new idea of the BOU," which "changed focus to ecology and behaviour…in the 1940s" so as to study birds in a more "scientfic" way, that is, "the 'scientification' of the pastime," in *Sec. 2.5.2*; the alternative novel challenge to the conventional wisdom about "the overkill hypothesis," by the new idea of "the comet hypothesis" which was "first publicly presented at the Spring 2007 joint assembly of the American Geophysical Union in Acapulco, Mexico," in that "the mass extinction was caused by a swarm of comets 12,900 years ago," in *Sec. 3.2.1*; the alternative novel challenge to the "traditional wisdom and indigenous knowledge for the…health and livelihood security," by the new idea of modern Western prescription medicines "in technologically advanced countries," where "the vast majority of this indigenous knowledge has been lost to society, and is of little consequence to modern 'Western' lifestyles," in *Sec. 3.2.2*; in orbital variation, there are the "Milankovitch cycles" which "drove the ice age cycles," so "CO2 followed temperature change," in *Sec. 3.3.1*; "barbed goatgrass [*Aegilops triuncialis*] was introduced to California" as an invasive species and, by novel uses of "resources…to live on previously uninhabited soil types," was able to "form dense stands on these soils, crowding out native species that have not adapted well to serpentine soils," in *Sec. 3.3.2*; the alternative novel challenge to the conventional practice of uncontrolled deforestation, by the new idea of forest management with such diverse mechanisms as "reducing emissions," "cyclic agriculture," "monitoring deforestation," "reforestation," "forest plantations," and so on, in *Sec. 3.4.1*; the alternative novel challenge to the conventional wisdom about about oil in the older days, by the new idea of "resource curse" nowadays, in that, "countries and regions with an abundance of natural resources, specifically point-source non-renewable resources like minerals and fuels, tend to have less economic growth and worse development outcomes than countries with fewer natural resources," in *Sec. 3.4.2*; the alternative novel challenge to the conventional wisdom about traditional knowledge on a naturalistic basis, by the new idea that "traditional knowledge is not a natural category, and may reflect power struggles and relationships for land, resources and social control rather than adherence to a claimed ancestry or heritage," in *Sec. 3.5.1*; and the alternative novel challenge to the conventional wisdom of rejecting "the idea of extinction" before 1750's, by the new idea of uniformitarianism in the 19th century, which accepts extinction events as part of evolution over a long period of time, in *Sec. 3.5.2*).

And the reverse direction also holds true.

Again, it should be stressed, however, that there are different shades of gray (or different degrees of truth) in the two opposites, and the classification is not necessarily mutually exclusive either. In fact, even when some combinations of the two occur, they only end up sharing the same dialectic relationship, but in a different degree.

Also, there are some other relationships (as a kind of family resemblance) which have something in common with the principle, but they are not exactly the same but only more or less comparable, which varies from case to case.

Good examples of family resemblance in relation to the principle are convergence-divergence (or convergent thinking vs. divergent thinking), normalness-nonnormalness, conformity-nonconformity, familiarity-freshness, inertia-newness, equilibrium-disturbance, and so on.

19th Thesis: The Evolution-Transformation Principle

The nineteenth thesis refers to the evolution-transformation principle (on the multiple kinds of agency). in the category of "agency" in existential dialectics (which was first worked out in *FAE* and then in other books of mine).

This principle—and the symmetry-asymmetry principle, for instance—are both about agency. More precisely, the word "agency," in a formal definition, refers to "a person or thing through which power is exerted or an end is achieved." (MWD 2007) It therefore does not have to necessarily involve an intelligent lifeform.

Because of this dual meaning in agency, the evolution-transformation principle is more concerned with *the kind of agency*, that is, both about the *evolution* in the state of nature (e.g., an object of natural beauty) and the *transformation* in the world of intelligent lifeforms (e.g., a work of art). There is no evolution without transformation—and vice versa.

In classical Darwinian evolutionary theory (as more detailedly analyzed in *BNN*), evolution is "blind." But in the human world, change often takes place because of the conscious intervention of humans in transforming society and culture, just to cite two instances.

And the transformative part of the principle precisely refers to the other dimension in the dual meaning of agency, in giving technology (as invented by intelligent lifeforms like humans and, soon, post-humans) a major role to play in the change of the world, which is something that I extensively analyzed in *FHC* in the context of the technophilic lifeworld, espeically though not exclusively since modern times. The same theme is also elaborated in many other books of mine.

For instance, on the one hand, there is evolution (e.g., the natural evolution of humans to benefit from biodiversity, like the impact on human health, in *Sec. 2.2.1*; the natural evolution of humans to hunt for food in the state of nature in a primitive way, in *Sec. 2.2.2*; the natural evolution of humans to use natural resources in the state of nature in a primitive way, in *Sec. 2.3.1*; the natural evolution of humans to use natural resources in the state of nature in a primitive way and on a relatively much smaller scale, in *Sec. 2.3.2*; the natural evolution of humans and animals to die in the state of nature, in *Sec. 2.4.1*; the natural evolution of humans to consume natural resources in the state of nature in a primitive way, in *Sec. 2.4.2*; the natural evolution of humans to "butcher" animals "for food and clothing" in the state of nature in a primitive way, in *Sec. 2.5.1*; the natural evolution of humans to deal with animals like fish and birds in the state of nature in a primitive way, in *Sec. 2.5.2*; the natural evolution of animals to become extinct in the state of nature over time, in *Sec. 3.2.1*; the natural evolution of different lifeforms to face the challenge of biodiversity loss in the state of nature, in *Sec. 3.2.2*; the natural evolution of climate change in the state of nature, especially in the context of natural causes, with its impact, say, on water deterioration, in *Sec. 3.3.1*; the natural evolution of some native species to become extinct or reduced due to invasive species in the state of nature, in *Sec. 3.3.2*; the natural evolution of deforestation, by natural causes, in the state of nature, like climate change which led to "the Carboniferous Rainforest Collapse," in *Sec. 3.4.1*; the natural evolution of petroleum compounds to form from "plankton," "bacteria," and "other plant and animal matter" over millions of years in the state of nature, in *Sec. 3.4.2*; the natural evolution of humans to form some understanding of the world in the state of nature in a primitive way, in *Sec. 3.5.1*; and the natural evolution of humans to emerge in the history of the Earth, in *Sec. 3.5.2*).

On the other hand, there is transformation (e.g., the technical transformation of human ability to benefit from biodiversity, by the invention of the pharmaceutical sector for "natural product exploration [bioprospecting']," "genomics and synthetic chemistry" for health improvement, in *Sec. 2.2.1*; the technical transformation of human ability to get food by the invention of different agricultural systems, like organic cropping, conventional, and permacultural systems, in *Sec. 2.2.2*; the technical transformation of human ability to use natural resoources by the invention of conservation biology for a more systematic planning which allows the use of natural resources to be sustainable in the long run, in *Sec. 2.3.1*; the technical transformation of human ability to manage natural resources by the invention of the techniques of adaptive management nowadays, in *Sec. 2.3.2*; the technical transformation of human ability to restore life by the invention of cloning, although there are still a lot of works needed to be done, in *Sec. 2.4.1*; the technical transformation of human ability to consumer natural resources by the invention of ecological economics, with its focus on natural capital for sustainable development, in *Sec. 2.4.2*; the technical transformation of human treatment of animals by the invention of anthropocentrism, which "gives rise to human duties…to treat animals humanely," in *Sec. 2.5.1*; the technical transformation of human ability to study animals by the invention of "birdwatching" and "fishkeeping" in a systematic way, as shown in the decision of the BOU," which "changed focus to ecology and behaviour…in the 1940s" so as to study birds in a more "scientfic" way, in *Sec. 2.5.2*; the technical transformation of human ability to understand extinction by the invention of "the overkill hypothesis," together with subsequent conservation efforts to save the endangered species, in *Sec. 3.2.1*; the technical transformation of human ability to deal with biodiversity loss by the invention of "internatonal efforts" and "collaboration across disciplines and across borders" to fight against biodiversity loss for the promotion of human health, such as the formation of "the United Nations Framework Convention on Climate Change [UNFCCC]," in *Sec. 3.2.2*; the technical transformation of human ability to deal with the impact of climate change, say, on water deterioration, by the invention of "water management," for the efforts like "preventative approaches…to avoid high costs of inefficiency," "innovations to decrease overall demand," etc., in *Sec. 3.3.1*; the technical transformation of human ability to deal with biodiversity loss by the invention of nature conservation campaigns, as shown in the efforts by "the Joint Nature Conservation Committee [JNCC] in the UK, which "contributes to maintaining and enriching biological diversity and educating about the effects of the introduction of invasive/non-native species," in *Sec. 3.3.2*; the technical transformation of human ability to deal with deforestation by the invention of forest management, with such diverse mechanisms as "cyclic agriculture," "monitoring deforestation," "reforestation," and so on, in *Sec. 3.4.1*; the technical transformation of human ability to deal with oil depletion by the invention of alternative solutions, like the use of "solar power and "wind power," in *Sec. 3.4.2*; the technical transformation of human ability to understand the world by the invention of "traditional knowledge" and "scientifc knowledge," in *Sec. 3.5.1*; and the technical transformation of human ability to understand the history of the Earth by the invention of different theories like catastrophism and uniformitarianism, in *Sec. 3.5.2*).

And the reverse direction also holds true.

Again, it should be stressed, however, that there are different shades of gray (or different degrees of truth) in the two opposites, and the classification is not necessarily mutually exclusive either. In fact, even when some combinations of the two occur, they only end up sharing the same dialectic relationship, but in a different degree.

Also, there are some other relationships (as a kind of family resemblance) which have something in common with the principle, but they are not exactly the same but only more or less comparable, which varies from case to case.

Good examples of family resemblance in relation to the principle are inorganicness-volition, naturalness-technologicalness, naturalness-nonnaturalness, nonwillingness-willingness, inorganicness-motivation, evolution-creation, naturalness-artificialness, and so on.

20th Thesis: The Symmetry-Asymmetry Principle

The twentieth thesis is labeled as the symmetry-asymmetry principle (on the relationships among existents) in the category of "agency" in existential dialectics (which was also already worked out in my previous books), in that there is no asymmetry without symmetry—and vice versa.

For instance, the Same can be symmetric and asymmetric towards the Others. But in case of asymmetry (as analyzed in *BDPD*), oppression and self-oppression can occur. So, when the Same is asymmetric towards the Others, the Same can also be relatively asymmetric towards itself in self-oppression, just as the Others can be likewise towards themselves. The subsequent oppressiveness is dualistic, as much by the Same against the Others and itself, as by the Others against the Same and themselves. Both oppression and self-oppression can be achieved by way of downgrading differences between the Same and the Others and of accentuating them.

This is true, even though not all forms of asymmetry have to be about oppression and self-oppression.

In addition, from Chapter Three of *FPHG*, symmetry is not perfect, to be understood in an approximate sense under many life circumstances. With this clarification in mind, hereafter are some empirical examples.

For instance, on the one hand, there is symmetry (e.g., the existence of different life forms, like microorganisms, animals, and humans in the world of biodiversity, in *Sec. 2.2.1*; the potato "blight ravaged potato crops throughout Europe" in the middle 19th century, in *Sec. 2.2.2*; the existence of voluntary international agreements for conservation in developed and developing countries, in *Sec. 2.3.1*; the existence of different models of decision making in *Sec. 2.3.2*; the existence of different lifeforms on Earth, like plants and humans, in *Sec. 2.4.1*; the existence of different lifeforms in nature, like humans and animals, in *Sec. 2.4.2*; the existence of different lifeforms like animals and humans, in *Sec. 2.5.1*; the existence of different names for observing birds, in *Sec. 2.5.2*; the existence of different lifeforms like fish and humans, in *Sec. 3.2.1*; the existence of different socioeconomic groups in a society to deal with biodviersity loss and health risks, like the rich and the poor, in *Sec. 3.2.2*; the existence of different ways to release carbon dioxide, like volcanic emissions and fossil fuel combustion, in *Sec. 3.3.1*; the existence of different plants, like

"*Bromus tectorum*, originally from Eurasia" and "native plants" in western North America, in *Sec. 3.3.2*; the existence of different countries to deal with deforestation around the world, in *Sec. 3.4.1*; the existence of different countries like "resource-rich countries of the Third World" and "powerful states such as the United States and other leading western countries," in *Sec. 3.4.2*; the existence of different countries, like "developing" and "developed" countries, in *Sec. 3.5.1*; and the existence of different lifeforms in the world like humans, animals, plants, and minerals, in *Sec. 3.5.2*).

On the other hand, there is asymmetry (e.g., humans can have a greater threat to biodiversity on Earth than animals, as shown in the problems like human overpopulation, overexploitation, etc., in *Sec. 2.2.1*; "although blight ravaged potato crops throughout Europe during the 1840s, the impact and human cost in Ireland…was exacerbated by a host of political, social and economic factors," so that Ireland suffered more than its neighbors, in *Sec. 2.2.2*; "in industrialized countries, voluntary environmental agreements often provide a platform for companies to be recognized for moving beyond the minimum regulatory standards and, thus, support the development of best environmental practice"—but "in developing countries, such as throughout Latin America, these agreements are more commonly used to remedy significant levels of non-compliance with mandatory regulation," in *Sec. 2.3.1*; the popularity of the decision making modeling with the features and processes of adaptive management among those in such organizations as DFG, CMR, LACPR, and so on—but the popularity of the decision making modeling which must be "economically viable" and "technically" feasible among "the majority of global companies operating in the markets," in *Sec. 2.3.2*; plants use "vegetative reproduction or apomixes" naturally—but humans use "somatic cell nuclear transfer" [SCNT] artificially, in *Sec. 2.4.1*; humans have damaged the environment on a larger scale than animals in the history of the Earth, as shown by the human impact on "global warming," in *Sec. 2.4.2*; humans can treat animals in a way on a larger scale that animals cannot do to humans, especially in the exploitative "treatment of the animals we harness as beasts of burden, that we butcher for food and clothing, or that we destroy as disease-bearing pests or as dangerous predators?," in *Sec. 2.5.1*; "in North America…the term *birder* is unknown to most lay people"—whereas the term "birdwatching" is more popular, in *Sec. 2.5.2*; humans can use "one effective approach" like "the individual transferable quotas [ITQs]" which "help prevent collapses and restore fisheries that appear to be in decline"—but fish do not have this ability to save others from extinction, in *Sec. 3.2.1*; "those who are poor or sick, or who experience low levels of livelihood security, are likely to be at greatest risk from such events, and are typically more dependent upon ecosystems for their well-being," in *Sec. 3.2.2*; "the US Geological Survey estimates are that volcanic emissions are at a much lower level than the effects of current human activities, which generate 100-300 times the amount of carbon dioxide emitted by volcanoes," in *Sec. 3.3.1*; "*Bromus tectorum*...is highly fire-adapted" and "not only spreads rapidly after burning, but also increases the frequency and intensity [heat] of fires, by providing large amounts of dry detritus during the fire season in western North America,...so much that native plants cannot survive the frequent fires, allowing *B. tectorum* to further extend and maintain dominance in its introduced range," in *Sec. 3.3.2*; "developing countries feel that some countries in the developed world, such as the United States of America, cut down their forests centuries ago and benefited greatly from this deforestation, that it is hypocritical to deny developing countries the same opportunities, and that the poor shouldn't have to bear the cost of preservation when the rich created the problem," in *Sec. 3.4.1*; "there is often international pressure on resource-rich countries of the Third World to avoid reinvesting resource revenues in social investment purposes, or even in developmental efforts towards economic diversification," and "this pressure comes from powerful states such as the United States and other leading western countries as well as pro-liberalization institutions such as the World Trade Organization and the International Monetary Fund," in *Sec. 3.4.2*; "these practices [of biopiracy] contribute to inequality between developing countries rich in biodiversity, and developed countries hosting companies that engage in 'biopiracy,'" in *Sec. 3.5.1*; and the "hierarchical structure of all matter and life" for those with the belief in the great chain of being, such that humans are ranked higher than animals, plants, and minerals in the chain, in *Sec. 3.5.2*).

And the reverse direction also holds true.

Again, for the last time, it should be stressed, however, that there are different shades of gray (or different degrees of truth) in the two opposites, and the classification is not necessarily mutually exclusive either. In fact, even when some combinations of the two occur, they only end up sharing the same dialectic relationship, but in a different degree.

Also, there are some other relationships (as a kind of family resemblance) which have something in common with the principle, but they are not exactly the same but only more or less comparable, which varies from case to case.

Good examples of family resemblance in relation to the principle are balance-extremity, reflexiveness-unreflexiveness, equality-inequality, harmoniousness-unharrmoniousness, center-periphery, compatibility-incompatibility, proportion-disproportion, balance-unbalance, and something like that.

21st Thesis: The Softness-Hardness Principle

The twenty-first thesis refers to the softness-hardness principle (on the force of change) in the category of "agency" in existential dialectics (which was first worked out in *ALD*), in that any change by an agent, be it organic (like humans) or non-organic (like natural objects), can occur in a forceful (aggressive) or gentle (pacific) way, which can come in all shapes and sizes, of course. There is no softness without hardness—and vice versa.

For instance, on the one hand, there is softness (e.g., the appreciation of the idea of the "latitudinal gradients" in biodiversity by the advocates, in *Sec. 2.2.1*; the defense of the "middleman system" by its advocates in early 19th century Ireland, in *Sec. 2.2.2*; "companies to be recognized for moving beyond the minimum regulatory standards" in developed countries, in *Sec. 2.3.1*; the appreciation by Kai Lee of the classification of adaptive management into two types by Holling and Walters, in *Sec. 2.3.2*; species can be reproduced in nature, as in "apomixes," in *Sec. 2.4.1*; the replacement of natural capital by man-made capital, in *Sec. 2.4.2*; humans can "treat animals humanely," in *Sec. 2.5.1*; the care of fish welfare by aquarian activists, in *Sec. 2.5.2*; the saving of fisheries by humans as shown in the use of "one effective approach" known as "the individual transferable quotas [ITQs]," in *Sec. 3.2.1*; "international efforts" among those in the United Nations Framework Convention on Climate Change [UNFCCC] to fight against biodiversity loss, in *Sec. 3.2.2*; the defense of the scientific evidences of climate change by the United States National Research Council, in *Sec. 3.3.1*; "invasive species often coexist with native species for an extended time," in *Sec. 3.3.2*; the defense of the forest transition theory by the advocates, in *Sec. 3.4.1*; the defense of the resource-curse thesis, as shown by those who argued that, unlike resource-rich countries, "resource-poor economies like Singapore, Taiwan or South Korea,...spent enormous efforts on education, and this contributed in part to their economic success," in *Sec. 3.4.2*; the occurrence of cooperation among parties to CBD, in *Sec. 3.5.1*; and the inclusion of extinction events by those in catastrophism, in *Sec. 3.5.2*).

On the other hand, there is hardness (e.g., the critique of the idea of the "latitudinal gradients" in biodiversity by the critics, in *Sec. 2.2.1*; the critique of the "middleman system" by some critical historians in the article aforecited, in *Sec. 2.2.2*; "significant levels of non-compliance with mandatory regulation" in developing countries, in *Sec. 2.3.1*; the critique by Hilbourne of the classification of adaptive management into two types by Holling and Walters, as Hilbourne proposed three types instead, in *Sec. 2.3.2*; species can be eliminated in nature, as in "extinction," in *Sec. 2.4.1*; the depletion of natural resources, in *Sec. 2.4.2*; humans can kill animals "as beasts of burden," "for food and clothing," as "disease-bearing pests or as dangerous predators," in *Sec. 2.5.1*; the killing of fish by inexperienced aquarists, in *Sec. 2.5.2*; the killing of "the dodo, a flightless bird from Mauritius" by humans, which is "another well known example of overexploitation," in *Sec. 3.2.1*; "some scientists and non-scientists question aspects of climate-change science," in *Sec. 3.2.2*; the critique of the scientific evidences of climate change by Dennis Dinger [2010], who argued that "the Bible has quite a lot to say" about global warming, because "the Bible says God created the heavens and the earth, the Bible teaches that God maintains the daily workings of the earth,...and finally, it says that God controls the wind and the rain," in *Sec. 3.3.1*; "an introduced species...can outcompete native species for resources, such as nutrients, light, physical space, water or food," in *Sec. 3.3.2*; the critique of the forest transition theory by the critics aforecited, in *Sec. 3.4.1*; the critique of the resource-curse thesis, as shown by those who argued that "natural resources generate easily taxable rents that more often than not result in increased spending on education" in resource-rich countries, in *Sec. 3.4.2*; the occurrence of biopiracy against traditional knowledge, in *Sec. 3.5.1*; and the exclusion of extinction events by those with the belief in the idea of the great chain of being, in *Sec. 3.5.2*).

And the reverse direction also holds true.

Again, it should be stressed, however, that there are different shades of gray (or different degrees of truth) in the two opposites, and the classification is not necessarily mutually exclusive either. In fact, even when some combinations of the two occur, they only end up sharing the same dialectic relationship, but in a different degree.

Also, there are some other relationships (as a kind of family resemblance) which have something in common with the principle, but they are not exactly the same but only more or less comparable, which varies from case to case.

Good examples of family resemblance in relation to the principle are peacefulness-violence, cooperation-competition, pacificity-aggression, dovishness-hawkishness, reward-punishment, peace-war, friendliness-unfriendliness, kindness-hostility, love-hatred, acceptance-discrimination, admiration-belittlement, inclusion-exclusion, and the like.

22nd Thesis: The Seriousness-Playfulness Principle

The twenty-second thesis refers to the seriousness-playfulness principle (on the extent of seriousness) in the category of "agency" in existential dialectics (which is first worked out in *FPHH*), in that, if there is seriousness, there is also playfulness, especially for any agent with some kind of intelligence life, be it about humans, animals, or, later, post-humans. There is no seriousness without playfulness—and vice versa.

Lest any misunderstanding occurs, the word "playfulness" here should not be confused with other terms like "laughter," "smile," "joke," "tickleness," and "laugh-like vocalization" (in some animals), for instance.

Being playful can come in all shapes and sizes and therefore is not necessarily tied up with any of these terms, alhough it can result from any of them, or a combination of them, or something else altogether.

This principle, together with some other principles (especially though not exclusively, like the formalness-informalness principle), have important implications for what constitutes rationality, be it about "rationality of application" at the practical level or "rationality of knowledge" at the meta-theoretical level (as summarized in *Sec. 4.20* of *FPHH*).

For instance, on the one hand, there is seriousness (e.g., the serious business of those who study the nature of biodiversity, in *Sec. 2.2.1*; the serious business of those who try to find out the causes of the Irish Potato Famine, in *Sec. 2.2.2*; the serious business of those trying to deal with conservation efforts, in *Sec. 2.3.1*; the serious business of those who try to make adaptive management successful, in *Sec. 2.3.2*; the serious business of those who try to understand the nature of cloning, in *Sec. 2.4.1*; the serious business of those who try to understand the nautre of economics, in *Sec. 2.4.2*; the serious business of those in anthropocentrism to understand man-nature interactions, in *Sec. 2.5.1*; the serious business of those who try to classify birdwatching activities, in *Sec. 2.5.2*; the serious business of those who try to explain the occurrence of extinction events in history, in *Sec. 3.2.1*; the serious business of those who try to make medicinal resources, in *Sec. 3.2.2*; the serious business of those who try to understand climate change, in *Sec. 3.3.1*; the serious business of those who try to understand genetic pollution and invasive species, in *Sec. 3.3.2*; the serious business of those who try to deal with the problem of deforestation, in *Sec. 3.4.1*; the serious business of those who try to predict the impacts of oil depletion in the future, in *Sec. 3.4.2*; the serious business of those who try to protect traditional knowledge against biopiracy, in *Sec.*

3.5.1; and the serious business of those who try to understand extinction events, in *Sec. 3.5.2*).

On the other hand, there is playfulness (e.g., the playful part of those who study the nature of biodiversity, when they play around with different ideas of biodiveristy over time, like "molecular diversity," "genetic diversity," "speices diversity," "ecosystem diversity," etc., in *Sec. 2.2.1*; in *Sec. 2.2.2*; the playful part of those who try to find out the causes of the Irish Potato Famine, when they play around with different possible contributing factors over time, like "laws that restricted the rights of Irish Catholics," the "middleman system," "tenants, subdivisions, and bankruptcy," "potato dependency," "potato blight," and so on, in *Sec. 2.2.2*; the playful part of those trying to deal with conservation efforts, when they play around with different possible approaches over time, such as "voluntary international agreements," "ecosystems approach," "international environmental agreements," etc., in *Sec. 2.3.1*; the playful part of those who try to make adaptive management successful, as they play around with different possibilities in regard to its features and processes over time, in order to make them better, in *Sec. 2.3.2*; the playful part of those who try to understand the nature of cloning, when they play around with different ways of molecular cloning over the years, like the traditional "four steps," "a number of alternative routes," and so on, in *Sec. 2.4.1*; the playful part of those who try to understand the nautre of economics, when they play around with different approaches to economics over the centures, like "neoclassical economics," "ecological economics," "environmental economics," etc., in *Sec. 2.4.2*; the playful part of those in anthropocentrism to understand man-nature interactions, when they play around with different justifications or roots over the decades, like the "philosophical," "religious," and "genetic" arguments aforecited, in *Sec. 2.5.1*; the playful part of those who try to classify birdwatching activities, when they play around with different conceptual distinctions over the decades, like "birding," "birdwatching," "twitching," "chasing," etc., in *Sec. 2.5.2*; the playful part of those who try to explain the occurrence of extinction events in history, when they play around with different possible explanations over the decades, namely, "the overkill hypothesis," "the comet hypothesis," "the climate change hypothesis," the "hyperdisease hypothesis," etc., in *Sec. 3.2.1*; the playful part of those who try to make medicinal resources, when they play around with different options over time, like "traditional medicines," "modern prescription medicines," etc., in *Sec. 3.2.2*; the playful part of those who try to understand climate change, when they play around with different explanations over the decades, like "ocean variability," "orbital variations," "solar output," "volcanism," "plate tectonics," "human influences," "the Bible," in *Sec. 3.3.1*; the playful part of those who try to understand genetic pollution and invasive species, when they play around with different terms over the decades like "genetic deterioration," "genetic pollution," "genetic swamping," "genetic takeover," "genetic mixing," "genetic aggression," and so on, in *Sec. 3.3.2*; the playful part of those who try to deal with the problem of deforestation, when they play around with different mechanisms over the decades, like "reducing emissions," "cyclic agriculture," "monitoring deforestation," "forest management," "reforestation," "forest plantations," and so on, in *Sec. 3.4.1*; the playful part of those who try to predict the impacts of oil depletion in the future, when they play around with different scenarios over time, like "catastrophe," "recession," "increasing food prices," and "replacement with renewable energy," in *Sec. 3.4.2*; the playful part of those who try to protect traditional knowledge against biopiracy, when they play around with different mechanisms over time, like "TRIPs," "CBD," "indegeneous intellectual property," etc., in *Sec. 3.5.1*; and the playful part of those who try to understand extinction events, when they play around with different stories to tell over time, like "the great chain of being," "creation theory," "catastrohism," and "uniformitarianis," in *Sec. 3.4.2*).

And the reverse direction also holds true.

Again, it should be stressed, however, that there are different shades of gray (or different degrees of truth) in the two opposites, and the classification is not necessarily mutually exclusive either. In fact, even when some combinations of the two occur, they only end up sharing the same dialectic relationship, but in a different degree.

Also, there are some other relationships (as a kind of family resemblance) which have something in common with the principle, but they are not exactly the same but only more or less comparable, which varies from case to case.

Good examples of family resemblance in relation to the principle are seriousness-jokingness, seriousness-humorousness, seriousness-wittiness, seriousness-nonseriousness, Apollo-Dionysus, seriousness-restingness, seriousness-relaxedness, seriousness-clownishness, and so forth.

23rd Thesis: The Activeness-Inactiveness Principle

The twenty-third thesis refers to the activeness-inactiveness principle (on the extent of activeness) in the category of "agency" in existential dialectics (which was first worked out in *FPHTRAN*), in that, if there is activeness, there is also inactiveness, especially (though not exclusively) for any agent with some kind of intelligence life, be it about humans, animals, or, later, post-humans. There is no activeness without inactiveness—and vice versa.

Being active can come in all shapes and sizes, just as a person (or an animal) can be inactive because he (or it) is tired, old, injured, sick, sleepy (or hibernating), uninterested, in a state of passive inertia (or disposition), etc.—and an object can be inactive because it is at rest, out of work (or order), etc.

This principle, together with some other principles (especially though not exclusively, like the formalness-informalness principle and seriousness-playfulness principle), have important implications for what constitutes rationality, be it about "rationality of application" at the practical level or "rationality of knowledge" at the meta-theoretical level (as summarized in *Sec. 4.20* of *FPHH*).

For instance, on the one hand, there is activeness (e.g., the relative activeness of those species that can survive in changing natural circumstances to "increase disease transmission," such as "that of West Nile Virus, Lyme disease and Hantavirus, according to a study" by Felicia Keesing and her colleagues, in *Sec. 2.2.1*; the relative activeness of those who were lucky enugh to not suffer from the Irish Potato Famine in the 1848-52 period and were able to work, in *Sec. 2.2.2*; the activeness of those who are willing to "mov[e]...beyond the minimum regulatory standards," so as to do more for conservation efforts, in *Sec. 2.3.1*; the relative activeness of those ordinary employees in "lateral decision making," where they are encouraged to engage in "affective participation," in *Sec. 2.3.2*; the activeness of those clones that survive birth, like Dolly the sheep which lived from 1996 to 2006, in *Sec. 2.4.1*; the relative activeness of those lands for food production, when they are healthy in being benefited from "organic farming" and do not suffer from soil problems like "erosion, an impending water crisis, soil salinity [from irrigation]," etc., in *Sec. 2.4.2*; the relative activeness of those animals which are treated humanely by humans and are able to survive and live healthily, in *Sec. 2.5.1*; "the most active times of the year for birding in temperate zones are during the spring or fall migrations when the greatest variety of birds may be seen," in *Sec. 2.5.2*; the relative activeness of those "non-dinosaurian archosaurs" which did not become sick or were lucky enough to survive in "the Triassic-Jurassic Extinction Event," in *Sec. 3.2.1*; the relative activeness of those who are rich and not sick, so they are not "at greatest risk from such events [like natural disasters]" and can engage in acitivities that they prefer, in *Sec. 3.2.2*; the relative activeness of those volcanoes which are still "active," in *Sec. 3.3.1*; the relative activeness of those plants like "*Bromus tectorum*, originally from Eurasia" which "is highly fire-adapted," so that not only they can survive fire but also they "spread…rapidly after burning" and "increase…the frequency and intensity [heat] of fires, by providing large amounts of dry detritus during the fire season in western North America," in *Sec. 3.3.2*; the relative activeness of biodiversity in in the Amazon basin, since "much of what remains of the world's rainforests is in the Amazon basin, where the Amazon Rainforest covers approximately 4 million square kilometres," in *Sec. 3.4.1*; the relative activeness of those in the U.S. nowadays, which remains one of the most stable countries on earth and attracts a lot of immigrants seeking for opportunities, when contrasted with those in "the Democratic Republic of the Congo," who "suffer" or die "from disease, civil war, and the resource curse, in *Sec. 3.4.2*; the relative activeness of those biopirates who are able to use, for profit, "indigenous knowledge…originating with indigenous people…without permission from and with little or no compensation or recognition to the indigenous people themselves," in *Sec. 3.5.1*; and the relative activeness of those animals which survived catastrophic events in history and were able to engage in different life activities in *Sec. 3.5.2*).

On the other hand, there is inactiveness (e.g., the relative inactiveness of those species that cannot survive in changing natural circumstances, so that they cannot "increase disease transmission," in *Sec. 2.2.1*; the relative inactiveness of those who were unfortunate enugh to get sick or die due to the Irish Potato Famine in the 1848-52 period, so they could not engage in productive work, in *Sec. 2.2.2*; the relative inactiveness of those who show "non-compliance with mandatory regulation" because of "poor governance," "corruption," etc., in *Sec. 3.2.1*; the relative inactiveness of those ordinary employees in "traditional…hierarchal decision making," where they are to follow what is decided by those at the top instead, in *Sec. 2.3.1*; the relative inactiveness of those clones that did not quite survive birth and died shortly after, like Bessie the cow, which died two days after birth in 2001, in *Sec. 2.4.1*; the relative inactiveness of those lands for food production, when they suffer from soil problems like "erosion, an impending water crisis, soil salinity [from irrigation]," etc., in *Sec. 2.4.2*; the relative inactiveness of those animals which are caught or killed by humans, because they are treated "as beasts of burden," "for food and clothing," as "disease-bearing pests or as dangerous predators," in *Sec. 2.5.1*; the most inactive times of the year for birding in temperate zones are outside the periods of the spring or fall migrations, when the least variety of birds may be seen, in *Sec. 2.5.2*; the relative inactiveness of those "non-dinosaurian archosaurs" which got sick or could not survive in "the Triassic-Jurassic Extinction Event," in *Sec. 3.2.1*; the relative inactiveness of those "who are poor or sick, or who experience low levels of livelihood security" and who "are likely to be at greatest risk from such events, and are typically more dependent upon ecosystems for their well-being," so that they are more likely to be unable to engage much in acitivities that they prefer, in *Sec. 3.2.1*; the relative inactiveness of those volcanoes which are either "dormant" or "extinct," in *Sec. 3.3.1*; the relative inactiveness of those plants like "native plants" in western North America, when they "cannot survive the frequent fires, allowing *B. tectorum* to further extend and maintain dominance in its introduced range," in *Sec. 3.3.2*; the relative inactiveness of biodiversity "in South Asia," where "about 88% of the rainforests have been lost," in *Sec. 3.3.2*; the relative inactiveness of those common folks like those in "the Democratic Republic of the Congo," who "suffer" or die "from disease, civil war, and the resource curse," in *Sec. 3.4.1*; the relative inactiveness of other biopirates who do not succeed to exploit traditional knowledge without permission or compensation, because of the mechanisms to protect traditional knowledge nowadays, in *Sec. 3.4.2*; the relative inactiveness of other biopirates who do not succeed to exploit traditional knowledge without permission or compensation, because of the mechanisms to protect traditional knowledge nowadays, in *Sec. 3.5.1*; and the relative inactiveness of those animals which did not survive catastrophic events in history and became so injured, sick, or hurt that they were unable to engage in different life activities, in *Sec. 3.5.2*).

And the reverse direction also holds true.

Again, it should be stressed, however, that there are different shades of gray (or different degrees of truth) in the two opposites, and the classification is not necessarily mutually exclusive either. In fact, even when some combinations of the two occur, they only end up sharing the same dialectic relationship, but in a different degree.

Also, there are some other relationships (as a kind of family resemblance) which have something in common with the principle, but they are not exactly the same but only more or less comparable, which varies from case to case.

Good examples of family resemblance in relation to the principle are alertness-unalertness, activeness-decrepidness, alertness-sleepiness, activeness-tiredness, alertness-unwareness, alertness-uninterestedness, soberness-inactiveness, alertness-inactiveness, soberness-unawareness, awakenness-hibernatedness, and so on.

24th Thesis: The Selfness-Otherness Principle

The twenty-fourth thesis refers to the selfness-otherness principle (on the magnitude of self-centeredness) in the category of "agency" in existential dialectics (which is first worked out in *FPHCA*), in that, if there is the tendency to be mostly centered on one's self or group, there is likewise the tendency to be mostly centered on others, especially (though not exclusively) for any agent with some kind of intelligence life, be it about humans, animals, or, later, post-humans. There is no selfness without otherness—and vice versa.

Being self-centered and being other-centered do not have to be mutually exclusive, as the two can come in all shapes and sizes, or in different degrees. And other-centeredness does not entail desirability, any more than self-centeredness implies non-desirability, as both can go either way.

In addition, whether self-centeredness constitutes a moral alternative or, conversely, an alternative to morality is of course debatable (as addressed in my other book titled *Beyond Ethics to Post-Ethics*).

For instance, on the one hand, there is selfness (e.g., the tendency to be mostly centered on one's self or group, as shown in "the latitudinal distribution of parasites," in *Sec. 2.2.1*; the tendency to be mostly centered on one's self or group, as shown by those who advocated the "middleman system" in 19th century Ireland, because they benefited from it at the expense of others, in *Sec. 2.2.2*; the tendency to be mostly centered on one's self or group, as shown by those who show "non-compliance with mandatory regulation" because of greed and corruption, in *Sec. 2.3.1*; the tendency to be mostly centered on one's self or group, as shown by "the majority of global companies operating in the markets," as they are reluctant to adopt adaptive management, because it is not "economically viable" and "technically" feasible, as their major concern is profit-making in *Sec. 2.3.2*; the tendency to be mostly centered on one's self or group, as shown in the greed of South Korea's top human cloning scientist Hwang Woo-suk, when his 2005 research on "the world's first cloned dog" was later found to be "fabricated," in *Sec. 2.4.1*; the tendency to be mostly centered on one's self or group, as shown in those profit-seeking companies which are interesting in exploiting "the rainforest" of "the Yasuni region," in *Sec. 2.4.2*; the tendency to be mostly centered on one's self or group, as shown by those "technocentrics, including imperialists," who "have absolute faith in technology and industry," so they go around to colonize and exploit nature for their own interests, in *Sec. 2.5.1*; the tendency to be mostly centered on one's self or group, as shown by those who try to win "goldfish as prizes at funfairs," in *Sec. 2.5.2*; the tendency to be mostly centered on one's self or group, as shown in the killing of "the mamo bird" to the point of overexploitation, for "the ceremonial cloaks worn by the Hawaiian kings," in *Sec. 3.2.1*; the tendency to be mostly centered on one's self or group, as shown by those organizatons like "the libertarian Competitive Enterprise Institute, conservative commentators, and some companies such as ExxonMobil" which "have challenged IPCC climate change scenarios" and pursue their own interests in regard to global warming and biodiversity, in *Sec. 3.2.2*; the tendency to be mostly centered on one's self or group, as shown in "carbon dioxide emissions from fossil fuel combustion," "aerosols [particulate matter in the atmosphere]," and "cement manufacture," by those companies which engaged in these activities for profit-seeking, in *Sec. 3.3.1*; the tendency to be mostly centered on one's self or group, as shown by "some conservation biologists and conservationists" which "have used genetic pollution for a number of years as a term to describe gene flow [which they disapprove of]," so that "the term 'genetic pollution'…is a propaganda term, not a technical or scientific term," for the pursuit of their own group agenda, in *Sec. 3.3.2*; the tendency to be mostly centered on one's self or group, as shown in those "countries in the developed world, such as the United States of America," which "cut down their forests centuries ago and benefited greatly from this deforestation" but now "deny developing countries the same opportunities," on the basis of the power politics of the North-South divide, in *Sec. 3.4.1*; the tendency to be mostly centered on one's self or group, as shown by those "authoritarian…high-ranking politicians and military leaders" in resource-cursed countries, who "rule without fear of consequence" for their own greed, in *Sec. 3.4.2*; the tendency to be mostly centered on one's self or group, as shown by the exploitation of traditional knowledge by biopirates, in *Sec. 3.5.1*; and the tendency to be mostly centered on one's self or group, as shown by some humans in "poaching," such that "species were deliberately driven to extinction," in *Sec. 3.5.2*).

On the other hand, there is otherness (e.g., the tendency to be mostly centered on others, as shown in those species that "most likely to disappear" but that can "buffer against infectious disease transmission" to benefit others like humans, according to a study" by Felicia Keesing and her colleagues, in *Sec. 2.2.1*; the tendency to be mostly centered on others, as shown by those critical historians who tried to provide a more objective analysis of the Irish Potato Famine for the advancement of historical knowledge, in *Sec. 2.2.2*; the tendency to be mostly centered on others, as shown by those who care a lot about "the tragedy of the commons" by participating in conservation efforts for the future benefits of humanity, in *Sec. 2.3.1*; the tendency to be mostly centered on others, as shown in those non-profit organizations like FOS and others, which "continued to actively work in promoting adaptive management for conservation projects and program," such that humanity can benefit from a more sustainable way to use natural resources, in *Sec. 2.3.2*; the tendency to be mostly centered on others, as shown in the proven cases of cloning like Dolly the sheep, which aim to provide "a viable outcome of the continuing advancements in our science and technology" for the human future, in *Sec. 2.4.1*; the tendency to be mostly centered on others, as shown in the effort of "the government of Ecuador" to "preserve the rainforest" of "the Yasuni region," together with the political support of "Bill Clinton, Paul Martin and other former world leaders," for the benefit of humanity, in *Sec. 2.4.2*; the tendency to be mostly centered on others, as shown by those ecocentrists, who "believe in the intrinsic value of nature," in *Sec. 2.5.1*; the tendency to be mostly centered on others, as shown by those aquarian activists who try to protect fish welfare by campaigning for banning the game of winning "goldfish as prizes at funfairs," since they succeeded to persuade "the United Kingdom," which "outlawed live-animal prizes such as goldfish in 2004," in *Sec. 2.5.2*; the tendency to be mostly centered on others, as shown in the saving of fisheries by those activists who used the "effective approach" known as "the individual transferable quotas [ITQs]," in *Sec. 3.2.1*; the tendency to be mostly centered on others, as shown by those parties to "the United Nations Framework Convention on Climate Change (UNFCCC)" at Cancún in 2010, which produced a voluntary, though not binding, agreement, for a global campaign against global warming and biodiversity loss, in *Sec. 3.2.2*; the tendency to be mostly centered on others, as shown in the efforts by the High Level Threat Panel of the United Nations to address environmental degradation caused by climate change, for the sustainable development of future humanity, in *Sec. 3.3.1*; the tendency to be mostly centered on others, as shown by some scientists like J. Rhymer and D. Simberloff [1996] who proposed a more neutral term like "gene mixing," for the enhancement of scientific knowledge, not for the propaganda of some groups, in *Sec. 3.3.2*; the tendency to be mostly centered on others, as shown in the efforts of some "international organizations including the United Nations and the World Bank," which "have begun to develop programs aimed at curbing deforestation," with different "monetary or other incentives to encourage developing countries to limit and/or roll back deforestation," for a better human future, in *Sec. 3.4.1*; the tendency to be mostly centered on others, as shown by "the Center for Global Development," which "advocates direct distribution of revenues," such that "a government would transfer some or all of the revenue from natural resource extraction to citizens in a universal, transparent, and regular payment,...thus forcing...the government to be transparent and accountable in its management of natural resource revenues," for the future of better global development, in *Sec. 3.4.2*; the tendency to be mostly centered on others, as shown by "the World Intellectual Property Organisation," which had pushed "to see the diverse wealth of this world's indigenous, intangible cultural heritage better valued and better protected against probable, ongoing misappropriation and misuse," so as to protect traditional knowledge as part of biodiversity which is endangered and requires protection, for the betterment of future humanity,in *Sec. 3.5.1*; and the tendency to be mostly centered on others, as shown in the design of creation by God, according to those with the belief in creation theory, in *Sec. 3.5.2*).

And the reverse direction also holds true.

Again, it should be stressed, however, that there are different shades of gray (or different degrees of truth) in the two opposites, and the classification is not necessarily mutually exclusive either. In fact, even when some combinations of the two occur, they only end up sharing the same dialectic relationship, but in a different degree.

Also, there are some other relationships (as a kind of family resemblance) which have something in common with the principle, but they are not exactly the same but only more or less comparable, which varies from case to case.

Good examples of family resemblance in relation to the principle are egoisticness-altruisticness, selfishness-unselfishness, selfness-selflessness, egocentricness-othercentricness, and so on.

In Relation to Outcome

The next five theses are in relation to outcome in existential dialectics, as shown below.

25th Thesis: The Regression-Progression Principle

The twenty-fifth thesis is called the regression-progression princple (on the direction of history) in the category of "outcome" in existential dialectics (which was also already worked out in my previous books), in that neither the cyclical nor the linear views are adequate for explaining many phenomena at all levels. There is no regression without progression—and vice versa.

History progresses to more advanced forms, but with a regressive touch. Examples include no freedom without unfreedom, no equality without inequality, and no civilization without barbarity. This is not an inevitable law, but merely a highly likely empirical trend.

For instance, on the one hand, there is regression (e.g., the regression made by the study of biodiversity and psychological benefits, as shown in the problems pointed out by the critics, in *Sec. 2.2.1*; agrobiodiversity can contribute to crop diseases resulting in human suffering, in *Sec. 2.2.2*; the regression made by the study on conservation biology and environmental protection , as shown in the problems pointed out by the critics aforecited, in *Sec. 2.3.1*; the regression made by adaptive management, as shown in the problems pointed out by the critics aforecited, in *Sec. 2.3.2*; the regression made by cloning, as shown in the problems pointed out by the critics, in *Sec. 2.4.1*; the regression made by ecological economics, as shown in the problems pointed out by the critics aforecited, in *Sec. 2.4.2*; the regression made by environmental ethics, as shown in the problems pointed out by the critics, in *Sec. 2.5.1*; the regression made by animal observing, as shown in the problems with fishkeeping and birdwatching pointed out by the critics, in *Sec. 2.5.2*; the regression made by the overkill hypothesis, as shown in the problems pointed out by the critics aforecited, in *Sec. 3.2.1*; the regression made by the study of biodiversity loss and health risks, as shown in the problems pointed out by the critics aforecited, in *Sec. 3.2.2*; the regression made by the study on climate change, as shown in the problems pointed out by the critics aforecited, in *Sec. 3.3.1*; the regression made by the study of genetic pollution and invasive species, as shown in the problems pointed out by the critics, in *Sec. 3.3.2*; the regression made by the study on deforestation, as shown in the problems pointed out by the critics aforecited, in *Sec. 3.4.1*; the regression made by the availiability of natural resources, as shown in the problems of "resource curse" pointed out by the critics aforecited, in *Sec. 3.4.2*; the regression made by the mechanisms against biopiracy, as shown in the problems pointed out by the critics, in *Sec. 3.5.1*; and the regression made by the study on extinction, as shown in the problems pointed out by the critics, in *Sec. 3.5.2*).

On the other hand, there is progress (e.g., the progress made by the study of biodiversity and psychological benefits, as shown in its contribution to a better understanding of the impact of biodiversity on human health aforecited, in *Sec. 2.2.1*; agrobiodiversity can contribute to food supply for human dietary health, in *Sec. 2.2.2*; the progress made by the study of conservation biology and environmental protection, as shown in its contribution to a better protection of natural resources aforecited, in

Sec. 2.3.1; the progress made by adaptive management, as shown in its contribution to a better way to make decision in the face of uncertainty, while addressing ecological concerns aforecited, in *Sec. 2.3.2*; in *Sec. 2.4.1*; the progress made by cloning, as shown in its contribution to the restoration of life for biodiversity aforecited, in *Sec. 2.4.1*; the progress made by ecological economics, as shown in its contribution to more investment in natural capital for sustainable development aforecited, in *Sec. 2.4.2*; the progress made by environmental ethics, as shown in its contribution to a more humane treatment of nature aforecited, in *Sec. 2.5.1*; the progress made by animal observing, as shown in its contribution to a better understanding of fish and bird behaviors aforecited, in *Sec. 2.5.2*; the progress made by the overkill hypothesis, as shown in its contribution to a better understanding of the causes of overexploitation aforecited, in *Sec. 3.2.1*; the progress made by the study of biodiversity loss and health risks, as shown in its contribution to a better understanding of the relationsip between biodiversity loss and health risks aforecited, in *Sec. 3.2.2*; the progress made by the study on climate change, as shown in its contribution to the understanding of its causes aforecited, in *Sec. 3.3.1*; the progress made by the study of genetic pollution and invasive species, as shown in the contribution to a better understanding about the relationship between invasive species and native species in a given habitat aforecited, in *Sec. 3.3.2*; the progress made by the study on deforestation, as shown in its contribution to a better understanding of the causes and impacts aforecited, in *Sec. 3.4.1*; the progress made by the availiability of natural resources, as shown in the contribution to "economic growth and…development" aforecited, in *Sec. 3.4.2*; the progress made by the mechanisms against biopiracy, as shown in the contribution to the protection of traditional knowledge aforecited, in *Sec. 3.5.1*; and the progress made by the study on extinction, as shown in the contribution to a better understanding of extinction events on Earth aforecited, in *Sec. 3.5.2*).

And the reverse direction also holds true.

Again, it should be stressed, however, that there are different shades of gray (or different degrees of truth) in the two opposites, and the classification is not necessarily mutually exclusive either. In fact, even when some combinations of the two occur, they only end up sharing the same dialectic relationship, but in a different degree.

Also, there are some other relationships (as a kind of family resemblance) which have something in common with the principle, but they are not exactly the same but only more or less comparable, which varies from case to case.

Good examples of family resemblance in relation to the principle are undesirability-desirability, risk-opportunity, badness-goodness, cost-benefit, dysfunctionality-functionality, cyclicity-linearity, and so on.

26th Thesis: The Same-Difference Principle

The twenty-sixth thesis refers to the same-difference principle (on the metamorphosis of change) in the category of "outcome" in existential dialectics (which was worked out in *ALD* and other books of mine), in that an entity, as it evolves over time, can be both different from and similar to its opposing alternatives and does not have to be solely more different from them over time. There is no similarity without difference—and vice versa.

Opposites are not absolute in a black-or-white fashion; so, an entity can become relatively more similar to (or more different from) its opposite over time.

This is further constrained by another principle, that is, the symmetry-asymmetry principle about the relationships among existents under the category about agency in existential dialectics, in that if there is symmetry (equality) between two entities, there is likewise asymmetry (inequality) emerging in a different way.

For instance, on the one hand, there is similarity in outcome (e.g., the contribution to the molding and control of beliefs, values, and behaviors, regardless of whether this be done by way of the evolutinoary theory to explain biodiversity, or by way of the creation theory to explain biodiversity, in *Sec. 2.2.1*; the contribution to the molding and control of beliefs, values, and behaviors, regardless of whether this be done by way of focusing on agrobiodiversity in understanding the Irish Potato Famine, or by way of focusing on "political, social and economic factors" in understanding the Irish Potato Famine, in *Sec. 2.2.2*; the contribution to the molding and control of beliefs, values, and behaviors, regardless of whether this be done by way of an extrinsic [utilitarian] conservation effort, or by way of intrinsic conservation effort, in *Sec. 2.3.1*; the contribution to the molding and control of beliefs, values, and behaviors, regardless of whether this be done by way of "lateral decision making" or by way of "hierarchal decision making," in *Sec. 2.3.2*; the contribution to the molding and control of beliefs, values, and behaviors, regardless of whether this be done by way of the restoration of life by cloning, or by way of the beginning of life only at conception in the Catholic tradition, in *Sec. 2.4.1*; the contribution to the molding and control of beliefs, values, and behaviors, regardless of whether this be done by way of ecological economics, or by way of mainstream economics, in *Sec. 2.4.2*; the contribution to the molding and control of beliefs, values, and behaviors, regardless of whether this be done by way of ecocentricism, or by way of technocentrism, in *Sec. 2.5.1*; the contribution to the molding and control of beliefs, values, and behaviors, regardless of whether this be done by way of banning the game of winning "goldfish as prizes at funfairs," or by way of winning "goldfish as prizes at funfairs," in *Sec. 2.5.2*; the contribution to the molding and control of beliefs, values, and behaviors, regardless of whether this be done by way of overexploitation, or by way of conservation, in *Sec. 3.2.1*; the contribution to the molding and control of beliefs, values, and behaviors, regardless of whether this be done by way of the UNFCCC, or by way of individual groups like "the libertarian Competitive Enterprise Institute, conservative commentators, and some companies such as ExxonMobil," in *Sec. 3.2.2*; the contribution to the molding and control of beliefs, values, and behaviors, regardless of whether this be done by way of using the report on climate change by the United States National Research Council, or by way of using the Bible on climate change, in *Sec. 3.3.1*; the contribution to the molding and control of beliefs, values, and behaviors, regardless of whether this be done by way of the use of a propaganda term like "genetic pollution," or by way of the use of a technical term like "genetic mixing," in *Sec. 3.3.2*; the contribution to the molding and control of beliefs, values, and behaviors, regardless of whether this be done by way of the North-South divide, or by way of the curbing mechanisms of forest management, in *Sec. 3.4.1*; the contribution to the molding and control of beliefs, values, and behaviors, regardless of whether this be done by way of the "high-ranking politicians and military leaders" in resource-cursed countries who "rule without fear of consequence," or by way of "the Center for Global Development" for "direct distribution of revenues," in *Sec. 3.4.2*; the contribution to the molding and control of beliefs, values, and behaviors, regardless of whether this be done by way of traditional knowledge, or by way of scientific knowledge, in *Sec. 3.5.1*; and the contribution to the molding and control of beliefs, values, and behaviors, regardless of whether this be done by way of creation theory, or by way of uniformitarianism, in *Sec. 3.5.2*).

On the other hand, there is difference in outcome (e.g., the contribution to the molding and control of beliefs, values, and behaviors by way of the evolutinoary theory to explain biodiversity for a relatively more scientific worldview—but the contribution to the molding and control of beliefs, values, and behaviors by way of the creation theory to explain biodiversity for a relatively more religious worldview, in *Sec.*

2.2.1; the contribution to the molding and control of beliefs, values, and behaviors by way of focusing on agrobiodiversity for a relatively more naturalistic worldview—but the contribution to the molding and control of beliefs, values, and behaviors by way of focusing on "political, social and economic factors" for a relatively more societal worldview, in *Sec. 2.2.2*; the contribution to the molding and control of beliefs, values, and behaviors by way of an extrinsic [utilitarian] conservation effort for a relatively more consequentialist lifestyle—but the contribution to the molding and control of beliefs, values, and behaviors by way of intrinsic conservation effort for a relatively more deontological lifestyle, in *Sec. 2.3.1*; the contribution to the molding and control of beliefs, values, and behaviors by way of "lateral decision making" for a relatively more egalitarian lifeworld—but the contribution to the molding and control of beliefs, values, and behaviors by way of "hierarchal decision making" for a relatively more inegalitarian lifeworld, in *Sec. 2.3.2*; the contribution to the molding and control of beliefs, values, and behaviors by way of the restoration of life by cloning for a relatively more technophilic lifeworld—but the contribution to the molding and control of beliefs, values, and behaviors by way of the beginning of life only at conception in the Catholic tradition for a relatively more religious lifeworld, in *Sec. 2.4.1*; the contribution to the molding and control of beliefs, values, and behaviors by way of ecological economics for a relatively more technologically pessimist worldview—but the contribution to the molding and control of beliefs, values, and behaviors by way of mainstream economics for a relatively more technologically optimist worldview, in *Sec. 2.4.2*; the contribution to the molding and control of beliefs, values, and behaviors by way of ecocentricism for a relatively more naturalistic lifeworld—but the contribution to the molding and control of beliefs, values, and behaviors by way of technocentrism for a relatively more technophilic lifeworld, in *Sec. 2.5.1*; the contribution to the molding and control of beliefs, values, and behaviors by way of banning the game of winning "goldfish as prizes at funfairs," for a relatively more compassionate lifeworld for animal welfare—but the contribution to the molding and control of beliefs, values, and behaviors by way of winning "goldfish as prizes at funfairs" for a relatively more competitive lifeworld for material gains, in *Sec. 2.5.2*; the contribution to the molding and control of beliefs, values, and behaviors by way of overexploitation for a relatively more overconsumeristic lifeworld—but the contribution to the molding and control of beliefs, values, and behaviors by way of conservation for a relatively more sustainable lifeworld, in *Sec. 3.2.1*; the contribution to the molding and control of beliefs, values, and behaviors by way of the UNFCCC for a relatively more cooperative lifeworld—but the contribution to the molding and control of beliefs, values, and behaviors by way of individual groups for a relatively more competitive lifeworld, in *Sec. 3.2.2*; the contribution to the molding and control of beliefs, values, and behaviors by way of using the report on climate change by the United States National Research Council for a relatively more scientific worldview—but the contribution to the molding and control of beliefs, values, and behaviors by way of using the Bible on climate change for a relatively more religious worldview, in *Sec. 3.3.1*; the contribution to the molding and control of beliefs, values, and behaviors by way of the use of a propaganda term like "genetic pollution" for a relatively more politicized worldview—but the contribution to the molding and control of beliefs, values, and behaviors by way of the use of a technical term like "genetic mixing" for a relatively more scientific worldview, in *Sec. 3.3.2*; the contribution to the molding and control of beliefs, values, and behaviors by way of the North-South divide for a relatively more politicized lifeworld—but the contribution to the molding and control of beliefs, values, and behaviors by way of the curbing mechanisms of forest management for a relatively more technocratic lifeworld, in *Sec. 3.4.1*; the contribution to the molding and control of beliefs, values, and behaviors by way of the "high-ranking politicians and military leaders" in resource-cursed countries for a relatively more authoritarian lifeworld—but the contribution to the molding and control of beliefs, values, and behaviors by way of "the Center for Global Development" for a relatively more egalitarian lifeworld, in *Sec. 3.4.2*; the contribution to the molding and control of beliefs, values, and behaviors by way of traditional knowledge for a relatively more spiritual worldview—but the contribution to the molding and control of beliefs, values, and behaviors by way of scientific knowledge for a relatively more secular worldview, in *Sec. 3.5.1*; and the contribution to the molding and control of beliefs, values, and behaviors by way of creation theory for a relatively more religious worldview—but the contribution to the molding and control of beliefs, values, and behaviors by way of uniformitarianism for a relatively more secular worldview, *Sec. 3.5.2*).

And the reverse direction also holds true.

Again, it should be stressed, however, that there are different shades of gray (or different degrees of truth) in the two opposites, and the classification is not necessarily mutually exclusive either. In fact, even when some combinations of the two occur, they only end up sharing the same dialectic relationship, but in a different degree.

Also, there are some other relationships (as a kind of family resemblance) which have something in common with the principle, but they are not exactly the same but only more or less comparable, which varies from case to case.

Good examples of family resemblance in relation to the principle are homogeneity-heterogeneity, we-they, similarity-contrast, and so forth.

27th Thesis: The Stability-Reaction Principle

The twenty-seventh thesis is called the stability-reaction princple (on the feedback mechanism for further changes) in the category of "outcome" in existential dialectics (which was first worked out in *FPHCHEM*), in that once an outcome becomes stabilized over time (which varies from case to case), it will undergo further changes in its multifaceted interactions with the outside world (e.g., with other entities, energy, space, and so on).

There is no stability without reaction for further changes—and vice versa.

Works on chemical reactions (as in *FPHCHEM*) and feedback mechanisms for learning in adaptive management (as in *BNR*), just to name a few of them, have contributed to the understanding of stability and reaction.

For instance, on the one hand, there is stability in outcome (e.g., the relatively more stable condition of the belief in the "latitudinal gradients" in biodiversity among the advocates, in *Sec. 2.2.1*; the relatively more stable condition of the potato dependency in 19th century Ireland, which had contribtued to the Irish Potato Famine, in *Sec. 2.2.2*; the relatively more stable condition of the "hotspot" priority approach, as shown by the fact that "the predominant approach to date has been to focus efforts on endangered species by conserving biodiversity hotspots," in *Sec. 2.3.1*; the relatively more stable condition of the "hierarchal decision making" in the older days, in *Sec. 2.3.2*; the relatively more stable condition of the idea about the beginning of life only at conception in the Catholic tradition in the older days, in *Sec. 2.4.1*; the relatively more stable condition of neoclassical economics in the older days, with its focus on "the Economic Man" as "a utility maximizing consumer," in *Sec. 2.4.2*; the relatively more stable condition of anthropocentrism in the older days, because of its being "an ethic reflected in the major interpretations of Western religions and the dominant economic paradigms of the industrialised world," in *Sec. 2.5.1*; the relatively more stable condition of winning "goldfish as prizes at funfairs" in the older days, as a "tradition…in many parts of the world," in

Sec. 2.5.2; the relatively more stable condition of the overkill hypothesis in the older days, when it was used to explain the extinction event in "the Quaternary period," in *Sec. 3.2.1*; the relatively more stable condition of "traditional wisdom and indigenous knowledge for the…health and livelihood security" of a local community, in *Sec. 3.2.2*; the relatively more stable condition of the scientific explanation of climate change in modern times, as shown in the view by the United States National Research Council that there is "a strong, credible body of evidence, based on multiple lines of research, documenting that climate is changing," in *Sec. 3.3.1*; the relatively more stable condition of "part of an ecosystem" which is not yet "disturbed" by human activities like land use, agricutlure, and so on, or by different species, in *Sec. 3.3.2*; the relatively more stable condition of deforestation in the pre-historical era, since there was only "small scale deforestation…practiced by some societies for tens of thousands of years before the beginnings of civilization," in *Sec. 3.4.1*; the relatively more stable condition of oil dependency in the older days, when, for example, "over 90% of transportation in the United States relies on oil," in *Sec. 3.4.2*; the relatively more stable condition of traditional knowledge in the older days, when it showed "the diversity and interactions among plant and animal species, landforms, watercourses, and other qualities of the biophysical environment in a given place,'" in *Sec. 3.5.1*; and the relatively more stable condition of the idea of the great chain of being in the older days, when "the possibility of extinction was not widely accepted before the 1800s," in *Sec. 3.5.2*).

On the other hand, there is reaction in outcome (e.g., the belief in the "latitudinal gradients" in biodiversity among the advocates led to further changes later on, when some critics nowadays suggest instead that "this characteristic [of latitudinal gradients] is unverified in aquatic ecosystems, especially in marine ecosystems," that "the latitudinal distribution of parasites does not follow this rule," and "other instances of great diversity in higher latitudes have also been recorded," in *Sec. 2.2.1*; the potato dependency in 19th century Ireland which had contribtued to the Irish Potato Famine then led to further changes later on, when the Irish became less dependent on potatoes over time by consuming other resources, even though this took many decades for them to overcome it, in *Sec. 2.2.2*; the "hotspot" priority approach then led to further changes later on, when nowadays, "some scientists" argue that "it is more cost effective, logical, and socially relevant to invest in *biodiversity coldspots*" instead, in *Sec. 2.3.1*; the "hierarchal decision making" in the older days then led to further changes later on, when nowadays, there are alternative models like "lateral decision making" in adaptive management for learning, in *Sec.*

2.3.2; the idea about the beginning of life only at conception in the Catholic tradition in the older days then led to further changes later on, when many scientists of cloning nowadays work on the restoration of life through the use of biotechnology, in *Sec. 2.4.1*; neoclassical economics in the older days then led to further changes later on, when nowadays "ecological economics" adds "natural capital…to the typical capital asset analysis of land, labor, and financial capital," since "ecological economists argue that neoclassical economics has ignored the environment," in *Sec. 2.4.2*; anthropocentrism in the older days then led to further changes later on, when it has to come to terms with its new rival like ecocentrism nowadays, in *Sec. 2.5.1*; the tradition of winning "goldfish as prizes at funfairs" in the older days then led to further changes later on, when, nowadays, for example, "the United Kingdom outlawed live-animal prizes such as goldfish in 2004," in *Sec. 2.5.2*; the stable condition of the overkill hypothesis in the older days then led to further changes later on, when nowadays, there are alternative hypotheses, like "the comet hypothesis" "the hyperdisease hypothesis," etc., in *Sec. 3.2.1*; traditional wisdom and indigenous knowledge for the…health and livelihood security" then led to further changes later on, when, nowadays, especially "in technologically advanced countries, the vast majority of this indigenous knowledge has been lost to society, and is of little consequence to modern 'Western' lifestyles," with their modern prescription medicines, in *Sec. 3.2.2*; the stable conditon of the scientific explanation of climate change in modern times then led to further changes later on, when nowadays,, there are some people like Dennis Dinger who argue that "the Bible has quite a lot to say" about global warming, in *Sec. 3.3.1*; the undistrubed part of an ecosystems can then lead to further changes later on, when "an ecosystem is disturbed as when land is cleared for agriculture," so that "the boundary between remaining undisturbed habitat and the newly cleared land itself forms a distinct habitat, creating new winners and losers and possibly hosting species that would not thrive outside the boundary habitat," in *Sec. 3.3.2*; "small scale deforestation" in the pre-historical era led to further changes later on, when in the industrial era, "global deforestation sharply accelerated around 1852, in *Sec. 3.4.1*; oil dependency in the older days then led to further changes later on, when, nowadays, there is an increasing call for the replacement with renewable resources, like "solar power and "wind power," *Sec. 3.4.2*; traditional knowledge then led to further changes later on, when nowadays, there is scientific knowledge which is "acquired by naturalistic trial and error," but not "through direct revelation through conversations with 'the creator,' spirits, or ancestors" as in traditional knowledge, in *Sec. 3.5.1*; and the idea of the great chain of being in the older days then led to further changes later on, when "today most geologists combine catastrophist and uniformitarianist standpoints, taking the view that Earth's history is a slow, gradual story punctuated by occasional natural catastrophic events that have affected Earth and its inhabitants," in *Sec. 3.5.2*).

And the reverse direction also holds true.

Again, it should be stressed, however, that there are different shades of gray (or different degrees of truth) in the two opposites, and the classification is not necessarily mutually exclusive either. In fact, even when some combinations of the two occur, they only end up sharing the same dialectic relationship, but in a different degree.

Also, there are some other relationships (as a kind of family resemblance) which have something in common with the principle, but they are not exactly the same but only more or less comparable, which varies from case to case.

Good examples of family resemblance in relation to the principle are stability-forwardness, normalcy-feedback, balance-forwardness, balance-reaction, consolidation-modification, normalcy-feedback, stability-breakup, rest-motion, and the like.

28th Thesis: The Functionality-Nonfunctionality Principle

The twenty-eighth thesis is called the functionality-nonfunctionality princple (on the presence of function) in the category of "outcome" in existential dialectics (which is first worked out in *FPHCA*), in that, if there is the presence of action for which something or someone is fitted, there is likewise the relative lack of action for which something or someone is fitted. There is no functionality without nonfunctionallity—and vice versa.

A *function* should not be confused with an *alternative function,* which is just another kind of function, just as *nonfunction* should not be confused with *dysfunction* (which has more to do with malfunction and is thus related more to the regression-progression principle).

In addition, being functional (or useful) and being nonfunctional (or useless) do not have to be mutually exclusive, as the two can come in all shapes and sizes, or in different degrees. And what is functional at one time may become nonfunctional at another time, and vice-versa.

For instance, on the one hand, there is functionality in outcome (e.g., the presence of action for which something is fitted, as shown in the trend that "contemporary biodiversity physics is 'firmly fixated on the visible [macroscopic] world,'" which serves the function of understanding biodiversity in mainstream society, in *Sec. 2.2.1*; the presence of action for which something is fitted, as shown in the potato dependency in 19th century Ireland, as it served the essential function of providing food supply for the Irish at the time, in *Sec. 2.2.2*; the presence of action for which something is fitted, as shown in the study that "Alaskan Tlingit peoples and the Haida of the Pacific Northwest had resource boundaries, rules, and restrictions among clans with respect to the fishing of Sockeye Salmon," as these rules served the function of resolving "communal resource conflict" at the time, in *Sec. 2.3.1*; the presence of action for which something is fitted, as shown in the "hierarchal decision making" in the older days, when it served the function of the inegalitarian social world at the time, in *Sec. 2.3.2*; the presence of action for which something is fitted, as shown in the idea about the beginning of life only at conception in the Catholic tradition in the older days, which served the function of religious beliefs at the time, in *Sec. 2.4.1*; the presence of action for which something is fitted, as shown in the popularity of neoclassical economics in the older days, when it served the function of economic rationality along the utilitarian line at the time, in *Sec. 2.4.2*; the presence of action for which something is fitted, as shown in anthropocentrism in the older days, which served the important function of being "an ethic reflected in the major interpretations of Western religions and the dominant economic paradigms of the industrialised world," in *Sec. 2.5.1*; the presence of action for which something is fitted, as shown in "the study of birds and natural history…in Britain during the Victorian Era," which served the function of "collection based taxonomy" at the time, in *Sec. 2.5.2*; the presence of action for which something is fitted, as shown in the killing of "the mamo bird" in Hawaii in the older days, which served the function of the provision of "the ceremonial cloaks worn by the Hawaiian kings" at the time; in *Sec. 3.2.1*; the presence of action for which something is fitted, as shown in the use of "traditional wisdom and indigenous knowledge" in the older days, which served the function of providing "the…health and livelihood security" of a local community at the time, in *Sec. 3.2.2*; the presence of action for which something is fitted, as shown in the use of the Bible to explain a lot of different things in medieval times, which served the function of understanding the world at the time, in *Sec. 3.3.1*; the presence of action for which something is fitted, as shown in the use of the term "invasive species" in the older days, which served the function of providing "a nomenclature term and categorization phrase used for flora and fauna, and for specific restoration-preservation processes in native habitats," in *Sec. 3.3.2*; the presence of action for which something is fitted, as shown in the "small scale deforestation" in the pre-historical era, which served the function of providing the needs of "some societies for tens of thousands of years before the beginnings of civilization" in that era, in *Sec. 3.4.1*; the presence of action for which something is fitted, as shown in the use of oil in modern times, which served the function of "economic growth and prosperity since the industrial revolution, in *Sec. 3.4.2*; the presence of action for which something is fitted, as shown in the belief in traditional knowledge in developing countries, which serves the function of providing a body of knowledge to understand "the diversity and interactions among plant and animal species, landforms, watercourses, and other qualities of the biophysical environment in a given place," in *Sec. 3.5.1*; and the presence of action for which something is fitted, as shown in the belief in the idea of the great chain of being in the older days, which served the function of explaining the "hierarchical structure of all matter and life" on the basis of the Platonic Forms at the time, in *Sec. 3.5.2*).

On the other hand, there is nonfunctionality in outcome (e.g., the relative lack of action for which something is fitted, as shown in the call by some critical scientists for an analysis of biodiversity not only in "the visible [macroscopic] world'" but also in the non-visible "microbial" world, because "the vast majority of Earth's species are microbial" and non-visible, such that the older way of being "fixated on the visible [macroscopic] world" is becoming more and more untenable and no longer functional in the coming future of research on biodiversity, in *Sec. 2.2.1*; the relative lack of action for which something is fitted, as shown in the less potato dependency in 21st century Ireland, when higher living standards allow better nutrition, so that the potato dependency is no longer functional for the Irish nowadays, because of the current higher consumption level, in *Sec. 2.2.2*; the relative lack of action for which something is fitted, as shown in different "international environmental agreements" nowadays, like "the Kyoto Protocol, Vienna Convention on the Protection of the Ozone Layer and Rio Declaration on Development and Environment," so the "resource boundaries, rules, and restrictions" among the Alaskan Tlingit peoples and the Haida of the Pacific Northwest are no longer adequately functional in a globalized world at our time, in *Sec. 2.3.1*; the relative lack of action for which something is fitted, as shown in the "lateral decision making" in adaptive management nowadays, so that the "hierarchal decision making" in the older days is no longer functional in our relatively more democratic time, in *Sec. 2.3.2*; the relative lack of action for which something is fitted, as shown in the modern view of using biotechnology to restore life through cloning, so that the idea about the beginning of life only at conception in the Catholic tradition in the older days can no longer serve the function of dealing with biodiveristy in our time, especially in light of extinct and endangered species, among other life forms, in *Sec. 2.4.1*; the relative lack of action for which something is fitted, as shown in the rise of ecological economics nowadays, with its focus on natural capital for sustainable development, so that the neoclassical utilitarian mindset in the older days becomes increasingly nonfunctional in our time, when the expectations are higher, in switching from "growth [quantitative increase in economic output]" to "development [qualitative improvement of the quality of life]," in *Sec. 2.4.2*; the relative lack of action for which something is fitted, as shown in the rise of ecocentrism in our time to challenge anthropocentrism, which is no longer functional as the dominant legitimate ethics nowadays, since a "2012 documentary *The Superior Human?* systematically analyzes anthropocentrism and concludes that value is fundamentally an opinion, and since life forms naturally value their own traits, some humans are misled to believe that they are actually more valuable than other species," in *Sec. 2.5.1*; and the relative lack of action for which something is fitted, as shown in the decision of the BOU," which "changed focus to ecology and behaviour…in the 1940s" so as to study birds in a more "scientfic" way, so that birdwatching solely for "collection based taxonomy" is no longer functional in our time, which has different expectations and needs, in *Sec. 2.5.2*; the relative lack of action for which something is fitted, as shown in the efforts by conservationists in our time to protect endangered species, so that killing a species to the point of overexploitation is no longer functional nowadays, because of the more conscious awareness of the need for conservation for sustainability, in *Sec. 3.2.1*; the relative lack of action for which something is fitted, as shown in the use of modern prescription medicines "in technologically advanced countries," where "the vast majority of this indigenous knowledge has been lost to society," so that the "traditional wisdom and indigenous knowledge" is no longer functional in these societies of our time, which has different scientific standards for "modern 'Western' lifestyles," in *Sec. 3.2.2*; the relative lack of action for which something is fitted, as shown in the use of science to explain the causes of climate change, such that the appeal to the Bible by Dennis Dinger does not draw a lot of scholarly support nowadays and is no longer much functional in our time, with its more scientific expectations instead, in *Sec. 3.3.1*; the relative lack of action for which something or someone is fitted, as shown in "the new nomenclature system based on biogeography rather than on taxa" to understand invasive species by R. Colautti and Hugh MacIsaac [2004], in that "the model evaluated individual populations rather than entire species" and "classified each population based on its success in that environment," so that the old usage of "invasive species" is no longer functional much in our time, because nowadays "the phrase *invasive species* is often criticized as an imprecise term for the scientific field of ecology," in *Sec. 3.3.2*; the relative lack of action for which something is fitted, as shown in the massive deforestation in the industrial era, like "the massive use of charcoal on an industrial scale in Early Modern Europe," such that the "small scale deforestation" in the pre-historical era is no longer function in our itme, because of different needs in the industrial era, in *Sec. 3.4.1*; the relative lack of action for which something is fitted, as shown in an increasing call for the replacement with renewable resources like "solar power," "wind power," and so on, such that the excessive oil dependency is no longer functional in our time, because of different needs and challenges, in *Sec. 3.4.2*; the relative lack of action for which something is fitted, as shown in the appeal to scientific knowledge in developed countries nowadays, such that traditional knowledge is no longer functional much in our time, because of different needs and expectations for a naturalistic basis of understanding, although it tries to make good use of traditional knowledge for some patents as shown in the mechanisms for its protection as part of endangered biodiversity, in *Sec. 3.5.1*; and the relative lack of action for which something is fitted, as shown in the belief in the combination of catastrophism and uniformitarianism nowadays, such that the idea of the great chain of being is no longer functional in our time, because of different scientific standards and expectations in *Sec. 3.5.2*).

And the reverse direction also holds true.

Again, it should be stressed, however, that there are different shades of gray (or different degrees of truth) in the two opposites, and the classification is not necessarily mutually exclusive either. In fact, even when some combinations of the two occur, they only end up sharing the same dialectic relationship, but in a different degree.

Also, there are some other relationships (as a kind of family resemblance) which have something in common with the principle, but they are not exactly the same but only more or less comparable, which varies from case to case.

Good examples of family resemblance in relation to the principle are usefulness-nonusefulness, valueness-nonvalueness, usefulness-uselessness, and the like.

29th Thesis: The Intentionality-Nonintentionality Principle

The twenty-ninth thesis is called the intentionality-nonintentionality princple (on the planning for outcomes) in the category of "outcome" in existential dialectics (which is first worked out in *FPHW*), in that, if there is the planning or design of something for a certain outcome, there is likewise the relative absence of planning or design of something for a certain outcome. There is no intentionality without nonintentionallity—and vice versa.

An intention for an outcome does not have to be certain, so the outcome can even be counter-intentional in being the opposite of what was originally planned. In this sense, counter-intentionality is a special case of non-intentionality; in other words, there are unintented consequences which can be just the opposite of what was originally planned.

An intention for an outcome is mostly relevant to intelligent life, but there are non-intelligent life and non-living matter in the universe too, so nonintentionality is also important.

In addition, an intention does not have to be obvious or straightforward, as William Shakspeare once famously said: "I must be cruel only to be kind."

In any event, on the one hand, there is intentionality in outcome (e.g., the planning or design of something for a certain outcome, as shown in the empirical study by Fuller and his colleagues for a confirmation of the causal relationship between psychological benefits and biodiversity, in *Sec. 2.2.1*; the planning or design of something for a certain outcome, as shown in the investigation by some scientists of the outbreak of "coffee rust" which "attacked coffee plantations in Sri Lanka, Brazil, and Central America in 1970," in *Sec. 2.2.2*; the planning or design of something for a certain outcome, as shown in the "voluntary international agreements" in developing countries, which "are more commonly used to remedy significant levels of non-compliance with mandatory regulation," in *Sec. 2.3.1*; the planning or design of something for a certain outcome, as shown in the processes of adaptive management, which are designed to deal with decision making in the face of uncertainty, in *Sec. 2.3.2*; the planning or design of something for a certain outcome, as shown in the advocation of cloning to resolve the threat to biodiversity, in *Sec. 2.4.1*; the planning or design of something for a certain outcome, as shown in the rise of ecological economics to focus on natural capital for sustainable development, in *Sec. 2.4.2*; the planning or design of something for a certain outcome, as shown in the use of ecocentrism to justify "the intrinsic value of nature while maintaining an understanding that 'human beings must use and even exploit nature to survive and live,'" in *Sec. 2.5.1*; the planning or design of something for a certain outcome, as shown by aquarists who keep fish alive so as to enjoy them as a leisure activity," in *Sec. 2.5.2*; the planning or design of something for a certain outcome, as shown in the use of "the hyperdisease hypothesis" as an alternative to "the overkill hypothesis," by claiming that "humans or animals traveling with them [e.g.,dogs]...introduced one or more highly virulent diseases into vulnerable populations of native mammals, eventually causing extinctions," in *Sec. 3.2.1*; the planning or design of something for a certain outcome, as shown in the use of "bioprospecting" for "the exploration and exploitation of chemical compounds from natural sources" which is important for human health and economic growth, in *Sec. 3.2.2*; the planning or design of something for a certain outcome, as shown in the study to consider "solar variability" as a cause of climate change, in *Sec. 3.3.1*; the planning or design of something for a certain outcome, as shown in "hybridization," when it is intended to "crossbreed...two genetically distinct varieties" in genetic engineering, in *Sec. 3.3.2*; the planning or design of something for a certain outcome, as shown in "a 2002 analysis of satellite imagery," which tried to find out the true rate of massive deforestation in our time, in *Sec. 3.4.1*; the planning or design of something for a certain outcome, as shown in the attempt by the Dutch, when they "discovered a massive natural gas field in Groningen in 1959" and tried to "tap this resource...to export the gas for profit," in *Sec. 3.4.2*; the planning or design of something for a certain outcome, as shown in the formation of traditional knowledge for a naturalistic understanding of the world, especially in regard to "the diversity and interactions among plant and animal species, landforms, watercourses, and other qualities of the biophysical environment in a given place," in *Sec. 3.5.1*; and the planning or design of something for a certain outcome, as shown in the intention of Georges Cuvier to use catastrophism "to explain the patterns of extinction and faunal succession that he and others were observing in the fossil record," but without "any reference to Noah's flood" or to "divine creation," in *Sec. 3.5.2*).

On the other hand, there is nonintentionality in outcome (e.g., the relative absence of planning or design of something for a certain outcome, such that the outcome can even be counter-intentional in being the opposite of what was originally planned, as shown in the findings by Fuller and his colleagues, when they came to realize that the relationship between psychological benefits and biodiversity may not be causal but can be illusionary, because people may simply "prefer...those that enhance their psychological well-being," so "selection by individuals of environments that enhance personal well-being makes interpretation of the causal structure of these relationships difficult," in *Sec. 2.2.1*; the relative absence of planning or design of something for a certain outcome, such that the outcome can even be counter-intentional in being the opposite of what was originally planned, as shown in the investigation by some scientists of the outbreak of "coffee rust" in Sri Lanka, Brazil, and Central America in 1970," when they discovered, to their surprise. that "the diseases are themselves a form of biodiversity," so agrobiodiversity has its dark side too, in *Sec. 2.2.2*; the relative absence of planning or design of something for a certain outcome, such that the outcome can even be counter-intentional in being the opposite of what was originally planned, as shown in the "voluntary international agreements" in developing countries, where, "due to the difficulties inherent in evaluating effectiveness, their use is often questioned and, indeed, the environment may well be adversely affected as a result" instead, in *Sec. 2.3.1*; the relative absence of planning or design of something for a certain outcome, such that the outcome can even be counter-intentional in being the opposite of what was originally planned, as shown in the processes of adaptive management, which can backfire at times, so the failures "can seriously compromise effective adaptive management decision making" instead, in *Sec. 2.3.2*; the relative absence of planning or design of something for a certain outcome, such that the outcome can even be counter-intentional in being the opposite of what was originally planned, as shown in the warning in 2006 by David Ehrenfeld who showed that "cloning…is likely to siphon funds from established and working projects and does not address any of the issues underlying animal extinction [such as habitat destruction, hunting or other overexploitation, and an impoverished gene pool]," contrary to the original goal, in *Sec. 2.4.1*; the relative absence of planning or design of something for a certain outcome, such that the outcome can even be counter-intentional in being the opposite of what was originally planned, as shown in the criticism that ecological economics evolves over time to become more mainstream and betrays itself, contrary to its original goal, as "this has been referred to as ecologists 'selling out on Nature,'" in *Sec. 2.4.2*; the relative absence of planning or design of something for a certain outcome, such that the outcome can even be counter-intentional in being the opposite of what was originally planned, as shown in the critique of ecocentrism, in that it "risks sacrificing human well-being for the sake of an ill-defined 'greater good'" in nature, contrary to its original claim, in *Sec. 2.5.1*; the relative absence of planning or design of something for a certain outcome, such that the outcome can even be counter-intentional in being the opposite of what was originally planned, as shown by "inexperienced aquarists," who "often keep too many fish in one tank, or add fish too quickly into an immature aquarium, killing many of them," contrary to the original goal of keeping them alive as a leisure activity, in *Sec. 2.5.2*; the relative absence of planning or design of something for a certain outcome, such that the outcome can even be counter-intentional in being the opposite of what was originally planned, as shown in the failure of "the hyperdisease hypothesis" as an alternative to "the overkill hypothesis," because, for the critics, "it can not account for several major extinction events, notably Australia and North America" when "dogs did not arrive in Australia until approximately 35,000 years after the first humans arrived and approximately 30,000 years after the megafaunal extinction was complete," contary to the origiinal goal of the hyperdisease hypothesis, in *Sec. 3.2.1*; the relative absence of planning or design of something for a certain outcome, such that the outcome can even be counter-intentional in being the opposite of what was originally planned, as shown in the overexploitation of bioprospecting," such that "unsustainable patterns of exploitation which impact on biodiversity can jeopardize these resources" and in turn negatively affect economic development—contrary to the original goal of the enhancement of human well-being, in *Sec. 3.2.2*; the relative absence of planning or design of something for a certain outcome, such that the outcome can even be counter-intentional in being the opposite of what was originally planned, as shown in "a 2010 study" which "suggests, 'that the effects of solar variability on temperature throughout the atmosphere may be contrary to current expectations,'" in *Sec. 3.3.1*; the relative absence of planning or design of something for a certain outcome, such that the outcome can even be counter-intentional in being the opposite of what was originally planned, as shown by the work of Jeremy Rifkin which showed that "intentional crossbreeding between two genetically distinct varieties" can end up producing "problems that might occur due to the unintentional process of [modernly] genetically modified organisms [GMOs] dispersing their genes into the natural environment by breeding with wild plants or animals," with extinction or reduction of species, contrary to the original goal of creating new ones, in *Sec. 3.3.2*; the relative absence of planning or design of something for a certain outcome, such that the outcome can even be counter-intentional in being the opposite of what was originally planned, as shown in the "2002 analysis of satellite imagery," which "suggested that the rate of deforestation in the humid tropics [approximately 5.8 million hectares per year] was roughly 23% lower than the most commonly quoted rates," contrary to the original expectation of a higher rate of deforestation in our time, in *Sec. 3.4.1*; the relative absence of planning or design of something for a certain outcome, such that the outcome can even be counter-intentional in being the opposite of what was originally planned, as shown in the "Dutch Disease," when the Dutch discovered, to their surprise, that, "when the gas began to flow out of the country so too did its ability to compete against other countries' exports," contrary to their original expectations, in *Sec. 3.4.2*; the relative absence of planning or design of something for a certain outcome, such that the outcome can even be counter-intentional in being the opposite of what was originally planned, as shown in the criticism that "the spiritual component of 'traditional knowledge' can be used to justify any activity, including the unsustainable harvesting of resources," contrary to its original naturalistic claims, in *Sec. 3.5.1*; and the relative absence of planning or design of something for a certain outcome, such that the outcome can even be counter-intentional in being the opposite of what was originally planned, as shown in the work of Robert Jameson, who "translated the introduction Cuvier wrote for a collection of his papers on fossil quadrupeds that discussed his ideas on catastrophic extinction into English" but "added extensive editorial notes to the translation that explicitly linked the latest of Cuvier's revolutions with the biblical flood," contrary to the original non-religious focus of Cuvier, in *Sec. 3.5.2*).

And the reverse direction also holds true.

Again, it should be stressed, however, that there are different shades of gray (or different degrees of truth) in the two opposites, and the classification is not necessarily mutually exclusive either. In fact, even when some combinations of the two occur, they only end up sharing the same dialectic relationship, but in a different degree.

Also, there are some other relationships (as a kind of family resemblance) which have something in common with the principle, but they are not exactly the same but only more or less comparable, which varies from case to case.

Good examples of family resemblance in relation to the principle are purposiveness-nonpurposiveness, intention-accidentalness, design-randomness, etc.

30th Thesis: The Post-Human Resilience

And the thirtieth thesis is about the role of "post-humans," which I originally proposed in my previous books, starting with the first book titled *The Future of Human Civilization* in 2000 and all others afterwards.

As already pointed out in *Sec. 1.8*, I need to emphasize, as this is something that I used to repeat (*almost verbatim*) from my previous books, two clarifications here about the term "post-human" as a neologism in my works.

Firstly, the word "post-human" here should *not* be confused with another term which looks similar but has a totally different meaning in the literature of postmodernism, namely, "post-humanism"—which constitutes a critique of "humanism" as tradionally understood (especially, though not exclusively, in relation to the idea of progress in science and reason in the Enlightenment project). (WK 2008)

My works reject the project of "postmodernism" and propose the future world of what I orignally called "after-postmodernity" in *FHC* and *FCD*, for instance.

And secondly, the word "post-human" here should also *not* be confused with a similar term which is used to champion the ideology of technology for the future co-existence between humans and cyborgs in "trans-humanism." (WK 2008a)

Instead, my term "post-human" in relation to "posthuman-ism" also rejects "transhumanism" (especially, though not exclusively, in relation to the promises of technology) and refers to something else altogether, that is, the future extinction of humans and its post-human successors in deep space and beyond unto multiverses.

My critique of "transhumanism" was more extensively elabroated in *Sec. 2.4.1* of *BEPE*—just as my critique of different futuristic scenarios of fictional artificial intelligence in science fiction was given in *Sec. 3.5.2* of *FPHCOM.*

With these two clarifications in mind (as summarized in *Table 1.31*)—the post-human resilience in the study of natural resources can be addressed in relation to five main directions, based on my *resilient theory of natural resources*, as analysed in this book (together with my visions as already worked out in my previous books).

Firstly, the post-human resilience in the study of natural resources will learn from my resilient theory of natural resources in relation to the dialectic context of no diversity without discontinuity (and vice versa), in the context of natural resources.

Secondly, the post-human resilience in the study of natural resources will learn from my resilient theory of natural resources in relation to the four perspectives of the mind, nature, society, and culture.

Thirdly, the post-human resilience in the study of natural resources will learn from my resilient theory of natural resources in relation to the non-privilege of any specific theory over others in the literature and the non-integration of them all (as they are not necessarily compatible with each other).

Fourthly, the post-human resilience in the study of natural resources will have to confront the new challenges as posed by the long-term civilizational development of intelligent life in the distant future, both here on this planet Earth and elsewhere in deep space until multiverses.

As this is something that I regularly pointed out in my previous books for background information (and summarize here, *almost verbatim*, with some updates), I already worked out, in my numerous books, what these new challenges will be and provide different original visions to meet them—especially in regard to the following future evolutions:

- of the mind (e.g., "the supersession computing," "metamorphic humor," "the contrarian personality," "the hyper-martial body," "the hyper-sexual body," "floating consciousness," "hyper-spatial consciousness," "thinking machines," "unfolding unconsciousness," "thinking robots," "genetically altered superior beings," "cyborgs," and others)
- of nature (e.g., "transdisciplinary architecture," "creational chemistry," "multilateral sound," "resettlement geology," "post-cosmology," "the alteration of space-time," "the creation of new matter-energy," "selective geometry," and the like)
- of society (e.g., "post-capitalism," "post-democracy," "transfigurative waste," "panoramic transportation," "cyclical-progressive migration," "reflective criminology," "interventive-reshaping geography," "detached gambling," "heterodox education," "multifaceted war and peace," "virtual organizations," the movement of "cyclical progression" at both structural and systemic levels, "ambivalent technology," and so forth)
- of culture (e.g., "inquisitive culinary art," "interactive semantics," "transdisciplinary performing arts," "comparative-impartial literature," "post-human mind games," "comparative-substitutive religion," "post-ethics," "post-civilization," "transformative aesthetic experience," "contrastive mathematical-logic," and the whatnot)
- in history (e.g., "multifold history," the age of "after-postmodernity," etc.)
- in methodology (e.g., "sophisticated methodologial holism," "critical-dialectic formal science," "complex data analysis," etc.)
- in ontology (e.g., "existential dialectics," "contrastive rationality," etc.)

Of course, the examples (as listed above) are not exhaustive but illustrative, since my numerous books have worked out many other visions (as already roughly summarized in *Sec.1.8* and *Sec. 1.9*). Many, though not all, of my visions on the mind, nature, society, culture, history, methodology, and ontology (as cited above) are summarized in the tables as shown in Chapter One and Chapter Four (especially in *Table 4.43* about my original theories on numerous topics).

And finally, or fifthly, the post-human resilience in the study of natural resources will learn from my resilient theory of natural resources in regard to the need to go beyond diversity and discontinuity (as further elaborated in the next section).

Towards the Post-Human Resilience

This "post-human" resilience in the study of natural resources to transcend both diversity and discontinuity can be understood in terms of nine great future transformations of natural resources, in relation to the mind, nature, society, and culture—as explained below (and summarized in *Table 1.2*).

Nine Great Future Transformations of Natural Resources

The nine great future transformations of natural resources (in relation to the mind, nature, society, and culture) can be explained hereafter, with the caveat, however, that the classification in terms of nine is solely aesthetic and utilitarian (illustrative), as there is no objective basis that it must be classified in term of nine (as this is something that I often stressed and repeated in my previous books, albeit in the context of different topics).

The First Great Future Transformation of Natural Resources

The first great future transformation of natural resources in the post-human era concerns what I originally call *new goals,* in relation to the mind.

This search for "new goals" will give future advanced humans and later their successors, namely, the post-humans, the capacity to be motivated to find natural resources in a way that current humans cannot.

There are already some indications that the future of natural resources is moving toward this direction.

Consider, for illustration, the different "goals" as already seen in the history of natural resources, such as the search for human health from biodiversity (*Sec. 2.2.1*), the quest for dietary health from agrobiodiversity (*Sec. 2.2.2*), the protection of natural resources (*Sec. 2.3.1*), the aim of adaptive management in the face of uncertainty (*Sec. 2.3.2*), the restoration of life (*Sec. 2.4.1*), the enhancement of life quality with nature (*Sec. 2.4.2*), the ethics of man-nature interactions (*Sec. 2.5.1*), the leisure activities of animal watching (*Sec. 2.5.2*), the fight against overexploitation and overconsumption (*Sec. 3.2.1*), the war on infectoius disease (*Sec. 3.2.2*), the concern with "anthropogenic factors" of climate change (*Sec. 3.3.1*), the awareness of genetic pollution (*Sec. 3.3.2*), the goal of reforestation (*Sec. 3.4.1*), the quest for renewable resources (*Sec. 3.4.2*), the fight against biopiracy (*Sec. 3.5.1*), and the goal of scientific objectivity (*Sec. 3.5.2*), for instance.

More importantly, this future quest for new goals should also be understood in relation to my previous works on the future "post-human" supersession of the human mind from the combined perspectives of the mind, nature, society, and culture—with good examples like "the supersession computing," "metamorphic humor," "the contrarian personality," "the hyper-martial body," "the hyper-sexual body," "floating consciousness," "hyper-spatial consciousness," "thinking machines," "unfolding unconsciousness," "thinking robots," "genetically altered superior beings," "cyborgs," and others.

In any event, all these are just some illustrative (relatively primitive) examples in our time, of course, but the important point to remember here is that the search for different new goals will further transform the study of natural resources at the more advanced level in the post-human future that the human world has never known.

The Second Great Future Transformation of Natural Resources

The second great future transformation of natural resources in the post-human era concerns what I originally call *new means,* in relation to the mind.

This quest for "new means" will give future advanced humans and later their successors, namely, the post-humans, the capacity to find novel options to achieve the goals in regard to natural resources in a way that current humans cannot.

There are already some indications that the future of natural resources is moving toward this direction.

Consider, for illustration, the different "means" as already seen in the history of natural resources, such as the use of biodiversity for drug discovery (*Sec. 2.2.1*), conservation efforts for sustainable food production (*Sec. 2.2.2*), voluntary environmental agreements for the protection of natural resources (*Sec. 2.3.1*), the use of decisions to improve learning (*Sec. 2.3.2*), cloning for the restoration of life (*Sec. 2.4.1*), investment in natural capital for the enhancement of life quality (*Sec. 2.4.2*), ecocentrism for enviromental ethics (*Sec. 2.5.1*), the use of air-travel for birdwatching (*Sec. 2.5.2*), the use of timber for housing (*Sec. 3.2.1*), the use of the monitoring of ecosystems for disaster prevention (*Sec. 3.2.2*), the use of the Milankovitch circles to study climate change (*Sec. 3.3.1*), the use of transport vehicles by organisms in species invasion (*Sec. 3.3.2*), the use of cyclic agriculture to fight deforestation (*Sec. 3.4.1*), the use of production cutbacks by OPEC to keep prices high (*Sec. 3.4.2*), the use of indigenous intellectual property to fight biopiracy (*Sec. 3.5.1*), and the use of Cuvier's work by Jameson for his religious translation (*Sec. 3.5.2*), for instance.

All these are just some illustrative (relatively primitive) examples in our time, of course, but the important point to remember here is that the search for different new means will further transform the study of natural resources at the more advanced level in the post-human future that the human world has never known.

The Third Great Future Transformation of Natural Resources

The third great future transformation of natural resources in the post-human era concerns what I originally call *new comparisons*, in relation to nature.

This search for "new comparisons" will give future advanced humans and later their successors, namely, the post-humans, the ability to learn from animals, plants, and the universe, through comparison and contrast, about the study of natural resources in a way that current humans cannot.

There are already some indications that the future of natural resources is moving toward this direction.

Consider, for illustration, the different forms of comparison and contrast, such as as already seen in the history of natural resources, such as the impact of potato disease on both crops and humans (*Sec. 2.2.2*), the impact of logging and mining on humans and forests (*Sec. 2.3.1*), the impact of climate change on humans and wildlife (*Sec. 2.3.2*), the occurrence of cloning in both nature and the human world (*Sec. 2.4.1*), the dependence of economic production on natural and non-natural resources (*Sec. 2.4.2*), the maintenance of a sustainable environment for human survival as opposed for its own sake (*Sec. 2.5.1*), the comparison of suffering from maltreatment in both animals and humans (*Sec. 2.5.2*), the effects of extinction on both animals and humans (*Sec. 3.2.1*), the impact of natural disasters on both animals and humans (*Sec. 3.2.2*), the impact of water deterioration on both animals and humans (*Sec. 3.3.1*), the impact of species invasion on both animals and humans (*Sec. 3.3.2*), the impact of deforestation on both animals and humans (*Sec. 3.4.1*), the impact of the use of pesticides and fertilizers on both animals and humans (*Sec. 3.4.2*), the impact of biopiracy on both humans and plants (*Sec. 3.5.1*), and the impact of extinction on all forms of life on Earth (*Sec. 3.5.2*), for instance.

All these are just some illustrative (relatively primitive) examples in our time, of course, but the important point to remember here is that the search for different new comparisons will further transform the study of natural resources at the more advanced level in the post-human future that the human world has never known.

The Fourth Great Future Transformation of Natural Resources

The fourth great future transformation of natural resources in the post-human era concerns what I originally call *new integrations*, in relation to nature.

This search for "new integrations" will give future advanced humans and later their successors, namely, the post-humans, the ability to live and interact with nature for natural resources in a way that current humans cannot.

There are already some indications that the future of natural resources is moving toward this direction.

Consider, for illustration, the different integrations for the study of natural resources as already seen in the history of natural resources, such as the relationship between biodiversity and human health (*Sec. 2.2.1*), the sustainable use of natural resources for agricultural production (*Sec. 2.2.2*), the preservaton of nature (*Sec. 2.3.1*), the linkage between human management and broad temporal and spatial scales (*Sec. 2.3.2*), ethical issues about all forms of cloning in nature and the human world (*Sec. 2.4.1*), the coevolution of human economies and natural ecosystems (*Sec. 2.4.2*), the close relationships between man and nature in deep ecology (*Sec. 2.5.1*), the call to treat animals with respect "as an integral part of the natural world" (*Sec. 2.5.2*), the call to not overexploit nature for long-term human well-being (*Sec. 3.2.1*), the call to "preserve traditional knowledge systems…in a manner that values and protects habitats and species" (*Sec. 3.2.2*), the call to address anthropogenic factors that negatively affect the climate (*Sec. 3.3.1*), the call to monitor species invasion that threaten native ecosystems (*Sec. 3.3.2*), the call to monitor illegal logging to preserve ecosystems (*Sec. 3.4.1*), the call for renewable energy to minimize the negative effects of oil depletion (and the extraction industries) on nature and society (*Sec. 3.4.2*), the treatment of traditional knowledge as part of endangered biodiversity for protection (*Sec. 3.5.1*), and the call to treat all forms of life as part of the great chain of being (*Sec. 3.5.2*), for instance.

More importantly, this future quest for new integrations should also be understood in relation to my previous works on the future "post-human" adventures into different dimensions of space-time unto multiverses from the combined perspectives of the mind, nature, society, and culture—with good examples like "the alteration of space-time" (*Table 1.3* and *Table 1.4*), "multiverses" (*Table 1.7* and *Table 1.8*), "hyperspace" (*Table 1.9* and *Table 1.25*), "time travel" (*Table 1.10* and *Table 1.11*), and others.

In addition, in my previous book titled *The Future of Post-Human Geology* (2010), I already envisioned the three great future transformations of geology, namely, "solar geology," "galactic geology," and "cosmic geology," in the context of my "resettlement theory of geology," in that humans and later post-humans will explore and migrate to other planets, solar systems, galaxies, and cosmos in the long haul.

Then, in another book of mine titled *The Future of Post-Human Geography* (2011), I linked the two books together, since geography has much to learn from geology (as in "physical geography," which, as already introduced in *Sec. 1.3* of *FPHGEOG*, relies heavily on the earth sciences) and discussed the search for "new explorations" in the context of the three forms of environments outside the planet Earth, namely, what I originally called "solar geography," "galactic geography," and "cosmic geography"—as a further elaboration of the three great future transformations of geology (in *FPHGEOL*).

Later, in still another book of mine titled *The Future of Post-Human Migration* (2012), I envisioned the future integrations in different forms, namely, what I originally called "solar migration," "galactic migration," "clustery migration," "cosmic migration," and "multiversal migration."

And then, in an additional book of mine titled *The Future of Post-Human Transportation* (2012), I envisioned the future integrations in different forms, namely, what I originally called "solar transportation," "galactic transportation," "clustery transportation," "cosmic transportation," and "multiversal transportation."

This extension is consistent with my other theory, namely, "the theory of the cyclical progression of system integration and fragmentation" in *FCD* (2002)—at the social-systemic level (as already clarified in *Sec. 1.8* and shown below):

Local → Regional → Global →Solar →
Galactic → Clustery…→ Multiversal

All these are just some illustrative (relatively primitive) examples in our time, of course, but the important point to remember here is that the search for different new integrations will further transform natural resources at the more advanced level in the post-human future that the human world has never known.

For instance, most advanced forms of future intelligent life in deep space until multiverses (e.g., "floating consciousness," "hyper-spatial consciousness," "unfolding unconsciousness," etc.) will need to adapt to different environments by seeking different natural resources (both renewable and non-renewable) through space colonization (as already discussed in *FPHMIG*, *FPHST*, *FPHC*, *FCD*, and *FHC*) in the distant future.

The Fifth Great Future Transformation of Natural Resources

The fifth great future transformation of natural resources in the post-human era concerns what I originally call *new praxes*, in relation to society.

This search for "new praxes" will give future advanced humans and later their successors, namely, the post-humans, the ability to engage in practice (like activism, politics, personal experience, etc.) for the future of natural resources in a way that current humans cannot.

There are already some indications that the future of natural resources is moving toward this direction.

Consider, for illustration, the different praxes as already seen in history of natural resources like the policy for biodiversity by the United Nations (*Sec. 2.2.1*), the efforts to avoid potato dependency after learning from the historical study of the Irish potato famine (*Sec. 2.2.2*), the conservation activities by the New York Zoological Society (*Sec. 2.3.1*), the applications of adaptive management to the business world (*Sec. 2.3.2*), the application of cloning for stem cell research (*Sec. 2.4.1*), the application of ecology to the study of the "flow diagram of income" (*Sec. 2.4.2*), the application of religion to the work on environmental management (*Sec. 2.5.1*), the application of animal welfare by PETA to the realm of fishkeeping (*Sec. 2.5.2*), the application of fisheries management in the form of ITQs (*Sec. 3.2.1*), the application of biodiversity awareness to the field of public policy for a legal standard to enforce it (*Sec. 3.2.2*), the application of preventive approaches to water management (*Sec. 3.3.1*), the application of the idea of genetic pollution for propaganda (*Sec. 3.3.2*), the application of different mechanisms to control deforestation (*Sec. 3.4.1*), the application of production cutbacks by OPEC to keep prices high (*Sec. 3.4.2*), the application of legal actions by the Indian government in the case of the Neem Tree (*Sec. 3.5.1*), and the application of the Bible for the study of extinction (*Sec. 3.5.2*), for instance.

All these are just some illustrative (relatively primitive) examples in our time, of course, but the important point to remember here is that the search for different new praxes will further transform the study of natural resources at the more advanced level in the post-human future that the human world has never known.

The Sixth Great Future Transformation of Natural Resources

The sixth great future transformation of natural resources in the post-human era concerns what I originally call *new technologies*, in relation to society.

This search for "new technologies" will give future advanced humans and later their successors, namely, the post-humans, the ability to come up with different techniques, materials, and experiments for natural resources in a way that current humans cannot.

There are already some indications that the future of natural resources is moving toward this direction.

Consider, for illustration, the use of different technologies as already seen in the history of natural resources, such as the technology to use biodiversity for drug production (*Sec. 2.2.1*), the technique of land clearing (*Sec. 2.2.2*), the technique of systematic planning in 6 stages (*Sec. 2.3.1*), the technique of computer modeling (*Sec. 2.3.2*), the technology of cloning (*Sec. 2.4.1*), the technique of forecasting in the Stem report (*Sec. 2.4.2*), the role of technology as in technocentrism (*Sec. 2.5.1*), the technique of using binoculars and telescopes for birdwatching (*Sec. 2.5.2*), the technique of ITQs (*Sec. 3.2.1*), the technique of relief, recovery, and redevelopment programmes (*Sec. 3.2.2*), the invention of the Milankovitch circles (*Sec. 3.3.1*), the technique of cattle gazing with impacts on trees (*Sec. 3.3.2*), the technique of data analysis like the FAO data (*Sec. 3.4.1*), the technique of production decline prediction (*Sec. 3.4.2*), the technique of patenting in biopiracy (*Sec. 3.5.1*), and the technique of finding fossil records (*Sec. 3.5.2*), for instance.

All these are just some illustrative (relatively primitive) examples in our time, of course, but the important point to remember here is that the search for new technologies will further transform the study of natural resources at the more advanced level in the post-human future that the human world has never known.

For instance, most advanced forms of future intelligent life in deep space until multiverses will make good use of their technological prowess (a) to restore life through cloning (as already discussed in *Sec. 2.4.1*—and also in *FCD* and *FHC*), (b) to extract different energy sources in outerspace through space colonization (as already discussed in *FPHMIG*, *FPHST*, *FPHC*, *FCD*, and *FHC*), (c) to create new forms of matter and energy in the distant future (as already discussed in *FPHST* and other works—and summarized in *Table 1.4*), and (d) to explore the technological frontiers of the micro-world (as already discussed in *FPHC*—and summarized in *Table 1.5*).

The Seventh Great Future Transformation of Natural Resources

The seventh great future transformation of natural resources in the post-human era concerns what I originally call *new stratifications*, in relation to society.

This emergence of "new stratifications" will lead future advanced humans and later their successors, namely, the post-humans, to face new inequalities for natural resources in a way that current humans cannot.

There are already some indications that the future of natural resources is moving toward this direction.

Consider, for illustration, the different inequalities as already seen in the history of natural resources, such as inadquate water supply in some developing countries but less so in developed countries (*Sec. 2.2.1*), the worst impact of potato disease on Ireland but not on other European countries (*Sec. 2.2.2*), more challenges facing African countries to implement environmental protection but less in developed countries (*Sec. 2.3.1*), the uneven acceptance of adaptive management in the corporate world (*Sec. 2.3.2*), the less successful rate of artificial cloning when contrasted with natural cloning (*Sec. 2.4.1*), the divide between the neoclassical approach and the ecological approach (*Sec. 2.4.2*), the divide between anthropocentrism and ecocentrism (*Sec. 2.5.1*), the divide between birders and birdwatchers (*Sec. 2.5.2*), the disagreement between those on the definition of extinction (*Sec. 3.2.1*), the loss of indegenous knowledge more in advanced countries than in developing countries (*Sec. 3.2.2*), the difference in powerful impact between short-term fluctuations and long-term alterations in ocean processes (*Sec. 3.3.1*), the competition between native species and invasive species (*Sec. 3.3.2*), the North-South divide in deforestation (*Sec. 3.4.1*), the tug of war between people and government in resource-cursed countries (*Sec. 3.4.2*), the divide between biopirates and their victims (*Sec. 3.5.1*), and the divide between religious catastrophism and scientific catastrophism (*Sec. 3.5.2*), for instance.

All these are just some illustrative (relatively primitive) examples in our time, of course, but the important point to remember here is that the emergence of new stratifications will further transform the quest for natural resources at the more advanced level in the post-human future that the human world has never known.

The Eighth Great Future Transformation of Natural Resources

The eighth great future transformation of natural resources in the post-human era concerns what I originally call *new sub-fields*, in relation to culture.

This search for "new sub-fields" will give future advanced humans and later their successors, namely, the post-humans, the ability to explore new specialties for the understanding of natural resources in a way that current humans cannot.

There are already some indications that the future of natural resources is moving toward this direction.

Consider, for illustration, the different sub-fields as already seen in the history of natural resources, such as the study of biodiversity and human health (*Sec. 2.2.1*), the field of agrobiodiversity (*Sec. 2.2.2*), the field of conservation biology (*Sec. 2.3.1*), the field of adaptive management (*Sec. 2.3.2*), the field of organism cloning (*Sec. 2.4.1*), the field of ecological economics (*Sec. 2.4.2*), the field of environmental ethics (*Sec. 2.5.1*), the field of animal watching (*Sec. 2.5.2*), the study of extinction (*Sec. 3.2.1*), the study of disaster prevention, relief, and recovery (*Sec. 3.2.2*), the study of water deterioration (*Sec. 3.3.1*), the field of genetic pollution (*Sec. 3.3.2*), the study of deforestation (*Sec. 3.4.1*), the study of oil depletion (*Sec. 3.4.2*), the study of biopiracy (*Sec. 3.5.1*), and the study of the history of scientific ideology (*Sec. 3.5.2*), for instance.

All these are just some illustrative (relatively primitive) examples in our time, of course, but the important point to remember here is that the search for new sub-fields will further transform the study of natural resources at the more advanced level in the post-human future that the human world has never known.

The Ninth Great Future Transformation of Natural Resources

And the ninth great future transformation of natural resources in the post-human era concerns what I originally call *new interpretations*, in relation to culture.

This search for "new interpretations" will give future advanced humans and later their successors, namely, the post-humans, the ability to come up with different evaluations and views about the study of natural resources in a way that current humans cannot.

There are already some indications that the future of natural resources is moving toward this direction.

Consider, for illustration, the different interpretations as already seen in the history of natural resources, such as the view on health risks of climate change (*Sec. 2.2.1*), the view on potato blight and the Irish potato famine (*Sec. 2.2.2*), the view on the tragedy of the commons (*Sec. 2.3.1*), the view on ecological uncertainty (*Sec. 2.3.2*), the religious objection to cloning (*Sec. 2.4.1*), the ideological view on "equity" in ecological economics (*Sec. 2.4.2*), the anthropocentric view of environmental management (*Sec. 2.5.1*), the view on birders as contrasted with birdwatchers (*Sec. 2.5.2*), the view on the overkill hypothesis (*Sec. 3.2.1*), the view on outbreaks of diseases by human impacts on ecosystems (*Sec. 3.2.2*), the view on the causes of climate change (*Sec. 3.3.1*), the view on species-based mechanisms in species invasion (*Sec. 3.3.2*), the view on the causes of deforestation (*Sec. 3.4.1*), the view on resource curse (*Sec. 3.4.2*), the view on the exploitatoin of traditional knowledge (*Sec. 3.5.1*), and (*Sec. 3.5.2*), the view on uniformitarianism, for instance.

All these are just some illustrative (relatively primitive) examples in our time, of course, but the important point to remember here is that the search for new interpretations will further transform the study of natural resources at the more advanced level in the post human future that the human world has never known.

Enormous Implications

These nine great future transformations of the study of natural resources have enormous implications for some soul-searching questions which have challenged some of the best minds in the history of natural resources (in the context of diversity and discontinuity).

For illustration, consider the following fundamental questions in the study of natural resources:

- Ex: are natural resources really so limited as to be unable to satisfy human greed?
- Ex: is the restoration of life possible?
- Ex: can new forms of matter and energy be created?
- Ex: are the ecosysems on Earth sustainable in the long run?

- Ex: is diversity really desirable, or is discontinuity really undesirable?

Questions and disagreements like this will receive new answers in the future history of natural resources in the advanced human and later post-human eras, at least in two fundamental ways.

The first fundamental way is to examine the relationships between diversity and discontinuity—such that their possibility (or impossibility) and desirability (or undesirability) are not to the extent that the respective defenders would like us to believe.

The different views about the study of natural resources, as already addressed in *Sec. 1.1*, *Sec. 1.5*, and the rest of the book, are just different ways to understand natural resources and their future, with each from its own vantage point.

Each views has its own merits and defects and is preferred over others by different proponents with their own different interests, but none of them is possible (or impossible) and desirable (or undesirable) to the extent that its respective advocates would like us to believe.

And the second fundamental way is to inquire about, and then transcend, in the long haul, the debate about diversity and discontinuity, such that it will be shifted to different directions, as shown in the nine great future transformations that I have envisioned and outlined above. To transcend them all is to move into these different directions, in conjunction with the various visions of my previous books (as already summarized in *Sec. 1.8* and also earlier in this chapter, with the further help of the tables at this end of Chapter One and this chapter).

Five good examples can be summarized below:

- to make good use of renewable energy (e.g., solar energy, wind energy, and so on) in the near future (as discussed in *Sec. 3.4.2*)
- to restore life through cloning (as already discussed in *Sec. 2.4.1*—and also in *FCD* and *FHC*)
- to exploit different energy sources in outerspace through space colonization (as already discussed in *FPHMIG*, *FPHST*, *FPHC*, *FCD*, and *FHC*)
- to create new forms of matter and energy in the distant future (as already discussed in *FPHST* and other works—and summarized in *Table 1.4*)
- to explore the technological frontiers of the micro-world (as already discussed in *FPHC*—and summarized in *Table 1.5*)

Of course, these examples are illustrative, not exhaustive.

In any event, lest any misunderstanding occurs, one must remember that these nine great future transformations of the study of natural resources will bring neither utopia nor dystopia, regardless of the current growing euphoria about the different versions of cloning (in *Sec. 2.4.1*), just to cite one obvious example—as this is something that I have warned time and again in all of my previous books, almost verbatim.

In any event, as I so often stressed in my previous books, it is not up to us current humans to decide whether or not, or to what extent, the post-human future of natural resources will be for better and for worse, because it will be judged by the values and beliefs of future post-humans on earth and in deep space unto multiverses—in a way that the entire history of the human world has never known.

Thus is the future of natural resources—and of its post-human fate.

Table 4.1. Sophisticated Methodological Holism (Part I)

• "My methodological holism implies the partiality-totality principle in the ontology of existential dialectics (see the table on the partiality-totality principle for summary), which is against the varieties of (a) reductionism and (b) reverse-reductionism, in relation to (i) concept, (ii) theory, (iii) methodology, and (iv) ontology." (*FC*)

• "[M]y methodological holism here is not opposed to methodological individualism but includes it (and, for that matter, other methodologies too)…." (*FPHC*) For this reason (and others too, as summarized hereafter), my version of methodological holism is sophisticated—not vulgar as sometimes used by inapt scholars using the same term. (*FC*)

• "[M]y methodological holism does not democratically presume that all levels are equally valid, as all levels are not created equal. In other words, in relation to issue X, level A may be more relevant than level B, but in relation to Y, level B can be more relevant than level A instead." (*FPHC*) One excellent example of this vulgar democratic presumption is what I called in *BNN* "the compromise fallacy." (*FC*)

• My methodological holism does not presume that a lower level of analysis is more important than a higher level, solely because the former serves as the foundation for the latter—and vice versa, for that matter. One excellent example of this reductionistic presumption is what I called in *FPHST* "the foundation fallacy." (*FPHST*, *FC*)

(continued on next page)

Table 4.1. Sophisticated Methodological Holism (Part II)

• "[M]y methodological holism does not make any a-priori postulation that there must be a definite (and, for that matter, indefinite) number of levels" in any analysis. (*FPHC*) Nor does it dogmatically require that there must be a certain combination of levels of analysis in a given inquiry. (*FC*)

• "[M]ethodological holism, in my usage, does not assume that all levels…can necessarily be integrated, since methodological holism is not aimed to search for the holy grail of 'an integral theory…' (as is the case for Wilber). In other words, it allows that sometimes some levels may experience irreducible gaps between them, to be understood, at best, as empirical correlations, not as causal relations…." (*FPHC*)

• "[D]ifferent levels may overlap and even interact with each other in a given context (but sometimes may not), and the fact that I even proposed different ways of re-classifying the levels (whenever needed) in *FDC* reinforces this point….The dual danger here is either forcefully making different levels interact when they are just different (or, metaphorically speaking, apples and oranges) or inappropriately ignoring their interactions when some situations instead require them." (*FPHC*)

• "[T]o understand different levels from their own (uniquc) perspectives (as required by my methodological holism) is not the same as trying to reduce them to a preferred level in the process of learning from other levels. This second kind of multidisciplinary work is not genuine and does no justice to the unique complexities and merits inherent at each level." (*FPHC*)

(continued on next page)

Table 4.1. Sophisticated Methodological Holism (Part III)

• "[My] methodological holism walks a fine line between the artificial classification (separation) of levels and the simultaneous incorporation of them, if only for the sake of human scholarly endeavor. It should be reminded that nature does not impose upon itself the academic classification of the levels of analysis as humans have. The enterprise of classification is therefore anthropocentric." (*FPHC*)

• "[M]y methodological holism advocates neither epistemic subjectivism nor epistemic non-subjectivism (e.g., realism, idealism, and historicism), neither epistemic relativism (e.g., subjectivism, historicism) nor epistemic absolutism (e.g., realism, positivism), neither epistemic reductionism nor epistemic emergencism, and neither epistemic objectivism (e.g., realism, idealism) nor epistemic historicism....Neither does methodological holism, in my usage, accept the false meta-conceptual dichotomy between nominalism and realism....These false dichotomies...are to be transcended. In other words, methodological holism does not fully accept epistemic realism, positivism (a form of epistemic idealism), historicism, subjectivism, and reductionism in epistemology and philosophy of science but learns from the strengths and weaknesses of all of the opposing approaches without siding with any of them...." (*FPHC*)

• "Sophisticated methodological holism is subject to the constraints as imposed by the syntax of existential dialectics (e.g., the partiality-totality principle and the predictability-unpredictability principle). Even in predictability, outcomes are subject to uncertainty, the degree of which varies from case to case." (*FC*)

(continued on next page)

Table 4.1. Sophisticated Methodological Holism (Part IV)

• "Sophisticated methodological holism—when applied, especially though not exclusively, as illustrated in my numerous works—can enrich the understanding of reality in some distinctive ways. Here are three examples (as revealed in each of my books)." (*FPHCT*; *FPHL*)

—"Firstly, it provides a comprehensive analysis of a subject matter, from which much can be learned about reality," "both in relation to the perspectives of the mind, nature, society, and culture and also in relation to a new classification of the subject matter."

—"Secondly, it suggests some visions of the future in relation to the subject matter in question."

—"And thirdly, it proposes some insights on meta-theory (e.g., methodology and ontology) in general—with the clear understanding, however, of the dilemma of specific vs. general ontology (as shown in the table on the syntax of existential dialectics in the context of the dilemma of ontology). For this very reason, all of these ways are important, without reducing one into the analysis of another."

• "In the end, my meta-theory (both sophisticated methodological holism and existential dialectics) serves as a foundation to unify all domains of knowledge into *an unified theory of everything* (by way of some ontological principles and the comprehensive perspectives of the mind, nature, society, and culture). This is so, without commiting the sins of reductionism and reverse-reductionism (as often seen in many holistic approaches, with the fad of sysems approach—be it about systems theory, chaos theory, complexity theory, or else—as a most recent notorious example, which I debunked in *The Future of Complexity* and also in *The Future of Post-Human Formal Science*). My distinctive approach makes good use of different schools of thought without favoring any of them nor trying to integrate them (as they are not necessarily compatible with each other), so as to adjust for subjectivity, diversity, conflict, and complexity, for example. In this sense, the word "unified" does not have to mean integrative approach, in a narrow sense, nor systems approach, in a broad one, both of which I reject while learning from them." (*FPHFS*; *FPHS*)

Sources: A summary of *Sec.1.2* in *FPHC*—and also from *BNN*, *FPHST*, *ALD*, *FC*, *FPHCT*, and the rest of all other books of mine. See the books for more detail.

Table 4.2. On Reductionism and Reverse-Reductionism (Part I)

- **The Partiality-Totality Principle**
  —The partiality-totality principle in the ontology of existential dialectics targets against the varieties of reductionism and reverse-reductionism (as already worked out in my previous books).

- **Against the Varieties of Reductionism**
  —*Conceptual Reductionism*
    - Some illustrative instances involve myriad dualities like mind vs. body, self vs. world, democracy vs. non-democracy, and the like (as already addressed in *FHC*, *FPHC*, and *BDPD*, for instance).

  —*Theoretical Reductionism*
    - A fascinating case study concerns what I originally called "the foundation fallacy" in *FPHST*, in any attempt to naively understand space-time from the physical perspective as the foundation and, consequently, to dangerously dismiss other perspectives.
    - In *FAE*, I elaborated further these versions of reductionism in the literature on aesthetics (e.g., form vs. content, representation vs. expression, critics vs. artists, and externalism vs. internalism).
    - In *FIA*, I revealed other forms of reductionism in the literature on information architecture (e.g., the constructivist argument).
    - In *FPHU*, I showed the persistent legacy of reductionism, this time, in the literature on anomalous experience (e.g., the obsession with physics, chemistry, and biology for explaining anomalous experience).
    - In *FPHE*, I examined another case of reductionism in action, in the context of engineering (e.g., technical constraints vs. normative constraints).
    - In *FPHMM*, I elaborated one more version of reductionism, in relation to the three domains of communication (e.g., the competing views on sending, connecting, and receiving).
    - In *FPHCT*, I explored another version of reductionism, in relation to invention and innovation (e.g., the bio-psychological argument vs. the socio-cultural arguments).

(continued on next page)

Table 4.2. On Reductionism and Reverse-Reductionism (Part II)

- **Against the Varieties of Reductionism *(cont'd)***
  - *—Theoretical Reductionism (cont'd)*
    - In *FPHG*, I identify another version of reductionism, in relation to infinity, symmetry, and dimensionality (e.g., the Euclidean argument vs. the non-Euclidean arguments).
    - In *FPHUP*, I analyzed another version of reductionism, in relation to density and void (e.g., the engineering argument and the ecology argument).
    - In *FPHL*, I examined another version of reductionism, in relation to structure and context (e.g., the structuralist argument and the contextualist argument).
    - In *PFHO*, I explored another version of reductionism, in relation to communcation, decision-making, and leadership (e.g., the rational-system argument vs. the natural-system argument).
    - In *PFHMA*, I revealed another version of reductionism, in relation to the martial body and spirit (e.g., the spiritual argument vs. the materialist argument vs. the defensive argument).
    - In *PFHS*, I analyzed another version of reductionism, in relation to the sexual body and spirit (e.g., the naturalist argument vs. the constructivist argument).
    - In *PFHLAW*, I examined another version of reductionism, in relation to law (e.g., the necessity argument vs. the contengency argument).
    - In *FPHWP*, I showed another version of reductionism, in relation to war and peace (e.g., the aggressivist argument vs. the pacifist argument).
    - In *BEPE*, I analyzed another version of reductionism, in relation to morality and immorality (e.g., the objectivist argument vs. the non-objectivist argument vs. the skeptical argument).
    - In *BCOS*, I examined another version of reductionism, in relation to the contested beginnings and speculative ends of the universe (e.g., the scientific argument vs. the religious argument vs. the esoteric argument vs. the metaphysical argument).
    - In *FPHP*, I show another version of reductionism, in relation to normality and abnormality (e.g., the natural argument vs. the social argument vs. the cultural argument vs. the mental argument).

(continued on next page)

Table 4.2. On Reductionism and Reverse-Reductionism (Part III)

- **Against the Varieties of Reductionism** ***(cont'd)***
  —*Theoretical Reductionism (cont'd)*
    - In *FPHGEOL*, I scrutinized another version of reductionism, in relation to statics and dynamics (e.g., the catastrophe argument vs. the uniformity argument vs. the revision argument).
    - In *FPHCHESS*, I showed another version of reductionism, in relation to tactics and strategy (e.g., the natural argument vs. the social argument vs. the cultural argument vs. the mental argument).
    - In *FPHR*, I revealed another version of reductionism, in relation to secularness and sacredness (e.g., the critical argument vs. the skeptical argument vs. the theist argument).
    - In *FPHEDU*, I examined another version of reductionism, in relation to teaching and learning (e.g., the teacher-centered argument vs. the student-centered argument vs. the balanced argument).
    - In *FPHH*, I explored another version of reductionism, in relation to joking and laughing (e.g., the natural argument vs. the social argument vs. the cultural argument vs. the mental argument).
    - In *FPHA*, I proposed another version of reductionism, in relation to sound and silence (e.g., the natural argument vs. the social argument vs. the cultural argument vs. the mental argument).
    - In *FPHGAM*, I proposed another version of reductionism, in relation to risk and caution (e.g., the natural argument vs. the social argument vs. the cultural argument vs. the mental argument).
    - In *FPHCOM*, I worked out another version of reductionism, in relation to hardware and software (e.g., the natural argument vs. the social argument vs. the cultural argument vs. the mental argument).
    - In *FPHLIT*, I proposed another version of reductionism, in relation to fiction and non-fiction (e.g., the natural argument vs. the social argument vs. the cultural argument vs. the mental argument).
    - In *FPHCHEM*, I proposed another version of reductionism, in relation to substances and their changes (e.g., the natural argument vs. the social argument vs. the cultural argument vs. the mental argument).

(continued on next page)

Table 4.2. On Reductionism and Reverse-Reductionism (Part IV)

- **Against the Varieties of Reductionism** ***(cont'd)***
  - *—Theoretical Reductionism (cont'd)*
    - In *FPHDA*, I proposed another version of reductionism, in relation to quantitative and qualitative research (e.g., the natural argument vs. the social argument vs. the cultural argument vs. the mental argument).
    - In *FPHGEOG*, I proposed another version of reductionism, in relation to environments and their interactions (e.g., the natural argument vs. the social argument vs. the cultural argument vs. the mental argument).
    - In *FPHPA*, I proposed another version of reductionism, in relation to the body and its presence (e.g., the natural argument vs. the social argument vs. the cultural argument vs. the mental argument).
    - In *FPHSEM*, I proposed another version of reductionism, in relation to internality and externality (e.g., the natural argument vs. the social argument vs. the cultural argument vs. the mental argument).
    - In *FPHARCH*, I proposed another version of reductionism, in relation to form and function (e.g., the natural argument vs. the social argument vs. the cultural argument vs. the mental argument).
    - In *FPHCRIM*, I proposed another version of reductionism, in relation to heroes and villains (e.g., the natural argument vs. the social argument vs. the cultural argument vs. the mental argument).
    - In *FPHHIST*, I proposed another version of reductionism, in relation to universality and relativity (e.g., the natural argument vs. the social argument vs. the cultural argument vs. the mental argument).
    - In *FPHMIG*, I proposed another version of reductionism, in relation to Sameness, Otherness, and identity (e.g., the natural argument vs. the social argument vs. the cultural argument vs. the mental argument).

(continued on next page)

Table 4.2. On Reductionism and Reverse-Reductionism (Part V)

- **Against the Varieties of Reductionism** ***(cont'd)***
  —*Theoretical Reductionism (cont'd)*
  - In *FPHTRAN*, I proposed another version of reductionism, in relation to networks and operations (e.g., the natural argument vs. the social argument vs. the cultural argument vs. the mental argument).
  - In *FPHCA*, I proposed another version of reductionism, in relation to ingredients and techniques (e.g., the natural argument vs. the social argument vs. the cultural argument vs. the mental argument).
  - In *FPHW*, I proposed another version of reductionism, in relation to uselessness and usefulness (e.g., the natural argument vs. the social argument vs. the cultural argument vs. the mental argument).
  - And in *BNR*, I propose another version of reductionism, in relation to diversity and discontinuity (e.g., the natural argument vs. the social argument vs. the cultural argument vs. the mental argument).

  —*Methodological Reductionism*
  - A good instance concerns the debate between different versions of qualitative and quantitative methods (as already analyzed in *FC* and also *FHC*). In *FPHML*, I examined similar reductionism, this time, in the literature on mathematical logic (e.g., the obsession with consistency, soundness, and completeness). And in *FPHFS*, I also explored the problems of reductionism in the context of formal science (e.g., the analytical argument).

  —*Ontological Reductionism*
  - An excellent example is the debate between emergentism and reductionism in complexity theory and also in psychology (as elaborated in *FPHC*, in the context of Being and Becoming).

- **Against the Varieties of Reverse-Reductionism**
  —*Conceptual Reverse-Reductionism*
  - Any concept of "art" (e.g., fine art, cave art, outsider art, junk art) is deemed acceptable in postmodernism (as already addressed in Ch.4 of *FHC*).

(continued on next page)

Table 4.2. On Reductionism and Reverse-Reductionism (Part VI)

- **Against the Varieties of Reverse-Reductionism *(cont'd)***

—*Theoretical Reverse-Reductionism*

- There are numerous art and literary theories co-exist. Take the case of literary studies, as there are now Literary Structuralism, Marxist Literary Criticism, New Criticism, Phenomenology, Hermeneutics, Language-Game Literary Criticism, Feminist Literary Criticism, Reception Theory, Reader Response Criticism, Poststructuralism, Semiotics, Pyschoanalytic Literary Criticism, just to cite some well-known ones, with no one being said to be better than any others (as detailedly analyzed in Ch.4 of *FHC*). (S. Raman 1997) In *BNN*, I even introduced "the compromise fallacy" as another good illustration of theoretical reverse-reductionism, in misleadingly treating both genetic and environmental approaches as equally valid.

—*Methodological Reverse-Reductionism*

- There is the "anything-goes" mentality in postmodernism (e.g., doing art without praxis, doing art with praxis, and doing art by sublation), as analyzed in Ch.4 of *FHC*. And in *FPHFS*, I also exposed the problems of reverse-reductionism in the context of formal science in relation to systems theory.

—*Ontological Reverse-Reductionism*

- There are likewise no privileged ontology, and the door is open for anything in postmodernism (e.g., the equal status of the ontology of Being vs. that of Becoming, as already addressed in Ch.4 of *FHC*—and also in *FPHC).* In *FAE*, I also introduced another version of reverse-reductionism, that is, "the pluralist fallacy," in the context of understanding aesthetic experience, for instance—although this fallacy has been committed not exclusively in relation to the ontological level (but also at the conceptual, theoretical, and methodological ones).

Sources: From my previous books.

Table 4.3. The Concepton of Existential Dialectics (Part I)

- **Sets and Elements**
  - —Sets
    - Ex: the Same
    - Ex: the Others
  - —Elements
    - Ex: whites in 20th century America (in the set of "the Same")
    - Ex: Iraq during the U.S. invasion in 2003 (in the set of "the Others")

- **Relations, Operations, Functions**
  - —Relations (e.g., "belongs," "equals to," "is greater than")
    - Ex: symmetric interactions within the Same (or the Others)
    - Ex: asymmetric interactions between the Same and the Others
  - —Operations (e.g., "and," "or," "not," "if…then")
    - Ex: if the Same oppresses the Others, it will also oppress itself.
    - Ex: the Same is not the Others.
  - —Functions (e.g., goals)
    - Ex: the Same is hegemonic in relation to the Others.

- **Truth Values**
  - —"1" if True (in Symbolic Logic)
    - Ex: the proposition that imperial Japan was hegemonic to China during WWII
  - —"0" if False (in Symbolic Logic)
    - Ex: the proposition that Grenada invaded France in 2003
  - —"1" & "0" if Both True and False (in Dialectic Logic)
    - Ex: the proposition that the rabbit-duck picture refers to a duck
  - —"~1" & "~0" if Neither True Nor False (or N/A, in Dialectic Logic)
    - Ex: the proposition that God really exists

(continued on next page)

Table 4.3. The Conception of Existential Dialectics (Part II)

- **Axioms and Postulates**
  - —Axioms
    - Ex: the reflexive axiom—"any quantity is equal to itself"
  - —Postulates
    - Ex: the SSS postulate—"if the three sides of a triangle are congruent to their corresponding parts, then the triangles are congruent"

- **Theorems and Principles**
  - —Theorems (and Principles) in Existential Dialectics
    - *In Relation to Method*
      - –#29: The formalness-informalness principle
      - –#28: The absoluteness-relativeness principle
      - –#27: The partiality-totality principle
      - –#26: The predictability-unpredictability principle
      - –#25: The explicability-inexplicability principle
      - –#24: The fiction-reality principle
      - –#23: The cognitiveness-noncognitiveness principle
    - *In Relation to Structure*
      - –#22: The finiteness-transfiniteness principle
      - –#21: The preciseness-vagueness principle
      - –#20: The simpleness-complicatedness principle
      - –#19: The openness-hiddenness principle
      - –#18: The denseness-emptiness principle
    - *In Relation to Process*
      - –#17: The change-constancy principle
      - –#16: The order-chaos principle
      - –#15: The slowness-quickness principle
      - –#14: The expansion-contraction principle

(continued on next page)

Table 4.3. The Conception of Existential Dialectics (Part III)

- **Theorems and Principles (*cont'd*)**
  —Theorems (and Principles) in Existential Dialectics
  - *In Relation to Agency*
    –#13: The theory-praxis principle
    –#12: The convention-novelty principle
    –#11: The evolution-transformation principle
    –#10: The symmetry-asymmetry principle
    –#9: The softness-hardness principle
    –#8: The seriousness-playfulness principle
    –#7: The activeness-inactiveness principle
    –#6: The selfness-otherness principle
  - *In Relation to Outcome*
    –#5: The regression-progression principle
    –#4: The same-difference principle
    –#3: The stability-reaction principle
    –#2: The functionality-nonfunctionality principle
    –#1: The intentionality-nonintentionality principle

Notes: The categories and examples in each are solely illustrative (not exhaustive). The comparison is also relative (not absolute), nor are they mutually exclusive. As generalities, they allow exceptions.
Sources: Starting from Ch.6 of *BCPC* and also from other books of mine

Table 4.4. The Syntax of Existential Dialectics I: The Principles (Part I)

- **In Relation to Method**

  —*The Formalness-Informalness Principle*

  (On the Formal Requirements of Logical Systems)

  - The formal requirements of a logical system (e.g., consistency, soundness, and completeness) have both usefulness and non-usefulness, to the extent that, if there are formal systems requiring them, there are alternative ones which do not. Thus, it does *not* exclude classical logics but simply goes beyond both classical and non-classical logics, while learning something from each. There is no formalness without informalness—and vice versa.
  - For instance, existential dialectics can make use of both classical logics under certain conditions (e.g., especially, though not exclusively, when they are clear-cut, etc.) and non-classical logics under alternative conditions (especially, though not exclusively, when they are "unknown," "irrelevant," "ambiguous," "possible," with "different degrees of truth," empirically inconsistent in a desirable way, etc.). (WK 2008u)
  - Family resemblance: e.g., logicalness-nonlogicalness, rationality-nonrationality, etc.
  - Sources: From *FPHML*, *FPHH*, *FPHGAM*, *FPHSFM*, and my later books.

(continued on next page)

Table 4.4. The Syntax of Existential Dialectics I: The Principles (Part II)

- **In Relation to Method** ***(cont'd)***

  —*The Absoluteness-Relativeness Principle*
  (On the Multiplicity of Things)

  - There is the multiplicity of things in reality, be they about entities, qualities (or properties), and relationships, such that what is acceptable from one standpoint may not be so from another. For instance, if there is something absolute, there is likewise something relative. There is no absoluteness without relativeness—and vice versa.
  - Both absoluteness and relativeness here are also relevant to different modalities often cited in the literature on ontology, such as possibility (e.g., something "can" happen) and its opposite (e.g., impossibility), probability (e.g., something "will" happen) and its opposite (e.g., improbability), and necessity (e.g., something "should" happen) and its opposite (e.g., contingency).
  - Family resemblance: e.g., independence-dependence, isolatedness-interconnectiveness, universalness-relativeness, immanence-transcendence, autonomy-interdependency, internalness-externalness, etc.
  - Sources: From *FPHK*. See also my later books.

(continued on next page)

Table 4.4. The Syntax of Existential Dialectics I: The Principles (Part III)

• **In Relation to Method** ***(cont'd)***

—*The Partiality-Totality Principle*

(On the Relationships between Whole and Parts)

- The whole is not the sum of the parts. There is no partiality without totality—and vice versa.
- Any inquiry about a phenomenon in the work is to guard against the varieties of (a) reductionism and (b) reverse-reductionism.
- Reductionism and reverse-reductionism can be (i) conceptual, (ii) theoretical, (iii) methodological, and (iv) ontological.
- Works on emergence (as in *FC*), gestalt psychology (as in *FPHGEOL*), the hermeneutic circle (as in *FPHDA*), and mental language (as in *FPHSEM*), to name just a few of them, have contributed to the understanding of whole and parts.
- Family resemblance: e.g., individualisticness-holisticness, partness-wholeness, analysis-synthesis, etc.
- Sources: Especially from *FC* & *FPHGEOL*. See also my later books.

(continued on next page)

Table 4.4. The Syntax of Existential Dialectics I: The Principles (Part IV)

- **In Relation to Method *(cont'd)***

  —*The Predictability-Unpredictability Principle (On the Occurrence of Events)*

  - Both predictability and unpredictability have a major role to play in the occurrence of things, so that neither determinism nor indeterminism wins the centuries-old fight. There is no predictability without unpredictability—and vice versa.
  - There are events which are predictable, just as there are those which are not. Or what is regarded as unpredictable at one point in time may turn out to be predictable later, and, conversely, what is deemed as predictable may turn out to not be so predictable. Even in predictability, outcomes are subject to uncertainty, the degree of which varies from case to case.
  - Works on complexity theory (as in *FC*), quantum mechanics (as in *FPHST* and *FPHG*), frequency probability, subjective (or Bayesian) probability, and "special causes" (as in *FPHGAM*), just to name a few of them, have contributed to the understanding of uncertainity.
  - Family resemblance: e.g., sureness-arbitrariness, certainty-uncertainty, deterministicness-indeterministicness, sureness-doubtfulness, likeliness-riskiness, deterministicness-randomness, etc.
  - Sources: Especially from *FC* & *FPHGAM.* See also my later books.

(continued on next page)

Table 4.4. The Syntax of Existential Dialectics I: The Principles (Part V)

• **In Relation to Method** ***(cont'd)***

—*The Explicability-Inexplicability Principle*
(On the Underlying Mechanisms of Things)

• Both explicability and inexplicability are part of the understanding of things. There is no explicability without inexplicability—and vice versa.
• This principle tells us the duality of the research dilemma, in that, if reality can be explained in some ways, it also has its other ways which are not quite explainable, at a given point in time.
• Family resemblance: e.g., underlyingness-regularness, causation-regularness, causation-correlation, explanation-description, explanation-narration, explanation-interpretation, etc.
• Sources: Especially from *FPHU*—and also *FPHC*. See also my later books.

—*The Fiction-Reality Principle*
(On the different faces of Reality)

• If there is reality, there is also fiction, especially for any agent with some kind of intelligence life, be it about humans, animals, or, later, post-humans. There is no reality without fiction—and vice versa.
• The distinction between reality and fiction is not absolute, since there can be a mixture of both. Fiction can have non-fictional elements, just as non-fiction can have fictional ones. So, the distinction is in degree, not in kind.
• Family resemblance: e.g., fiction-nonfiction, fiction-fact, imagination-fact, subjectivity-objectivity, imagination-nonfiction, etc.
• Sources: First from *FPHLIT*.

(continued on next page)

Table 4.4. The Syntax of Existential Dialectics I: The Principles (Part VI)

- **In Relation to Method *(cont'd)***
  —*The Cognitiveness-Noncognitiveness Principle*
  (On the Understanding of Meaning)
  - If a view of things is based on reason and evidences, it is also affected by non-cognitive factors (such that it is neither true nor false in the latter case)—and vice versa. In other words, there is no cognitiveness without noncognitiveness—and vice versa.
  - Noncognitive factors can come in all shapes and sizes, and good examples include envy, jealousy, power, nationality, race, gender, age, class, greed, lust, status, faith, anger, sadness, joy, fear, wish, etc.
  - Communication, even in its ideal type, is seldom solely based reason and evidences, as falsely assumed during the Age of the Enlightenment and its continued supporters nowadays. What is understood about something is often twisted on the basis of something else too.
  - Family resemblance: e.g., proposition-attitude, cognition-emotion, cognition-prescription, theory-extratheory, etc.
  - Sources: First from *BNR*.

(continued on next page)

Table 4.4. The Syntax of Existential Dialectics I: The Principles (Part VII)

- **In Relation to Structure**

—*The Finiteness-Transfiniteness Principle*

(On the Nature of Numbers)

- If there are finite things, there are likewise transfinite ones. There is no finiteness without transfiniteness—and vice versa.
- To avoid confusion, my usage of the word "transfinite" here differs radically from the one used by Cantor (and other mathematicians) for "relative" infinity—and is more limited, in light of the problems confronting any attempt to understand the idea of infinity, be it by intuition, imagination, and conception (as detailedly analyzed in *Sec. 2.2.3* of *FPHG*).
- Instead, by "transfinity," I allow numbers which can be many times larger—or smaller, for that matter—than the finite things that we encounter in daily life, but they do not have to be related to the idea of infinity at all (which may not exist).
- Of course, there may be some *borderline* cases, in which it is not clear whether the number in question is transfinite (in my usage) or simply a mathematical convenience. A good example of a borderline case is the Planck unit of length for "the smallest space possibly measured in nature," which is "less than billionths of trillionths of trillionths of an inch" (or something like 1.6×10^{-35} meters).
- That said—my usage of "transfinity" can also resolve (or better, dissolve) an age-old problem in philosophy known as "Zeno's paradoxes" (as already explained in *Sec. 2.2.3* of *FPHG*).
- Family resemblance: e.g., boundedness-quasiunboundedness, smallness-largeness, boundedness-unboundedness, etc.
- Source: From *FPHG*. See also my later books.

(continued on next page)

Table 4.4. The Syntax of Existential Dialectics I: The Principles (Part VIII)

- **In Relation to Structure** ***(cont'd)***
  —*The Preciseness-Vagueness Principle*
  (On the Refinement of Things)
  - Both preciseness and vagueness are important, not that one is better than the other, but that both are used, in different degrees of preference, in accordance to the contextual application from the perspectives of nature, the mind, culture, and society. Even when both are used in a combination, the dilemma is shifted instead to one of combinational concern. There is no preciseness without vagueness—and vice versa.
  - In relation to taxonomy, preciseness has its taxonomic clarity, just as vagueness has its classificatory flexibility, for instance. And in relation to network, vagueness has its explorative liberty, just as preciseness has its conceptual definitiveness, for instance.
  - Family resemblance: e.g., clarity-ambiguity, directness-indirectness, quantitativeness-qualitativeness, describability-nondescribability, specificity-obscurity, specificity-generality, thickness-thinness, translatability-untranslatability, clearcutness-transitionalness, concreteness-abstractness, etc.
  - Sources: Especially from *FIA*. See also my later books.

(continued on next page)

Table 4.4. The Syntax of Existential Dialectics I: The Principles (Part IX)

- **In Relation to Structure** ***(cont'd)***

  —*The Simpleness-Complicatedness Principle*
  (On the Interconnection among Things)

  - Both simpleness and complicatedness are vital, without favoring one over the other, in that each is utilized, depending on the basis of the perspectives of nature, the mind, culture, and society. And even when a combination of them is preferred, the dilemma is only shifted to a combinational degree of concern. There is no simpleness without complicatedness—and vice versa.
  - In relation to taxonomy, simpleness has its heuristic usefulness, just as complicatedness has its realistic representation, for instance. And in relation to network, simpleness has its economical attractiveness, just as complicatedness has its practical reliability, for instance.
  - Family resemblance: e.g., inflexibility-flexibility, standardization-specialization, imperfectness-perfectness, superficiality-depth, shallowness-deepness, economicalness-elaboratedness, plainness-circumspection, onesidedness-multisidedness, succinctness-duplicatedness, etc.
  - Sources: Especially from *FIA*. See also my later books.

(continued on next page)

Table 4.4. The Syntax of Existential Dialectics I: The Principles (Part X)

- **In Relation to Structure** ***(cont'd)***

  —*The Openness-Hiddenness Principle*

  (On the Detection of Things)

  - Reality has its hidden face, just as it is open to outside view in some other ways. There is no openness without hiddenness—and vice versa.
  - For instance, in the context of anomalous experience, certain aspects of reality can be open for examination (e.g., the experiment with SPECT images, the transmission of telepathemic bit, and the use of metonymies and metaphors). Other aspects, however, remain hidden, and examples include the elusive deeper nature of spiritual reality in the intangible realm, the undetectability of different branched universes, the underlying mechanisms of signals traveling between different folds of physical space-time, and the dependence on language and the bias for science and logic).
  - Family resemblance: e.g., overtness-covertness, publicness-privateness, openness-closedness, transparency-secrecy, openness-biasedness, participation-obstruction, exposition-illusion, surface-depth, etc.
  - Sources: Especially from *FPHU*. See also my later books.

(continued on next page)

**Table 4.4. The Syntax of Existential Dialectics I:
The Principles
(Part XI)**

• **In Relation to Structure** ***(cont'd)***
—*The Denseness-Emptiness Principle*
(On the Distribution of Entities in Space)
- Both density and void are needed, in relation to the mind, nature, culture, and society, albeit in different ways. There is no denseness without emptiness—and vice versa.
- For clarity, the term "void" is used here only as an approximation of emptiness (depending on the degree of the lack of density), since, in physics, it is well known that "empty" space is not really empty all the way, because it can be full of energy (e.g., random quantum fluctuations at the sub-atomic level, and, for that matter, dark energy in the universe) and matter (e.g., different versions of sub-atomic particles, and, for that matter, dark matter in the universe).
- Family resemblance: e.g., fullness-voidness, nearness-farness, concentration-dispersion, denseness-nothingness, presence-absence, focus-unfocus, solidness-void, something-nothing, heaviness-lightness, composition-emptiness, emphasis-disemphasis, etc.
- Sources: First worked out in *FPHUP*. See also my later books.

• **In Relation to Process**
—*The Change-Constancy Principle*
(On the Alteration of Things)
- Change occurs over time, although constancy is also allowed. There is no change without constancy—and vice versa.
- Asymmetry undergoes changes over time, so does symmetry.
- Old players fade away, and new ones emerges, with ever new causes and ever new forms.
- Family resemblance: e.g., dynamicness-staticness, instability-stability, etc.
- Sources: First named in *BCPC*. Especially from *FHC*, *FCD*, and *FPHC*. See also other books of mine (from that point on).

(continued on next page)

Table 4.4. The Syntax of Existential Dialectics I:
The Principles
(Part XII)

- **In Relation to Process *(cont'd)***
  —The Order-Chaos Principle
  (On the Pattern of Things)
  - Both order and chaos are vital in the process of change in the world. The preference for order is biased, since it does not give sufficient attention to the vital role of chaos in the transformation of the world (without somehow reducing it for the understanding of order). There is no order without chaos—and vice versa.
  - The scientific search for order in the world is often a hidden bias in its ontological obsession with order, since chaos is often treated as the "bad" guy, with order as the "good" guy (for the end goal of science).
  - Neither order nor chaos is the final end of the world, and one is not to be treated as the means for the other in the transformation of things. Both are fundamental in their recurrent dialectical interactions with each other over time, without reducing one for the other.
  - Family resemblance: e.g., lawfulness-disorder, order-disorder, linearity-nonlinearity, linearity-irreversalness, pattern-chaos, unity-disunity, etc.
  - Sources: Especially from *FC*. See also my later books.

(continued on next page)

Table 4.4. The Syntax of Existential Dialectics I: The Principles (Part XIII)

- **In Relation to Process *(cont'd)***

  —*The Slowness-Quickness Principle*
  (On the Speed of Change)
  - Both slowness and quickness co-exist, with their own internal tension, to the extent that each fights for its own relevance with the other, in accordance to the perspectives of nature, the mind, culture, and society, without one being the victor and the other being the vanquished in the long haul. Even when both are chosen in other cases, this dilemma is only transferred into something else with a combinational character. There is no slowness without quickness—and vice versa.
  - In relation to taxonomy, quickness has its efficient usability, just as slowness has its aesthetic appeal, for instance. And in relation to network, quickness has its adventurous readiness, just as slowness has its risk-adverse convenience, for instance.
  - Family resemblance: e.g., inconvenience-convenience, passiveness-activenessness, gradualness-abruptness, deceleration-acceleration, difficulty-ease, nonresilience-resilience, etc.
  - Sources: Especially from *FIA*. See also my later books.

  —*The Expansion-Contraction Principle*
  (On the Growth of Things)
  - Entities in the world can both expand in some ways and contract in other ones, as part of their nature. There is no expansion without contraction—and vice versa.
  - For instance, in the context of anomalous experience, one of its most fascinating expressions is none other than the analysis of different causes of the slow but steady expansion (and, furthermore, if put in a different way, unfolding) of unconsciousness.
  - Family resemblance: e.g., conquest-autarky, rise-fall, spread-shrink, extendingness-shorteningness, widening-narrowing, expansion-destruction, expansion-decline, construction-demolition, creation-destruction, construction-deconstruction, etc.
  - Sources: Especially from *FPHU*. See also my later books.

(continued on next page)

Table 4.4. The Syntax of Existential Dialectics I: The Principles (Part XIV)

- **In Relation to Agency**
  —*The Theory-Praxis Principle*
  (On the Duality of Knowledge))
  - This has to do with the duality of knowledge in existential dialectics, in that, if there is theoretical construction, there is likewise its practical application, both technical and normative. There is no theory without praxis—and vice versa.
  - Family resemblance: e.g., discovery-application, knowledge-action, invention-innovation, knowledge-motivation, theory-intentionality, universality-particularity, discourse-action, knowledge-interest, potentiality-actuality, etc.
  - Source: Especially from *FPHE*. See also my later books.

(continued on next page)

Table 4.4. The Syntax of Existential Dialectics I: The Principles (Part XV)

- **In Relation to Agency *(cont'd)***
  —*The Convention-Novelty Principle*
  (On the Nature of Creative Thinking)
  - If there is conventional wisdom, there is likewise novel challenge, to the extent that both convergent and divergent thinking are part of life. There is no convention without novelty—and vice versa.
  - In addition, in this context of the convention-novelty principle, there are, in the absence of better words, what I want to call (a) *creative techniques* and (b) *creative traits*, which, when satisfied—in relation to the larger context of the mind, nature, society, and culture—can be used to enhance creative works. The list of creative techniques and traits (as summarized in *Table 4.1* of *FPHCT*) are not exhaustive, of course, but the examples here constitute a great beginning of understanding the structure of creative thinking.
  - Yet, lest the reader gets carried away by the euphoria about creative thinking in our time, it should be reminded that creative thinking has its own possibilities and limits (as shown in *Table 2.1* of *FPHCT* on invention), just as it has its own promises and pitfalls (as shown in *Table 3.1* of *FPHCT* on innovation). In the end, it should also be stressed that creative thinking has its own desirability and dark sides (as shown in *Table 4.2* of *FPHCT*, *BNN*, and *FPHMIG*).
  - This has important daily implications for an average individual, and the word "average" here allows different degrees of variation from one individual to another in a population, at any given point in history.
  - Family resemblance: e.g., normalness-nonnormalness, conformity-nonconformity, convergence-divergence, familiarity-freshness, inertia-newness, equilibrium-disturbance, etc.
  - Sources: First from *FPHCT*. See also my later books.

(*continued on next page*)

**Table 4.4. The Syntax of Existential Dialectics I:
The Principles
(Part XVI)**

• **In Relation to Agency *(cont'd)***

—*The Evolution-Transformation Principle*
(On the Multiple Kinds of Agency)

• This principle (and the symmetry-asymmetry principle) are both about *the agency of change*. The word "agency," in a formal definition, refers to "a person or thing through which power is exerted or an end is achieved." (MWD 2007b) It therefore does not have to necessarily involve an intelligent lifeform.

• Because of this dual meaning in agency, the evolution-transformation principle is more concerned with *the multiple kinds of agency*, that is, both about the *evolution* in the state of nature (e.g., an object of natural beauty) and the *transformation* in the world of intelligent lifeforms (e.g., a work of art, an air-conditioner, etc.). There is no evolution without transformation—and vice versa.

• And the transformative part of the principle precisely refers to the other dimension in the dual meaning of agency, in giving technology a major role to play in the change of the world, which is something that I extensively analyzed in *FHC* in the context of the technophilic lifeworld, espeically though not exclusively since modern times. The same theme is also elaborated in other books of mine.

• Family resemblance: e.g., nonwillingness-willingness, naturalness-technologicalness, evolution-creation, naturalness-nonnaturalness, inorganicness-volition, naturalness-artificialness, inorganicness-motivation, etc.

• Sources: Especially from *FAE*. See also other books of mine.

(*continued on next page*)

Table 4.4. The Syntax of Existential Dialectics I: The Principles (Part XVII)

- **In Relation to Agency** ***(cont'd)***

  —*The Symmetry-Asymmetry Principle*
  (On the Relationships among Existents)

  - There is no symmetry without asymmetry—and vice versa.
  - For instance, the Same can be symmetric and asymmetric towards the Others. But in case of asymmetry, oppression and self-oppression can occur. So, when the Same is asymmetric towards the Others, the Same can also be relatively asymmetric towards itself in self-oppression, just as the Others can be likewise towards themselves.
  - The subsequent oppressiveness is dualistic, as much by the Same against the Others and itself, as by the Others against the Same and themselves.
  - Both oppression and self-oppression can be achieved by way of downgrading differences between the Same and the Others and of accentuating them.
  - This is true, even though not all forms of asymmetry have to be about oppression and self-oppression.
  - In addition, from Chapter Three of *FPHG*, symmetry is not perfect, to be understood in an approximate sense under many life circumstances.
  - Family resemblance: e.g., balance-extremity, reflexiveness-unreflexiveness, balance-unbalance, equality-inequality, center-periphery, compatibility-incompatibility, proportion-disproportion, harmoniousness-unharrmoniousness, etc.
  - Sources: From all my books, starting with *FHC*. First named in *BCPC*. See, for instance, *FCD* for more details on the Same and the Others.

(*continued on next page*)

Table 4.4. The Syntax of Existential Dialectics I: The Principles (Part XVIII)

- **In Relation to Agency** ***(cont'd)***
  —*The Softness-Hardness Principle*
  (On the Force of Change)
  - This has to do with *the force of change* under the category about agency in existential dialectics, in that any change by an agent, be it organic (like humans) or non-organic (like natural objects), can occur in a forceful (aggressive) or gentle (pacific) way, which can come in all shapes and sizes, of course. There is no softness without hardness—and vice versa.
  - Family resemblance: e.g., peacefulness-violence, cooperation-competition, acceptance-discrimination, love-hatred, peace-war, pacificity-aggression, dovishness-hawkishness, reward-punishment, kindness-hostility, admiration-belittlement, inclusion-exclusion, friendliness-unfriendliness, etc.
  - Sources: Especially from *ALD*. See also my later books.

(continued on next page)

Table 4.4. The Syntax of Existential Dialectics I: The Principles (Part XIX)

- **In Relation to Agency *(cont'd)***
  - *—The Seriousness-Playfulness Principle*
    (On the Extent of Seriousness)
    - If there is seriousness, there is also playfulness, especially for any agent with some kind of intelligence life, be it about humans, animals, or, later, post-humans. There is no seriousness without playfulness—and vice versa.
    - Lest any misunderstanding occurs, the word "playfulness" here should not be confused with other terms like "laughter," "smile," "joke," "tickleness," and "laugh-like vocalization" (in some animals), for instance.
    - Being playful can come in all shapes and sizes and therefore is not necessarily tied up with any of these terms, alhough it can result from any of them, or a combination of them, or something else altogether.
    - This principle, together with some other principles (especially though not exclusively, like the formalness-informalness principle), have important implications for what constitutes rationality, be it about "rationality of application" at the practical level or "rationality of knowledge" at the meta-theoretical level (as summarized in *Sec. 4.20* of *FPHH*).
    - Family resemblance: e.g., seriousness-jokingness, seriousness-humorousness, seriousness-wittiness, seriousness-nonseriousness, seriousness-relaxedness, seriousness-restingness, Apollo-Dionysus, etc.
    - Sources: First from *FPHH* & *FPHGAM* See also my later books.

(*continued on next page*)

Table 4.4. The Syntax of Existential Dialectics I: The Principles (Part XX)

- **In Relation to Agency** ***(cont'd)***
  —*The Activeness-Inactiveness Principle*
  (On the Extent of Activeness)
  - If there is activeness, there is also inactiveness, especially (though not exclusively) for any agent with some kind of intelligence life, be it about humans, animals, or, later, post-humans. There is no activeness without inactiveness—and vice versa.
  - Being active can come in all shapes and sizes, as a person can be inactive because he is tired, old, sick, sleepy, uninterested, etc., just as an object can be inactive because it is at rest, out of work, etc.
  - This principle, together with some other principles (especially though not exclusively, like the formalness-informalness principle and seriousness-playfulness principle), have important implications for what constitutes rationality, be it about "rationality of application" at the practical level or "rationality of knowledge" at the meta-theoretical level (as summarized in *Sec. 4.20* of *FPHH*).
    - Family resemblance: e.g., alertness-unalertness, activeness-decrepidness, alertness-sleepiness, awakenness-hibernatedness, activeness-tiredness, alertness-uninterestedness, alertness-unwareness, soberness-unawareness, soberness-inactiveness, alertness-inactiveness, etc.
  - Sources: First from *FPHTRAN*.

(*continued on next page*)

Table 4.4. The Syntax of Existential Dialectics I: The Principles (Part XXI)

- **In Relation to Agency** ***(cont'd)***
  —*The Selfness-Otherness Principle*
  (On the Magnitude of Self-Centeredness)
  - If there is the tendency to be mostly centered on one's self or group, there is likewise the tendency to be mostly centered on others, especially (though not exclusively) for any agent with some kind of intelligence life, be it about humans, animals, or, later, post-humans. There is no selfness without otherness—and vice versa.
  - Being self-centered and being other-centered do not have to be mutually exclusive, as the two can come in all shapes and sizes, or in different degrees. And other-centeredness does not entail desirability, any more than self-centeredness implies non-desirability, as both can go either way.
  - In addition, whether self-centeredness constitutes a moral alternative or, conversely, an alternative to morality is of course debatable (as addressed in my other book titled *Beyond Ethics to Post-Ethics*).
  - Good examples of family resemblance in relation to the principle are egoisticness-altruisticness, selfishness-unselfishness, selfness-selflessness, egocentricness-othercentricness, and so on.
  - Source: First from *FPHCA*

(*continued on next page*)

Table 4.4. The Syntax of Existential Dialectics I: The Principles (Part XXII)

- **In Relation to Outcome**

  —*The Regression-Progression Principle*
  (On the Direction of History)

  - Neither the cyclical nor the linear views are adequate for explaining many phenomena at all levels. There is regression without progression—and vice versa.
  - History progresses to more advanced forms, but with a regressive touch. Examples include no freedom without unfreedom, no equality without inequality, and no civilization without barbarity.
  - This is not an inevitable law, but merely a highly likely empirical trend.
  - Family resemblance: e.g., cost-benefit, undesirability-desirability, badness-goodness, risk-opportunity, dysfunctionality-functionality, cyclicity-linearity, etc.
  - Sources: From all my books, starting with *FHC*. First named in *BCPC*.

  —*The Same-Difference Principle*
  (On the Metamorphosis of Change)

  - An entity, as it evolves over time, can be both different from and similar to its opposing alternatives and does not have to be solely more different from them over time. There is no similarity without difference—and vice versa.
  - Opposites are not absolute in a black-or-white fashion; so, an entity can become relatively more similar to (or more different from) its opposite over time.
  - Family resemblance: e.g., homogeneity-heterogeneity, we-they, similarity-contrast, etc.
  - Sources: Especially from ALD. See also my later books

(*continued on next page*)

Table 4.4. The Syntax of Existential Dialectics I: The Principles (Part XXIII)

- **In Relation to Outcome *(cont'd)***
  —The Stability-Reaction Principle
  (On the Feedback Mechanism for Further Changes)
  - Once an outcome becomes stabilized over time (which varies from case to case), it will undergo further changes in its multifaceted interactions with the outside world (e.g., with other entities, energy, space, and so on).
  - There is no stability without reaction for further changes—and vice versa.
  - Family resemblance: e.g., stability-forwardness, balance-reaction, balance-forwardness, rest-motion, normalcy-feedback, stability-breakup, consolidation-modification, etc.
  - Sources: First from *FPHCHEM.*

(*continued on next page*)

Table 4.4. The Syntax of Existential Dialectics I: The Principles (Part XXIV)

- **In Relation to Outcome *(cont'd)***
  —*The Functionality-Nonfunctionality Principle*
  (On the Presence of Function)
  - If there is the presence of action for which something or someone is fitted, there is likewise the relative lack of action for which something or someone is fitted. There is no functionality without nonfunctionallity—and vice versa.
  - A *function* should not be confused with an *alternative function*, which is just another kind of function; just as *nonfunction* should not be confused with *dysfunction* (which has more to do with malfunction and is thus related more to the regression-progression principle).
  - In addition, being functional (or useful) and being nonfunctional (or useless) do not have to be mutually exclusive, as the two can come in all shapes and sizes, or in different degrees. And what is functional at one time may become nonfunctional at another time, and vice-versa.
  - Good examples of family resemblance in relation to the principle are usefulness-nonusefulness, usefulness-uselessness, valueness-nonvalueness, and the like.
  - Source: First from *FPHCA*

(*continued on next page*)

Table 4.4. The Syntax of Existential Dialectics I: The Principles (Part XXV)

- **In Relation to Outcome** ***(cont'd)***

  —The Intentionality-Nonintentionality Principle
  (On the Planning for Outcomes)

  - If there is the planning or design of something for a certain outcome, there is likewise the relative absence of planning or design of something for a certain outcome. There is no intentionality without nonintentionallity—and vice versa.
  - An intention for an outcome does not have to be certain, so the outcome can even be counter-intentional in being the opposite of what was originally planned. In this sense, counter-intentionality is a special case of non-intentionality; in other words, there are unintented consequences which can be just the opposite of what was originally planned.
  - An intention for an outcome is mostly relevant to intelligent life, but there are non-intelligent life and non-living matter in the universe too, so nonintentionality is also important.
  - In addition, an intention does not have to be obvious or straightforward, as William Shakspeare once famously said: "I must be cruel only to be kind."
  - Good examples of family resemblance in relation to the principle are purposiveness-nonpurposiveness, intention-accidentalness, design-randomness, etc.
  - Source: First from *FPHW*

Notes: The features in each principle are solely illustrative (not exhaustive). The comparison is also relative (not absolute), nor are they mutually exclusive. As generalities, they allow exceptions.

Table 4.5. The Syntax of Existential Dialectics II: The Principles as Short Cuts (Part I)

• The principles should be treated with caution, lest misunderstanding occurs, since they do not constitute rigid dualities (or dichotomies).

• The reason is that each pair in an ontological principle consists of two opposites, which are, however, merely short cuts both for multiple variations and degrees, as well as for different interactions with multiple other entities unlike them. In this light, each pair can end up having hundred (or even thousand, if not more) different versions, which interact with hundred (or even thousand, if not more) other entities. There are two clarifications here.

• Firstly, this conception of shortcuts is not mutually exclusive nor absolute, in that the opposites can come in all shapes and sizes, with different degrees. For instance, by analogy, just as there are different degrees of the two colors "white" and "black"—there are likewise different degrees of the opposites in each ontological pair, to the extent that there can be multiple entities (not only two) interacting with each other in each pair.

• And secondly, this conception of shortcuts do not ignore other possible entities in interacting with the two opposites (with their different versions) in each pair. For instance, by analogy, there are not only the two colors "white" and "black" as opposites, since there are other colors too besides them like "yellow," "green," "purple," or else. The same logic can be applied to each ontological pair, in that they also interact with other entities, not with only two of them (with their different versions).

• Therefore, with these two clarifications in mind—each pair in an ontological principle serves only as an abbreviation for something more complicated and, therefore, although it contains two opposites, it should not be confused as a duality (dualism).

(continued on next page)

Table 4.5. The Syntax of Existential Dialectics II: The Principles as Short Cuts (Part II)

• Existential dialectics rejects any dualism (or dichotomy) as too rigid and instead allows the multiplicity of entities, to the extent that between the two opposites in each pair exist many other alternatives to choose from. They are named in that short form for aesthetic elegance, instead of listing all possible entities between the two opposites in the title.

• To be dialectic is to go beyond any rigid dichotomy and transcend into something different altogether in the long haul. One may be tempted to call the dialectic logic here with a different name like existential "multilectics" (instead of "dialectics"), but this naming is incorrect (or even misleading), for the two reasons aforestated.

• After all, each ontological pair come in all shapes and sizes (with different degrees) and do not exist by themselves but also interact with other entities unlike them. The virtue of revealing an ontological pair is to show how they relate within themselves (in multiple versions) and also interact with others unlike them (also in multiple versions).

• This will be clear in the table on "the dialectic constraints imposed by the principles."

Source: From Ch.1 of *FPHK*

Table 4.6. The Syntax of Existential Dialectics III: The Principles as Family Resemblances

• Each ontological principle is generic, with some other comparable ontological pairs to be put in the same family (like a *family resemblance*). Ludwig Wittgenstein (1953) once suggested the idea of "family resemblance" in explaining different games classified under the same family called *games*.

• Why should, for instance, playing football and chess as playing "games," when it is well understood that football is not the same as chess? The answer is that, although each game is different and has different rules, many of them (though not all) share, more or less, some commonalites (e.g., scoring as necessary for winning).

• And this is so, even though some games share more than some others in any given selection of critieria, and no two games are exactly identical. So, his point here is that there is no essential core which is common to all games, and the best that one can look for is some characteristics which are common to many (but not all) games. (A. Biletzki 2006)

• By the same logic—in the previous section on selection criteria, the flexibility-inflexibility pair can be put in the family resemblance of the simpleness-complicatedness principle, although the two pairs are not exactly identical. Likewise, the directness-indirectness pair can be put in the family resemblance of the preciseness-vagueness principle, although, again, the two pairs are not exactly identical.

• In this sense, which specific pair in a family should be used to designate the name of the family can be at times a bit arbitrary, but with good reason.

• For illustration, in the context of method, the partiality-totality principle can take the different form like individualisticness-holisticness, just as the explicability-inexplicability principle can take the different form like underlyingness-regularness—although each two pairs are not exactly identical and have slightly different meanings and usages. A more comprehensive listing of this family resemblance for all other principles is shown in the table on the syntax of existential dialectics.

Source: From Ch.1 of *FPHK*

Table 4.7. The Syntax of Existential Dialectics IV: The Dialectic Constraints Imposed by the Principles (Part I)

• **Co-Existent and Asymmetric**

—The principles, as they constitute the syntax of existential dialectics, are dialectic in character, such that, when they are applied, they impose dialectic constraints on how reality is to be understood. Consider, say, the symmetry-asymmetry principle as an illustration here, in order to summarize two main characters of the dialectic constraints in question.

—Firstly, to be dialectic here is to go beyond the narrow dichotomies (and, for that matter, any rigid multi-dimensional classificatory scheme), be they about "self" vs. "world," "freedom" vs. "unfreedom," "barbarity" vs. "civilization," "individuality" vs. "communality," and so on.

—One way to do so (to go beyond) is to consider them all in terms of co-existence (without favoring one over the rest). For instance, my theory of "post-civilization" (to be summarized later in the section on the pragmatics of existential dialectics) is to go beyond barbarity and civilization in terms of understanding barbarity and civilization as being co-existent. And the same logic can be said in relation to my theories of "post-democracy," "post-capitalsm," and others (also to be introduced later in the section on the pragmatics of existential dialectics), in regard to freedom vs. unfreedom, equality vs. inequality, communality vs. individuality, spirituality vs. materiality, and so on.

—But to consider them all (in the dichotomies—and, for that matter, in any rigid multi-dimensional classifactory schme) as co-existent is not the same as to imply that the opposites in any classificatory scheme are all equal, since, in accordance to the symmetry-asymmetry principle (as an illustration here), if they are equal in terms of being considered as co-existent, they are asymmetric in terms of being unequal in dialectic interaction (e.g., X can be more dominant than Y in case A, or Y is more dominant than X in case B).

(continued on next page)

Table 4.7. The Syntax of Existential Dialectics IV: The Dialectic Constraints Imposed by the Principles (Part II)

—For this reason, there are different versions of "post-democracy" and "post-capitalism" in my theories. As an illustration, in version I of the theory of post-democracy, freedom is more dominant than equality, whereas in version II of the theory of post-democracy, equality is more so than freedom.

—But this "X more than Y" has to be understood in the context of dialectic logic (not in conventional logic), in that both "X" and "Y" are important in post-democracy (in the context of dialectic logic), but in an asymmetry way. By contrast, in conventional logic, it often favors one over the other—be it in regard to privileging freedom over equality in Fascism, favoring freedom relatively more than equality in Liberal Democracy, or favoring equality relatively more than freedom in Socialist Democracy. In the latter two cases (about Liberal Democracy and Socialist Democracy), the difference between dialectic logic and conventional logic can be one in degree, not in kind—in this sense, albeit not in other senses.

—The same logic can be said about the relatonships between individuality and communality, between spirituality and materiality, and between formal legalism and informal legalism in the different versions of my theory of post-capitalism.

• **Transcendent**

—Secondly, to be dialectic is to go beyond the narrow dichotomies (and, for that matter, any rigid multi-dimensional classificatory scheme) in another way, this time, in a transcendent way, that is, in exploring other possibilities or even other issues not considered within the narrow confines of narrow dichotomies (and, for that matter, any rigid multi-dimensional classificatory scheme).

(continued on next page)

Table 4.7. The Syntax of Existential Dialectics IV: The Dialectic Constraints Imposed by the Principles (Part III)

—As an analogy, to go beyond the narrow color dichotomy of "black" and "white" is not just to choose both "black" and "white" (as in the first meaning) but also to explore other color options (e.g., "green," "purple," "blue," etc.—and, alternatively, "shade," "line," "curve," etc.). By the same logic, to go beyond "democracy" is to transcend democracy (as in version III of the theory of "post-democracy") and to explore other possibilties of lifeforms (e.g., floating consciousness, hyper-spatial consciousness, etc., to live beyond the narrow obsession with freedom and equality).

—This dialectic character of the principles in existential dialectics has important implications for the pragmatics of existential dialectics (as will be clear shortly, in the section on the pragmatics of existential dialectics).

Notes: The examples here are solely illustrative (not exhaustive). The comparison is also relative (not absolute), nor are they mutually exclusive. As generalities, they allow exceptions.

Sources: First explicitly stated in *ALD*. Also from all other books of mine.

Table 4.8. The Syntax of Existential Dialectics V: Further Clarifications (Part I)

- **Pioneering**
  —Firstly, the total number of ontological principles is unknown, to be discovered later, as our knowledge of the world becomes more advanced.
  —Consequently, the principles as introduced in my books are not exhaustive, with new ones being added, whenever more of them are discovered in later research.
  —At least, future generations can pick up where I leave off and continue the discovery.
  —In this sense, my work should be treated as a pioneering effort for the development of a systematic, comprehensive analysis of a new general ontology for the future of knowledge.

- **Flexible**
  —Secondly, the principles are not rigidly classified, as they can be reclassified in a different way.
  —For instance, the preciseness-vagueness principle is classified under the category of "structure" but can be reclassified under the category of "method," although in so doing, it has a different meaning in the context of method.
  —The same logic applies to the same-difference principle under the category of "outcome," which can be reclassified under the category of "structure," although in so doing, once more, it has a different meaning in the context of structure.

- **Mutually Constraining**
  —Thirdly, the principles are to be understood together, not that each prinicple is to be analyzed independently of others. After all, the principles are mutually constraining, in that they work together as a whole.

(continued on next page)

Table 4.8. The Syntax of Existential Dialectics V: Further Clarifications (Part II)

- **Selectively Useful**
  —Fourthly, the principles are relevant to all subject matters, but some principles are more useful to some subject matters than others—as implied in the symmetry-asymmetry principle. This is true, even if different studies of the same kind can yield different views about the degree of relevance for each ontological principle, depending on the specific nature of a research in question, needless to say

- **Anti-Reductionistic**
  —And fifthly, the principles constitute only two levels of analysis, this time, at the ontological and methodological levels—while other levels of analysis (from the perspectives of the mind, nature, society, and culture) are also needed, in order to understand reality in its totality (as explained in the section on "sophisticated methodological holism," so as to avoid the dual dangers of reductionism and reverse-reductionism).

Notes: The examples here are solely illustrative (not exhaustive). The comparison is also relative (not absolute), nor are they mutually exclusive. As generalities, they allow exceptions.
Sources: From my previous books like *FHC*, *FCD*, *FPHC*, *BDPD*, *BCPC*, *BNN*, *FC*, *FAE*, and so on.

Table 4.9. The Syntax of Existential Dialectics VI: The Dilemma of Specific vs. General Ontology (Part I)

• Ontology is often more complicated and imposes some seemingly insurmountable difficulties concerning what constitutes a kind of ontology which can be valid enough to be accepted by the wider intellectual community in question. An excellent example concerns what I want to call *the dilemma of ontology* in relation to the relationship between "specific ontology" and "general ontology." (M. Bunge 1999; R. Corazzon 2007)

• On the one hand, "general ontology," as Mario Bunge (1999) put it, "studies all existents." But, on the other hand, "special ontology studies one genus of thing or process—physical, chemical, biological, social, etc." (M. Bunge 1999; R. Corazzon 2007)

• Consequently, the relationship between the former and the latter is that, as an illustration, "whereas general ontology studies the concepts of space, time, and event, the ontology of the social investigates such general sociological concepts as those of social system, social structure, and social change." (M. Bunge 1999; R. Corazzon 2007)

• With this formal definition in mind, the relationship between the two forms of ontology only poses *the dilemma of ontology*, in that general ontology is possible to the extent that its general contructs must be broad enough to serve as the "lowest common denominator" for all disciplines, but this character of the lowest common denominator is not only hard to find but also its usefulness is limited. (WK 2008b)

• But specific ontology fares no better either, in that its specific constructs primarily serve a specific discipline in question, although other disciplines can appropriate them for their own purposes, but in a different context (especially, though not exclusively, when used as a metaphor from one field to another). For instance, the Kantian category of understanding under the heading of "quantity" constitutes a type of specific ontology, which is specific to the discipline of mathematics, although other disciplines can make use of the categories in their own context.

(continued on next page)

Table 4.9. The Syntax of Existential Dialectics VI: The Dilemma of Specific vs. General Ontology (Part II)

• A solution lies in my proposal of "sophisticated methodological holism," which requires all levels of analysis in any subject matter. See the tables on sophisticated methodological holism, and on reductionism and reverse-reductionism, for more details.

Source: From *Sec. 1.4* of *FPHK*

Table 4.10. The Syntax of Existential Dialectics VII: Types of Inappropriate Family Resemblances (Part I)

- **Inadequate Family Resemblances**
  - —Ex: consistency-inconsistency: the formalness-informalenss principle
  - —Ex: cognition-noncognition: the formalness-informalness principle

- **Compound Family Resemblances**
  - —Ex: reality-nonreality: the formalness-informalness principles, the absoluteness-relativeness principle
  - —Ex: normality-abnormality: the symmetry-asymmetry principle, the density-emptiness principle, the convention-novelty principle, the same-difference principle
  - —Ex: structure-context: the simpleness-complicatedness principle, the theory-praxis principle
  - —Ex: freedom-unfreedom: the expansion-contraction principle
  - —Ex: morality-immorality: the denseness-emptiness principle, the symmetry-asymmetry, the same-difference principle
  - —Ex: particularness-universalness: the preciseness-vagueness principle, the simpleness-complicatedness principle
  - —Ex: pureness-mixedness: the absoluteness-relativeness principle, the simpleness-complicatedness principle, the same-difference principle
  - —Ex: shortsightedness-foresightedness: the partiality-totality principle, the simpleness-complicatedness principle
  - —Ex: shorttermness-longtermness: the partiality-totality principle, the simpleness-complicatedness principle, the slowness-quickness principle

- **One-Sided Family Resemblances**
  - —Ex: sound-silence: asymmetry (in the symmetry-asymmetry principle)
  - —Ex: vision-blindness: asymmetry (in the symmetry-asymmetry principle)

(continued on next page)

Table 4.10. The Syntax of Existential Dialectics VII: Types of Inappropriate Family Resemblances (Part II)

- **Vague Family Resemblances**
  —Ex: shortness-longness, narrowness-broadness, smallness-largeness, smallness-greatness, microness-macroness, and littleness-muchness: three different possibilities, namely, (a) finiteness (in the finiteness-transfiniteness principle), if both opposites are defined on a finite scale, (b) transfiniteness (in the finiteness-transfiniteness principle) if both opposites are defined on a transfinite scale, and (c) the finiteness-transfiniteness principle if one opposite is defined on a finite scale and the other opposite is defined on a transfinite scale.

- **One-Sided and Compound Family Resemblances**
  —Ex: extroversion-introversion: one-sided, in being part of the side (category) of "normality" in the normality-abnormality pair. And compounded, because the normality-abnormality pair is in turn related to the symmetry-asymmetry principle, the density-emptiness principle, the convention-novelty principle, the same-difference principle
  —Ex: organizedness-spontaneousness: one-sided, in being part of the side (category) of "normality" in the normality-abnormality pair. And compounded, because the normality-abnormality pair is in turn related to the symmetry-asymmetry principle, the density-emptiness principle, the convention-novelty principle, the same-difference principle
  —Ex: risk-caution: one-sided, in being part of the side (category) of "normality" in the normality-abnormality pair. And compounded, because the normality-abnormality pair is in turn related to the symmetry-asymmetry principle, the density-emptiness principle, the convention-novelty principle, the same-difference principle

Notes: The examples here are solely illustrative (not exhaustive). The comparison is also relative (not absolute), nor are they mutually exclusive. As generalities, they allow exceptions.
Source: First summarized in *BEPE* (and updated in later books)

Table 4.11. The Semantics of Existential Dialectics (Part I)

• **Abstract Structure vs. Specific Meanings**

—The syntax of existential dialectics so understood in terms of ontological principles only gives us the structure of ontology in the world, in an abstract (general) sense. These principles by themselves do not tell us the specific meanings in a given context.

—In order to grasp the specific meanings of the principles in a given context, it is necessary to study the semantics of existential dialectics. And the analysis of the ontological principles as family resemblances in the section on syntax is only a starting point (and thus overlaps a bit with the semantics of existential dialectics here).

—With this caveat in mind, the reason that I have often gone in great lengths in my previous books on different subjects is to explain the specific meanings of the principles when applied in different contexts.

• **Specific Meanings in Specific Fields**

—For instance, in *FPHST*, I used the first three principles (i.e., the change-constancy principle, the regression-progression principle, and the symmetry-asymmetry principle) to propose "the perspectival theory of space-time," for a better way to understand space and time—especially, though not exclusively, in relation to future post-human history (as summarized in *Table 3.6*, *Table 3.7*, *Table 3.8*, and *Table 3.9*). In so doing, I had to introduce concepts and theories specific to the field of physics and other related fields (e.g., "absolute space" and "absolute time" in "classical mechanics" and "relative space-time" in "the theory of relativity").

—In *BNN*, I also exploited the three principles to propose the "transcendent" approach to the study of genes and memes as a new way to understand the interaction between nature and nurture. In so doing, I had to explore concepts and theories in the world of evolutionary theory (e.g., "mutation," "variation," "adaptation," "selection," and "inheritance" in Darwinian evolutionary theory) and neural biology (e.g., "chromosome," "gene," "DNA," "RNA," "protein," "neuron," "neural network," and "behavior").

(continued on next page)

Table 4.11. The Semantics of Existential Dialectics (Part II)

- **Two Types of Translation (Interpretation)**
  —In the technical jargons of semantics, these exercises in my books perform two essential functions of translation (or interpretation), namely, (a) "intensive" interpretation (about the properties or relations of the principles in existential dialectics), and (b) "extensive" interpretation (about the classes or cases to which the properties and relations of the principles in existential dialectics are applied). (WK 2011oo)

- **Limits of Translation (Interpretation)**
  —But there are limits of translation (or interpretation), and an excellent example is that "there is a problem with the 'objectivity' of translation, as it is relative to different cultural (or sub-cultural) contexts. For instance, one of the still unresolved problems in linguistics is: 'Is there an objective gauge for the quality of translation?'" (*FPHCOM* & WK 2011pp)
  —This is one serious problem about the logic of translation and interpretation that I already addressed in *Sec. 3.5.1* of *FPHCOM (*and also, for that matter, in *FHC*, in the context of hermeneutics).

Notes: The categories and examples in each are solely illustrative (not exhaustive). The comparison is also relative (not absolute), nor are they mutually exclusive. As generalities, they allow exceptions.
Sources: First explicitly stated in *ALD* (and later revised in *FPHCOM*). Also from all other books of mine.

Table 4.12. The Pragmatics of Existential Dialectics (Part I)

- **The Two-Way Street Connecting Theory and Meta-Theory**
  - —The pragmatics of existential dialectics is not a one-way street (that is, using the ontological principles for theoretical insights in praxis) but a two-way one, that is, (a) from meta-theory to theory, and (b) from theory to meta-theory.
  - —(a) On one side of the street, the ontological principles can inspire some theoretical insights in praxis, that is, in relation to some specific fields.
  - —(b) On the other (opposing) side of the street, however, the study of a subject matter in the specific fields in turn reveals some more hitherto unknown ontological principles to be discovered and identified. For this reason, three new principles were added in *FC*, one in *FAE*, and two in *ALD*, on top of the original three in *BCPC*—after some research on the specific subject matters.

- **Direct and Indirect Applications**
  - —*Direct*
    - The logic of existential dialectics can shed some theoretical insights on diverse phenomena in the world, and good instances are the pertinent use of the principles of existential dialectics for the theoretical insights on the freedom/unfreedom dialectics, the equality/inequality dialectics, and the wealth/poverty dialectics in my previous works.
    - My latest books like *FPHST* and *BNN* also use the principles to reveal some theoretical insights on the perspectives of space and time (as in *FPHST*) and of nature and nurture (as in *BNN*).
  - —*Indirect*
    - The theoretical insights can further be used to reveal other phenomena directly from them (viz., the theoretical insights) and therefore indirectly from the principles themselves. A good instance is the use of the theoretical insights on the freedom/unfreedom and equality/inequality dialectics for the understanding of the civilization/barbarity dialectics.

(continued on next page)

Table 4.12. The Pragmatics of Existential Dialectics (Part II)

• **Direct and Indirect Applications (*cont'd*)**

  • Even in indirect applications, however, a phenomenon under study can still be directly related back to the principles themselves. In the example as cited above, the civilization/barbarity dialectics can be directly related to the principles of existential dialectics without the intermediate role of the freedom/unfreedom and equality/inequality dialectics.

• **Multiple Levels of Application**

—The theoretical insights can be applied to different levels of analysis, even though in a given example, it may refer to one level only. For instance, in the example concerning the freedom/unfreedom dialectics, it can be used at the structural level (e.g., in relation to the theory of cyclical progression of hegemony), but it can be exploited as well for other levels (e.g., the theory of post-capitalism at the institutional level).

Notes: The categories and examples in each are solely illustrative (not exhaustive). The comparison is also relative (not absolute), nor are they mutually exclusive. As generalities, they allow exceptions.
Sources: From Ch.6 of *BCPC*. See also other books of mine.

Table 4.13. The Freedom-Unfreedom Dialectics (Part I)

- **On Having**
  - —*In Relation to the Technological*
    - (1) if freer from submission to Nature, then less free from ecological degradation (Deep and Social Ecology), even if in a hi-tech form
    - (2) if freer from technological inconvenience / backwardness, then less free from technological control and the loss of privacy
    - (3) if freer from technological (material) backwardness, then less free from the abusive (barbaric) maltreatment of the primitive Others
  - —*In Relation to the Everyday*
    - (1) if freer from abject poverty, then less free from artificial needs/discontents (Frankfurt School)
    - (2) if freer from sensual suppression, then less free from violent sublimation (Freud)
    - (3) if freer from the snobbishness of high culture, then less free from the shabbiness (leveling-off effect) of mass culture (Tocqueville)
    - (4) if freer from the inefficiency of traditional "compassionate economy," then less free from the bondage of a "ruthless [competitive] economy" (Keynes)
    - (5) if freer from anarchy in the state of nature (system fragmentation), then less free from government regulations and controls in system integration

(continued on next page)

Table 4.13. The Freedom-Unfreedom Dialectics (Part II)

- **On Belonging**
  - *—In Relation to the Good and the Just*
    - (1) if freer from disciplinary society, then less free from society of control (Foucault)
    - (2) if freer from the tyranny of one or a few, then less free from the tyranny of the majority (or sometimes, minority veto)
    - (3) if freer from elitist decision making, then less free from political gridlock/cleavage
    - (4) if freer from arbitrary (discretionary) administration, then less free from bureaucratic irrationality (Weber) and legal trickery (loopholes)

- **On Being**
  - *—In Relation to the True*
    - (1) if freer from unscientific dogmas, then less free from instrumental abyss (nihilism). Or conversely, if freer from meaninglessness, then less free from dogmas.
    - (2) if freer from the bondage of partiality/partisanship (e.g., prejudice, discrimination), then less free from the danger of impartiality and neutrality (e.g., opportunism, unrealisticness, lack of compassion, inaction)
    - (3) if freer from making generalizations, then less free from being unable to understand much of anything
  - *—In Relation to the Holy*
    - (1) if freer from collective conscience, then less free from social loneliness
    - (2) if freer from religious absoluteness, then less free from spiritual emptiness
  - *—In Relation to the Beautiful/Sublime*
    - (1) if freer from artistic non-autonomy, then less free from aesthetic disillusion (deconstruction)

(continued on next page)

Table 4.13. The Freedom-Unfreedom Dialectics (Part III)

Notes: The examples in each category are solely illustrative (not exhaustive), and the comparison is relative (not absolute), nor are they necessarily mutually exclusive. And some can be easily re-classified elsewhere. As generalities, they allow exceptions.
Sources: A reconstruction from Ch.10 of *FCD*, based on *FHC*

Table 4.14. The Equality-Inequality Dialectics (Part I)

- **On Having**
  - *—In Relation to the Technological*
    - (1) if more equal in treating Nature with spiritual unity, then less equal in suppressing the dominant drive to transcend it altogether
  - *—In Relation to the Everyday*
    - (1) if more equal in building social plurality, then less equal in leveling-off effects (e.g., the subsequent relative intolerance of high/intellectual ethos in mass culture industry)
    - (2) if more equal in socioeconomic distribution beyond a certain point, then less equal in efficiency (e.g. resentment, the erosion of work ethics)
    - (3) if more equal in urging an affirmative action program, then less equal in creating victim mentality (in oneself), stigma (from others), reverse discrimination (against the once privileged), and mediocracy (against the more able)

- **On Belonging**
  - *—In Relation to the Good and the Just*
    - (1) if more equal in banning monarchic/oligarchic exclusion, then less equal in producing "the tyranny of the majority" or of "minority veto"
    - (2) if more equal in encouraging participatory decision making, then less equal in inducing political divisiveness (gridlock/cleavage in power blocs) and organizational oligarchy
    - (3) if more equal in institutionalizing a decentralized bureaucracy, then less equal in falling into more territorial/turf politics (intrigues)

(continued on next page)

Table 4.14. The Equality-Inequality Dialectics (Part II)

- **On Being**
  - —In Relation to the Beautiful / Sublime
    - (1) if more equal in accepting diverse styles ("anything goes" mentality), then less equal in artistic good quality (in leveling-off effects against the best)
  - —In Relation to the True
    - (1) if more equal in tolerating multiple viewpoints (no matter how extreme), then less equal in epistemic standards
  - —In Relation to the Holy
    - (1) if more equal in celebrating any cults and sects (no matter how questionable), then less equal in spiritual depth and authenticity

Notes: The examples in each category are solely illustrative (not exhaustive), and the comparison is relative (not absolute), nor are they mutually exclusive. And some can be easily reclassified elsewhere. As generalities, they allow exceptions.
Sources: A reconstruction from Ch.10 of *FCD*, based on *FHC*

Table 4.15. The Duality of Oppression in Existential Dialectics: Oppression and Self-Oppression (Part I)

- **From the Same to the Others and Itself**
  —The Oppression by the Same against the Others
  - *By way of downgrading differences*
    –Ex: on judiciary caprice for corporate crimes (*Sec.2.2.1.2.1*)
    –Ex: on the deceptive politics of liberation (*Sec.3.5*)
    –Ex: on the humanitarian mystique (*Sec.4.4*)
    –Ex: on the fad of emotional intelligence (*Sec.5.3*)
  - *By way of accentuating differences*
    –Ex: on the legal sophistry of self-defense (*Sec.2.3*)
    –Ex: on the legal semantics of proportionality (*Sec.2.4*)
    –Ex: on the tricky politics of external threat (*Sec.3.4*)
    –Ex: on the appeal of the Far Right for democracy (*Sec.5.4*)
    –Ex: on the democratic axis of evil (*Sec.5.5*)
    –Ex: on the democratic way of brutality and revenge (*Sec.5.6*)
    –Ex: on democratic autocracy (*Sec.6.4*)

  —The Oppression by the Same against Itself
  - *By way of downgrading differences*
    –Ex: on the politics of fear (*Sec.2.2*)
    –Ex: on the trickery of compassionate conservatism (*Sec.3.2*)
    –Ex: on the deceptive politics of patriotism (*Sec.3.3*)
  - *By way of accentuating differences*
    –Ex: on the caprice of due process on domestic suspects (*Sec.2.2*)
    –Ex: on the false security/freedom dilemma (*Sec.6.5.2*)

(continued on next page)

Table 4.15. The Duality of Oppression in Existential Dialectics: Oppression and Self-Oppression (Part II)

- **From the Others to the Same and Themselves**
  - —The Oppression by the Others against the Same
    - *By way of downgrading differences*
      - –Ex: on judiciary caprice in the reverse direction (*Sec.2.2.1.2.2*)
      - –Ex: on equal pay (*Sec.6.2.1.1*)
      - –Ex: on equal representation (*Sec.6.2.1.2*)
      - –Ex: on affirmative action program (*Sec.6.3.1.1*)
      - –Ex: on same-sex marriage (*Sec.6.3.1.2*)
    - *By way of accentuating differences*
      - –Ex: on sexual harassment (*Sec.6.2.2.1*)
      - –Ex: on physical violence (*Sec.6.2.2.2*)
      - –Ex: on sexual exploitation (*Sec.6.2.2.3*)
  - —The Oppression by the Others against Themselves
    - *By way of downgrading differences*
      - –Ex: on the reverse-class mystique (*Sec.4.2*)
      - –Ex: on the reverse-black mystique (*Sec.4.3*)
      - –Ex: on self-discrimination by downgrading (*Sec.6.3.2.2*)
    - *By way of accentuating differences*
      - –Ex: on self-discrimination by accentuating (*Sec.6.3.2.1*)

Notes: The examples are solely illustrative (not exhaustive), nor are they mutually exclusive. As generalities, they allow exceptions. Also, both forms of oppression co-exist in all of the examples, so the listing of them are only meant in a relative, not absolute, sense.
Source: A summary of the sections (as cited) in Chs.2-6 of *BDPD*. See text for more info and references.

Table 4.16. The Structure of Existential Dialectics I: The Freedom/Unfreedom and Equality/Inequality Dialectics

• Each freedom and equality produces its own unfreedom and inequality, regardless of whether the pair occurs in political society (with the nation-state), in civil society (with some autonomy from the state), or elsewhere (e.g., in the private sphere of individual homes)—and regardless of whether freedom and equality are understood as "negative" or "positive."

• Oppression is dualistic, as much by the Same against the Others and itself, as by the Others against the Same and themselves.

• Both forms of oppression and self-oppression can be achieved by way of downgrading differences (between the Same and the Others) and of accentuating them.

• The relationships are relatively asymmetric between the Same and the Others and relatively symmetric within them. This is true, even when the Same can be relatively asymmetric towards itself in self-oppression, just as the Others can be likewise towards themselves.

• Symmetry and asymmetry change over time, with ever new players, new causes, and new forms.

Notes: The examples in each category are solely illustrative (not exhaustive) not necessarily mutually exclusive, and the comparison is relative (not absolute). As generalities, they allow exceptions. "Negative" freedom is freedom "from" (e.g., freedom from poverty), whereas "positive" freedom is freedom "to" (e.g., freedom to the state of enlightenment). "Negative" equality is "procedural" equality (e.g., equality of opportunity), while "positive" equality is "substantive" equality (e.g., equality of outcome). Existential dialectics impose constraints on freedom and equality in democracy, non-democracy, and post-democracy. There is no utopia, in the end; even should there be one, dystopia would exist within it.

Sources: From *Table 1.5* of *BDPD*—and also from *FHC*, *FCD*, and *FPHC*

Table 4.17. The Structure of Existential Dialectics II: The Wealth/Poverty Dialectics

• There is no wealth without poverty, just as there is no poverty without wealth.

• The wealth/poverty dialectics occurs in the realms of having, belonging, and being, in relation to the material, relational, and spiritual.

• The wealth/poverty dialectics also expresses itself at the multiple levels of analysis in accordance to methodological holism, be they about the micro-physical, the chemical, the biological, the psychological, the organizational, the institutional, the structural, the systemic, the cultural, and the cosmological.

• The wealth/poverty dialectics is a different manifestation of existential dialectics in general, subject to the principles in its logic of ontology—just as the freedom/unfreedom and equality/inequality dialectics are likewise.

• There is no economic utopia, in the end; even should there be one, dystopia would exist within it.

Notes: The main points here are solely illustrative (not exhaustive) nor necessarily mutually exclusive, and the comparison is relative (not absolute). As generalities, they allow exceptions.
Sources: From *BCPC*. See also *FCD* and *FHC*.

Table 4.18. The Structure of Existential Dialectics III: The Civilization/Barbarity Dialectics

- There is no civilization without barbarity.

- The civilization/barbarity dialectics applies in the four civilizing processes (e.g., the rationalizing process, the pacifying process, the stewardizing process, and the subliming process).

- The civilization/barbarity dialectics is another (different) manifestation of existential dialectics in general, subject to the principles in its logic of ontology—just as the freedom/unfreedom and equality/inequality dialectics and the wealth/poverty dialectics are likewise.

- There is no utopia, in the end; even should there be one, dystopia would exist within it.

Notes: The main points here are solely illustrative (not exhaustive) nor necessarily mutually exclusive, and the comparison is relative (not absolute). As generalities, they allow exceptions.
Sources: From *BCIV*. See also *FCD*, *FHC*, and *BDPD*.

Table 4.19. The Double Sides of Virtual Organizations (Part I)

- **Psychological**
  —Ex: virtual psychosis
  —Ex: impersonality and loneliness in quaternary social relations

- **Organizational**
  —Ex: the race for power and interests
  —Ex: the world of unequal successes
  —Ex: the bureaucratic life of its own
  —Ex: lesser accountability and transparency

- **Economic**
  —Ex: the economic divides
  —Ex: the erosive impact of commercialization

- **Political**
  —Ex: the anti-authoritarian myth
  —Ex: different power struggles among groups (e.g., Conservative, Reformist, Radical)

- **Structural**
  —Ex: different localities (e.g., climates, scenery)
  —Ex: divided domains (e.g., core, peripheral, sub-peripheral)
  —Ex: substitution effect of social capital

- **Systemic**
  —Ex: uneven advances in transportation and communications, and geographical migration

- **Cultural**
  —Ex: conflicting civilizational fabrics (e.g., Confucian, Islamic, Western)
  —Ex: the bias of the Liberal Democratic agenda
  —Ex: the addiction to moral fanaticism

(continued on next page)

Table 4.19. The Double Sides of Virtual Organizations (Part II)

Notes: These categories and examples are solely illustrative (not exhaustive), and some of the items can be reclassified somewhere else. Nor are they mutually exclusive. Since they are generalities, exceptions are expected.
Source: From Ch.7 of *FCD*

Table 4.20. Beyond the World of Titans, and the Remaking of the World Order

- **Hyper-Empires**
  - —Ex: The Chinese Union
  - —Ex: The Indian Union
- **Meso-Empires**
  - —Ex: The European Union
  - —Ex: The North American Union
- **Micro-Empires**
  - —Ex: The Latin American Union
  - —Ex: The Middle Eastern Union
- **The Rest of the World**
  - —<u>Odd Powers</u>
    - Ex: Japan
    - Ex: Russia
  - —<u>The Poor Club</u>
    - Ex: The African Union
  - —<u>Ambivalent Regions</u>
    - Ex: Southeast Asia
    - Ex: Oceania
    - Ex: South Asia
    - Ex: Central Asia
    - Ex: Southern/Eastern Europe
    - Ex: North Africa
    - Ex: Central America
    - Ex: Others (e.g., the Korean peninsula)

Source: A summary of Chs.2-5 (of *BWT*)

Table 4.21. The Origins of Authoritarian Liberal Democracy

- **The Geopower of Nature (Ch.4 of *ALD*)**
  —Ex: Power Character and Geographical Strategy
  —Ex: Living Space and Territorial Expansion
  —Ex: Strategic Heartland and Containment

- **The Biopsychology of the Mind (Ch.5 of *ALD*)**
  —Ex: The Bell Curve and Mass Intelligence
  —Ex: Group Analysis and Mass Knowledge
  —Ex: Groupthink and Elite (Mis)calculation

- **The Disciplinary Control of Society (Ch.6 of *ALD*)**
  —Ex: Social Organizations of Ruthlessness
  —Ex: Social Institutions of Greed
  —Ex: Social Structure of Exclusion
  —Ex: Social Systems of Violence

- **The Molding Force of Culture (Ch.7 of *ALD*)**
  —Ex: The Tradition of Conquest
  —Ex: The Rationalization of Unreason

Notes: The examples in each category are solely illustrative (not exhaustive), and they are also relative (not absolute), nor are they mutually exclusive. As generalities, they allow exceptions.
Source: A summary of Chs.4-7 of *ALD*

Table 4.22. The Theory of Post-Democracy I: The Priority of Freedom over Equality (Part I)

- **Differences**
  - —*For the aggressive Lions (the strong Elitists)*
    - •Setting up rank distinctions among unequals (e.g., between inferior humans and superior post-humans, or later among inferior post-humans and superior ones, relatively speaking)
    - •Yearning for being not only distinguished from unequals, but also the first among equals (the best of the very best)
    - •Soul-searching for a high spiritual culture (not the trashy one for the masses). Mass culture is a dirty joke for them.
  - —*For the manipulative Foxes (the weak Counter-Elitists)*
    - •Seeking a gentle hegemony by way of more communitarian concerns (for inferior humans and, later, inferior post-humans)
    - •Being more sympathetic to less formal-legalistic institutions and values

- **Similarities**
  - —*For both Lions and Foxes*
    - •Exploring different spheres of non-human consciousness in the cosmos (something vastly superior than the human one)
    - •Recognizing the democratic illusions (e.g., no freedom without unfreedom, no equality without inequality, or simply no justice without injustice, and vice versa)

(continued on next page)

Table 4.22. The Theory of Post-Democracy I: The Priority of Freedom over Equality (Part II)

Notes: The two callings and examples in each category are solely illustrative (not exhaustive), since there will be many different post-human value ideals in the distant future of post-human civilization. The comparison is also relative (not absolute) towards post-democracy, so this is not just a version of free-market democracy (nor Fascism/Nazism, as shown in the table later on democracy, non-democracy, and post-democracy). Nor are they mutually exclusive. As generalities, they allow exceptions. And the specific forms of post-human post-democratic ideals need to be further developed in future after-postmodern history, as they will be different from the ones we now know. The point here is to solely give an extremely rough picture of a small part of the world to come that we have never known.
Source: From Ch.10 of *FCD*. Refer to text for more info and references.

Table 4.23. The Theory of Post-Democracy II: The Priority of Equality over Freedom

- **Hybrid Versions of**
  —Ex: the Trans-Feminine Calling
  —Ex: the Trans-Sinitic Calling
  —Ex: the Trans-Islamic Calling
  —Ex: the Trans-Outerspace Calling

- **Qualifications**
  —These four versions of post-capitalist value ideals need not automatically be post-democratic, just as capitalism does not necessarily mean democracy. They are two different entities—though closely related.
  —But up to a certain threshold of elevating equality at the farther expense of freedom, the democratic ideals will be overcome and cease to exist.
  —The overcome will not be socialist or communist, but post-democratic with no freedom without unfreedom and no equality without inequality, subject to the constraints of existential dialectics.

Notes: The callings are solely illustrative (not exhaustive), since there will be many different post-human value ideals in the distant future of post-human lifeforms. The comparison is also relative (not absolute), nor are they mutually exclusive. As generalities, they allow exceptions. And the specific forms of post-human post-democratic ideals need to be further developed in future after-postmodern history, as they will be different from the ones we now know. The point here is to solely give an extremely rough picture of a small part of the world to come that we have never known.
Source: From Ch.10 of *FCD*. Refer to text for more info and references.

Table 4.24. The Theory of Post-Democracy III: The Transcendence of Freedom and Equality (Part I)

• **Transcending Freedom in Floating Existence**

—*Freedom*: seeking an ultimate elimination of the body. Being without the body. The aim is to transcend freedom in the end into a metaphysical state (i.e., beyond the physique).

—*Unfreedom*: yet facing difficult trade-offs. The sacrifice of bodily existence and its joyfulness. An eternal boredom in floating existence in dark deep space, though with alternative pleasures. There is no free lunch even in the state of transcending freedom.

• **Transcending Equality in the Rivalry of Cosmic Hegemony**

—*Inequality*: competing to outlast other lifeforms in floating existence, or just marginalizing them for one's hegemonic expansiveness in the rest of the cosmos (and even beyond). Universalism is only for the mediocre.

—*Equality*: accepting only those of one's rank as equal partners in the vast spacetime for cosmic supremacy. Even here, the aim is to transcend equality into a metaphysical state.

(continued on next page)

k

Table 4.24. The Theory of Post-Democracy III: The Transcendence of Freedom and Equality (Part II)

Notes: Do not confuse this transcendence of freedom and equality (as one version of post-democracy) with the naïve temptation to transcend the freedom/unfreedom and equality/inequality dialectics. Existential dialectics hold true for freedom and equality in all cultures and societies—past, present, or future (i.e., democracy, non-democracy, and post-democracy), regardless of whether freedom and equality are conventionally understood as "negative" or "positive."

Also, the two features and examples in each are solely illustrative (not exhaustive), since there will be many different post-human value ideals in the distant future of post-human lifeforms. The comparison is also relative (not absolute), nor are they mutually exclusive. As generalities, they allow exceptions. And the specific forms of post-human ideals even for these radically alien floating lifeforms (and others unknown to us) need to be further developed in future after-postmodern history, as they will likely be different from the ones herein illustrated. The point here is to solely give a very rough picture of a small part of the extremely alien world to come that we have never known.

Source: From Ch.10 of *FCD*. Refer to text for more info and references.

Table 4.25. Democracy, Non-Democracy, and Post-Democracy (Part I)

• **Democracy**

—*Theoretical Constructs*

•The pursuit of freedom and equality (in various degrees), regardless of whether freedom and equality can be understood as "negative" or "positive"

(1) more equality than freedom: The relative priority of the good over the right

(2) more freedom than equality: The relative priority of the right over the good

—*Types*

•Only (1): Different versions of communitarian moral universalism

•Only (2): Different versions of liberal moral universalism

•(1) or (2): Different versions of anarchic (non-nation-state) moral universalism

•(1) or (2): Different versions of postmodern moral localism

• **Non-Democracy**

—*Theoretical Constructs*

•The focus on (1') equality or (2') freedom, but not both, regardless of whether freedom and equality can be understood as "negative" or "positive"

—*Types*

•Only (1'): Different versions on the Far Left (e.g., Stalinism, Robespierrianism)

•Only (2'): Different versions on the Far Right (e.g., Nazism, absolute monarchism)

(continued on next page)

Table 4.25. Democracy, Non-Democracy, and Post-Democracy (Part II)

- **Post-Democracy**
  —*Theoretical Constructs*
  •The priority of (1'') equality over freedom, or (2'') freedom over equality, or (3'') the transcendence of freedom and equality, regardless of whether freedom and equality are "negative" or "positive." In degree, (1'') or (2'') is less than (1') or (2') but more than (1) or (2)—respectively.
  •Like democracy and non-democracy, post-democracy is also subject to the freedom/unfreedom and equality/inequality dialectics (or existential dialectics in general). Unlike them, post-democracy acknowledges the constraints of existential dialectics and no longer value freedom and equality as sacred virtues. There is no utopia, in the end; even were there one, dystopia would exist within it.
  —*Types*
  •(1''): Different versions of trans-Sinitic value ideals
  •(1''): Different versions of trans-feminine value ideals
  •(1''): Different versions of trans-Islamic value ideals
  •(1''): Different versions of trans-outerspace value ideals
  •(2''): Different versions of post-human elitist value ideals
  •(3''): Different versions of the value ideals of floating consciousness (etc.)

(continued on next page)

Table 4.25. Democracy, Non-Democracy, and Post-Democracy (Part III)

Notes: The examples are solely illustrative (not exhaustive), nor are they mutually exclusive. As generalities, they allow exceptions. "Negative" freedom is freedom "from" (e.g., freedom from poverty), whereas "positive" freedom is freedom "to" (e.g., freedom to the state of enlightenment). "Negative" equality is "procedural" equality (e.g., equality of opportunity), while "positive" equality is "substantive" equality (e.g., equality of outcome). Existential dialectics impose constraints on freedom and equality in democracy, non-democracy, and post-democracy, regardless of whether freedom and equality can be understood as "negative" or "positive" in conventional discourse. Therefore, do not confuse the transcendence of freedom and equality in (3'') with the naïve temptation to transcend existential dialectics. There is no utopia, in the end; even should there be one, it would not exist without dystopia embedded within it.
Sources: A summary, based on my previous works, especially Ch.5 of *FHC*, Chs.5-10 of *FCD*, Chs.2-4 of *FPHC*, and Chs.1 & 7 of *BDPD*. The reader should consult the books for more analysis, as this is only a summary here.

Table 4.26. Multiple Causes of the Emergence of Post-Democracy (Part I)

- **At the Micro-Physical Level**
  —Ex: intelligent life without the human physical-chemical system
  —Sources: Ch.7 of *FHC*; Chs.9-10 of *FCD*; Ch.1 of *FPHC*

- **At the Chemical Level**
  —Ex: space radiation and toxins
  —Sources: Ch.7 of *FHC*; Chs.9-10 of *FCD*

- **At the Bio-Psychological Level**
  —Ex: exo-biological evolution in deep space
  —Ex: genetic engineering of new beings
  —Ex: limits of cognitive partiality
  —Ex: illusions of emotional neutrality
  —Ex: human biological inequality
  —Ex: the rise of unfolding unconsciousness
  —Sources: Ch.2 & Chs.9-10 of *FCD*; Ch.7 of *FHC*; Ch.4 of *BCPC*; *FPHU*

- **At the Institutional Level**
  —Ex: the flawed logic of equality
  —Ex: the conflicting nature of governance
  —Sources: Ch. 5 of *FHC*; Chs. 6 & 10 of *FCD*; Ch. 3 of *FPHC*; Chs.2-5 of *BDPD*

- **At the Organizational Level**
  —Ex: e-civic alienation
  —Ex: the dark sides of formal-legalistic routines
  —Sources: Ch.3 of *FHC*; Ch.7 of *FCD*; Ch.3 of *FPHC*

- **At the Structural Level**
  —Ex: ever new forms of inequities, at home and abroad
  —Ex: the emergence of China, women, and Islam as major actors
  —Sources: Chs.5-6 of *FHC*; Chs.7, 9 & 10 of *FCD*; Chs.4-5 of *BDPD*

(continued on next page)

Table 4.26. Multiple Causes of the Emergence of Post-Democracy (Part II)

• **At the Cultural Level**
—Ex: freedom/unfreedom dialectics
—Ex: equality/inequality dialectics
—Ex: system fragmentation and integration
—Sources: Ch.5 of *FHC*; Chs. 3, 9 & 10 of *FCD*; Ch.4 of *FPHC*; Ch.1 of *BDPD*; Ch.4 of *BCPC*

• **At the Systemic Level**
—Ex: space habitats (in zero-gravity) and colonization
—Ex: ultra advanced future info systems
—Ex: qualitative demography
—Sources: Ch.7 of *FHC*; Chs.9 &10 of *FCD*

• **At the Cosmological Level**
—Ex: the colonization of multiverses
—Ex: the alteration of space-time and the creation of new matter-energy
—Ex: the expansion of floating consciousness
—Ex: the spread of hyper-spatial consciousness
—Sources: Ch.7 of *FHC*; Chs.9 &10 of *FCD*; Ch.4 of *FPHC*; *FPHST*

Notes: The examples in each category are solely illustrative (not exhaustive), and some of the items can be reclassified somewhere else. Nor are they always mutually exclusive. Since they are generalities, exceptions are expected.
Sources: Especially from *FHC*, *FCD*, *FPHC*, *BCPC*, *BDPD*, *FPHST*, and *FPHU*. See also other books and my perspectives on civilizational holism.

Table 4.27. Some Clarifications about Post-Capitalism and Post-Democracy (Part I)

• The prefix "trans-" in the first category of post-capitalism (with its four versions) refers to something "going beyond" (not "uniting" or "combining"). Ex: *Sec.10.3.3* of *FCD*; *Sec.2.4* & *Sec.4.4* of *FPHC*; *Sec.7.2* of *BCPC*

• Such terms like "post-democracy," "post-capitalism," "post-human elitist," "trans-feminine calling," and the like as used in my works are more for our current intellectual convenience than to the liking of future humans and post-humans, who will surely invent more tasteful neologisms to call their own eras, entities, and everything else, for that matter. But the didactic point here is to use the terms to foretell what the future might be like, not that its eras and entities must be called so exactly and permanently.
Ex: *Sec.11.1* of *FCD*; *Sec.7.2* of *BCPC*

• The four versions in the first category of post-capitalist value ideals need not automatically be post-democratic, just as capitalism does not necessarily mean democracy. They are two different entities—though closely related. But up to a certain threshold of elevating equality at the farther expense of freedom, the democratic ideals will be overcome and cease to exist. The same is true for the post-human elitist calling in the second category of post-capitalism in relation to post-democracy, depending on the extent to which freedom is elevated at the expense of equality.
Ex: *Sec.10.4.3.3* of *FCD*; *Table 3.9* of *FPHC*; *Table 7.6* of *BDPD*

• The comparison in each of the three realms of existence in all forms of post-capitalism is not absolute, but relative. Examples include "communal" vs. "individualistic," and the like.
Ex: Notes in *Table 10.8*, *Table 10.9*, *Table 10.10*, & *Table 10.11* of *FCD*; Chs.2-4 of *FPHC*; *Sec.7.2* of *BCPC*

(continued on next page)

Table 4.27. Some Clarifications about Post-Capitalism and Post-Democracy (Part II)

• The emergence of post-capitalism (and post-democracy, for that matter) has multiple causes (to not be reduced to one or only a few).
Ex: Ch.10 of *FCD*, Chs.2-4 of *FPHC*; *Sec.1.3* & *Sec.7.2* of *BCPC* (or *Table 1.8* & *Table 7.11*)

• The specific forms of post-capitalism (and post-democracy, for that matter) need to be further developed in future after-postmodern history, as they will be different from the ones we now know. The point here is to solely give an extremely rough sketch of a world to come that we have never known.
Ex: *Sec.10.3.3* & *Sec.10.4.3.3* of *FCD*; *Table 10.14* & *Table 10.15* of *FCD*; *Sec.7.2* of *BCPC*

• All forms of post-capitalism are not part of a "teleological law," but of "historical trends" only. The same is also true for all forms of post-democracy.
Ex: *Sec.7.1* of *FHC*; *Sec.9.5.3.2* & *Sec.10.3.4.2* of *FCD*; *Sec.7.2* of *BCPC*

• Post-capitalism is not better than capitalism in an "absolute" sense but only fits in better, on the basis of the historical contingency of culture, society, nature, and the mind in some future eras. The same is true for post-democracy in relation to democracy. The term "better" is historically relative.
Ex: *Sec.10.3.3* of *FCD*; *Sec.1.7* of *BDPD*; *Sec.1.5* of *BCPC*

(continued on next page)

Table 4.27. Some Clarifications about Post-Capitalism and Post-Democracy (Part III)

• All forms of post-capitalism and post-democracy are subject to the constraints of existential dialectics. In the process, the dialectic direction is to go beyond the conventional "either-or" dichotomies (e.g., freedom vs. unfreedom, equality vs. inequality, freedom vs. equality, individuality vs. communality, spirituality vs. materiality, formal legalism vs. informal legalism, etc.). As is true in post-civilization, to go beyond the dichotomies is to acknowledge the co-existence of both in each dichotomy, although the degree of scaling one over the over varies from case to case (e.g., the theory of post-capitalism I, the theory of post-capitalism II, the theory of post-democracy I, the theory of post-democracy II, etc.)—but is not to be extreme in largely favoring one over the other, *on average* (all things considered). There is no utopia to be had in the end; even should there be one, dystopia would exist within it.
Ex: Ch.5 of *FHC*; *Sec.10.4.4.2* of *FCD*; *Sec.1.5* of *BDPD*; *Sec.1.3* of *BCPC*; *BCIV*

• All forms of post-capitalism, however different from each other though they are, share one common feature, in that they all inspire for a higher spiritual culture. The same is also true for post-democracy.
Ex: *Sec.10.3*, *Sec.10.4* & *Sec.10.5* of *FCD*; *Chs.2-4* of *FPHC*; *Sec.7.2* of *BCPC*

• All forms of post-capitalism try to avoid the excess in capitalist consumerism by favoring more basic than artificial needs in having, but the quality and quantity of these "basic" needs will be measured by future standards, not by our current ones. Standards are historically relative.
Ex: *Sec.10.3*, *Sec.10.4* & *Sec.10.5* of *FCD*; Ch.2 of *FPHC*; *Sec.7.2* of *BCPC*

(continued on next page)

Table 4.27. Some Clarifications about Post-Capitalism and Post-Democracy (Part IV)

• All forms of post-capitalism make use of a different degree of political authority with advanced info systems in future history and strives for higher spiritual cultures (especially in the post-human age), while acknowledging the constraints of existential dialectics and no longer valuing free market (as in capitalism) and economic control (as in non-capitalism) as sacred virtues.
Ex: *Sec.10.3.4.2*, *Sec.10.3*, *Sec.10.4* & *Sec.10.5* of *FCD*; Chs.2-4 of *FPHC*; *Sec.1.5* of *BDPD*; *Sec.7.2* of *BCPC*

Notes:: The main points here are solely illustrative (not exhaustive) nor necessarily mutually exclusive, and the comparison is relative (not absolute). As generalities, they allow exceptions. The sections as cited are only illustrative (not exhaustive).
Sources: From *FHC*, *FCD*, *FPHC*, and *BDPD*

Table 4.28. The Theory of Post-Capitalism I.1: By Group— Ex: Spiritual/Communal in the Trans-Feminine Calling

- **More Communal Than Individual**
  —*Sharing*: learning from others, as different ideas mutually enrich
  —*Cooperative*: encouraging a sense of shared leadership and teamwork

- **More Informal-Legalistic Than Formal-Legalistic**
  —*Specific*: listening more from the heart than from the head, to know a person as a concrete, not as an abstract, unit
  —*Affective*: thinking and acting with others on a more affective tone. Business can mix with an emotional touch.
  —*Ascriptive*: hiring (or firing) can be done on the basis of merit (or lack of it), but deep solidarity (sisterhood) is important too.
  —*Particularistic*: making decisions on the basis of cost-benefit analysis, but a given group relationship is vital

- **More Spiritual Than Secular**
  —*Long-Term Looking*: sharing for a long-term relationship (e.g., love, friendship), not just for a short-term gain
  —*Loving/Caring*: showing compassion for the sufferings of others, without quickly blaming and pre-judging
  —*Respectful*: showing acceptance about others' feelings (and thoughts)

Notes: The categories and examples are solely illustrative, since there can be different versions, and the comparison is relative (not absolute), nor are they mutually exclusive. As generalities, they allow exceptions. The specific forms of the trans-feminine version need to be further developed in future after-postmodern history, as they will be different from the ones we now know, since the prefix "trans-" here means going beyond or deconstructing the feminine values, while using them as the inspirational point at the beginning.
Source: From Ch.10 of *FCD*. Refer to text for more info and references.

Table 4.29. The Theory of Post-Capitalism I.2: By Nation-State— Ex: Spiritual/Communal in the Trans-Sinitic Calling

- **More Communal Than Individualistic**
  —*Centralized*: being more top-down in management
  —*Collective*: encouraging more group cooperation
  —*Social*: investing in trust and connection

- **More Informal-Legalistic Than Formal-Legalistic**
  —*Specific*: knowing more of those related or connected
  —*Affective*: behaving in a paternalistic, hierarchical way
  —*Ascriptive*: favoring family members and those related
  —*Particularistic*: building connection (guanxi) as imperative

- **More Spiritual Than Secular**
  —*Expansionist*: diffusing civilizational values (e.g., the superiority complex of civilizationalism)
  —*Holistic*: synthesizing things into a panoramic horizon
  —*Historical*: learning from the lessons of the ancient past
  —*Respectful*: deferential to elders and superiors

Notes: The categories and examples are solely illustrative, since there can be different versions, and the comparison is relative (not absolute), nor are they mutually exclusive. As generalities, they allow exceptions. The specific forms of the trans-Sinitic version need to be further developed in future after-postmodern history, as they will be different from the ones we now know, since the prefix "trans-" here means going beyond or deconstructing the Sinitic values, while using them as the inspirational point at the beginning.
Source: From Ch.10 of *FCD*. Refer to text for more info and references.

Table 4.30. The Theory of Post-Capitalism I.3: By Region— Ex: Spiritual/Communal in the Trans-Islamic Calling

- **More Communal Than Individualistic**
  —*Collective*: building the webs of relationships to bind individuals
  —*Sharing*: cultivating the established "wisdom" through common experience
  —*Cooperative*: stressing harmony, solidarity, and commonality

- **More Informal-Legalistic Than Formal-Legalistic**
  —*Specific*: making efforts to know well the participants (family and larger community) in matters of common concern
  —*Affective*: mixing work with language and ritual on explicit religious (Islamic) ideals, texts, stories, and examples
  —*Ascriptive*: privileging local history and custom on relationships among kinship groups
  —*Particularistic*: preferring an unbiased insider with ongoing connections to all parties

- **More Spiritual Than Secular**
  —*Historical*: learning from the lessons of the past as a source of stability and guidance
  —*Deferential*: showing respect for age, experience, status, and leadership in communal affairs
  —*Honorable*: emphasizing face, dignity, prestige, and fairness
  —*Compassionate*: giving mercy and charity ("Zahah") to others

Notes: The categories and examples are solely illustrative (not exhaustive), and the comparison is relative (not absolute), nor are they mutually exclusive. As generalities, they allow exceptions. The specific forms of the trans-Islamic version need to be further developed in future after-postmodern history, as they will be different from the ones we now know, since the prefix "trans-" here means going beyond or deconstructing the Islamic values, while using them as the inspirational point at the beginning.
Sources: From Ch.10 of *FCD*. Refer to text for more info and references, especially from the works by George Irani (2000) and C. Murphy (September 19, 2001).

Table 4.31. The Theory of Post-Capitalism I.4: By Universe— Ex: Spiritual/Communal in the Trans-Outerspace Calling

- **More Communal Than Individual**
  —*Cooperative*: requiring teamwork in small space habitats
  —*Sharing*: learning from, and enjoying being with, each other in a small group in outer space

- **More Informal-Legalistic Than Formal-Legalistic**
  —*Specific*: knowing more about each other to facilitate living and working together in space, both as fellow astronauts and space-mates
  —*Affective*: being friendly and social to each other as vital to working and living in small space quarters
  —*Ascriptive*: nurturing comaraderie among fellow astronauts as if they are family members over time
  —*Particularistic*: building work relationship with enduring memory in a space mission

- **More Spiritual Than Secular**
  —*Long-Term*: looking beyond selfish materialistic concerns in a precarious space environment with potential life or death
  —*Loving/Caring*: cultivating deep bondage for the success of a long term space mission
  —*Transcendent*: searching for life meaning in outer space

Notes: The calling and examples in each category are solely illustrative (not exhaustive), since there will be many different outer-space value ideals in the distant future of space colonization. The comparison is also relative (not absolute), nor are they mutually exclusive. As generalities, they allow exceptions. And the specific forms of trans-outer-space calling need to be further developed in future after-postmodern history, as they will be different from the ones we now know, since the prefix "trans-" here means going beyond or deconstructing the current outer-space values, while using them as the inspirational point at the beginning. The point here is to solely give an extremely rough picture of a small part of the world to come that we still do not know much about.
Source: From Ch.10 of *FCD*. Refer to text for more info and references.

Table 4.32. The Theory of Post-Capitalism II: Spiritual/Individualistic in the Post-Human Elitist Calling (Part I)

- **More Individualistic Than Communal**
  —Setting up rank distinctions among unequals (e.g., between inferior humans and superior post-humans, or later among inferior post-humans and superior ones, relatively speaking)
  —Yearning for being not only distinguished from unequals, but also the first among equals (the best of the very best)
  —Recognizing the constraints of equality/inequality dialectics (or existential dialectics in general)

- **More Spiritual Than Secular**
  —Soul-searching for a high spiritual culture (not the trashy one for the masses). Mass culture is a dirty joke for them.
  —Exploring different spheres of non-human consciousness in the cosmos (something vastly superior than the human one)
  —Recognizing the constraints of freedom/unfreedom dialectics (or existential dialectics in general)

- **Qualifications**
  —Although post-human elitist post-democracy is comparable to post-human elitist post-capitalism in some respects, the former does not necessarily imply the latter (post-human elitist post-capitalism), just as democracy does not have to entail capitalism. They are two different (though related) entities.
  —But up to a certain threshold of incorporating government intervention with advanced info systems in future civilizations for higher spiritual concerns at the expense of the free market and materialist pursuit, the capitalist ideal will be overcome.
  —The overcome will not be Fascist or feudalistic, but post-capitalist, subject to the constraints of existential dialectics.

(continued on next page)

Table 4.32. The Theory of Post-Capitalism II: Spiritual/Individualistic in the Post-Human Elitist Calling (Part II)

Notes: The calling and examples in each category are solely illustrative (not exhaustive). The comparison is also relative (not absolute), nor are they mutually exclusive. As generalities, they allow exceptions. And the specific forms of post-human elitist post-capitalism need to be further developed in future after-postmodern history, as they will be different from the ones we now know, while using them as the inspirational point at the beginning. The point here is to solely give an extremely rough picture of a small part of the world to come that we still do not know much about.
Sources: From Ch.10 of *FCD* (and also *FPHC*, *BDPD*, and *BCPC*). Refer to the text for more info and references.

Table 4.33. Capitalism, Non-Capitalism, and Post-Capitalism (Part I)

- **Capitalism**
  - —*Theoretical Constructs*
    - •Allocation of scarce resources among alternative wants largely by free market for competition (whose characteristics in its ideal form include, for instance, no barrier to entry or exit, homogeneity, perfect information, a large number of buyers/sellers, and perfect factor mobility)
    - •More formal-legalistic than informal-legalistic, more individualistic than communal, and more material (secular) than spiritual
    - •Either (1) minimal government or (2) relatively active government
  - —*Types*
    - •Only (1): Different versions of market capitalism (e.g., the U.S.)
    - •Only (2): Different versions of welfare capitalism (e.g., Sweden)

- **Non-Capitalism**
  - —*Theoretical Constructs*
    - •Allocation of scarce resources among alternative wants mainly by political authority for policies (which can be regulative, redistributive, symbolic, and participatory)
    - •More informal-legalistic than formal-legalistic
    - •Either (1') more individualistic (for the elites), often (though not always) for material (secular) concerns, or (2') more communal (for the masses), often (though not always) for spiritual concerns
  - —*Types*
    - •Only (1'): Different versions on the Right (e.g., Fascist corporate-state economy for the glory of the new Rome, medieval lord-vassal-serf economy for the power of the feudalistic order)
    - •Only (2'): Different versions on the Left (e.g., Soviet command economy for the creation of the New Socialist Man)

(continued on next page)

Table 4.33. Capitalism, Non-Capitalism, and Post-Capitalism (Part II)

- **Post-Capitalism**

—*Theoretical Constructs*

- •Allocation of scarce resources among alternative wants largely by political authority with advanced info systems in future civilizations, subject to existential dialectics. In degree of allocating by authority, post-capitalism is more than capitalism but less than non-capitalism.
- •More spiritual than secular (material)
- •Either (1'') more individualistic or (2") more communal
- •Like capitalism and non-capitalism, post-capitalism is also subject to the freedom/unfreedom and equality/inequality dialectics (or existential dialectics in general). There is no utopia, in the end; even were there one, dystopia would exist within it.
- •Unlike capitalism and non-capitalism, post-capitalism makes use of a different degree of political authority with advanced info systems in future civilizations and strives for higher-spiritual cultures (especially in the post-human age), while acknowledging the constraints of existential dialectics and no longer valuing free market (as in capitalism) and economic control (as in non-capitalism) as sacred virtues.

—*Types*

- •Only (1"): Different versions of post-human elitist value ideals
- •Only (2''): Different versions of trans-Sinitic value ideals
- •Only (2''): Different versions of trans-feminine value ideals
- •Only (2''): Different versions of trans-Islamic value ideals
- •Only (2''): Different versions of trans-outerspace value ideals

---

(continued on next page)

Table 4.33. Capitalism, Non-Capitalism, and Post-Capitalism (Part III)

Notes: The calling and examples in each category are solely illustrative (not exhaustive). The comparison is also relative (not absolute), nor are they mutually exclusive. As generalities, they allow exceptions. And the specific forms of each calling need to be further developed in future after-postmodern history, as they will be different from the ones we now know, while using them as the inspirational point at the beginning. The point here is to solely give an extremely rough picture of a small part of the world to come that we still do not know much about.
Source: From Ch.10 of *FCD*. Refer to the text for more info and references.

Table 4.34. Multiple Causes of the Emergence of Post-Capitalism (Part I)

• **At the Micro-Physical Level**
—Ex: intelligent life without the human physical-chemical system
—Ex: mastering of quantum mechanics, electromagnetism, and other fields for the understanding of a broad range of anomalous experiences and the application for artificial intelligence for spiritual quest
—Sources: Ch.7 of *FHC*; Chs.9-10 of *FCD*; Ch.1 of *FPHC*

• **At the Chemical Level**
—Ex: space radiation and toxins
—Sources: Ch.7 of *FHC*; Chs.9-10 of *FCD*

• **At the Bio-Psychological Level**
—Ex: exo-biological evolution in deep space
—Ex: genetic engineering of new beings
—Ex: limits of human cognition
—Ex: the rise of unfolding unconsciousness
—Sources: Ch.2 & Chs.9-10 of *FCD*; Ch.7 of *FHC*; *FPHU*

• **At the Institutional Level**
—Ex: the flawed logic of the free market
—Ex: the need of a post-autistic economics
—Sources: Ch.10 of *FCD*

• **At the Organizational Level**
—Ex: the dark sides of formal-legalistic routines
—Sources: Ch.3 of *FHC*; Ch.7 of *FCD*; Ch.3 of *FPHC*

• **At the Structural Level**
—Ex: ever new forms of inequities, at home and abroad
—Ex: the emergence of China, women, and Islam as major actors
—Sources: Chs.5-6 of *FHC*; Chs.7, 9 & 10 of *FCD*; Chs.4-5 of *BDPD*

(continued on next page)

Table 4.34. Multiple Causes of the Emergence of Post-Capitalism (Part II)

- **At the Cultural Level**
  —Ex: freedom/unfreedom dialectics
  —Ex: equality/inequality dialectics
  —Sources: Ch.5 of *FHC*; Chs.3 & 10 of *FCD*; Ch.4 of *FPHC*; Ch.1 of *BDPD*

- **At the Systemic Level**
  —Ex: space habitats (in zero-gravity) and colonization
  —Ex: ultra advanced future info systems
  —Ex: qualitative demography
  —Sources: Ch.7 of *FHC*; Chs. 9 & 10 of *FCD*

- **At the Cosmological Level**
  —Ex: the colonization of multiverses
  —Ex: the alteration of space-time and the creation of new matter-energy
  —Ex: the expansion of floating consciousness
  —Ex: the spread of hyper-spatial consciousness
  —Sources: Ch.7 of *FHC*; Chs. 9 & 10 of *FCD*; Ch.4 of *FPHC*; *FPHST*

Notes: The examples in each category are solely illustrative (not exhaustive), and some of the items can be reclassified somewhere else. Nor are they always mutually exclusive. Since they are generalities, exceptions are expected.
Sources: Especially from *FHC*, *FCD*, *FPHC*, *BCPC*, *BDPD*, *FPHST*, and *FPHU*. See also other books and my perspectives on civilizational holism.

Table 4.35. The Theoretical Debate on Civilization

• **The Progressive Theory of Civilization**
—*Thesis*: The "civilizing" process is "good," as opposed to the "barbarizing" process as something "bad," relatively speaking.
—*Discourse*: Especially, though not exclusively, in the Enlightenment era and a bit before. Example: Thomas Hobbes—in that the tribes in primitive societies were "savages."

• **The Romantic Theory of Barbarity**
—*Thesis*: The "civilizing" process is "bad," as opposed to the "barbarizing" process as something "good," relatively speaking.
—*Discourse*: Especially, though not exclusively, in the Counter-Enlightenment circle. Example: Jean-Jacques Rousseau—in that civilization "corrupts" men, and the "savages" are in fact "noble."

• **The Moderate Theory of Civilization**
—*Thesis*: The "civilizing" process is "good," but there is a price to pay, especially in systematic (compulsive) self-control.
—*Discourse*: Especially, though not exclusively, in some late modern and postmodern circles. Example: Norbert Elias—in that social manners become more refined in the civilizing process, but self-control also becomes more systematic.

• **The Theory of Post-Civilization**
—*Thesis*: The civilizing process is as evil and good as barbarity, and each cannot exist without the other, to be eventually superseded by post-civilization unto the post-human age.
—*Discourse*: Proposed by Peter Baofu. See the rest of *BCIV* for more analysis.

Source: From *BCIV* on the theoretical debate

Table 4.36. No Freedom Without Unfreedom in the Civilizing Processes (Part I)

- **The Rationalizing Process (at the Level of Culture)**
  —if freer from the dominance of unreason (as in barbarism) in the civilizing process, then less free from the rationalizing process (be it in the form of the principle of either transcendence or immanence)
  —if freer from the principle of immanence in the rationalizing process, then less free from the inclination to commit terror in the name of reason and the relative underdevelopment of non-reason (e.g., in relation to yoga and meditation)
  —if freer from the principle of transcendence in the rationalizing process, then less free from the relative underdevelopment of reason (e.g., in relation to systematic methodology) and the occurrence of oppression in the name of non-reason

- **The Pacifying Process (at the Level of Society)**
  —if freer from the dominance of pillage (as in savagery) in the civilizing process, then less free from the pacifying process (be it in the form of external control or self-control)
  —if freer from self-control in the pacifying process, then less free from the temptation of expansionist oppression and rebellious mindset in external control
  —if freer from external control in the pacifying process, then less free from the gruesome psychological self-torture and conformism in self-control

- **The Stewardizing Process (at the Level of Nature)**
  —if freer from the dominance of nature (as in the state of nature) in the civilizing process, then less free from the stewardizing process (be it in the form of the stewardship of creation or the covenant with nature)
  —if freer from the stewardship of creation in the stewardizing process, then less free from material underdevelopment, relatively speaking, and spiritual exclusion in the covenant with nature
  —if freer from the covenant with nature in the stewardizing process, then less free from ecological degradation and spiritual disconnection from nature in the stewardship of creation

(continued on next page)

Table 4.36. No Freedom Without Unfreedom in the Civilizing Processes (Part II)

- **The Subliming Process (at the Level of the Mind)**
  —if freer from the dominance of spontaneity (as in the wild state of the mind) in the civilizing process, then less free from the subliming process, be it in the form of (cyclical-centric) self-refinement or (linear-centric) self-discipline
  —if freer from (cyclical-centric) self-refinement in the subliming process, then less free from the (linear-centric) self-regimen (as a form of neurosis)
  —if freer from (linear-centric) self-discipline in the subliming process, then less free from the (cyclical-centric) self-torture (equally as a form of neurosis)

Notes: The examples in each category are solely illustrative (not exhaustive), and the comparison is relative (not absolute), nor are they necessarily mutually exclusive. And some can be easily re-classified elsewhere. As generalities, they allow exceptions.
Sources: From *BCIV*. See also *FHC*, *FCD*, *FPHC*, *BDPD*, and *BCPC*.

Table 4.37. No Equality Without Inequality in the Civilizing Processes (Part I)

- **The Rationalizing Process (at the Level of Culture)**
  —if more equal for the role of rationalization in the rationalizing process (of civilizational making), then less equal for that of mythicization (as in barbarism)
  —if more equal for the principle of transcendence in (linear-centric) rationalizing process, then less equal for the principle of immanence
  —if more equal for the principle of immanence in (cyclical-centric) rationalizing process, then less equal for the principle of transcendence

- **The Pacifying Process (at the Level of Society)**
  —if more equal for pacification in civilizational making, then less equal for the institution of pillaging and others (as in savagery)
  —if more equal for external control, relatively speaking, in pacifying process, then less equal for self-control
  —if more equal for self-control, relatively speaking, in pacifying process, then less equal for external-control

- **The Stewardizing Process (at the Level of Nature)**
  —if more equal for stewardship in the stewardizing process (of civilizational making), then less equal for reverent (submissive) existence (as in barbarism)
  —if more equal for the stewardship of creation in (linear- centric) stewardizing process, then less equal for the (cyclical-centric) covenant with nature for harmonious co-existence
  —if more equal for the (cyclical-centric) covenant with nature in the stewardizing process, then less equal for the (linear-centric) stewardship of nature for domination

(continued on next page)

Table 4.37. No Equality Without Inequality in the Civilizing Processes (Part II)

- **The Subliming Process (at the Level of the Mind)**
  - —if more equal for the role of reason in the subliming process, then less equal for that of unreason (as in the natural state of wildness)
  - —if more equal for the primacy of reason in (linear-centric) subliming process, then less equal for other faculties (e.g., intuition, existential feelings, and analogous thinking) in cyclical-centric one
  - —if more equal for the exercise of other faculties (e.g., intuition, existential feelings, and analogous thinking) in cyclical-centric subliming process, then less equal for the role of reason in linear-centric counterpart

Notes: The examples in each category are solely illustrative (not exhaustive), and the comparison is relative (not absolute), nor are they mutually exclusive. And some can be easily reclassified else-where. As generalities, they allow exceptions.
Sources: From *BCIV*. See also *FHC*, *FCD*, *FPHC*, *BDPD*, and *BCPC*.

Table 4.38. Five Theses on Post-Civilization

• Post-civilization no longer treats civilization as good and barbarity as evil (relatively speaking), nor does it nostalgically regard barbarity as good and civilization as evil (relatively speaking again). Civilization is as evil and good as barbarity.

• Post-civilization also no longer accepts the dichotomy between civilization and barbarity. Civilization cannot exist without barbarity. It is no longer necessary to preserve civilization, any more than it is imperative to destroy barbarity. To go beyond civilization and barbarity is to acknowledge the co-existence of both, although the degree of scaling one over the over varies from case to case—but is not to be extreme in largely favoring one over the other, *on average* (subject to the constraints of existential dialectics).

• Post-civilization is thus subject to the constraints of existential dialectics. There is no freedom without unfreedom, and no equality without inequality, for instance. There will be no utopia; even should there be one, there would be dystopia embedded within it.

• Post-civilization will eventually replace civilization (as a form of life settlement), to be dominated by post-capitalist and post-democratic lifeforms here on earth and in deep space (besides other alien lifeforms that we have never known), unto the post-human age in multiverses. Those few post-humans who keep civilization will live in a "post-human civilization," while the rest (the majority), who choose post-civilization, will evolve towards the state of "post-human post-civilization." One therefore should not confuse "post-human civilization" with "post-human post-civilization," as the two are not the same.

• Post-civilization will confront psychosis as a primary problem in the culture of virtuality unto the post-human age, just as civilization has neurosis as a primary one of its own (although both neurosis and psychosis are major problems in both).

Notes: The comparison in each category is relative (not absolute), nor are they necessarily mutually exclusive. And some can be easily re-classified elsewhere. As generalities, they allow exceptions.
Sources: From *BCIV*. See also *FHC*, *FCD*, *FPHC*, *BDPD*, and *BCPC*.

Table 4.39. Barbarity, Civilization, and Post-Civilization

- **The Rationalizing Process (at the Level of Culture)**
  —*Barbarity*
    - More mythicizing than rationalizing, relatively speaking
  —*Civilization*
    - More rationalizing than mythicizing, relatively speaking
  —*Post-Civilization*
    - Beyond the dichotomy, subject to existential dialectics

- **The Pacifying Process (at the Level of Society)**
  —*Barbarity*
    - More pillaging than pacifying, relatively speaking
  —*Civilization*
    - More pacifying than pillaging, relatively speaking
  —*Post-Civilization*
    - Beyond the dichotomy, subject to existential dialectics

- **The Stewardizing Process (at the Level of Nature)**
  —*Barbarity*
    - More revering than stewardizing, relatively speaking
  —*Civilization*
    - More stewardizing than revering, relatively speaking
  —*Post-Civilization*
    - Beyond the dichotomy, subject to existential dialectics

- **The Subliming Process (at the Level of the Mind)**
  —*Barbarity*
    - More impulsing than subliming, relatively speaking
  —*Civilization*
    - More subliming than impulsing, relatively speaking
  —*Post-Civilization*
    - Beyond the dichotomy, subject to existential dialectics

Notes: The comparison in each category is relative (not absolute), nor are they necessarily mutually exclusive. And some can be easily re-classified elsewhere. As generalities, they allow exceptions.
Sources: From *BCIV*. See also *FHC*, *FCD*, *FPHC*, *BDPD*, and *BCPC*.

Table 4.40. Types of Super Civilization in the Cosmos (Part I)

- **Type I**
  —a civilization which gains control of and uses the total energy output "falling on its planet from its sun for interstellar communication" (or, in general, space colonization). For N. Kardashev, who proposed the first three types, human civilization is currently Type Zero (Type O), which is below even Type I, since its present energy consumption for all purposes, let alone for interstellar communication, is still 10,000 times less.

- **Type II**
  —a civilization which gains control of and uses directly the total energy output of its sun for interstellar communication (or, in general, space colonization).

- **Type III**
  —a civilization which gains control of and uses the total energy output of its galaxy for interstellar communication (or, in general, space colonization).

- **Type IV**
  —a civilization which gains control of and uses the total energy output of its cluster of galaxies for interstellar communication (or, in general, space colonization).

- **Type V**
  —a civilization which gains control of and uses the total energy output of its supercluster of galaxies for interstellar communica-tion (or, in general, space colonization).

(continued on next page)

Table 4.40. Types of Super Civilization in the Cosmos (Part II)

- **Type…n**
  —So continues the series in what I call the cyclical progression of hegemony in the cosmos and beyond.

Notes: The Russian astrophysicist Nikolai Kardashev proposed the first three types of super civilization in terms of total energy out-put for interstellar communication. (CSM 1979) I extend his argument further to propose Type IV, Type V, Type VI, and Type…n, in the context of my claim about the cyclical progression of he-gemony in the cosmos and beyond.
Sources: From *Table 9.4* of *FCD.* See *FHC*, *FCD*, and *FPHC* for more info.

Table 4.41. The Civilizational Project from Pre-Modernity to After-Postmodernity (Part I)

	Pre-Modern	*Modern*	*Postmodern*	*After-Postmodern*
Main narratives	•Sacralness •Courtliness •Vitalism •Animism	•Freedom •Equality •Fraternity	•Multiplicity •Hybridization	•Naked contingency •Cyclical progres-sion of hegemony
Main institutions	•Monarchy •Aristocracy •Feudalism •Holy order •Primitivism	•Capitalism •Liberalism •Socialism •Nazism •Fascism	•Capitalism •Liberalism •Postmodern politics of difference	•Post-Capitalism •Post-Democracy •Others
Main techno-logical and economic revolutions	•Agricultural	•Service •Industrial	•Informational	•Biological •Material •Energy •Space •Others

(continued on next page)

Table 4.41. The Civilizational Project from Pre-Modernity to After-Postmodernity (Part II)

	Pre-Modern	*Modern*	*Postmodern*	*After-Postmodern*
Main agents	•Males •Upper strata •Mini-states	•Males •Upper strata •Whites •Empires	•Males •Upper strata •Whites •Others •Supra-states •IO's	•Post-humans •Humans •Others
Main impacts	•Local	•Inter-national	•Global	•Outer-space •Multiverse
Main outcomes	•Towards moderntiy •Rise of linear- & cyclical-centric civiliza-tions	•Towards post-modernity •Dominance of linear-centric civilization	•Towards after-post-moderntiy •Linear-centric civiliza-tion in crisis	•Towards human (& maybe post-human) extinction •Rise of post-civiliza-tion, especially in post-human forms of space-time

Notes: The examples in each category are solely illustrative (not exhaustive) nor necessarily mutually exclusive, and the comparison is relative (not absolute). As generalities, they allow exceptions.

Sources: From *Table 10.16* of *FCD*—and also from *BCIV* on post-civilization (and *FPHST*)

Table 4.42. Civilizational Holism (Part I)

- **At the Micro-Physical Theoretical Level**
  —Ex: Mastering of quantum mechanics, electromagnetism, and other fields for the understanding of a broad range of anomalous experiences and theapplication for artificial intelligence (*Sec.1.4.1* of *FPHC*)

- **At the Chemical Theoretical Level**
  —Ex: Unprecedented expansion of (and violence to) the mind through ever new forms of drugs (and virtual technologies, for that matter) (Ch.9 of *FCD*)

- **At the Biological Theoretical Level**
  —Ex: Humans are not biologically equal, on the basis of race, gender, ethnicity, age, and whatnot. (*Sec.2.6* & Ch.10 of *FCD*; *BNN*) And post-humans will experience the same fate, in an even more amazing way.

- **At the Psychological Theoretical Level**
  —Ex: Human cognitive impartiality and emotional neutrality are quite limited. (*Secs.2.4-2.5* of *FCD*)
  —Ex: Rise of Floating Consciousness (Ch.10 of *FCD*; Chs.1 & 4 of *FPHC*) and Unfolding Unconsciousness (*FPHU*)

- **At the Organizational Theoretical Level**
  —Ex: Administrative colonization of deep space, with less legal-formalism in some corners. (Chs.9-10 of *FCD*)

- **At the Institutional Theoretical Level**
  —Ex: Both capitalism and democracy will not last, to be superseded by different versions of post-capitalism and post-democracy in after-postmodernity. (Ch.10 of *FCD*)

(continued on next page)

Table 4.42. Civilizational Holism (Part II)

- **At the Structural Theoretical Level**
  —Ex: Social stratification reappears in ever new forms, also with new causes and new players in the cyclical progression of hegemony. (Chs.8-10 of *FCD*)
  —Ex: The world of hyper-empires, and the union of the unions (*BWT*)

- **At the Systemic Theoretical Level**
  —Ex: Outerspace expansion: local → regional → global → solar → galactic → clustery → multiversal (Ch.9 of *FCD*)
  —Ex: Demographic transition: human extinction, and the rise of post-humans (e.g., cyborgs, thinking machines, thinking robots, genetically altered superior beings, floating consciousness, hyper-spatial consciousness) (Ch.4 of *FPHC*; Ch.10 of *FCD*; & Ch.7 of *FHC*)
  —Ex: New technological forces in material sciences, electronic and communication sciences, energy sciences, biosciences, manufacturing and engineering sciences, and space sciences (Ch.10 of *FCD* & Ch.7 of *FHC*)
  —Ex: Systematic dominance towards nature for space colonization (Chs.9-10 of *FCD*; Chs.2 & 7 of *FHC*)

- **At the Cultural Theoretical Level**
  —Ex: The post-human transcendence of freedom and equality (Ch.10 of *FCD*)
  —Ex: Methodological Holism (Ch.1 of *FCD*; Ch.1 of *FPHC*; *Sec.2.1* & *Sec.2.5* of *BCPC*)
  —Ex: The Evolution from Barbarity to Post-Civilization (*BCIV*)

- **At the Cosmological Theoretical Level**
  —Ex: Mastering of dark matter and dark energy, and the exploration of multiverses (Ch.4 of *FPHC*; Ch.10 of *FCD*; & Ch.7 of *FHC*)
  —Ex: Alternation of space-time (*FPHST*)
  —Ex: The emergence of hyper-spatial consciousness (*FPHC*)

(continued on next page)

Table 4.42. Civilizational Holism (Part III)

- **At Other Levels**
  —Ex: Historical: pre-modernity → modernity → postmodernity → after-postmodernity (human distinction, and the rise of post-humans, including floating consciousness, hyper-spatial consciousness, and unfolding unconsciousness) (Ch.7 of *FHC*; Ch.10 of *FCD*; *FPHC*; *FPHU*)

Notes: These examples are solely illustrative (not exhaustive), and some of the items can be reclassified somewhere else. Nor are they always mutually exclusive. Since they are generalities, exceptions are expected. And the comparison is relative, not absolute.
Sources: From *Table 5.1* of *FPHC*—with details from *FHC*, *FCD*, and the rest of my books. Not every aspect in each category as discussed in all my books are presented here, since there are too many issues. For more info, also consult the table on theories on civilizational holism and, of course, the books themselves.

Table 4.43. Theories on Civilizational Holism (Part I)

I. Theories in Relation to Nature

—*At the Chemical Theoretical Level*

- 63. Creational Theory of Chemistry (Peter Baofu) (*FPHCHEM*)
- (• 58). Resilient Theory of Natural Resources (Peter Baofu) (*BNR*)

—*At the Macro-Physical (Cosmological) Theoretical Level*

- 62. Resettlement Theory of Geology (Peter Baofu) (*FPHGEOL*)
- 61. Theory of Post-Cosmology (Peter Baofu) (*BCOS*)
- 60. Theory of Hyper-Spatial Consciousness (Peter Baofu) (Ch.4 of *FPHC*; *FPHG*)
- (• 58). Resilient Theory of Natural Resources (Peter Baofu) (*BNR*)
- (• 57). Panoramic Theory of Transportation (Peter Baofu) (*FPHTRAN*)
- (• 55). Multilateral Theory of Acoustics (Peter Baofu) (*FPHA*)
- (• 54). Selective Theory of Geometry (Peter Baofu) (*FPHG*)
- (• 53). Perspectival Theory of Space-Time (Peter Baofu) (*FPHST*)
- (• 52). Dialectic Theory of Complexity (Peter Baofu) (*FC*)
- (• 30). Theory of Floating Consciousness (Peter Baofu) (Ch.10 of *FCD*; Chs.1 & 4 of *FPHC*)
- 59. Theory of the Geopower of Nature (Ch.4 of *ALD*)—refined as the Interventive-Reshaping Theory of Geography (Peter Baofu) (*FPHGEOG*)

(continued on next page)

Table 4.43. Theories on Civilizational Holism (Part II)

I. Theories in Relation to Nature (*cont'd*)

—*At the Micro-Physical Theoretical Level*

- 58. Resilient Theory of Natural Resources (Peter Baofu) (*BNR*)
- 57. Panoramic Theory of Transportation (Peter Baofu) (*FPHTRAN*)
- 56. Transcendent Theory of Architecture (Peter Baofu) (*FPHARCH*)
- 55. Multilateral Theory of Acoustics (Peter Baofu) (*FPHA*)
- 54. Selective Theory of Geometry (Peter Baofu) (*FPHG*)
- 53. Perspectival Theory of Space-Time (Peter Baofu) (*FPHST*)
- 52. Dialectic Theory of Complexity (Peter Baofu) (*FC*)

II. Theories in Relation to Culture

—*At the Cultural Theoretical Level*

- 51. Inquisitive Theory of Culinary Art (Peter Baofu) (*FPHCA*)
- 50. Interactive Theory of Semantics (Peter Baofu) (*FPHSEM*)
- 49. Transdicsiplinary Theory of Performing Arts (Peter Baofu) (*FPHPA*)
- 48. Comparative-Impartial Theory of Literature (Peter Baofu) (*FPHLIT*)

(continued on next page)

Table 4.43. Theories on Civilizational Holism (Part III)

II. Theories in Relation to Culture (*cont'd*)

—At the Cultural Theoretical Level (*cont'd*)

- • 47. Mediative-Variative Theory of Chess (Peter Baofu) (*FPHCHESS*)
- • 46. Theory of Post-Ethics (Peter Baofu) (*BEPE*)
- • 45. Dualistic Theory of Mass Culture (Peter Baofu) (Ch.2 of *FHC*)
- • 44. Comparative Theory of Religion—also known as the Comparative-Substitutive Theory of Religion (Peter Baofu) (Ch.3 of *FHC*; Ch.9 of *FCD*; Ch.1 of *FPHK*; *FPHR*)
- • 43. Theory of Post-Civilization (Peter Baofu) (*BCIV*)
- • 42. Theory of the Trinity of Modernity to Its After-Postmodern Counterpart (Peter Baofu) (*FHC*; Ch.10 of *FCD*)
- • 41. Transformative Theory of Aesthetic Experience (Peter Baofu) (*FAE*)
- (• 20). Theory of Post-Capitalism (Peter Baofu) (Ch.10 of *FCD*; Chs.2 & 4 of *FPHC*; *BCPC*)
- (• 19). Theory of Post-Democracy (Peter Baofu) (Ch.10 of *FCD*; Chs.3 & 4 of *FPHC*; *BDPD*)
- (• 5). Theory of Methodological Holism (Peter Baofu) (Ch.1 of *FCD*; Ch.1of *FPHC*; *Sec.2.1* & *Sec.2.5* of *BCPC*; *FC*; *FPHK*; etc.)

(continued on next page)

Table 4.43. Theories on Civilizational Holism
(Part IV)

II. Theories in Relation to Culture (*cont'd*)

—*At the Cultural Theoretical Level* (*cont'd*)

(• 4). Theory of Existential Dialectics, or the Holistic Theory of Knowledge (Peter Baofu) (*FHC*; *FCD*; *FPHC*; *BDPD*; *FC*; *FAE*; *ALD; FIA; FPHK*; etc,)

(• 3). Contrastive Theory of Rationality (Peter Baofu) (*FPHML*)

III. Theories in Relation to the Mind

—*At the Biological Theoretical Level*

• 40. Theory of Contrastive Advantages (Peter Baofu) (*Sec.2.6* & Ch.10 of *FCD; BNN*)

(• 30).Theory of Floating Consciousness (Peter Baofu) (Ch.10 of *FCD*; Chs.1 & 4 of *FPHC*)

—*At the Psychological Theoretical Level*

• 39. Complex Theory of Data Analysis (Peter Baofu) (*FPHDA*)

• 38. Supersession Theory of Computing (Peter Baofu) (*FPHCOM*)

• 37. Metamorphic Theory of Humor (Peter Baofu) (*FPHH*)

• 36. Contrarian Theory of Personality (Peter Baofu) (*FPHP*)

• 35. Theory of Virtual Sexuality (Peter Baofu) (*FPHS*)

• 34. Expansive-Contractive Theory of Martial Arts (Peter Baofu) (*FPHMA*)

• 33. Multilogical Theory of Learning (Peter Baofu) (*FPHL*)

• 32. Comprehensive Theory of Creative Thinking (Peter Baofu) (*FPHCT*)

(continued on next page)

Table 4.43. Theories on Civilizational Holism (Part V)

III. Theories in Relation to the Mind (*cont'd*)

—*At the Psychological Theoretical Level* (*cont'd*)

- 31. Theory of Unfolding Unconsciousness—also known as the Unfolding Theory of Anomalous Experience (Peter Baofu) (*FPHU*)
- 30. Theory of Floating Consciousness (Peter Baofu) (Ch.10 of *FCD*; Chs.1 & 4 of *FPHC*)
- 29. Theory of Cognitive Partiality (Peter Baofu) (*Sec.2.4* of *FCD*; *Sec.4.5.1.1* of *BCPC*)
- 28. Theory of Emotional Non-Neutrality (Peter Baofu) (*Sec.2.5* of *FCD*; *Sec.4.5.2* of of *BCPC*)
- 27. Theory of Behavioral Alteration (Peter Baofu) (*Sec.4.5.3* of *BCPC*)

IV. Theories in Relation to Society

—*At the Organizational Theoretical Level*

- 26. Theory of E-Civic Alienation (Peter Baofu) (Ch.7 of *FCD*)
- 25. Combinational Theory of Organization (Peter Baofu) (*FPHO*; Ch.6 of *ALD*)

—*At the Institutional Theoretical Level*

- 24. Reflective Theory of Criminality (Peter Baofu) (*FPHCRIM*)
- 23. Detached Theory of Gambling (Peter Baofu) (*FPHGAM*)
- 22. Heterodox Theory of Education (Peter Baofu) (*FPHEDU*)
- 21. Reconstruction Theory of Law (Peter Baofu) (*FPHLAW*)
- 20. Theory of Post-Capitalism (Peter Baofu) (Ch.10 of *FCD*; Chs.2 & 4 of *FPHC*; *BCPC*)
- 19. Theory of Post-Democracy (Peter Baofu) (Ch.10 of *FCD*; Chs.3 & 4 of *FPHC*; *BDPD*)

(continued on next page)

Table 4.43. Theories on Civilizational Holism (Part VI)

IV. Theories in Relation to Society (*cont'd*)

—*At the Institutional Theoretical Level (cont'd)*

- 18. Dynamic Theory of Comparative Political Systems (Peter Baofu) (*ALD*)
- (• 8.) Theory of the Cyclical Progression of Empire-Building (Peter Baofu) (*BWT*)

—*At the Systemic Theoretical Level*

- 17. Transfigurative Theory of Waste (Peter Baofu) (*FPHW*)
- 16. Theory of the Cyclical Progression of Migration (Peter Baofu) (*FPHMIG*)
- (• 62). Resettlement Theory of Geology (Peter Baofu) (*FPHGEOL*)
- 15. Contingent Theory of Urban Planning (Peter Baofu) (*FPHUP*)
- 14. Totalistic Theory of Communication (Peter Baofu) (*FPHMM*; *FCD*; *FHC*)
- 13. Ambivalent Theory of Technology (Peter Baofu) (*FPHE*; *FCD*; *FHC*)
- 12. Multifaceted Theory of War and Peace (Peter Baofu) (Ch.9 of *FCD*; Ch.1 of *FPHK*)
- 11. Theory of Post-Humanity (Peter Baofu) (Ch.7 of *FHC*; Chs.3, & 10 of *FCD*; Chs.1, 3 & 4 of *FPHC*; and other books of mine)
- 10. Theory of the Cyclical Progression of System Integration and Fragmentation (Peter Baofu) (Chs.9-10 of *FCD*)
- 9. Synthetic Theory of Information Architecture (Peter Baofu) (*FIA*)

—*At the Structural Theoretical Level*

- 8. Theory of the Cyclical Progression of Hegemony (Peter Baofu) (Chs.9-10 of *FCD*; Chs.1, 3 & 4 of *FPHC*; *BDPD*)
- 7. Theory of the Cyclical Progression of Empire-Building (Peter Baofu) (*BWT*)

(continued on next page)

Table 4.43. Theories on Civilizational Holism (Part VII)

V. Meta-Theories (in Relation to Theories)

—*At the Methodological Meta-Theoretical Level*

(• 50).Interactive Theory of Semantics (Peter Baofu) (*FPHSEM*)

(• 39). Complex Theory of Data Analysis (Peter Baofu) (*FPHDA*)

• 6. Critical-Dialectic Theory of Formal Science (Peter Baofu) (*FPHFS*)

• 5. Theory of Methodological Holism (Peter Baofu) (Ch.1 of *FCD*; Ch.1of *FPHC*; *Sec.2.1* & *Sec.2.5* of *BCPC*; *FC*; *FPHK*; etc.)

—*At the Ontological Meta-Theoretical Level*

• 4. Theory of Existential Dialectics, or the Holistic Theory of Knowledge (Peter Baofu) (*FHC*; *FCD*; *FPHC*; *BDPD*; *FC*; *FAE*; *ALD; FIA; FPHK*; etc,)

• 3. Contrastive Theory of Rationality (Peter Baofu) (*FPHML*)

VI. Theories in Relation to the Rest

—*At the Meta-Historical Level*

• 2. Multifold Theory of History (Peter Baofu) (*FPHHIST*)

—*At the Empirical-Historical Level*

• 1. Theory of the Evolution from Pre-Modernity to After-Postmodernity (Peter Baofu) (*FHC*; Ch.9-10 of *FCD*; *FPHC*)

Notes: All these theories are my constructions, as some of the main contributions of my grant project on civilization and its future. These examples are solely illustrative (not exhaustive), and some of the items can be reclassified somewhere else. Nor are they always mutually exclusive. Since they are generalities, exceptions are expected.

Sources: From my previous books.

Table 4.44. Three Great Future Transformations of Mind Games

• **Virtual Games**
  —Ex: online chess
  —Ex: virtual experience

• **Novel Games**
  —Ex: new chess variants
  —Ex: new chess engines

• **Post-Human Mind Games**
  —Ex: the quest for broader/deeper mental benefits of chess playing
  —Ex: games designed for the evolution of the mind into different body-less forms (e.g., "floating consciousness," "hyper-spatial consciousness," "unfolding unconsciousness")

Notes: The examples in the categories are solely illustrative (not exhaustive), and the comparison is relative (not absolute), nor are they necessarily mutually exclusive. And some can be easily re-classified elsewhere. As generalities, they allow exceptions.
Sources: From *Sec. 4.16* of *FPHCHESS*. See text for more info. Also, consult *FCD* for strategy and tactics in warfare and *FPHO* for strategy and tactics in organization.

BIBLIOGRAPHY

Achard, Frederic; Eva Hugh; Hans-Stibig Jurgen; Mayaux Philippe; et al. 2002. "Determination of Deforestation Rates of the World's Humid Tropical Forests." *Science*, 297 (5583): 999–1003.

Adams, David. 2006. "Royal Society Tells Exxon: Stop Funding Climate Change Denial" (September 18). *The Guardian*. <http://www.guardian.co.uk/environment/2006/sep/20/oilandpetrol.business>.

Adams, Nancy K.; Bruce Houghton; Sarah Fagents; & W. Hildreth. 2006. "The Transition from Explosive to Effusive Eruptive Regime: The Example of the 1912 Novarupta Eruption, Alaska." *Geological Society of America Bulletin*, 118 (5–6): 620.

Aggie Horticulture (AGG). 2012. "Asexual Propagation" (August 14). <http://aggie-horticulture.tamu.edu/archives/parsons/misc/asexualpropaga tion.html>.

Adler, Mortimer J. 1993. *The Difference of Man and the Difference It Makes*. New York: Fordham University Press.

Allen, A. P.; J. Gillooly; V. Savage; & J. Brown. 2006. "Kinetic Effects of Temperature on Rates of Genetic Divergence and Speciation." *PNAS*, 103 (24): 9130–9135. <http://www.ncbi.nlm.nih.gov/pmc/articles/PMC1474011/?tool=pmcentrez>.

Alroy, J; C. Marshall; R. Bambach; K. Bezusko; M. Foote; F. Fursich; T. Hansen; S. Holland, et al. 2001. "Effects of Sampling Standardization on Estimates of Phanerozoic Marine Diversification." *Proceedings of the National Academy of Sciences of the United States of America*, 98 (11): 6261–6. <http://www.ncbi.nlm.nih.gov/pmc/articles/PMC33456/?tool=pmcentrez>.

American Geophysical Union (AGU). 2011. "Human Activities Emit Way More Carbon Dioxide Than Do Volcanoes." <http://www.agu.org/news/press/pr_archives/2011/2011-22.shtml>.

Angelsen, Arild; & David Kaimowitz. 1999. "Rethinking the Causes of Deforestation: Lessons from Economic Models." *The World Bank Research Observer*, 14: 1.

Arvai, J., R. Gregory; & D. Ohlson. 2006. "Deconstructing Adaptive Management: Criteria for Applications to Environmental Management." *Ecological Applications*, 16 (6): 2411-2425.

Ayres, D., et al. 2004. "Spread of Exotic Cordgrasses and Hybrids (Spartina sp.) in the Tidal Marshes of San Francisco Bay, California." B*iological Invasions*, 6 (2): 221–231.

Baez, John. 2006. "Extinction" (April 08). <http://math.ucr.edu/home/baez/extinction/>.

Bailey, Ronald. 2010. "Peak Everything?" (April 27). *Reason.com*. <http://reason.com/archives/2010/04/27/peak-everything>.

Bannon, Ian; & Paul Collier, eds. 2003. *Natural Resources and Violent Conflict: Options and Actions*. Washington, DC: World Bank.

Baofu, Peter. 2012. *The Future of Post-Human Waste: Towards a New Theory of Uselessness and Usefulness*. Cambridge, England: Cambridge Scholars Publishing.

_____.2012a. *The Future of Post-Human Culinary Art: Towards a New Theory of Ingredients and Techniques*. Cambridge, England: Cambridge Scholars Publishing.

_____.2012b. *The Future of Post-Human Transportation: A Preface to a New Theory of Networks and Operations*. Cambridge, England: Cambridge Scholars Publishing.

_____.2012c. *The Future of Post-Human Migration: A Preface to a New Theory of Sameness, Otherness, and Identity*. Cambridge, England: Cambridge Scholars Publishing.

_____.2012d. *The Future of Post-Human History: A Preface to a New Theory of Universality and Relativity*. Cambridge, England: Cambridge Scholars Publishing.

_____.2012e. *The Future of Post-Human Criminality: A Preface to a New Theory of Heroes and Villains*. Cambridge, England: Cambridge International Science Publishing, Ltd.

_____.2012f. *The Future of Post-Human Architecture: A Preface to a New Theory of Form and Function*. Cambridge, England: Cambridge International Science Publishing, Ltd.

_____.2012g. *The Future of Post-Human Semantics: A Preface to a New Theory of Internality and Externality*. Cambridge, England: Cambridge Scholars Publishing.

_____.2012h. *The Future of Post-Human Performing Arts: A Preface to a New Theory of the Body and its Presence*. Cambridge, England: Cambridge Scholars Publishing.

_____.2011. *The Future of Post-Human Geography: A Preface to a New Theory of Environments and their Interactions*. Cambridge, England: Cambridge International Science Publishing, Ltd.

_____.2011a. *The Future of Post-Human Data Analysis: A Preface to a New*

Theory of Quantitative and Qualitative Research. Cambridge, England: Cambridge International Science Publishing, Ltd.

_____.2011b. *The Future of Post-Human Chemistry: A Preface to a New Theory of Substances and their Changes*. Cambridge, England: Cambridge Scholars Publishing.

_____.2011c. *The Future of Post-Human Literature: A Preface to a New Theory of Fiction and Non-Fiction*. Cambridge, England: Cambridge International Science Publishing, Ltd.

_____.2011d. *The Future of Post-Human Computing: A Preface to a New Theory of Hardware, Software, and the Mind*. Cambridge, England: Cambridge International Science Publishing, Ltd.

_____.2011e. *The Future of Post-Human Gambling: A Preface to a New Theory of Risk and Caution*. Cambridge, England: Cambridge International Science Publishing, Ltd.

_____.2011f. *The Future of Post-Human Acoustics: A Preface to a New Theory of Sound and Silence*. Cambridge, England: Cambridge International Science Publishing, Ltd.

_____.2011g. *The Future of Post-Human Humor: A Preface to a New Theory of Joking and Laughing*. Cambridge, England: Cambridge International Science Publishing, Ltd.

_____.2011h. *Beyond Ethics to Post-Ethics: A Preface to a New Theory of Morality and Immorality*. Charlotte, NC: Information Age Publishing.

_____.2011i. *The Future of Post-Human Education: A Preface to a New Theory of Teaching and Learning*. Cambridge, England: Cambridge International Science Publishing, Ltd.

_____.2010. *The Future of Post-Human Religion: A Preface to a New Theory of Spirituality*. Cambridge, England: Cambridge International Science Publishing, Ltd.

_____.2010a. *The Future of Post-Human Chess: A Preface to a New Theory of Tactics and Strategy*. Cambridge, England: Cambridge International Science Publishing, Ltd.

_____.2010b. *The Future of Post-Human Geology: A Preface to a New Theory of Statics and Dynamics*. Cambridge, England: Cambridge International Science Publishing, Ltd.

_____.2010c. *Beyond Cosmology to Post-Cosmology: A Preface to a New Theory of Different Worlds*. Cambridge, England: Cambridge International Science Publishing, Ltd.

_____.2010d. *The Future of Post Human Personality: A Preface to a New Theory of Normality and Abnormality*. Cambridge, England: Cambridge International Science Publishing, Ltd.

_____.2010e. *The Future of Post-Human War and Peace: A Preface to a New Theory of Aggression and Pacificity*. Cambridge, England: Cambridge Scholars Publishing.

_____.2010f. *The Future of Post-Human Formal Science: A Preface to a New Theory of Abstraction and Application*. Cambridge, England: Cambridge Scholars Publishing.

_____.2010g. *The Future of Post-Human Law: A Preface to a New Theory of Necessity, Contingency, and Justice*. Cambridge, England: Cambridge Scholars Publishing.

_____.2010h. *The Future of Post-Human Sexuality: A Preface to a New Theory of the Body and Spirit of Love Makers*. Cambridge, England: Cambridge Scholars Publishing.

_____,2009. *The Future of Post-Human Martial Arts: A Preface to a New Theory of the Body and Spirit of Warriors*. Cambridge, England: Cambridge Scholars Publishing.

_____.2009a. *The Future of Post-Human Organization: A Preface to a New Theory of Communication, Decision-Making, and Leadership*. Cambridge, England: Cambridge Scholars Publishing.

_____.2009b. *The Future of Post-Human Language: A Preface to a New Theory of Structure, Context, and Learning*. Cambridge, England: Cambridge Scholars Publishing.

_____.2009c. *The Future of Post-Human Urban Planning: A Preface to a New Theory of Density, Void, and Sustainability*. Cambridge, England: Cambridge Scholars Publishing.

_____.2009d. *The Future of Post-Human Geometry: A Preface to a New Theory of Infinity, Symmetry, and Dimensionality*. Cambridge, England: Cambridge Scholars Publishing.

_____.2009e. *The Future of Post-Human Creative Thinking: A Preface to a New Theory of Invention and Innovation*. Cambridge, England: Cambridge Scholars Publishing.

_____.2009f. *The Future of Post-Human Mass Media: A Preface to a New Theory of Communication*. Cambridge, England: Cambridge Scholars Publishing.

_____.2009g. *The Future of Post-Human Engineering: A Preface to a New Theory of Technology*. Cambridge, England: Cambridge Scholars Publishing.

_____.2008. *The Future of Post-Human Mathematical Logic: A Preface to a New Theory of Rationality*. Cambridge, England: Cambridge Scholars Publishing.

_____.2008a. *The Future of Post-Human Knowledge: A Preface to a New Theory of Methodology and Ontology*. Oxford, United Kingdom: Chandos Publishing (Oxford) Ltd.

_____.2008b. *The Future of Post-Human Unconsciousness: A Preface to a New*

Theory of Anomalous Experience. Cambridge, England: Cambridge Scholars Publishing.

_____.2008c. *The Future of Information Architecture: Conceiving a New Way to Understand Taxonomy, Network, and Intelligence*. Oxford, United Kingdom: Chandos Publishing (Oxford) Ltd.

_____.2007. *The Rise of Authoritarian Liberal Democracy: A Preface to a New Theory of Comparative Political Systems*. Cambridge, England: Cambridge Scholars Publishing.

_____.2007a. *The Future of Aesthetic Experience: Conceiving a Better Way to Beauty, Ugliness, and the Rest*. Cambridge, England: Cambridge Scholars Publishing.

_____.2007b. *The Future of Complexity: Conceiving a Better Way to Understand Order and Chaos*. London, United Kingdom: World Scientific Publishing Co.

_____.2007c. *Beyond the World of Titans, and the Remaking of World Order: A Preface to a New Logic of Empire-Building*. Cambridge, England: Cambridge Scholars Publishing.

_____.2006. *Beyond Nature and Nurture: Conceiving a Better Way to Understand Genes and Memes*. Cambridge, England: Cambridge Scholars Publishing.

_____.2006a. *The Future of Post-Human Space-Time: Conceiving a Better Way to Understand Space and Time*. New York: Peter Lang Publishing, Inc.

_____.2006b. *Beyond Civilization to Post-Civilization: Conceiving a Better Model of Life Settlement to Supersede Civilization*. New York: Peter Lang Publishing, Inc.

_____.2005. *Beyond Capitalism to Post-Capitalism: Conceiving a Better Model of Wealth Acquisition to Supersede Capitalism*. New York: The Edwin Mellen Press.

_____.2004. *The Future of Post-Human Consciousness*. New York: The Edwin Mellen Press.

_____.2004a. *Beyond Democracy to Post-Democracy: Conceiving a Better Model of Governance to Supersede Democracy*. Volume 1. New York: The Edwin Mellen Press.

_____.2004b. *Beyond Democracy to Post-Democracy: Conceiving a Better Model of Governance to Supersede Democracy*. Volume 2. New York: The Edwin Mellen Press.

_____.2002. *The Future of Capitalism and Democracy*. Maryland: The University Press of America.

_____.2000. *The Future of Human Civilization*. Volume 1. New York: The Edwin Mellen Press.

_____.2000a. *The Future of Human Civilization*. Volume 2. New York: The

Edwin Mellen Press.

Bard E.; G. Raisbeck; Yiou Francoise; & Jouzel Jean. 2000. "Solar Irradiance during the Last 1200 Years Based on Cosmogenic Nuclides." *Tellus B*, 52 (3): 985–992. <http://adsabs.harvard.edu/abs/2000TellB..52..985B>.

Barrow, John D. 1998. *Impossibility: The Limits of Science and the Science of Limits*. Oxford: Oxford University Press.

Barrow, Mark. 1998. *A Passion for Birds*. NJ: Princeton University Press.

BBC. 2007. "The Roman Way" (March 5) <http://www.bbc.co.uk/radio4/history/romanway.shtml>.

_____.2007a. "R.I.P. Prize Goldfish in a Bag" (April 22). <http://news.bbc.co.uk/2/hi/uk_news/magazine/3893889.stm>.

_____.2005. "South Korea Cloning Research was Fake" (December 23). <http://news.bbc.co.uk/2/hi/asia-pacific/4554422.stm>.

_____.2003. "Scientists 'to Clone Mammoth'" (August 18). <http://news.bbc.co.uk/2/hi/asia-pacific/3075381.stm>.

_____.2003a. "The Amazon Rainforest" (February 14). <http://www.h2g2.com/approved_entry/A925913>.

_____.1999. "Details of Hybrid Clone Revealed" (June 18). <http://news.bbc.co.uk/2/hi/science/nature/371378.stm>.

Begley, Sharon. 2007. "The Truth About Denial" (August 13). *Newsweek*. <http://www.thedailybeast.com/newsweek/2007/08/13/the-truth-about-denial.html>.

Bengtsoon, J., et al. 2005. "The Effects of Organic Agriculture on Biodiversity and Abundance: A Meta-Analysis." *Journal of Applied Ecology*, 42: 261–269.

Benton, M. J. 2001. "Biodiversity on Land and in the Sea." *Geological Journal*, 36 (3–4): 211–230. <http://onlinelibrary.wiley.com/doi/10.1002/gj.877/abstract>.

Berman, M. 1981. *The Reenchantment of the World*. Ithaca, NY: Cornell University Press.

Berns, M.; A. Townend; Z. Khayat; B. Balagopal; M. Reeves; M. Hopkins; & N. Kruschqwitz. 2009. "The Business of Sustainability: Imperatives, Advantages, and Actions." *The Boston Consulting Group Report 2009*. <http://www.bcg.com/documents/file29480.pdf>.

Biletzki, Anat; & Anat Matar. 2006. "Ludwig Wittgenstein" (Nov. 17). *Stanford Encyclopedia of Philosophy*. <http://plato.stanford.edu/entries/wittgenstein/>.

Black, Richard. 2008. "Nature Loss 'to Hurt Global Poor'" (May 29). *BBC News*, <http://news.bbc.co.uk/2/hi/science/nature/7424535.stm>.

Blackman, A. 2008. "Can Voluntary Environmental Regulation Work in Developing Countries? Lessons from Case Studies." *Policy Studies Journal*, 36 (1): 119-141.

Blackwell Dictionary of Western Philosophy (BDWP). 2004. "The Great Chain of

Being."

Boffey, Philip. 1983. "Italy's Wild Dogs Winning Darwinian Battle" (December 13). *The New York Times*. <http://www.nytimes.com/1983/12/13/science/italy-s-wild-dogs-winning-darwinian-battle.html>.

Bolden, E. G.; & W. Robinson. 1999. *Wildlife Ecology and Management*. Upper Saddle River, NJ: Prentice-Hall, Inc..

Bookchin, Murray. 1987. "Social Ecology versus Deep Ecology: A Challenge for the Ecology Movement." Green Perspectives/Anarchy Archives. <http://dwardmac.pitzer.edu/Anarchist_Archives/bookchin/socecovdeepeco.html>.

Bormann, B. T. ; F. Wagner; G. Wood; J. Algeria; P. Cunningham; M. Brooks, P, Friesema, P.; J. Berg; & J. Henshaw. 1999. *Ecological Stewardship: A common reference for ecosystem management*. Amsterdam: Elsevier.

Bosselman, Fred. 2004. "A Dozen Biodiversity Puzzles." *Environmental Law Journal*, 364. <http://www1.law.nyu.edu/journals/envtllaw/issues/vol12/bosselman-for%20web.pdf>.

Bräutigam, Deborah 2008. "Development Policy Outlook 'Taxation and Governance in Africa.'" Aei.org.

Brian, Kirk. 1986-87. "Man's Great Geophysical Experiment." *Oceanus*, 29 (4). <http://www.gfdl.noaa.gov/bibliography/related_files/kb8602.pdf?PHPSESSID=3ce1d32ab89d5b593b86c95ce91614df>.

Brooks, M. L.; C. M. D'Antonio; D. M. Richardson; J. B. Grace; J. E. Keeley; J. M. DiTomaso; R. J. Hobbs; M. Pellant; & D. Pyke. 2004. "Effects of Invasive Alien Plants on Fire." *BioScience*, 54 (54): 677–688.

Brown, Lester. 2006. "How Food and Fuel Compete for Land" (February 01). *The Globalist*. <http://www.theglobalist.com/StoryId.aspx?StoryId=5077>.

Brown, Tony. 1997. "Clearances and Clearings: Deforestation in Mesolithic/Neolithic Britain." *Oxford Journal of Archaeology*, 16 (2): 133. <http://onlinelibrary.wiley.com/doi/10.1111/1468-0092.00030/abstract>.

Bruckschen, Peter; Susanne Oesmanna; & Ján Veizer. 1999. "Isotope Stratigraphy of the European Carboniferous: Proxy Signals for Ocean Chemistry, Climate and Tectonics." *Chemical Geology*, 161 (1–3): 127. <http://www.whoi.edu/oceanus/viewArticle.do?id=2508>.

Burney, D. A.; & T. Flannery. 2005. "Fifty Millennia of Catastrophic Extinctions after Human Contact." *Trends in Ecology & Evolution* (Elsevier), 20 (7): 395–401. <http://www.anthropology.hawaii.edu/Fieldschools/Kauai/Publications/Publication%204.pdf>.

Bunge, Mario. 1999. *Dictionary of philosophy*. Amherst, MA: Prometheus Books.

Butler, D. 1994. "Bid to Protect Wolves from Genetic Pollution." Nature, 370: 497. <http://www.nature.com/nature/journal/v370/n6490/pdf/370497a0.pdf>.

Butler, Rhett. 2007. "Destruction of Renewable Resources" (July 22). <http://rainforests.mongabay.com/0905.htm>.

Butt, Kyle. 2001. "Apparent Age." <http://www.apologeticspress.org/apcontent.aspx?category=9&article=548>.

Byers, J. E. 2002. "Impact of Non-Indigenous Species on Natives Enhanced by Anthropogenic Alteration of Selection Regimes." *Oikos*, 97 (3): 449–458. <http://onlinelibrary.wiley.com/doi/10.1034/j.1600-0706.2002.970316.x/abstract;jsessionid=3F48B59AE42BEE19208ABC349AE5997A.d04t01>.

California Institute of Public Affairs (CIPA). 2001. "An Ecosystem Approach to Natural Resource Conservation in California." CIPA Publication No. 106. InterEnvironment Institute.

Campbell, A. K. 2003. "Save Those Molecules: Molecular Biodiversity and Life." *Journal of Applied Ecology*, 40 (2): 193–203. <http://onlinelibrary.wiley.com/doi/10.1046/j.1365-2664.2003.00803.x/abstract;jsessionid=E3E1E388E2ADB340BE73F77B897ECF15.d03t01>.

Carr, Marjorie Harris; & Archie Fairly Carr. 1994. *A Naturalist in Florida: A Celebration of Eden*. New Haven, CO: Yale University Press.

Cassey, P; T. M. Blackburn; R. P. Duncan; & S. L. Chown. 2005. "Concerning Invasive Species: Reply to Brown and Sax." *Austral Ecology*, 30 (4): 475. <http://onlinelibrary.wiley.com/doi/10.1111/j.1442-9993.2005.01505.x/abstract>.

cbd.int (CBD). 2009. "Climate Change and Biological Diversity." Convention on Biological Diversity. <http://www.cbd.int/climate/>.

Chamberlin, J. E. 2003. *If This Is Your Land, Where Are Your Stories? Finding Common Ground*. Toronto, Canada: Alfred A. Knopf.

Chan, K. M. 2008. "Value and Advocacy in Conservation Biology: Crisis Discipline or Discipline in Crisis?." *Conserv. Biol.*, 22 (1): 1–3.

Chase, Scott. 1993. "Tachyons" (March). *Physics FAQ*. <http://math.ucr.edu/home/baez/physics/ParticleAndNuclear/tachyons.html>.

Chapin, F. Stuart; Pamela A. Matson; & Harold A. Mooney. 2002. *Principles of Terrestrial Ecosystem Ecology*. New York: Springer.

Chivian, Eric; & Aaron Bernstein, eds. 2008. Sustaining Life: How Human Health Depends on Biodiversity. New York: Oxford University Press.

Climate Institute (CLI). 2012. "Water" (August 20). <http://www.climate.org/topics/water.html>.

Cloyd, E. L. 1972. *James Burnett, Lord Monboddo*. New York: Oxford University Press.

Cohab Initiative (COH). 2012. "Emerging Infectious Disease" (August 19). <http://www.cohabnet.org/en_issue3.htm>.

_____.2012a. "Natural Products and Medicinal Resources" (August 19).

<http://www.cohabnet.org/en_issue4.htm>.

_____.2012b. “Disaster Prevention, Relief, and Recovery” (August 19). <http://www.cohabnet.org/en_issue5.htm>.

Cohen, David.2007. “Earth's Natural Wealth: An Audit” (May 23). <http://www.science.org.au/nova/newscientist/027ns_005.htm?id=mg19426051.200&print=true>.

Colautti, Robert I.; & Hugh MacIsaac. 2004. “A Neutral Terminology to Define 'Invasive' Species.” *Diversity and Distributions*, 10 (2): 135–141.

Colby, M. E. 1991. “Environmental Management in Development: The Evolution of Paradigms.” *Ecological Economics*, 3: 193-213.

Collier, Paul. 2003. “Natural Resources, Development and Conflict: Channels of causation and Policy Interventions.” Washington, DC: World Bank.

Columbia Encyclopedia (COL). 2007. “Uniformitarianism.” NY: Columbia University Press.

Connor, Steve. 2009. “Cloned Goat Dies after Attempt to Bring Species back from Extinction.” *The Independent* (London). <http://www.independent.co.uk/news/science/cloned-goat-dies-after-attempt-to-bring-species-back-from-extinction-1522974.html>.

Cooper, C. B.; & J. A. Smith. 2010. “Gender Patterns in Bird-Related Recreation in the USA and UK.” *Ecology and Society*, 15 (4): 4.

Corazzon, Raul. 2007. “Definitions of Ontology: From Nicolai Hartmann to the Present Time.” <http://www.formalontology.it/section_4_two.htm>.

Corbett, Dan. 2008. “Introduction to String Theory” (November 23). *ThinkQuest*. <http://library.thinkquest.org/27930/stringtheory1.htm>.

Corvalan, Carlos; Simon Hales, & Anthony McMichael. 2005. *Ecosystems and Human Well-Being: Halth Synthesis*. Geneva, Switzerland: World Health Organization.

Costanza, R. 2003. “Early History of Ecological Economics and ISEE.” *Internet Encyclopaedia of Ecological Economics*. <http://www.ecoeco.org/pdf/costanza.pdf.>

_____.1989. “What is Ecological Economics?” *Ecological Economics*, 1: 1-7.

Currie, D. J.; G. G. Mittelbach; H. V. Cornell; D. M. Kaufman; J. T. Kerr; T. Oberdorff; & J.-F. Gugan. 2004. “A Critical Review of Species-Energy Theory.” *Ecology Letters*, 7: 1121-1134.

Daily, B. F. & S. Huang. 2001. Achieving Sustainability through Attention to Human Resource Factors in Environmental Management.” *International Journal of Operations & Production Management*, 21 (12): 1539 – 1552.

Daily, G. C. 1997. *Nature's Services: Societal Dependence on Natural Ecosystems*. Washington, D.C.. Island Press.

Dandekar, Abhijit Y. 2006. *Petroleum Reservoir Rock and Fluid Properties*. US: CRC Press.

Davis, M. A.; J. P. Grime; & K. Thompson. 2000. "Fluctuating Resources in Plant Communities: A General Theory of Invisibility." *Journal of Ecology*, 88 (3): 528–534. <http://onlinelibrary.wiley.com/doi/10.1046/j.1365-2745.2000.00473.x/abstract>.

Diggles, Michael. 2005. "The Cataclysmic 1991 Eruption of Mount Pinatubo, Philippines" (February 28). *U.S. Geological Survey Fact Sheet 113-97*. United States Geological Survey. <http://pubs.usgs.gov/fs/1997/fs113-97/>.

Dinger, Dennis. 2010. "Global Climate Change, the Bible, And Science—A Book Review." <http://ezinearticles.com/?Global-Climate-Change,-the-Bible,-And-Science---A-Book-Review&id=5470434>.

Dunlap, Thomas R. 2011. *In the Field, Among the Feathered: A History of Birders & Their Guides*. Oxford: Oxford University Press.

Dunne, Pete. 2003. *Pete Dunne on Bird Watching*. Boston: Houghton Mifflin.

Dunphy, D.; A. Griffitths; & S. Benn. 2007. *Organizational Change for Corporate Sustainability*. UK: Taylor & Francis, Inc.

Economist, the (ECO). 2008. "A Rising Tide: Scientists Find Proof that Privatising Fishing Stocks Can Avert a Disaster" (September 18). <http://www.economist.com/node/12253181?source=hptextfeature&story_id=12253181>.

_____.2007. "The Economist—The End Of Cheap Food" (December 6). <http://www.economist.com/node/10252015?subjectid=7216688&story_id=10252015>.

Ehrenfeld, David. 2006. "Transgenics and Vertebrate Cloning as Tools for Species Conservation." *Conservation Biology*, 20 (3): 723–732. <http://onlinelibrary.wiley.com/doi/10.1111/j.1523-1739.2006.00399.x/abstract;jsessionid=0C06671B31616C3422D6E0C0C32973E1.d03t04>.

Eldredge, Niles. 1985. "Time Frames: Rethinking of Darwinian Evolution and the Theory of Punctuated Equilibria." New York: Simon and Schuster.

Ellstrand, Norman C. 2001. "When Transgenes Wander, Should We Worry?" *Plant Physiol*, 125: 1543-1545.

Elton, C. S. 2000. *The Ecology of Invasions by Animals and Plants*. Chicago, IL: University of Chicago Press.

Elzinga, C. L.; D. W. Salzer; & J. W. Willoughby. 1998. *Measuring and Monitoring Plant Populations*. Denver, CO: Bureau of Land Management.

Encyclopædia Britannica (ENC). 2012. "Hand Tool: Neolithic Tools" (August 21). <http://www.britannica.com/EBchecked/topic/254115/hand-tool/39205/Neolithic-tools>.

e-Notes.com (EN). 2012. "Reagan Promotes Supply-Side Economics" (July 21). <http://www.enotes.com/topic/Supply-side_economics>.

Evans, David. 1997. *A History of Nature Conservation in Britain*. New York: Routledge.

Everett, Hugh. 1957. "'Relative State' Formulation of Quantum Mechanics" (July). *Reviews of Modern Physics*, vol. 29, no.3. <http://www.univer.omsk.su/

omsk/Sci/Everett/paper1957.html>.

Ewell, J. J.; D. J. O'Dowd; J. Bergelson; C. C. Daehler; C. M. D'Antonio; L. D. Gomez; D. R. Gordon; R. J. Hobbs; A. Holt; K. R. Hopper; C. E. Hughes; M. LaHart; R. R. B. Leakey; W. G. Wong; L. L. Loope; D. H. Lorence; S. M. Louda; A. E. Lugo; P. B. McEvoy; D. M. Richardson; P. M. Vitousek. 1999. "Deliberate Introductions of Species: Research Needs-Benefits Can be Reaped, but Risks are High." *BioScience*, 49 (8): 619–630. <http://www.jstor.org/stable/1313438?origin=crossref&>.

Falanrue, M. 1984. "People Pressure and Management of Limited Resources on Yap." *National Parks, Conservation and Development: The Role of Protected Areas in Sustaining Society*. J. McNeely and K. Miller, eds. Washington DC: The Smithsonian Institution Press.

Farber, Paul Lawrence. 2000. *Finding Order in Nature: The Naturalist Tradition from Linnaeus to E. O. Wilson*. Baltimore, MA: Johns Hopkins University Press.

Farley, M. A. 2001. *The Human Embryonic Stem Cell Debate: Science, Ethics, & Public Policy*. K. L. Suzanne Holland, ed. Cambridge, MA: MIT Press.

Fastovsky, D. E.; & P. Sheehan. 2005. "The Extinction of the Dinosaurs in North America." *GSA Today*, 15 (3): 4–10.

Feinholz-Klip, Dafna; Luis García Barrios; & Julie Cook Lucas. 2009. "The Limitations of Good Intent: Problems of Representation and Informed Consent in the Maya ICBG Project in Chiapas, Mexico." *Indigenous Peoples, Consent and Benefit Sharing*. Rachel Wynberg, Doris Schroeder, & Roger Chennells, eds. Netherlands: Springer.

Firehammer, Reginald. 2005. "Basic Principles of Ontology" (January 05). <http://usabig.com/autonomist/philosophy/ontology2.html>.

Flannery, T. 1994. *The Future Eaters*. Melbourne: Reed Books.

Forest, C. E.; J. Wolfe; P. Molnar; & K. Emanuel. 1999. "Paleoaltimetry Incorporating Atmospheric Physics and Botanical Estimates of Paleoclimate." *Geological Society of America Bulletin*, 111 (4): 497–511. <http://mr.crossref.org/iPage/?doi=10.1130%2F0016-7606%281999%29111%3C0497%3APIAPAB%3E2.3.CO%3B2>.

Frederick, Kenneth D.; & David C. Major. 1997. "Climate Change and Water Resources." *Climatic Change*, 37 (1): 7-23.

Freundel, Barry. 2004. "Evolution and Life on Other Planets." *Contemporary Orthodox Judaism's Response to Modernity*. Jersey City, NJ: KTAV Publishing House.

Fryer, Jonathan. 2002. "Bringing the Dodo back to Life." *BBC News*. <http://news.bbc.co.uk/2/hi/programmes/from_our_own_correspondent/2255991.stm>.

Fuller, Richard; Katherine Irvine; Patrick Devine-Wright; Philip Warren; & Kevin

Gaston. 2007. "Psychological Benefits of Greenspace Increase with Biodiversity." *Biol Lett.*, 3(4): 390–394.

Gale, Andrew S. 1989. "A Milankovitch Scale for Cenomanian Time." *Terra Nova*, 1 (5): 420. <http://onlinelibrary.wiley.com/doi/10.1111/j.1365-3121.1989.tb00403.x/abstract;jsessionid=E2D1FD9FCEDFE89C8B8C95913392E41A.d03t01>.

Gao, L. Z. 2003. "The Conservation of Rice Biodiversity in China: Significance, Genetic Erosion, Ethnobotany and Prospect." *Genetic Resources and Crop Evolution*, 50: 17-32.

Gaston, Kevin J.; Philip H. Warren; Patrick Devine-Wright; Katherine Irvine; & Richard Fuller. 2007. "Psychological Benefits of Greenspace Increase with Biodiversity." *Biology Letters*, 3 (4): 390–394. <http://www.ncbi.nlm.nih.gov/pmc/articles/PMC2390667/?tool=pmcentrez>.

_____; & John I. Spicer. 2004. *Biodiversity: An Introduction*. Malden, MA: Blackwell Publishing.

Geist, Helmut J.; & Eric F. Lambin. 2002. "Proximate Causes and Underlying Driving Forces of Tropical Deforestation." *BioScience*, 52 (2): 143–150.

Gómez de Silva, Héctor; & Ernesto Alvarado Reyes. 2010. "Breve Historia de la Observación de Aves en México en el Siglo XX y Principios del Siglo XXI." *Huitzil*, 11: 9–20.

Gott, J. Richard. 2002. *Time Travel in Einstein's Universe*. Boston : Houghton Mifflin.

Grafton, R. Q.; T. Kompas; & R. Hilborn. 2007. "Economics of Overexploitation Revisited." *Science*, 318 (5856): 1601.

Gray, Richard; & Roger Dobson. 2009. "Extinct ibex is Resurrected by Cloning" (January 31). *Telegraph*. <http://www.telegraph.co.uk/science/science-news/4409958/Extinct-ibex-is-resurrected-by-cloning.html>.

Grey, W. 1993. '"Anthropocentrism and Deep Ecology." *Australasian Journal of Philosophy*, 71: 463-475.

Guardian, the. (GUARD). 2008. "Map Reveals Extent of Deforestation in Tropical Countries" (July 1). <http://www.guardian.co.uk/environment/2008/jul/01/forests.conservation>.

Guarino, L., 1995. "Assessing the Threat of Genetic Erosion." *Collecting Plant Genetic Diversity: Technical Guidelines*. L. Guarino; V. Rao, & R. Reid, eds. Wallingford: CAB International.

Gunderson, L.; & C. S. Holling, eds. 2002. *Panarchy: Understanding Transformations in Human and Natural Systems*. Washington, DC: Island Press.

Guth, Alan. 1997. *The Inflationary Universe: The Quest for a New Theory Of Cosmic Origins*. New York: Helix Books.

Hagemann, Steffen G.; Musie Gebre-Mariam; David Groves. 1994. "Surface-Water Influx in Shallow-Level Archean Lode-Gold Deposits in Western,

Australia." *Geology*, 22 (12): 1067. <http://adsabs.harvard.edu/abs/1994 Geo....22.1067H>.

Haigh, Joanna D.; Ann R. Winning; Ralf Toumi; & Jerald W. Harder. 2010. "An Influence of Solar Spectral Variations on Radiative Forcing of Climate." *Nature*, 467 (7316): 696–9. <http://adsabs.harvard.edu/abs/2010Natur. 467..696H>.

Halim, N. 2002. "Extensive New Study Shows Abnormalities in Cloned Animals." M.I.T. <http://web.mit.edu/newsoffice/2002/cloning-0911.html>.

Hamilton, E., & H. Cairns, eds. 1961. *Plato: The Collected Dialogues*. Princeton, NJ: Princeton University Press.

Haug, Gerald H. 2004. "How the Isthmus of Panama Put Ice in the Arctic" (March 22). *WHOI: Oceanus*. <http://www.whoi.edu/oceanus/viewArticle.do?id= 2508>.

Hardin, G. 1968. "The Tragedy of the Commons" (December). *Science*, 162 (3859): 1243–8.

Harris, J. 2006. *Environmental and Natural Resource Economics: A Contemporary Approach*. Boston, MA: Houghton Mifflin Company.

Hawkes, C.V.; I. F. Wren; D. J. Herman; & M. K. Firestone. 2005. "Plant Invasion Alters Nitrogen Cycling by Modifying the Soil Nitrifying Community." *Ecology Letters*, 8 (9): 976–985.

Hawkins, E. S.; & M. Reich. 1992. "Japanese-Originated Pharmaceutical Products in the United States from 1960 to 1989: An Assessment of Innovation." *Clin Pharmacol Ther*, 51 (1): 1–11. <http://www.ncbi.nlm.nih.gov/pubmed/ 1732073>.

Hawksworth, D. L., ed 1996. *Biodiversity: Measurement and Estimation*. NY: Springer.

Hayden, Cori. 2003. *When Nature Goes Public: The Making and Unmaking of Bioprospecting in Mexico*. NJ: Princeton University Press.

Health & Science (HEA). 2001. "Cloning with Stem Cells Causes Abnormalities in Mice" (July 05). <http://www.usatoday.com/news/health/2001-07-05-cloning.htm>.

Hilbourne, R. 1992. "Can Fisheries Agencies Learn from Experience?." *Fisheries*, 4 (17): 6–14.

Hillebrand, H. 2004. "On the Generality of the Latitudinal Diversity Gradient." *The American Naturalist*, 163 (2): 192–211. <http://www.jstor.org/discover/10.1086/381004?uid=3738528&uid=2&uid=4 &sid=21100974085723>.

Holdaway, R. N.; & C. Jacomb. 2000. "Rapid Extinction of the Moas (Aves: Dinornithiformes): Model, Test, and Implications." *Science*, 287 (5461): 2250–2254. <http://www.esf.edu/efb/gibbs/efb413/moa.pdf>.

Holling, C. S., ed. 1978. *Adaptive Environmental Assessment and Management.* Chichester: Wiley.

Hubbert, M. King. 1956. "Nuclear Energy and the Fossil Fuels 'Drilling and Production Practice.'" *API.* <http://www.hubbertpeak.com/hubbert/1956/1956.pdf>.

Hunter, Malcolm L. 1996. *Fundamentals of Conservation Biology.* Oxford: Blackwell Science.

Illmensee, K, M. Levanduski; A. Vidali; N. Husami; & V. Goudas. 2009. "Human Embryo Twinning with Applications in Reproductive Medicine." *Fertil. Steril,* 93 (2): 423–7. <http://www.fertstert.org/article/S0015-0282%2808%2904795-X/abstract>.

InterAcademcy Panel (IAP). 2006. "IAP Statement on the Teaching of Evolution." The Interacademy Panel on international issues. <http://www.interacademies.net/10878/13901.aspx>.

International Council for Science (ICS). 2002. "Science and Traditional Knowledge: Report from the ICSU Study Group on Science and Traditional Knowledge" (March). <http://www.icsu.org/publications/reports-and-reviews/science-traditional-knowledge/Science-traditional-knowledge.pdf>.

International Monetary Fund (IMF). 2007. "Angola: Selected Issues and Statistical Appendix" (October). <http://www.imf.org/external/pubs/ft/scr/2007/cr07355.pdf>.

IPCC. 2007. "Box 6.2: What Caused the Low Atmospheric Carbon Dioxide Concentrations During Glacial Times?." <http://www.ipcc.ch/publications_and_data/ar4/wg1/en/ch6s6-4.html>.

Jaenisch, R. 2001. "Cloning with Stem Cells Causes Abnormalities in Mice" (July 25). *Health & Science.* <http://www.usatoday.com/news/health/2001-07-05-cloning.htm>.

James-Griffiths, P. 2012. "Creation Days and Orthodox Jewish Tradition." <http://creation.com/creation-days-and-orthodox-jewish-tradition>.

Jha, Alok. 2005. "Amazon Rainforest Vanishing at Twice Rate of Previous Estimates" (October 21). *The Guardian.* <http://www.guardian.co.uk/environment/2005/oct/21/brazil.conservationandendangeredspecies>.

Johnson, F.A.; B. Williams; J. Nichols; J. Hines; W. Kendall; G. Smith; & D. Caithamer. 1993. "Developing an Adaptive Management Strategy for Harvesting Waterfowl in North America." *Trans N Am Wildl Nat Resour Conf*, 58: 565–583.

Johnson, Kristin. 2004. "The Ibis: Transformations in a Twentieth Century British Natural History Journal." *Journal of the History of Biology*, 37 (3): 515–555.

Johnson, Paul M.; & Paul Kenny. 2010. "Addiction-Like Reward Dysfunction and Compulsive Eating in Obese Rats: Role for Dopamine D2 Receptors." *Nature*

Neuroscience, 13 (5): 635–41. <http://www.ncbi.nlm.nih.gov/pmc/articles/PMC2947358/?tool=pmcentrez>.

Kacowicz, Arie M. 2007. "Globalization, Poverty, and the North–South Divide." *International Studies Review*, 9.4 (2007): 565-580.

Kaku, Michio. 1994. *Hyperspace: A Scientific Odyssey Through Parallel Universes, Time Warps, and The Tenth Dimension.* Illustrations by Robert O'Keefe. Oxford: Oxford University Press.

Karasov, C. 2001. "Who Reaps the Benefits of Biodiversity?." *Environmental Health Perspectives (Environmental Health Perspectives*, 109 (12): A582–A587. <http://www.ncbi.nlm.nih.gov/pmc/articles/PMC1240518/?tool=pmcentrez>.

Kareiva, P.; M. Marvier. 2003. "Conserving Biodiversity Coldspots." *American* Scientist, 91: 344–51.

Karamanos, P. 2001. "Voluntary Environmental Agreements: Evolution and Definition of a New Environmental Policy Approach." *Journal of Environmental Planning and Management*, 44 (1): 67-84.

Kay, C. E. 1997. "The Ultimate Tragedy of Commons." *Conserv. Biol.*, 11 (6): 1447-8.

Kheel, Marti. 1990. "Ecofeminism and Deep Ecology; Reflections on Identity and Difference." *Reweaving the World; The Emergence of Ecofeminism.* Irene Diamond. Gloria Orenstein, eds. San Francisco, CA: Sierra Club Books.

Kinealy, Christine. 1995. *This Great Calamity: The Irish Famine 1845–52.* Dublin: Gill & Macmillan.

King, Clarence. 1877. "Catastrophism and Evolution." *The American Naturalist*, 11 (8): 449–470.

Koerner, Lisbet. 1999. "God's Endless Larder." *Linnaeus: Nature and Nation.* Cambridge, MA: Harvard University Press.

Kommers, N.; & P. Mackie. 2005. "Journalist Guide to World Resources 2005." World Resources Institute. <http://pdf.wri.org/worldresources_2005_journalist_guide.pdf>.

Konikow, Leonard F. 2011. "Contribution of Global Groundwater Depletion since 1990 to Sea-level Rise." *Geophysical Research Letters*, 38 (17).

Kurz, W. A.; C. Dymond; G. Stinson, et al. 2008. "Mountain Pine Beetle and Forest Carbon Feedback to Climate Change." *Nature*, 452 (7190): 987–90.

Labandeira, C.; & J. Sepkoski. 1993. "Insect Diversity in the Fossil Record." *Science*, 261 (5119): 310–5. <http://adsabs.harvard.edu/abs/1993Sci...261..310L>.

Larsson, Tor-Björn. 2001. *Biodiversity Evaluation Tools for European Forests.* Oxford: Wiley-Blackwell.

Laurance, William F. 1999. "Reflections on the Tropical Deforestation Crisis"

(December). *Biological Conservation*, 91 (2-3):.109–117. <http://studentresearch.wcp.muohio.edu/BiogeogDiversityDisturbance/ReflectionsDeforestCrisis.pdf>.

Lavery, James V. 2007. "Case 1: Community Involvement in Biodiversity Prospecting in Mexico." *Ethical Issues in International Biomedical Research: A Casebook*. England: Oxford University Press.

Le Billon, Philippe. 2006. "Fuelling War: Natural Resources and Armed Conflicts." *Adelphi Paper*, 373. IISS.

Leopold, A. 1949. *A Sand County Almanac*. New York: Oxford University Press.

Leroux, S. J.; & F. Schmiegelow. 2007. "Biodiversity Concordance and the importance of Endemism." *Conserv. Biol.*, 21 (1): 266–8; discussion 269–70.

Liep, John. 2001. "Airborne Kula: The Appropriation of Birds by Danish Ornithologists." *Anthropology Today* 17(5):10-15.

Lindenmeyer, D. & M. Burgman. 2005. *Practical Conservation Biology*. Collingwood, Australia: CSIRO Publishing.

Lovejoy, Arthur O. 1964. *The Great Chain of Being: A Study of the History of an Idea*. Cambridge, MA: Harvard University Press.

Maas, Peter. 2007. "Extinct in the Wild." The Extinction Website. <http://www.petermaas.nl/extinct/wilduk.htm>.

MacDonald, Glen M. 2010. "Water, Climate Change, and Sustainability in the Southwest." PNAS, 107 (50): 56-62.

Macdonald, H. 2002. "What Makes you a Scientist is the Way you Look at Things: Ornithology and the Observer 1930–1955." *Studies in History & Philosophy of Biological & Biomedical Sciences*, 33 (1): 53.

Mack, R.; D. Simberloff; W.M. Lonsdale; H. Evans; M. Clout; & F.A. Bazzazf. 2000. "Biotic Invasions: Causes, Epidemiology, Global Consequences, and Control." *Ecological Applications*, 10 (3): 689–710.

MacKenzie, Debora. 2008. "'Shares' in Fish Stocks Halt Commercial Free-for-All" (September 18). <http://www.newscientist.com/article/dn14762-guaranteed-fish-quotas-halt-commercial-freeforall.html?feedId=online-news_rss20>.

Macleod, N.; P. Rawson; P. Forey; F. T. Banner; M. K. Boudagher-Fadel; P. R. Bown; J. A. Burnett; P. Chambers; S. Culver; S. E. Evans; C. Jeffery; M. A. Kaminski; A. R. Lord; A. C. Milner; A. R. Milner; N. Morris; E. Owen; B. R. Rosen; A. B. Smith; P. D. Taylor; E. Urquhart; & J. R. Young. 1997. "The Cretaceous-Tertiary biotic transition." *Journal of the Geological Society*, 154 (2): 265–292. <http://mr.crossref.org/iPage/?doi=10.1144%2Fgsjgs.154.2.0265>.

Maharaj, V. J.; J. Senabe; & R. Horak. 2008. "Hoodia, a Case Study at CSIR. Science real and Relevant." 2nd CSIR Biennial Conference, CSIR International Convention Centre Pretoria (November, 17 &18). <http://researchspace.csir.co.za/dspace/bitstream/10204/2539/1/Maharaj_2008.pdf>.

Margules, C. R.: & R. Pressey. 2000. "Systematic Conservation Planning." *Nature*, 405 (6783): 243–53.

Marty, B. 2006. "Water in the Early Earth." *Reviews in Mineralogy and Geochemistry*, 62: 421. <http://rimg.geoscienceworld.org/content/62/1/421>.

Mason, Rachel; & Judith Ramos. 2004. "Traditional Ecological Knowledge of Tlingit People concerning the Sockeye Salmon Fishery of the Dry Bay Area, A Cooperative Agreement Between Department of the Interior National Park Service and the Yakutat Tlingit Tribe, Final Report (FIS) Project 01-091." Yakutat, Alaska.

Matisoo-Smith, E.; R. M. Roberts; G. J. Irwin; J. S. Allen; D. Penny; & D.M. Lambert. 1998. "Patterns of Prehistoric Human Mobility in Polynesia Indicated by mtDNA from the Pacific rat." *Proceedings of the National Academy of the Sciences*, 95 (25): 15145–15150. <http://www.ncbi.nlm.nih.gov/pmc/articles/PMC24590/?tool=pmcentrez>.

Maugeri, Leonardo. 2012. "Oil: The Next Revolution.' Discussion Paper 2012-10." Belfer Center for Science and International Affairs, Harvard Kennedy School. <http://belfercenter.ksg.harvard.edu/files/Oil-%20The%20Next%20Revolu tion.pdf>.

McAleer, Neil. 1987. *The Mind-Boggling Universe*. New York: Doubleday & Co.

McAnallen, Samuel. 2012. "The Superior Human?." Produced by Dr Jenia Meng. <http://thesuperiorhuman.ultraventus.info/>.

McCauley, D. J. 2006. "Selling out on Nature." Nature, 443 (7): 27-28.

McElwain, J. C.; & S. Punyasena. 2007. "Mass Extinction Events and the Plant Fossil Record." *Trends in Ecology & Evolution,* 22 (10): 548–557. <http://www.sciencedirect.com/science/article/pii/S016953470700256X>.

McGowan, Christopher. 2001. *The Dragon Hunters*. Cambridge, MA: Persus Publishing.

McKee, Jeffrey K. 2004. *Sparing Nature: The Conflict Between Human Population Growth and Earth's Biodiversity*. New Jersey: Rutgers University Press.

Medline Plus (MED). 2011. "Fetal Development." <http://www.nlm.nih.gov/ medlineplus/ency/article/002398.htm>.

Meffe, Gary K.; & Martha J. Groom. 2006. *Principles of Conservation Biology*. Sunderland, MA: Sinauer Associates.

Merriam-Webster's Collegiate Dictionary Online (MWD). 2012. "Diversity" (August 08). <http://www.merriam-webster.com/dictionary/diversity>.

_____.2012a. "Discontinuity" (August 08). <http://www.merriam-webster.com/dictionary/discontinuity>.

_____.2007. "Agency" (March 20). <http://www.m-w.com/dictionary/agency>.

Meyfroidt, P.; & E. Lambin. 2011. "Global Forest Transition: Prospects for an End to Deforestation." *Annual Review of Environment and Resources*, 36: 343-371.

Millennium Ecosystem Assessment (MEA). 2005. *Ecosystems and Human Well-Being: Biodiversity Synthesis*. Washington, D.C.: World Resources Institute

Miller, G. T.; & S. Spoolman. 2011. *Living in the Environment: Principles, Connections, and Solutions*. Belmont, CA: Brooks-Cole.

Mitchell, R. B. 2003. "International Environmental Agreements: A Survey of Their Features, Formation, and Effects." *Annual Review of Environment and Resources*, 28 (1543-5938): 429-461.

Mongabay.com (MON). 2012. "Chart—Tropical Deforestation by Country & Region" (August 22). <http://www.mongabay.com/deforestation_rate_ tables.htm>.

Moore, Alana L.; & A. Michael. 2009. "On Valuing Information in Adaptive-Management, Models." *Conservation Biology*, 24 (4): 984–993.

Moore, Mick; & Sue Unsworth. 2007. "How Does Taxation Affect the Quality of Governance?." IDS Policy Briefing.

Morand, Serge; & Boris R. Krasnov. 2010. *The Biogeography of Host-Parasite Interactions*. Oxford: Oxford University Press.

Moss, Stephen. 2004. *A Bird in the Bush: A Social History of Birdwatching*. London: Aurum Press.

_____.1995. *Birds and Weather A Birdwatcher's Guide*. London: Hamlyn.

Moustakas, A.; & I. Karakassis. 2005. "How Diverse is Aquatic Biodiversity Research?." *Aquatic Ecology*, 39: 367-375. <https://springerlink3.metapress.com/content/p2q719335u606034/resource-secured/?target=fulltext.pdf&sid=njghlarsyopgjsfx354addlq&sh=www.spring erlink.com>.

MSNBC. 2007. "Exxon Cuts Ties to Global Warming Skeptics" (January 12). <http://www.msnbc.msn.com/id/16593606>.

_____.2006. "FDA Says Cloned Animals are OK to Eat" (December 28). <http://wifi-login-11.starnet.md/login?dst=http%3A%2F%2Fwww.msnbc. msn.com%2Fid%2F16372490>.

Myers, N. 1990. "The Biodiversity Challenge: Expanded Hot-Spots Analysis." *Environmentalist*, 10 (4): 243–256. <http://www.springerlink.com/content/ f84g623376724173/>.

_____.1988. "Threatened Biotas: 'Hot Spots' in Tropical Forests." *Environmentalist*, 8 (3): 187–208. <http://www.springerlink.com/content/ y66976j1n1201251/?MUD=MP>.

Naess, A. 1973. "The Shallow and the Deep, Long-Range Ecology Movement." Inquiry, 16: 95-100.

Naidoo, R.; A. Balmford; R. Costanza, et al. 2008. "Global Mapping of Ecosystem Services and Conservation Priorities." *Proc. Natl. Acad. Sci.*, 105 (28): 9495–500.

NASA. 2011. "What's in a Name? Global Warming vs. Climate Change" (July 23). <http://www.nasa.gov/topics/earth/features/climate_by_any_other_name. html>.

_____.2003. "NASA Study Finds Increasing Solar Trend That Can Change Climate." <http://www.nasa.gov/centers/goddard/news/topstory/2003/0313 irradiance.html>.

_____.2003a. "Panama: Isthmus that Changed the World" (December 31). NASA Earth Observatory. <http://earthobservatory.nasa.gov/IOTD/view.php?id =4073>.

National Geographic (NAT). 2012. "Forest Holocaust" (August 21). <http://www.nationalgeographic.com/eye/deforestation/effect.html>.

Nature Conservancy, the (NAT). 2006. "Rainforests—Facts and Information about the Rainforest" (October 06). <http://www.nature.org/ourinitiatives/ urgentissues/rainforests/rainforests-facts.xml>.

Nee, S. 2004. "More than Meets the Eye." *Nature*, 429 (6994): 804–805. <http://www.nature.com/nature/journal/v429/n6994/full/429804a.html>.

New York Times (NYT). 2010. "Restaurants Selling Exotic Meats" (September 3).

Newman, Mark. 1994. "A Mathematical Model for Mass Extinction" (May 20). Cornell University. <http://www.lassp.cornell.edu/newmme/science/ extinction.html>.

Nichols, J. D.; M. Runge; F. Johnson; & B. Williams. 2007. "Adaptive Harvest Management of North American Waterfowl Populations: A Brief History and Future Prospects." *Journal of Ornithology*, 148 (148): 343.

Nielsen, Ron. 2006. *The Little Green Handbook: Seven Trends Shaping the Future of Our Planet*. NY: Picador.

Norman, C. S.. 2009. "Rule of Law and the Resource Curse." *Environmental and Resource Economics*, 43 (2): 183–207. <http://www.springerlink.com/ content/573331512g41v538/?MUD=MP>.

Normile, Dennis. 2010. "Saving Forests to Save Biodiversity." *Science*, 329 (5997): 1278–1280. <http://www.sciencemag.org/content/329/5997/1278. summary?sid=3d8a15d7-279e-4177-a5f0-46743a80212a>.

Nova Online (NO). 2000. "Time Can Vary." <http://www.pbs.org/ wgbh/nova/time/thinktime.html>.

Ó Gráda, Cormac. 2006. *Ireland's Great Famine: Interdisciplinary Perspectives*. Dublin: Dublin Press.

O'Neil, Patrick. 2004. *Essentials of Comparative Politics*. New York: Norton & Company, Inc.

Oddie, Bill. 1980. *Bill Oddie's Little Black Bird Book*. London: Butler & Tanner Ltd.

Odendaal, L. J.; T. Haupt; & C. Griffiths. 2008. "The Alien Invasive Land Snail Theba Pisana in the West Coast National Park: Is There Cause for Concern?."

Koedoe—African Protected Area Conservation and Science, 50(1): 93-98. <http://www.koedoe.co.za/index.php/koedoe/article/view/153>.

Okasha, S. 2010. "Does Diversity Always Grow?." *Nature*, 466 (7304): 318. <http://adsabs.harvard.edu/abs/2010Natur.466..318O>.

Oppenheimer, Clive. 2003. "Climatic, Environmental and Human Consequences of the Largest Known Historic Eruption: Tambora Volcano (Indonesia) 1815." *Progress in Physical Geography*, 27 (2): 230. <http://ppg.sagepub.com/content/27/2/230>.

Otchet, Amy. 1998. "Jeremy Rifkin: Fears of a Brave New World." An interview hosted by The United Nations Educational, Scientific and Cultural Organization (UNESCO) <http://wifi-login-11.starnet.md/login?dst=http %3A%2F%2Fwww.unesco.org%2Fcourier%2F1998_09%2Fuk%2Fdires%2F txt1.htm>.

Overton, William R. 1982. "McLean v. Arkansas Board of Education." McLean v. Arkansas. TalkOrigins Archive. <http://www.talkorigins.org/faqs/mclean-v-arkansas.html>.

Owen, Nick: Oliver R. Inderwildi; & David A. King. 2010. "The Status of Conventional World Oil Reserves—Hype or Cause for Concern?." *Energy Policy*, 38 (8): 4743. <http://www.sciencedirect.com/science/article/pii/S0301421510001072>.

Pahl-Wost, C. 2006 'The Implications of Complexity for Integrated Resource Management." *Environmental Modelling and Software*, 22: 561-569.

Pallangyo, D. M. 2007. "Environmental Law in Tanzania; How Far Have We Gone?." *LEAD: Law, Environment & Development Journal*, 3 (1).

Palley, Thomas I. 2003. "Lifting the Natural Resource Curse." Globalpolicy.org. <http://www.globalpolicy.org/component/content/article/198/40112.html>.

Pallottinia, L.; E. Garciab; J. Kamic; G. Barcacciaa; & P. Gepts. 2004. "The Genetic Anatomy of a Patented Yellow Bean." *Crop Science*, 44 (3): 968–977.

Parrish, Judith T. 1993. "Climate of the Supercontinent Pangea." *Chemical Geology*, 101 (2): 215–233.

Passmore, J. 1974. *Man's Responsibility for Nature*. London: Duckworth.

PBS. 2003. "Bloodlines Timeline." <http://www.pbs.org/bloodlines/timeline/text_timeline.html>.

Peña, Devon G. 2005. *Mexican Americans and the Environment*. AZ: The University of Arizona Press.

People for the Ethical Treatment of Animals (PETA). 2007. "Fish in Tanks: No, Thanks!" (April 22). <http://www.peta.org/issues/Companion-Animals/fish-in-tanks-no-thanks.aspx>.

Pepper, D. 1984. *The Roots of Modern Environmentalism*. Dover, NH: Croom

Helm.

Perlman, Heidi B. 2000. “Scientists Close on Extinct Cloning.” *The Washington Post*. <http://www.washingtonpost.com/wp-srv/aponline/20001008/aponline 171938_000.htm>.

Pence, Gregory E. 2005. *Cloning After Dolly: Who's Still Afraid?*. Lanham, MD: Rowman & Littlefield.

Plumwood, V. 1996. “Androcentrism and Anthrocentrism: Parallels and Politics.” *Ethics and the Environment*.

_____.1993. *Feminism and the Mastery of Nature*. London: Routledge.

Potts, B. M.; R. Barbour; A. Hingston; & R. Vaillancourt. 2003. “*Corrigendum to*: TURNER REVIEW No. 6 Genetic Pollution of Native Eucalypt Gene Pools—Identifying the Risks.” *Australian Journal of Botany*, 51, 333–333. <http://www.publish.csiro.au/?paper=BT02035_CO>.

PIB. 2010. “Know Instances of Patenting on the UES of Medicinal Plants in India” (May 06). Ministry of Environment and Forests. <http://pib.nic.in/newsite/erelease.aspx?relid=61511>.

Posey, Darrel Addison. 2008. “Indigenous Management of Tropical Forest Ecosystems: The case of the Kayapo Indians of the Brazilian Amazon.” *Environmental Anthropology: A Historical Reader*. Oxford: Blackwell Publishing.

Primack, Richard B. 2004. *A Primer of Conservation Biology*. Sinauer Associates.

Prudhomme, Christel, et al. 2002. “Soil and Water: Climate Change and Water Resources Management in Arid and Semi-Arid Regions: Prospective Challenges for the 21st Century.” *Biosystems Engineering*, 81 (1): 3-34.

Purser, R.; & A. Montuori. 1996. “Ecocentrism is in the Eye of the Beholder.” *Academy of Management Review*, 21: 611-613.

Quince, C. et al. 2006. “Deleting Species from Model Food Webs.” <http://web.archive.org/web/20060925092540/http://theory.ph.man.ac.uk/~ajm/qui05a.pdf>.

Rabinowitz, A. R. 1986. *Jaguar: One Man's Battle to Establish the World's First Jaguar Preserve*. New York: Arbor House.

Raman, Shankar. 1997. Lectures on Literary Theory. Cambridge, MA: MIT (spring).

Ramanujan, Krishna. 2010. “Study: Loss of Species is Bad for your Health” (December 02). Cornell Chronicle. <http://www.news.cornell.edu/stories/Dec10/BiodiversityHealth.html>.

Raup, David M. 1991. *Extinction: Bad Genes or Bad Luck?* New York: W.W. Norton and Company.

Revington, John. 2012. “The Causes of Tropical Deforestation.” *New Renaissance* Magazine. <http://www.ru.org/ecology-and-environment/the-causes-of-

tropical-deforestation.html>.

Rhymer, J. M.; & D. Simberloff. 1996. "Extinction by Hybridization and Introgression." *Annual Review of Ecology and Systematics*, 27 (27): 83–109.

Ricklefs, R. E. 2005. *The Economy of Nature*. New York: WH Freeman.

Rifkin, Jeremy. 1998. *The Biotech Century: Harnessing the Gene and Remaking the World*. J P. Tarcher.

Roopesh, J., et al. 2008. "Marine Organisms: Potential Source for Drug Discovery." *Current Science*, 94 (3): 292.

Røpke, I. 2005. "Trends in the Development of Ecological Economics from the Late 1980s to the Early 2000s." *Ecological Economics*, 55 (2): 262-290.

_____.2004. "The Early History of Modern Ecological Economics." *Ecological Economics*, 50 (3-4): 293-314.

Ross, David. 2002. *Ireland: History of a Nation*. New Lanark: Geddes & Grosset.

Routley, R.. V. 1980. "Human Chauvinism and Environmental Ethics." *Environmental Philosophy*. D.S. Mannison, M. McRobbie, and R. Routley, eds. Canberra, Australia: ANU Research School of Social Sciences.

Rowe, Stan J. 1994."Ecocentrism: the Chord that Harmonizes Humans and Earth." *The Trumpeter*, 11 (2): 106-107.

Royal Society (ROY). 2005. "Letter from The Royal Society: A GUIDE TO FACTS AND FICTIONS ABOUT CLIMATE CHANGE: Misleading arguments: Many scientists do not think that climate change is a problem. Some scientists have signed petitions stating that climate change is not a problem" (April 13). <http://www.publications.parliament.uk/pa/ld200506/ldselect/ldeconaf/12/12we24.htm>.

Roussel, B. 2003. "The CBD: Local Knowledge Systems at the Heart of International Debates." *Synthèse*, 2. Institute for Sustainable Development and International Relations. <http://www.iddri.org/Publications/The-CBD-Local-knowledge-systems-at-the-heart-of-international-debates>.

Rudel, T. K. 2005 "Tropical Forests: Regional Paths of Destruction and Regeneration in the Late 20th Century." NY: Columbia University Press.

Rudwick, Martin J.S. 1972. *The Meaning of Fossils*. Chicago, IL: The University of Chicago Press.

Running, S. W. 2008. "Climate Change. Ecosystem Disturbance, Carbon, and Climate." *Science*, 321 (5889): 652–3.

Russel, Peter J. 2005. *Genetics: A Molecular Approach*. San Francisco, CA: Pearson Education.

Sagan, C.; & C. Chyba. 1997. "The Early Faint Sun Paradox: Organic Shielding of Ultraviolet-Labile Greenhouse Gases." *Science*, 276 (5316): 1217–21. <http://www.sciencemag.org/content/276/5316/1217>.

_____.& G. Mullen. 1972. "Earth and Mars: Evolution of Atmospheres and

Surface Temperatures." <http://www.sciencemag.org/content/177/4043/52.abstract?ck=nck>.

Sahney, S.; M. Benton; & P. Ferry. 2010. "Links between Global Taxonomic Diversity, Ecological Diversity and the Expansion of Vertebrates on Land." *Biology Letters*, 6 (4): 544–547. <http://rsbl.royalsocietypublishing.org/content/6/4/544.full.pdf+html>.

_____; M. Benton; & H. Falcon-Lang. 2010a. "Rainforest Collapse Triggered Pennsylvanian Tetrapod Diversification in Euramerica." *Geology*, 38 (12): 1079–1082. <http://geology.geoscienceworld.org/content/38/12/1079.abstract>.

_____.2008. "Recovery from the Most Profound Mass Extinction of all Time." *Proceedings of the Royal Society: Biological*, 275 (1636): 759–65. <http://rspb.royalsocietypublishing.org/content/275/1636/ 759.full.pdf+html>.

Salvati, L.; & Z. Marco. 2008. "Natural Resource Depletion and Economic Performance of Local Districts: Suggestions from a Within-Country Analysis." *Journal of Sustainable Development and World Ecology*, 15 (6): 518-523.

Sandell, Clayton. 2007. "Report: Big Money Confusing Public on Global Warming" (January 3). *ABC*. <http://abcnews.go.com/Technology/Business/story?id=2767979&page=1>.

Schilling, M.; & L. Chiang. 2011. "The Effect of Natural Resources on Sustainable Development Policy: The Approach of Non-Sustainable Externalities." *Energy Policy*, 39: 990-998.

Science & Environmental Policy Project (SEPP). 2008. "Frequently Asked Questions About Climate Change" (May 11). <http://web.archive.org/web/20080511011611/http://www.sepp.org/FAQ/faq.html>.

Scott, Eugenie C. 2004. *Evolution vs. Creationism: An Introduction*. Berkley, CA: University of California Press.

Selin. S.; & D. Chavez. 1995. "Developing a Collaborative Model for Environmental Planning and Management." *Environmental Management*, 19: 189-195.

Shearin, Gloria. 2008. "Neltje Blanchan." *Early American Nature Writers*. D. Patterson, R. Thompson, S. Bryson, et al. Westport, CT: Greenwood Press.

Shiva, Vandana. 2012. "The Neem Tree—A Case History of Biopiracy." <http://www.twnside.org.sg/title/pir-ch.htm>.

Shlain, Leonard. 1991. *Art and Physics: Parallel Visions in Space,Time, and Light*. New York: William Morrow and Co., Inc.

Smith, J. P., Jr.; & K. Berg. 1988. *Inventory of Rare and Endangered Vascular Plants of California*. Sacramento, CA: California Native Plant Society.

Smithsonian Centre for Education and Museum Studies (SCEMS). 2012. "A Traditional Brew Leads to Cancer Cure" (August 24). <http://www.smithsonianeducation.org/migrations/zoofood/rosper.html>.

Soderbaum P. 2004. "Politics and Ideology in Ecological Economics." *Internet Encyclopaedia of Ecological Economics.* <http://www.ecoeco.org/pdf/politics_ideology.pdf>.

Solomon, Susan; Gian-Kasper Plattner; Reto Knutti; & Pierre Friedlingstein. 2009. "Irreversible Climate Change due to Carbon Dioxide Emissions." *Proceedings of the National Academy of Sciences of the United States of America*, 106 (6): 1704–9. <http://www.pnas.org/content/early/2009/01/28/0812721106.full.pdf+html>.

Soule, Michael E. 1986. "What is Conservation Biology?." *BioScience* (American Institute of Biological Sciences), 35 (11): 727–34. <http://www.michaelsoule.com/resource_files/85/85_resource_file1.pdf>.

_____.1986a. *Conservation Biology: The Science of Scarcity and Diversity.* Sunderland, MA: Sinauer Associates.

Spash, C. L. 1999. "The Development of Environmental Thinking in Economics." *Environmental Values*, 8 (4): 413-435.

Spotts, Peter. 2005. "Satellite Images Reveal Amazon Forest Shrinking Faster" (October 21). *CSMonitor.com.* <http://www.csmonitor.com/2005/1021/p04s01-sten.html>.

Steinfeld, H.; P. Gerber; T. Wassenaar; V. Castel; M. Rosales; & C. de Haan. 2006. "Livestock's Long Shadow." <http://www.fao.org/docrep/010/a0701e/a0701e00.HTM>.

Sustainable Environment (SUS). 2012. "Earth: Ecocentrism & Technocentrism." <http://www.sustainable-environment.org.uk/Earth/Ecocentrism_and_Technocentrism.php>.

Svensmark, H.; T. Bondo; & J. Svensmark. 2009. "Cosmic Ray Decreases Affect Atmospheric Aerosols and Clouds." *Geophys. Res. Lett.*, 36 (15): L15101. <http://www.agu.org/pubs/crossref/2009/2009GL038429.shtml>.

Stachowicz, J. J.; & D. Tilman. 2005. "Species Invasions and the Relationships between Species Diversity, Community Saturation, and Ecosystem Functioning." *Species Invasions: Insights into Ecology, Evolution, and Biogeography*. D. F. Sax, J. J. Stachowicz, and S. D. Gaines, eds. Sunderland, MA: Sinauer Associates.

Stijns, Jean-Philippe. 2006. "Natural Resource Abundance and Human Capital Accumulation." *World Development*, 34 (6). <http://www.sciencedirect.com/science/article/pii/S0305750X06000453>.

Stohlgren, T. J.; D. Binkley; G.W. Chong; M.A. Kalkhan; L.D. Schell; K.A. Bull; Y. Otsuki; G. Newman; M. Bashkin; & Y. Son. 1999. "Exotic Plant Species Invade Hot Spots of Native Plant Diversity." *Ecological Monographs*, 69:

25–46.

Swanepoel, Esmarie. 2007. "Food versus Fuel Debate Escalates" (November 02). <http://www.engineeringnews.co.za/article/food-versus-fuel-debate-escalates-2007-11-02>.

Theel, Heather J.: E. Dibble; & J. Madsen. 1948. "Differential Influence of a Monotypic and Diverse Native Aquatic Plant Bed on a Macroinvertebrate Assemblage; an Experimental Implication of Exotic Plant Induced Habitat." <http://cat.inist.fr/?aModele=afficheN&cpsidt=20189959>.

ThinkExist.com (TE). 2012. "Mahatma Gandhi" (August 07). <http://thinkexist.com/quotation/earth_provides_enough_to_satisfy_every_m an- s_need/181709.html>.

_____.2012a. "Samuel Butler" (August 18). <http://thinkexist.com/search/ searchquotation.asp?search=extinction&page=2>.

_____.2012b. "Fred Ketchen" (August 25). <http://thinkexist.com/search/ searchquotation.asp?search=natural%20resources&page=2>.

Tilman, D. 2004. "Niche Tradeoffs, Neutrality, and Community Structure: A Stochastic Theory of Resource Competition, Invasion, and Community Assembly." *Proceedings of the National Academy of Sciences*, 101 (30): 10854–10861. <http://www.ncbi.nlm.nih.gov/pmc/articles/PMC503710/?tool =pmcentrez>.

Turner, W. R.; K. Brandon, T. M. Brooks; R. Costanza; G. A. B. Da Fonseca; and R. Portela. 2007. "Global Conservation of Biodiversity and Ecosystem Services." *BioScience*, 57 (10): 868-873.

Underwood E. C.; M. Shaw; K. Wilson, et al. 2008. "Protecting Biodiversity when Money Matters: Maximizing Return on Investment." *PLoS ONE*, 3 (1): e1515.

_____.2012a. "Sri Sathya Sai Baba" (August 10). <http://thinkexist.com/ search/searchquotation.asp?search=diversity&page=4>.

United Nations (UN). 2011. "Core Publications, Agenda 21, Section Two: Conservation and Management of Resources for Development" (September 12). <http://www.un.org/esa/dsd/agenda21/res_agenda21_09.shtml>.

_____.2011a. "Our Common Future, Chapter 2: Towards Sustainable Development" (September 12). <http://www.un-documents.net/ocf-02.htm>.

_____.2010. "Blueprint for Corporate Sustainability Leadership." UN Global Compact 2010. UN Global Compact Office.

_____.2007. "Investment and Financial Flows to Address Climate Change." UNFCCC. <http://unfccc.int/files/essential_background/background_publi cations_htmlpdf/application/pdf/pub_07_financial_flows.pdf>.

_____.2007a. "Indigenous peoples." Office of the United Nations High Commissioner of Human Rights. Geneva, Switzerland.

<http://web.archive.org/web/20071107144031/http://www.ohchr.org/english/issues/indigenous/>.

_____.2002. "Integrating Environment and Development: 1972–2002." United Nations Environmental Program.

_____.2002a. "Gene Flow from GM to Non-GM Populations in the Crop, Forestry, Animal and Fishery Sectors." Background Document to Conference: May 31-July 6. Electronic Forum on Biotechnology in Food and Agriculture, Food and Agriculture Organization of the United Nations (FAO),

_____.1997. "The State of the World's Plant Genetic Resources for Food and Agriculture. Food and Agriculture Organization of the United Nations." <http://www.fao.org/WAICENT/FAOINFO/AGRICULT/AGP/AGPS/Pgrfa/pdf/swrfull.pdf>.

_____.1996. "Global Plan of Action for the Conservation and Sustainable Utilization of Plant Genetic Resources for Food and Agriculture." Food and Agriculture Organization of the United Nations. <http://www.fao.org/ag/AGP/AGPS/GpaEN/gpatoc.htm>.

United States (US). 2010. "Our Invaluable Invertebrate Collections." USDA. <http://www.ars.usda.gov/is/AR/archive/jan10/insects0110.htm>.

_____.2006. "Volcanic Gases and Their Effects." U.S. Department of the Interior. <http://volcanoes.usgs.gov/hazards/gas/index.php>.

_____.1973. "U.S. Endangered Species Act (7 U.S.C. § 136, 16 U.S.C. § 1531 et seq.) of 1973." Washington, DC: U.S. Government Printing Office.

United States Water News (USWN). 2006. "World's Largest Aquifer Going Dry" (February). <http://www.uswaternews.com/archives/arcsupply/6worllarg2.html>.

University of California Museum of Paleontology (UCMP). 2012. "The Cambrian Period." <http://www.ucmp.berkeley.edu/cambrian/cambrian.php>.

University of Montana (UM). 2011. "Milankovitch Cycles and Glaciation." <http://web.archive.org/web/20110716144130/http://www.homepage.montana.edu/~geol445/hyperglac/time1/milankov.htm>.

University of Washington (UW). 2001. "Aquatic Invasive Species. A Guide to Least-Wanted Aquatic Organisms of the Pacific Northwest." <http://www.wsg.washington.edu/mas/pdfs/leastwanted.pdf>.

van den Bergh, Jeroen C. J. M. 2001. "Ecological Economics: Themes, Approaches, and Differences with Environmental Economics." *Regional Environmental Change*, 2 (1): 13-23.

van Dyke, Fred. 2008. *Conservation Biology: Foundations, Concepts, Applications*. Springer Verlag.

Verling, E.; G. M. Ruiz; L. D. Smith; B. Galil; A.W. Miller; & K. R. Murphy. 2005. "Supply-Side Invasion Ecology: Characterizing Propagule Pressure in

Coastal Ecosystems." *Proceedings of the Royal Society of London, Ser. B: Biological Science*, 272 (1569): 1249–1256. <http://www.ncbi.nlm.nih.gov/pmc/articles/PMC503710/?tool=pmcentrez>.

Wade, Nicholas. 2008. "Regenerating a Mammoth for $10 Million" (November 19). *The New York Times*.

Walters, C. J. 1986. *Adaptive Management of Renewable Resources*. New York, NY: Mc Graw Hill.

Wasongo, V. O.; D. Kambewa; & I. Bekalo. 2011. "Community-Based Natural Resource Management." *Managing Natural Resources for Development in Africa. A Resource Book*. W.O. Ochola, P.C. Sanginga, and I. Bekalo, eds. Nairobi, Kenya: Univ. of Nairobi Press.

Watson, Eb; & T. Harrison. 2005. "Zircon Thermometer Reveals Minimum Melting Conditions on Earliest Earth" (May). *Science*, 308 (5723): 841–4. <http://www.sciencemag.org/content/308/5723/841>.

Watson, Irene. 1992. "1993: International Year for Indigenous Peoples." *Aboriginal Law Bulletin*.

Weart, S. 2009. "The Public and Climate Change (cont.–since 1980). Section: After 1988" (July). American Institute of Physics. <http://www.aip.org/history/climate/public2.htm>.

Weber, Max. 1930. *The Protestant Ethic and the Spirit of Capitalism*. Talcott Parsons, tr. New York: Scribner.

Weirum, Brian K. 2007. "Will Traditional Chinese Medicine Mean the End of the Wild Tiger?" (November 11). *San Francisco Chronicle*. <http://www.sfgate.com/travel/article/Will-traditional-Chinese-medicine-mean-the-end-of-3236621.php>.

White, Gilbert. 1909. *Natural History and Antiquities of Selborne, in the County of Southampton* London: Dent.

White, L. 1967. "The Historical Roots of our Ecologic Crisis." Science, 155: 1203-1207.

Wignall, P. 2001. "Large Igneous Provinces and Mass Extinctions." *Earth-Science Reviews*, 53: 1. <http://adsabs.harvard.edu/abs/2001ESRv...53....1W>.

Wikipedia (WK). 2012. "Resource" (August 07). <http://en.wikipedia.org/wiki/Resource>.

_____.2012a. "Natural Resource" (August 07). <http://en.wikipedia.org/wiki/Natural_resource>.

_____.2012b. "Diversity" (August 08). <http://en.wikipedia.org/wiki/Diversity>.

_____.2012c. "Conservation Biology" (August 08). <http://en.wikipedia.org/wiki/Conservation_biology>.

_____.2012d. "Ecosystem" (August 08). <http://en.wikipedia.org/wiki/Ecosystem>.

_____.2012e. "Extinction" (August 08). <http://en.wikipedia.org/wiki/Extinction>.

_____.2012f. "Habitat Fragmentation" (August 08). <http://en.wikipedia.org/wiki/Habitat_fragmentation>.

_____.2012g. "Range (Biology)" (August 08). <http://en.wikipedia.org/wiki/Range_(Biology)>.

_____.2012h. "Earth Science" (August 08). <http://en.wikipedia.org/wiki/Earth_science>.

_____.2012i. "Biodiversity" (August 09). <http://en.wikipedia.org/wiki/Biodiversity>.

_____.2012j. "Overexploitation" (August 09). <http://en.wikipedia.org/wiki/Overexploitation>.

_____.2012k. "Environmental Protection" (August 09). <http://en.wikipedia.org/wiki/Environmental_protection>.

_____.2012l. "Invasive Species" (August 09). <http://en.wikipedia.org/wiki/Invasive_species>.

_____.2012m. "Cloning" (August 09). <http://en.wikipedia.org/wiki/Cloning>.

_____.2012n. "Environmental Management" (August 09). <http://en.wikipedia.org/wiki/Environmental_management>.

_____.2012o. "North-South Divide" (August 09). <http://en.wikipedia.org/wiki/North-South_divide>.

_____.2012p. "Human Impact on the Environment" (August 09). <http://en.wikipedia.org/wiki/Human_impact_on_the_environment>.

_____.2012q. "Fishkeeping" (August 09). <http://en.wikipedia.org/wiki/Fishkeeping>.

_____.2012r. "Bioprospecting" (August 09). <http://en.wikipedia.org/wiki/Bioprospecting>.

_____.2012s. "Extinction Event" (August 10). <http://en.wikipedia.org/wiki/Mass_extinction>.

_____.2012t. "Cambrian Explosion" (August 10). <http://en.wikipedia.org/wiki/Cambrian_explosion>.

_____.2012u. "Agricultural Biodiversity" (August 11). <http://en.wikipedia.org/wiki/Agricultural_biodiversity>.

_____.2012v. "Great Famine (Ireland)" (August 11). <http://en.wikipedia.org/wiki/Great_Famine_(Ireland)>.

_____.2012w. "Junk Food" (August 11). <http://en.wikipedia.org/wiki/Junk_food>.

_____.2012x. "Punctuated Equilibrium" (August 12). <http://en.wikipedia.org/wiki/Punctuated_equilibrium>.

_____.2012y. "Ecological Economics " (August 12). <http://en.wikipedia.org/

wiki/Ecological_economics>.

_____.2012z. "Adaptive Management" (August 12). <http://en.wikipedia.org/wiki/Adaptive_management>.

_____.2012aa. "Anthropocentrism" (August 15). <http://en.wikipedia.org/wiki/Anthropocentrism>.

_____.2012bb. "Ecocentrism" (August 15). <http://en.wikipedia.org/wiki/Ecocentrism>.

_____.2012cc. "Deep Ecology" (August 15). <http://en.wikipedia.org/wiki/Deep_ecology>.

_____.2012dd. "Birdwatching" (August 17). <http://en.wikipedia.org/wiki/Birdwatching>.

_____.2012ee. "Quaternary Extinction Event" (August 17). <http://en.wikipedia.org/wiki/Quaternary_extinction_event>.

_____.2012ff. "Global Warming" (August 19). <http://en.wikipedia.org/wiki/Global_warming>.

_____.2012gg. "Climate Change" (August 20). <http://en.wikipedia.org/wiki/Climate_change>.

_____.2012hh. "Environmental Degradation" (August 20). <http://en.wikipedia.org/wiki/Environmental_degradation>.

_____.2012ii. "Environmental Issues in India" (August 20). <http://en.wikipedia.org/wiki/Environmental_issues_in_India>.

_____.2012jj. "Volcano" (August 20). <http://en.wikipedia.org/wiki/Volcano>.

_____.2012kk. "Genetic Pollution" (August 20). <http://en.wikipedia.org/wiki/Genetic_pollution>.

_____.2012ll. "Deforestation" (August 21). <http://en.wikipedia.org/wiki/Deforestation>.

_____.2012mm. "Forest Transition" (August 21). <http://en.wikipedia.org/wiki/Forest_transition>.

_____.2012nn. "Oil Depletion" (August 23). <http://en.wikipedia.org/wiki/Oil_depletion>.

_____.2012oo. "Resource Curse" (August 23). <http://en.wikipedia.org/wiki/Resource_curse>.

_____.2012pp. "Oil Resources" (August 23). <http://en.wikipedia.org/wiki/Oil_resources>.

_____.2012qq. "Hubbert Curves" (August 23). <http://en.wikipedia.org/wiki/Hubbert_curves>.

_____.2012rr. "Traditional Knowledge" (August 24). <http://en.wikipedia.org/wiki/Traditional_knowledge>.

_____.2012ss. "Ideology" (August 25). <http://en.wikipedia.org/wiki/Ideology>.

_____.2012tt. “Great Chain of Being” (August 25). <http://en.wikipedia.org/wiki/Great_Chain_of_Being>.

_____.2012uu. “Young Earth Creationism” (August 25). <http://en.wikipedia.org/wiki/Young_Earth_creationism>.

_____.2012vv. “Catastrophism” (August 25). <http://en.wikipedia.org/ wiki/Catastrophism>.

_____.2011nn. “Design Elements and Principles” (September 30). <http://en.wikipedia.org/wiki/Design_elements_and_principles>.

_____.2011oo. “Interpretation (Logic)” (April 13). <http://en.wikipedia.org/wiki/Interpretation_(Logic)>.

_____.2011pp. “Unsolved Problems in Linguistics” (February 20). <http://en.wikipedia.org/wiki/Unsolved_problems_in_linguistics>.

_____.2008. “Posthumanism” (October 20). <http://en.wikipedia.org/wiki/Posthumanism>.

_____.2008a. “Transhumanism” (October 20). <http://en.wikipedia.org/wiki/Transhumanism>.

_____.2008b. “Lowest Common Denominator” (March 26). <http://en.wikipedia.org/ wiki/Least_common_denominator>.

_____.2008c. “Infinity” (November 11). <http://en.wikipedia.org/wiki/Infinity>.

_____.2008d. “Transfinite Number” (November 10). <http://en.wikipedia.org/wiki/Transfinite_ number>.

_____.2008e. “Quantum Mechanics” (November 26). <http://en.wikipedia.org/wiki/Quantum_mechanics>.

_____.2008f. “Copenhagen Interpretation” (November 26). <http://en.wikipedia.org/wiki/Copenhagen_interpretation >.

_____.2008g. “Many Worlds Interpretation” (November 26). <http://en.wikipedia.org/wiki/Many_Worlds_Interpretation>.

_____.2008h. “Hyperspace (Science Fiction)” (November 23). <http://en.wikipedia.org/wiki/Hyperspace_(science_fiction)>.

_____.2008i. “Time Travel” (November 26). <http://en.wikipedia.org/wiki/Time_travel#Time_dilation>.

_____.2008j. “Wormhole” (November 25). <http://en.wikipedia.org/wiki/Wormhole>.

_____.2007. “Linguistics” (March 04). <http://en.wikipedia.org/wiki/Linguistics>.

Wikiversity (WKV). 2008. “Why 10 Dimensions” (November 23). <ttp://en.wikiversity.org/wiki/Why_10_dimensions>.

Wilcox, Bruce A. 1984. “In Situ Conservation of Genetic Resources: Determinants of Minimum Area Requirements.” *National Parks, Conservation and Development, Proceedings of the World Congress on National Parks*. J.A. McNeely and K.R. Miller. DC: Smithsonian Institution Press.

_____; & Michael E. Soulé. 1980. *Conservation Biology: An Evolutionary-Ecological Perspective*. Sunderland, MA: Sinauer Associates.

Williams, J. D.; & G. K. Meffe. 1998. “Nonindigenous Species.” *Status and Trends of the Nation's Biological Resources*. Reston, Virginia: United States Department of the Interior, Geological Survey.

Wilson, David Alec. 2002. *Darwin's Cathedral: Evolution, Religion, and the Nature of Society*. Chicago, IL: University of Chicago Press.

Wilson, Edward Raymond. 2007. “Rethinking the Theoretical Foundation of Sociobiology.” *Q Rev Biol*, 82 (4): 327–48

_____.2002. *The Future of Life*. Boston, MA: Little, Brown.

_____.2000. “On the Future of Conservation Biology.” *Conservation Biology*, 14 (1): 1–3. <http://onlinelibrary.wiley.com/doi/10.1046/j.1523-1739.2000.00000-e1.x/abstract;jsessionid=762F3086E3B42C9E01449567F91AEE71.d03t02>.

Willson, Richard C.; & Alexander V. Mordvinov. 2003. “Secular Total Solar Irradiance Trend during Solar Cycles 21–23.” *Geophys. Res. Lett*. 30 (5): 1199. <http://adsabs.harvard.edu/abs/2003GeoRL..30e...3W>.

_____.& Hugh S. Hudson. 1991. “The Sun's Luminosity over a Complete Solar Cycle.” *Nature*, 351 (6321): 42–44. <http://www.nature.com/nature/journal/v351/n6321/abs/351042a0.html>.

Wittgenstein, Ludwig.1953. *Philosophical Investigations*. G. E. M. Anscombe and R. Rhees, eds. G.E.M. Anscombe, tr. Oxford: Blackwell.

Wood, C. C.; & M. Gross. 2008. “Elemental Conservation Units: Communicating Extinction Risk without Dictating Targets for Protection.” *Conserv. Biol*., 22 (1): 36–47.

Woodham-Smith, Cecil. 1991. *The Great Hunger*. London: Penguin.

World Bank (WB). 2011. “Poverty Reduction and Equity” (Sept 10). <http://web.worldbank.org/WBSITE/EXTERNAL/TOPICS/EXTPOVERTY/0,,contentMDK:23003457~pagePK:148956~piPK:216618~theSitePK:336992,00.htm>.

Worldwatch.org (WW). 2011. “The Hidden Shame of the Global Industrial Economy.” <http://www.worldwatch.org/node/543>.

Xepapadeas, Anastasios. 2008. “Ecological Economics.” *The New Palgrave Dictionary of Economics*. UK: Palgrave MacMillan.

Young, Gordon J.; James Dooge; & John C. Rodda. 2004. *Global Water Resource Issues*. Cambridge, England: Cambridge UP.

Zitner, A. 2000. “Cloned Goat Would Revive Extinct Line” (December 24). *Los Angeles Times*. <http://articles.latimes.com/2000/dec/24/news/mn-4250/2>.

INDEX

•A•

•C•

•D•

•E•

•F•

•G•

•H•

•J•

•K•

•L•

•M•

•N•

•O•

•P•

•T•

•U•

•V•

•W•

•X•

•Y•